Soil and Environmental Chemistry

Soil and Environmental Chemistry

William F. Bleam
University of Wisconsin, Madison

AMSTERDAM • BOSTON • HEIDELBERG • LONDON • NEW YORK • OXFORD
PARIS • SAN DIEGO • SAN FRANCISCO • SINGAPORE • SYDNEY • TOKYO
Academic Press is an imprint of Elsevier

Academic Press is an imprint of Elsevier
30 Corporate Drive, Suite 400, Burlington, MA 01803, USA
The Boulevard, Langford Lane, Kidlington, Oxford, OX5 1GB, UK

Notices
Knowledge and best practice in this field are constantly changing. As new research and
experience broaden our understanding, changes in research methods, professional practices, or
medical treatment may become necessary.

   Practitioners and researchers must always rely on their own experience and knowledge in
evaluating and using any information, methods, compounds, or experiments described herein. In
using such information or methods they should be mindful of their own safety and the safety of
others, including parties for whom they have a professional responsibility.

   To the fullest extent of the law, neither the Publisher nor the authors, contributors, or editors,
assume any liability for any injury and/or damage to persons or property as a matter of product
liability, negligence or otherwise, or from any use or operation of any methods, products,
instructions, or ideas contained in the material herein.

Library of Congress Cataloging-in-Publication Data
Bleam, W. F. (William F.)
   Soil and environmental chemistry / William F. Bleam.
      p. cm.
   Reprinted with corrections.
   Includes bibliographical references and index.
   ISBN 978-0-12-415797-2 (alk. paper)
1. Soil chemistry. 2. Environmental chemistry. I. Title.
   S592.5.B565 2012b
   631.4'1—dc23

                                                                                2011028878

British Library Cataloguing-in-Publication Data
A catalogue record for this book is available from the British Library.

For information on all Academic Press publications,
visit our website: www.elsevierdirect.com
Transferred to Digital Printing in 2012

To Wilbert F. Bleam, my father and so much more.

Contents

2. Soil Moisture and Hydrology   41

3. Clay Mineralogy and Clay Chemistry   85

The Earth, atmosphere, water, and living things abide intimately and inseparably in a place we call the *soil*. Any attempt to describe the chemistry of this peculiar corner of the environment inevitably sets boundaries. Hydrology sits beyond the boundary for many soil chemists; but certain chemical processes take decades to develop, making residence time an important variable, to say nothing of belowground water transport. Again, soil microbiology may lay "beyond the black stump" separating chemistry from its cousin biology; yet respiratory processes drive redox chemistry wherever microbes are found in nature. Is there sufficient chemistry in risk assessment to warrant its inclusion despite differences in scientific dialect?

Moving the boundaries outward to include *Soil Moisture & Hydrology* and *Risk Assessment* forces compensating choices. Herein *Clay Mineralogy & Clay Chemistry* embraces layer silicates and swelling behavior, but demurs the broader discussion of structural crystallography. Mineral weathering makes its appearance in *Acid-Base Chemistry* as the ultimate source of basicity, but detailed chemical mechanisms receive scant attention. *Ion exchange* is stripped to its bare essentials—the exchange isotherm and the factors determining its appearance. *Natural Organic Matter & Humic Colloids* adds content related to carbon turnover and colloidal behavior, but does not take on the humification process or the primary structure of humic molecules. *Water Chemistry* downplays algebraic methods in favor of model validation, reflecting my experience that students attain more success applying their chemical knowledge to validating simulation results than slogging through mathematically complex and symbolically unfamiliar coupled equilibrium expressions. The hallmark of *Acid-Base Chemistry* is integration—pulling together water chemistry, ion exchange, and fundamental acid-base principles to develop an understanding of two chemically complex topics: exchangeable acidity and sodicity.

Each chapter includes insofar as possible: actual experimental data plotted in graphic form, one or more simple models designed to explain the chemical behavior manifest in experimental data, and quantitative examples designed to develop problem-solving skills. Examples and problems rely on actual experimental data when available, supplemented by data gleaned from the USDA Natural Resources Conservation Service *Soil Data Mart*, the United States Environmental Protection Agency *PBT Profiler* and *Integrated Risk Information System*, and the *National Atmospheric Deposition Program*, to name a few. These examples are designed to develop key problem-solving skills and to demonstrate methods for making sound estimates and predictions using basic chemistry principles, proven chemical models, and readily accessible soil, environmental, and chemical data.

I wish to acknowledge several individuals who contributed to the planning, writing, and completion of this book. Philip Helmke and Phillip Barak, my chemistry colleagues at Madison, influenced uncounted choices of content and emphasis. Birl Lowery, John Norman, and Bill Bland, my soil physics colleagues, guided my choice to include hydrology. Robin Harris and Bill Hickey fed my interest and passion in environmental microbiology. Dr. Beat Müller of *EAWAG (Das Schweizer Eidgenössischen Technischen Hochschulen und Forschung)* was most helpful on all questions regarding ChemEQL. Robert W. Taylor, a friend dating from my Philadelphia days, exemplified encouragement. My graduate advisors—Roscoe Ellis, Murray B. McBride, and Roald Hoffman—guided my development as a young scientist, setting me on a course that ultimately led to writing this book. Edna J. Cash, a dear friend, diligently edited final drafts and proofs. Sharon—my wife—was always patient, always supportive.

William F. "Will" Bleam,
October 2010

Elements: Their Origin and Abundance

1.1. INTRODUCTION

Popular Science published an electronic version of the *Periodic Table of the Elements* in November 2006 (http://www.popsci.com/files/periodic_popup.html) that contained pictures of virtually all of the elements in pure form. Notably absent are the radioactive elements: promethium *Pm*, astatine *At*, radon *Rn*, francium *Fr*, actinium *Ac*, protactinium *Pa*, and elements beyond uranium *U*.

There is a reason images of these elements are absent: every one of them is unstable and, therefore, extremely rare. The Periodic Table of the Elements asserts that all elements exist *in principle*, but this particular table correctly implies that all of the elements from hydrogen to uranium exist on planet Earth. In fact, every sample of water, rock, sediment, and soil contains every *stable* element—and probably most of the *unstable* elements—from hydrogen to uranium.

The Environmental Working Group published an article in October 2008 entitled "Bottled Water Quality Investigation: 10 Major Brands, 38 Pollutants" (Naidenko, Leiba et al., 2008). Among the contaminants found in bottled water sold in the United States were the radioactive strontium isotope Sr-90 (0.02 $Bq \cdot L^{-1}$), radioactive radium (isotopes Ra-286 plus Ra-288; $0.02\,Bq \cdot L^{-1}$), boron (60–$90\,\mu g \cdot L^{-1}$), and arsenic ($1\,\mu g \cdot L^{-1}$). These concentrations, combined with a commentary listing the potential and actual toxicity of these substances, can be alarming. The important question is not *whether* drinking water, food, air, soil, or dust contains toxic elements; it most certainly does! The important question is: is the level of any toxic element in drinking water, food, or soils harmful?

The United States Environmental Protection Agency (USEPA) has established a *maximum contaminant level* (MCL) for beta emitters—such as strontium-90 or radium—in public drinking water: $0.296\,Bq \cdot L^{-1}$ ($8\,pCi \cdot L^{-1}$). The current USEPA drinking water MCL for arsenic is $10\,\mu g \cdot L^{-1}$; the mean arsenic concentration in U.S. groundwater (based on over 20,000 samples) is $2\,\mu g \cdot L^{-1}$. The USEPA does not have a drinking water standard for boron, but the World Health Organization (WHO) recommends boron levels in drinking water less than $500\,\mu g \cdot L^{-1}$, and in 1998 the European Union adopted a drinking water standard of $100\,\mu g \cdot L^{-1}$. Typical boron concentrations in U.S. groundwater fall below $100\,\mu g \cdot L^{-1}$ and 90% below 40 $\mu g \cdot L^{-1}$. Evaluated in context, the contaminants found in bottled water are less threatening and raise the question of whether it is reasonable to entirely eliminate trace elements from water and food.

1.2. A BRIEF HISTORY OF THE SOLAR SYSTEM AND PLANET EARTH

The gravitational collapse of a primordial cloud of gas and dust gave birth to the present-day Solar System. The conservation of angular momentum in the primordial dust cloud explains the rotation of the Sun (25-day rotation period at the equator) and a primordial accretion disk that spawned the planets and other bodies that orbit the Sun. The primordial gas and dust cloud was well mixed and uniform in composition. Gravitational accretion and radioactive decay released sufficient heat to melt the early Earth, leading segregation into a solid metallic core, a molten mantle, and a crystalline crust. Planetary formation and segregation altered the composition of Earth's crust relative to the primordial cloud and imposed variability in the composition of each element as a direct consequence of each separation process.

Throughout its entire history, planet Earth has experienced continual transformation as plate tectonics generate new inner (oceanic) crust and shift plates of outer (continental) crust around like pieces of a jigsaw puzzle. Plate tectonics and the hydrologic cycle drive a rock cycle that reworks portions of the outer crust through weathering, erosion, and sedimentation. The rock cycle imposes a new round of geochemical separation processes that alters the composition of the terrestrial land surface relative to the outer crust from which it derives. The rock cycle and soil development, much like planetary formation and segregation early in Earth's history scale, impose additional variability in the composition of each element. Transformations in the *overall composition* of planet Earth and the imposition of *composition variability* relative to the primordial gas cloud are the central themes of this chapter.

1.3. THE COMPOSITION OF EARTH'S CRUST AND SOILS

Most books on geology, soil science, or environmental science include a table listing the composition of Earth's crust or soil. There are several reasons for listing the elemental composition of Earth materials. The composition of soil constrains the biological availability of each element essential for living organisms. Soils develop from the weathering of rocks and sediments and inherit much of their composition from the local geology. The most common elements—those accounting for 90–95% of the total composition—determine the dominant mineralogy of rocks and materials that form when rocks weather (residuum, sediments, and soils).

The soil at a particular location develops in starting or *parent* material—rock or sediments—under a variety of local influences—landform relief, climate, and biological community over a period of decades or centuries. Not surprisingly, the composition of soil is largely inherited from the parent material. The rocks or sediments in which a soil develops are usually not the basement rocks that compose the bulk of Earth's crust. Regardless of geologic history, virtually every rock exposed at the Earth's terrestrial surface owes its composition to the crystalline igneous rock of the continental crust. The Earth's outer crust inherits its composition from the overall composition of the planet and, by extension, the gas and dust cloud that gave rise to the Solar System as a whole.

How much do the composition of soil, Earth's crust, and the Solar System have in common? What can we learn about the processes of planetary formation, the rock cycle, and soil development from any differences in composition? What determines the relative abundance of elements in soil?

1.4. THE ABUNDANCE OF ELEMENTS IN THE SOLAR SYSTEM, EARTH'S CRUST, AND SOILS

We are looking for patterns, and, unfortunately, data in tabular form usually do not reveal patterns in their most compelling form. Patterns in the elemental abundance of the Solar System, Earth's crust, and soils are best understood when abundance is plotted as a function of atomic number Z. Astronomers believe the composition of the Sun's photosphere is a good representation of the primordial gas and dust cloud that gave rise to the Solar System. Usually the composition of the Solar System is recorded as the atom mole fraction of each element, normalized by the atom mole fraction of silicon.

The composition of Earth's crust and soil is usually recorded as the mass fraction of each element. If we are to compare the compositions of the Solar System, Earth's crust, and soils, the abundance data must have the same units. Since atoms combine to form compounds based on atomic mole ratios, the atom mole fraction is the most informative. The atom mole fraction is found by dividing the mass fraction of each element—in crust or soil—by its atomic mass and then normalized by the atom mole fraction of silicon.

An abundance plot of the Solar System, Earth's crust, and soil using a linear atomic mole fraction scale reveals little because 99.86% of the Solar System consists of hydrogen and helium. A plot using a *logarithmic* atom mole fraction scale for the Solar System, Earth's crust, and soil reveals three important features. First, the abundance of the elements decreases exponentially with increasing atomic number Z (the decrease appears roughly linear when plotted using a logarithmic scale). Second, a zigzag pattern is superimposed on this general tread, known as the *even-odd effect*: elements with an *even* atomic number are consistently more abundant than elements with an *odd* atomic number. Third, while the data set for soil composition is missing many of the elements recorded for the Solar System and Earth's crust, the exponential decrease in abundance with atomic number (Figure 1.1) and the even-odd effect (Figure 1.2) appear in all three data sets.

Cosmological processes, beginning with the so-called *Big Bang*, determined Solar System composition. The cosmological imprint—the exponential decrease in abundance with atomic number and the even-odd effect—is clearly discernible in the composition of the Earth's crust and terrestrial soils despite the profound changes resulting from planetary formation, the rock cycle, and soil development.

1.5. ELEMENTS AND ISOTOPES

The Periodic Table of the Elements organizes all known elements into *groups* and *periods*. Elements in the same *chemical group* have the same number of electrons in the outermost *electronic shell* but differ in the number of occupied electronic shells. Elements in the same *chemical period* have the same number of occupied electronic shells but differ in the number of electrons in the partially filled outermost electronic shell. Period 1 contains elements *H* and *He*, the filling of atomic shell *1s*. Atomic shell *2s* is filled in groups 1 and 2 of period 2—elements *Li* and *Be*—while *2p* is filled in groups 13–18 of the same period. Elements of the same period exhibit similar chemical properties because their outermost or *valence* electronic shell has the same number of electrons. It is this characteristic—the *valence electron configuration*—that has the greatest influence on the chemistry of each element.

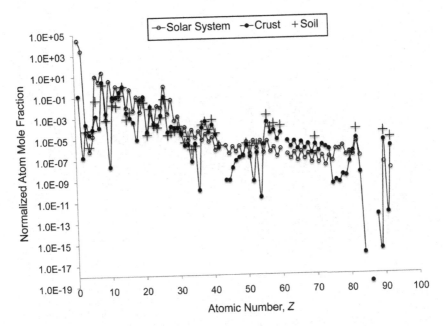

FIGURE 1.1   Elemental abundance of the Solar System, Earth's crust, and soil decreases exponentially with atomic number Z (Shacklette, 1984; Lide, 2005).

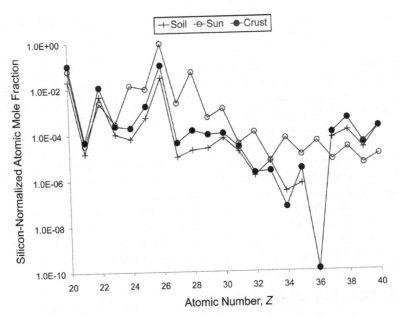

FIGURE 1.2   Elemental abundance of the Solar System, Earth's crust, and soil from calcium (atomic number Z = 20) to zirconium (atomic number Z = 40), showing the even-odd effect (Shacklette, 1984; Lide, 2005).

FIGURE 1.3 Periodic Table of the Elements: the abundance of each element in the bulk silicate Earth (i.e., mantle plus crust) plotted on the vertical scale in logarithmic units. *Source: Reproduced with permission from Anders, E., and N. Grevesse, 1989. Abundance of the elements: meteoric and solar. Geochim. Cosmochim. Acta. 53, 197–214.*

The Periodic Table of the Elements always lists the symbol, and usually the *atomic number* Z and *atomic weight,* of each element. The atomic number Z is an integer equal to the number of positively charged protons in the nucleus of that element. The atomic weight, unlike Z, is not an integer but a decimal number larger than Z. The atomic weight is the mass of one mole of the pure element and accounts for the relative abundance of the stable and long-lived radioactive *isotopes* for that element (see Appendix 1A).

Figure 1.3 is an unusual Periodic Table of the Elements because it plots the relative abundance of each element in the Earth's outermost layers—the mantle and the crust—on a logarithmic vertical scale. We will have more to say about the segregation of the Earth later, but for now we are most concerned with the processes that determine why some elements are more abundant than others. You will notice that the abundance of elements in a given period—say, period 4—fluctuate; *Ca* is more abundant than *K* or *Sc,* and *Ti* is more abundant than *Sc* or *V.* This is the even-odd effect mentioned earlier and results from the *nuclear stability* of each element, a topic discussed further in Appendix 1A.

1.6. NUCLEAR BINDING ENERGY

The even-odd effect fails to explain the abundance of iron ($Z = 26$) or, to be more precise, $^{56}_{26}Fe$. Figure 1.4 is a plot of the *nuclear binding energy* per nucleon for all known isotopes. It clearly shows $^{56}_{26}Fe$ is the isotope with the

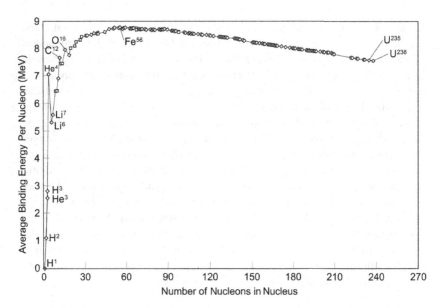

FIGURE 1.4   Binding energy per nucleon of each isotope as a function of the mass number A.

greatest binding energy per nucleon. The fusion of lighter nuclei to form heavier nuclei is exothermic when the product has a mass number $A \leq 56$ but is endothermic for all heavier isotopes. You will also notice that the energy release from fusion reactions diminishes rapidly as the mass number increases to 20 and then gradually in the range from 20 to 56.

The mass of an electron (0.000549 unified atomic mass units u) is negligible compared to that of protons (1.007276 u) and neutrons (1.008665 u). The *mass number* A of each isotope is an integral sum of protons Z and neutrons N in the nucleus, while the *isotope mass* is the mass of one mole of pure isotope. The mass of an isotope with mass number A is always lower than the mass of Z-independent protons and N-independent neutrons. The nuclear binding energy of hydrogen isotope $_{1}^{2}H$ (Example 1.1) illustrates the so-called *mass defect*.

Example 1.1 Calculate the Nuclear Binding Energy for Deuterium $_{1}^{2}H$, an Isotope of Hydrogen.

The nucleus of a deuterium atom consists of one proton and one neutron. The masses of the constituents, in unified atomic mass units u (1 $u = 1.66053886 \times 10^{-27}$ kg):

$$m_{proton} = 1.007276 \ u$$
$$m_{neutron} = 1.008665 \ u$$
$$m_{proton} + m_{neutron} = 2.015941 \ u$$

The mass of one mole of pure $_{1}^{2}H$, however, is 2.014102 u—0.001839 u less than the mass found by adding the rest mass of a proton and a neutron. The mass difference, multiplied by 931.494 $MeV \ u^{-1}$ and divided by 2 to account for the two nucleons of $_{1}^{2}H$, is the binding energy per nucleon for $_{1}^{2}H$: 0.8565 MeV.

Exothermic fusion reactions are the source of the energy output from stars, beginning with ^1_1H nuclei and other nuclei produced during the early stages of the universe following the Big Bang. Figure 1.4 shows that exothermic fusion becomes increasingly inefficient as an energy source. Simply put: a star's nuclear fuel is depleted during its lifetime and is eventually exhausted. Appendix 1B describes the nuclear processes that formed every isotope found in our Solar System and the Earth today. Some of those processes occurred during the early moments of the universe under conditions that no longer exist. Other processes continue to occur throughout the universe because the necessary conditions exist in active stars or in the interstellar medium. Nucleosynthesis is an accumulation process that gradually produces heavier nuclides from lighter nuclides, principally by exothermic fusion or neutron capture. Lighter nuclides are generally more abundant than heavier nuclides. The extraordinary energies required for exothermic nuclear fusion and the complex interplay of neutron-capture and radioactive decay, the principal nucleosynthesis processes, translate into an abundance profile (see Figure 1.1) that embodies the relative stability of each nuclide.

1.7. ENRICHMENT AND DEPLETION DURING PLANETARY FORMATION

Gravitational collapse of a primordial cloud of gas and dust and subsequent planetary formation in the Solar System was a process of *differentiation*— the transformation from a homogeneous state to a heterogeneous state—on scales ranging from the scale of the Solar System itself to the smallest microscopic scale we can characterize experimentally.

Homogeneity in the primordial gas cloud that ultimately became the Solar System resulted from the random motion of gas molecules and dust particles. If we collected samples from the primordial cloud for elemental analysis, the sample population would yield a frequency distribution consistent with the conditions existing in the gas cloud. A homogenous, well-mixed cloud of gas and dust would yield a *normal distribution* of concentrations for each element (Figure 1.5). The average (or *central tendency*) concentration of each element is estimated by calculating the *arithmetic mean* concentration for the sample population.

A consequence of the various separation processes during planetary accretion and segregation, the rock cycle on planet Earth, and, ultimately, soil formation is more than the enrichment and depletion of individual elements, altering the abundance profile (see Figures 1.2 and 1.3). Elemental analysis shows that the weathered rock materials and soils that blanket the Earth's continental crust bear the imprint of numerous separation processes and, as we will presently discover, yield abundance distributions that differ markedly from the normal distribution shown in Figure 1.5.

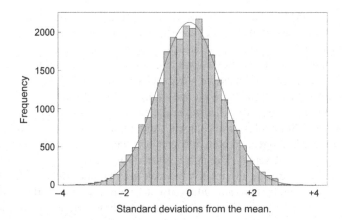

FIGURE 1.5 The frequency of concentrations from a sample population (bar graph) closely approximates a normal probability density function (line). The mean (or central tendency) for a normal probability distribution equals the arithmetic mean for the sample population. *Source: Image generated using "Sample versus Theoretical Distribution" (Wolfram Demonstration Project) using 40 bins and a sample size* n = 4118.

1.8. PLANETARY ACCRETION

The Solar System formed during the gravitational collapse of a gas cloud that ultimately became the Sun. Preservation of angular momentum within the collapsing gas cloud concentrated dust particles into a dense rotating central body—the Sun—and a diffuse disc (Figure 1.6, left) perpendicular to the axis of the Sun's rotation. Gravity led to the accretion of planetesimals within the disc that grew in size as they collided with one another.

The Solar System formed about 4 billion years ago and contains 8 major planets and 153 confirmed moons, asteroids, meteors, comets, and uncounted planetesimals at its outer fringe. The innermost planets—Mercury, Venus, Earth, and Mars—are rocky (density ≈ 5 *Mg* m^{-3}), while the outermost

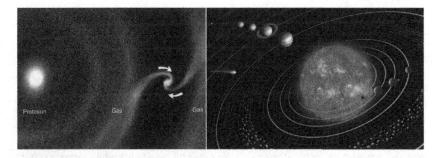

FIGURE 1.6 Planetesimal formation in the accretion disc of a protosun (left) and the present Solar System (right) resulted in the differentiation and segregation of the primordial gas cloud.

planets—Jupiter, Saturn, Uranus, and Neptune—are icy (density <2 Mg m^{-3}). Earth is the largest of the rocky planets, with a radius of nearly 6370 km. Mars, mean radius 3390 km, has the most moons of the rocky planets: two. The icy planets are much larger—ranging from roughly 4 to 11 times the radius of Earth; are more massive—from 17 to 318 times the mass of Earth; and are surrounded by far more moons—ranging from 13 to 63.

Goldschmidt (1937) identified several element groups based on geochemical behavior: lithophiles, chalcophiles, atmophiles, and siderophiles. The atmophiles are hydrogen, carbon, nitrogen, and the noble gases helium, neon, argon, krypton, and xenon. The noble gases do not combine with other elements and easily escape the relatively weak gravitational field of the rocky planets. Hydrogen, carbon, and nitrogen react with oxygen, but the products tend to be volatile gases. The gaseous planets, however, are cold enough and massive enough to capture atmophilic gases. Earth and the other rocky planets are depleted of the noble gas elements, hydrogen and other volatile elements relative to the Solar System, as shown in Figure 1.7.

Planet Earth began segregating into layers as the heat released by accretion and radioactive decay triggered melting. Modern Earth consists of a metallic core composed of *ferromagnetic* elements and an outer silicate-rich layer—sometimes called the *basic silicate Earth*—that further separated into a

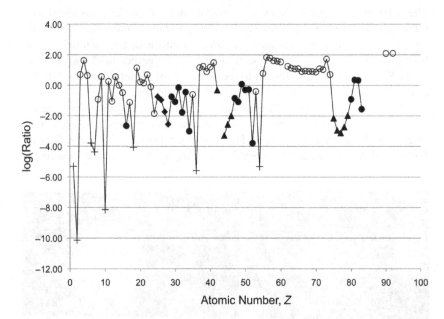

FIGURE 1.7 Segregation of the Earth's crust relative to the Solar System: atmophilic elements (crosses), ferromagnetic metals (diamonds), noble metals (triangles), chalcophilic elements (filled circles), and lithophilic elements (open circles). Ferromagnetic transition metals and noble metals are grouped as siderophilic elements (Goldschmidt, 1937; Lide, 2005).

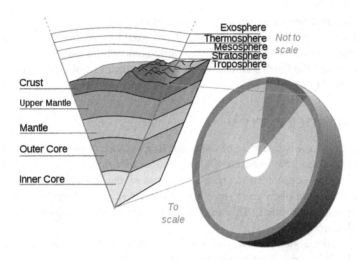

FIGURE 1.8   Segregated Earth: core, mantle, and crust.

molten mantle and crust of crystalline silicate rock (Figure 1.8). The silicate-rich crust is further segregated into an inner or oceanic crust and an outer or continental crust. The inner crust has a composition and density very similar to the mantle, while the outer crust is less dense, thicker, and far more rigid than the inner crust. The composition of the outer crust reflects the effect of segregation on the planetary scale but bears little imprint from the rock cycle because the mean age of the outer crust is roughly the same as the age of planet Earth.

The ferromagnetic elements comprise four transition metals from the fourth period: manganese Mn, iron Fe, cobalt Co, and nickel Ni. These four metals are depleted in Earth's crust (see Figure 1.7), having become major components of the Earth's metallic core. The noble metals—ruthenium Ru, rhodium Rh, palladium Pd, silver Ag, rhenium Re, osmium Os, iridium Ir, platinum Pt, and gold Au—have little tendency to react with either oxygen or sulfur. These elements are depleted from the basic silicate Earth, enriching the metallic core. Goldschmidt (1937) grouped the ferromagnetic transition metals and noble metals together as *siderophilic* metals.

The lithophilic elements (Goldschmidt, 1937) consist of all elements in the periodic table characterized by their strong tendency to react with oxygen. The chalcophilic elements—which include sulfur and the elements from periods 4 (copper through selenium), 5 (silver through tellurium), and 6 (mercury through polonium)—combine strongly with sulfur. The chalcophilic and lithophilic elements would be enriched in basic silicate Earth after the segregation of the metallic core.

Given the atomic mass of oxygen and sulfur and the relative atomic mass of lithophilic chalcophilic elements in the same period; it should not be surprising that sulfide minerals are significantly denser than oxide minerals. Buoyancy in the Earth's gravitational field would cause lithophilic magma to migrate toward the Earth's surface, while chalcophilic magma would tend to sink toward the Earth's core, thereby depleting the crust of chalcophilic elements (see Figure 1.7).

1.9. THE ROCK CYCLE

The Earth's crust is in a state of continual flux because convection currents in the mantle form new inner crust along midoceanic ridges. The spreading of newly formed oceanic crust drives the edges of the inner crust underneath the more buoyant outer crust along their boundaries, propelling slabs of outer crust against each other. This movement—*plate tectonics*—drives the rock cycle (Figure 1.9).

If we take a global view, the outer crust is largely composed of granite, while the inner crust is primarily basalt. Granite is a coarse-grained rock that crystallizes slowly and at much higher temperatures than basalt. Basalt rock has a density comparable to the mantle; it solidifies along midoceanic ridges, where convection currents in the mantle carry *magma* to the surface. Geologists classify granite, basalt, and other rocks that solidify from the molten state as *igneous* rocks.

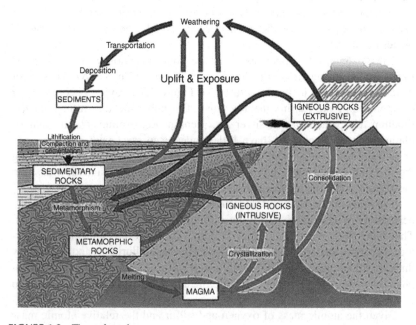

FIGURE 1.9 The rock cycle.

A more detailed view of the crust on the scale of kilometers to tens of kilometers reveals other rocks that owe their existence to volcanic activity, weathering, erosion, and sedimentation. Magma welling up from the mantle melts through the crust. Rock formations along the margin of the melted zone are annealed; minerals in the rock recrystallize in a process geologists call *metamorphism*. Both the mineralogy and texture of metamorphic rocks reveal radical transformation under conditions just short of melting. Metamorphic rocks provide geologists with valuable clues about Earth history, but they are far less abundant than igneous rocks.

Inspection on the kilometer scale reveals a zone of rock weathering and sediment deposits covering both the continental and oceanic crusts. Stress fractures, caused by thermal expansion and contraction, create a pathway for liquid water to penetrate igneous and metamorphic rock formations. Freezing and thawing accelerate fracturing, exfoliating layers of rock and exposing it to erosion by flowing water, wind, gravity, and ice at the Earth surface. The water itself, along with oxygen, carbon dioxide, and other compounds dissolved in water, promotes reactions that chemically degrade minerals formed at high temperatures in the absence of liquid water, precipitating new minerals that are stable in the presence of liquid water.

Burial, combined with chemical cementation by the action of pore water, eventually transforms sedimentary deposits into sedimentary rocks. *Primary* minerals—minerals that crystallize at high temperatures where liquid water does not exist—comprise igneous and metamorphic rocks. *Secondary* minerals—those that form through the action of liquid water—form in surface deposits and sedimentary rocks.

The rock cycle (see Figure 1.9) is completed as plate tectonics carry oceanic crust and their sedimentary overburden downward into the mantle at the continental margins, or magma melting upward from the mantle returns rock into its original molten state.

1.10. SOIL FORMATION

The rock cycle leaves much of the outer crust untouched. Soil development occurs as rock weathering alters rocks exposed on the terrestrial surface of the Earth's crust. Weathered rock materials, sediments and saprolite, undergo further differentiation during the process of soil formation. Picture freshly deposited or exposed geologic formations (sediments deposited following a flood or landslide, terrain exposed by a retreating glacier, fresh volcanic ash and lava deposits following a volcanic eruption). This fresh parent material transforms through the process of chemical weathering, whose intensity depends on climate (rainfall and temperature), vegetation characteristic of the climate zone, and topography (drainage and erosion). Over time these processes and unique characteristics of the setting lead to the development of soil.

FIGURE 1.10   A northern temperate forest soil developed in glacial till (Northern Ireland).

The excavation of a pit into the soil reveals a sequence of layers or *soil horizons* with depth (Figure 1.10). This sequence of soil horizons is known as the *soil profile* for that site. Soil horizons vary in depth, thickness, composition, physical properties, particle size distribution, color, and other properties. Soil formation takes hundreds to thousands of years, and, provided the site has remained undisturbed for that length of time, the soil profile reflects the natural history of the setting.

Plotting the abundance of each element in the Earth's crust divided by its abundance in the Solar System reveals those elements enriched or depleted during formation of planet Earth and the formation of the outer crust during the segregation of the Earth (see Figure 1.7). Plotting the abundance of each element soil divided by its abundance in the Earth's crust reveals those elements enriched or depleted during the rock cycle and soil development. When both are displayed together on the same graph (Figure 1.11), plotting the logarithm of the abundance ratio of crust to the Solar System and the ratio of soil to crust, an important and not surprising result is immediately apparent: enrichment and depletion during planetary formation and segregation are orders of magnitude greater than those that take place during the rock cycle and soil formation. This explains the findings (Helmke, 2000) that a characteristic soil composition cannot be related to geographic region or soil taxonomic group.

Enrichment from biological activity is clearly apparent when plotting the logarithm of the atom mole fraction ratio soil-to-crust (Figure 1.12). Carbon and nitrogen, two elements strongly depleted in the Earth's crust relative to the Solar System, are the most enriched in soil relative to the crust. Both

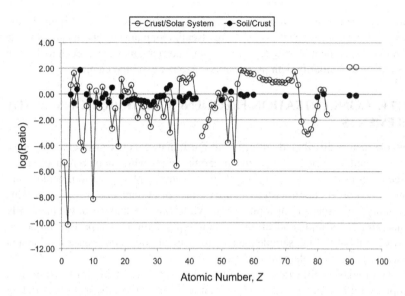

FIGURE 1.11   Elemental abundance of the Solar System relative to the Earth's crust and soil relative to the Earth's crust reveal the magnitude of planetary formation, the rock cycle, and soil formation (Shacklette, 1984; Lide, 2005).

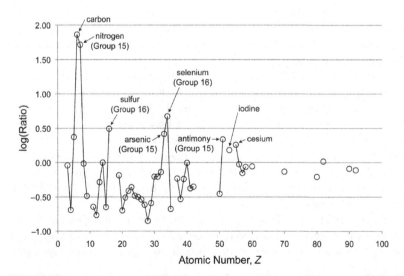

FIGURE 1.12   Elemental abundance of soil relative to the Earth's crust, showing enrichment of carbon, nitrogen, and sulfur caused by biological activity (Shacklette, 1984).

are major constituents of biomass. Selenium, sulfur, and boron—among the top six most enriched elements in soil—are essential for plant and animal growth. The relative enrichment of arsenic, which is not a nutrient, may have much to do with its chemical similarity with phosphorus—another essential element.

1.11. CONCENTRATION FREQUENCY DISTRIBUTIONS OF THE ELEMENTS

No element is distributed uniformly in Earth materials because separation processes tend to enrich an element in certain zones while depleting it in others (see Figures 1.10 and 1.11). Geologists, soil scientists, environmental chemists, and others often wish to identify zones of enrichment or depletion. This requires collecting samples representative of the site under study, chemically analyzing the sample collection (or sample population), and performing a statistical analysis of the sample data set to estimate the *mean*[1] concentration and variability of each element analyzed.

Geologists delineate zones of enrichment (or depletion) using this approach while prospecting for economic mining deposits or reconstructing the geologic history of rock formations. Soil scientists rely on chemical analysis to classify certain soil horizons characterized by either the loss or accumulation (see Figure 1.10). Environmental chemists map contamination and assess the effectiveness of remediation treatments by analyzing sample populations. In short, statistical analysis of a representative sample population will establish whether enrichment or depletion is significant.

Natural variation at the field site results in an abundance distribution for each element in the sample population. Understanding the abundance distribution typical of environmental samples enables us to identify appropriate statistical methods for estimating the *central tendency* and variation for each element in the sample population.

Figures 1.13 and 1.14 illustrate typical abundance distributions for environmental samples. The data in Figure 1.13 represent samples where geochemical processes dominate, while Figure 1.14 shows samples where the dominant processes are biological. Table 1.1 lists a number of studies reporting abundance distributions from the chemical analysis of rock, soil, water, and atmospheric samples. The results in Figures 1.13 and 1.14, along with the studies listed in Table 1.1, clearly show a characteristic property of frequency distributions for environmental samples: they tend to be skewed rather than symmetric (see Figure 1.8).

Ahrens (1954) discovered that the skewed frequency distributions usually encountered by geochemists tend to follow a *logarithmic-transformed normal* (or more commonly, *log-normal*) distribution. Biologists consider log-normal

---

1. The *mean* is one measure of the *central tendency* of a distribution. *Median* and *mode* are other statistics that serve as central tendency indicators.

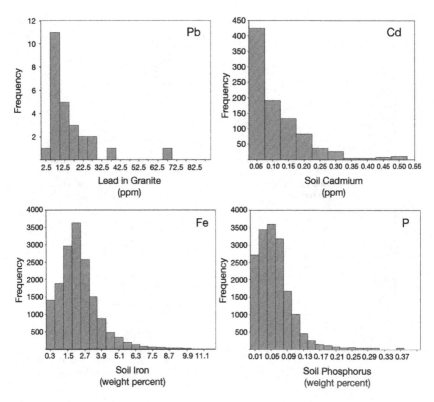

FIGURE 1.13   Abundance distributions for the elements Pb in Canadian granite (Ahrens, 1954), soil Cd from Taiwan (Yang and Chang, 2005), and Fe and P from the USGS National Geochemical Survey data set of U.S. soils and sediments (Zhang, Manheim et al., 2005). The number of samples n in each study is, respectively, 32 (Pb), 918 (Cd), and 16,511 (Fe and P).

distributions typical of biological processes (Limpert, Stahel et al., 2001). A logarithmic-transformed normal concentration distribution is a skewed distribution that under a logarithmic-transformation (i.e., a plot of *logarithm of sample concentrations*) becomes a symmetric *normal* distribution. Many environmental chemists, geochemists, and biologists estimate the *geometric mean* concentration (i.e., a *central tendency* statistic for log-normal distributions) and *geometric standard deviation* for their data sets.

Some of the studies listed in Table 1.1 assume their data set follows a log-normal distribution (Dudka, Ponce-Hernandez et al., 1995; Chen, Ma et al., 1999; Budtz-Jørgensen, Keiding et al., 2002), while others perform a statistical analysis to verify that the concentration distribution is best described by a log-normal distribution function (Hadley and Toumi, 2003). In some cases the concentration distributions may not conform to a true log-normal distribution, yet they clearly deviate from a normal distribution (Yang and Chang, 2005; Zhang, Manheim et al., 2005).

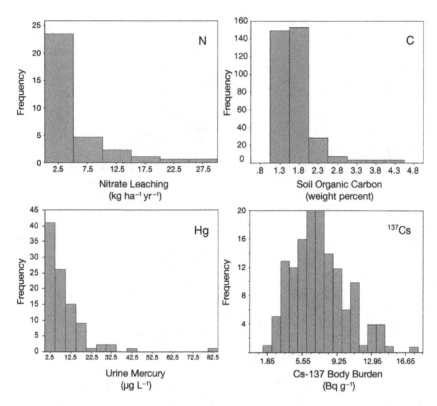

FIGURE 1.14 Frequency distributions for nitrate nitrogen leaching from German soils (Borken and Matzner, 2004), soil organic carbon from China (Liu, Wang et al., 2006), urinary mercury levels in a human population in Rio-Rato, Brazil (Silva, Nyland et al., 2004), and Cs-127 levels in a species of tree frogs (*Hyla cinerea*) near a nuclear facility in South Carolina (Dapson and Kaplan, 1975). The number of samples *n* in each study is, respectively, 57 (nitrate-nitrogen), 354 (soil organic carbon), 98 (Hg), and 141 (Cs-137).

Appendix 1E describes two models of environmental processes that result in log-normal concentration frequency distributions. The simplest to understand is the *sequential random dilution* model (Ott, 1990). The second model, known as the *law of proportionate effect* (Kapteyn, 1903), is completely general.

1.12. ESTIMATING THE MOST PROBABLE CONCENTRATION AND CONCENTRATION RANGE USING THE LOGARITHMIC TRANSFORMATION

Each type of frequency distribution results from processes with specific characteristics. For now we will rely on the same assumption as Ahrens (1954); a log-normal distribution is a reasonable representation of the abundance distribution of trace elements in the environment.

TABLE 1.1 Studies Reporting Frequency Distributions for Environmental Samples

Medium	Elements Analyzed	Sample Size	Reference
Rock	K, Rb, Sc, V, Co, Ga, Cr, Zr, La, Cs, F, Mo	158	(Ahrens, 1966)
Surface Material	Al, Ca, Fe, K, Mg, Na, P, Ti, Ba, Ce, Co, Cr, Cu, Ga, La, Li, Mn, Nb, Nd, Ni, Pb, Sc, Sr, Th, V, Y, Zn	16,511	(Zhang, Manheim et al., 2005)
Soil	Cd, Co, Cu, Cr, Fe, Mn, Ni, S, Zn	73	(Dudka, Ponce-Hernandez et al., 1995)
Soil	Ag, As, Ba, Be, Cd, Cr, Cu, Hg, Mn, Mo, Ni, Pb, Sb, Se, Zn	448	(Chen, Ma et al., 1999)
Soil	Cd	918	(Yang and Chang, 2005)
Soil	C	354	(Liu, Wang et al., 2006)
Groundwater	Ag, As, Ba, Be, Cd, Co, Cr, Cu, Hg, Ni, Pb, Sb, Se, Tl, Sn, V, Zn	104,280	(Newcomb and Rimstidt, 2002)
Groundwater	N	57	(Borken and Matzner, 2004)
Atmosphere	SO_2	>365	(Hadley and Toumi, 2003)
Human Tissue	Hg	98–140	(Silva, Nyland et al., 2004)
Human Tissue	methylmercury	1022	(Budtz-Jørgensen, Keiding et al., 2002)
Green Tree Frogs	Cs-137	141	(Dapson and Kaplan, 1975)

The best way to understand log-normal distributions is to consider a real example. The example we will use is the data set published by Ahrens (1954) consisting of Pb concentrations from a sample population of Canadian granites (Table 1.2; see Figures 1.13 and 1.15). The upper portion of Table 1.2 lists the original data set based on Pb concentrations in parts per million ($mg \cdot kg^{-1}$). The lower portion of Table 1.2 lists the transformed data set based on the natural logarithm of Pb concentrations in the original data set. The

TABLE 1.2 Estimates of the Central Tendency and the Variation Encompassing 95.5% of the Pb Concentrations in Canadian Granite Samples

Original Data Set		
Data Set $(mg \cdot kg^{-1})$	{2.0, 3.8, 5.5, 6.0, 6.0, 6.5, 7.5, 7.7, 8.0, 8.0, 8.5, 8.5, 9.0, 9.0, 9.0, 10.0, 10.0, 11.0, 12.0, 14.0, 14.0, 15.0, 17.0, 18.0, 20.0, 20.0, 22.0, 24.0, 26.0, 29.0, 39.0, 68.0}	
Arithmetic Mean	$\bar{x} = \left(\frac{1}{n}\right) \cdot \sum_i x_i$	$\bar{x} = 14.8$
Arithmetic Standard Deviation	$\sigma = \sqrt{\left(\frac{1}{n}\right) \cdot \sum_i (x_i - \bar{x})^2}$	$\sigma = 12.7$
95.5% Range	$[(\bar{x} - 2 \cdot \sigma) \leq x \leq (\bar{x} + 2 \cdot \sigma)]$	$[2.1\ mg \cdot kg^{-1} \leq x \leq 27.5\ mg \cdot kg^{-1}]$
Logarithmic-Transformed Data Set		
Transformed Data Set	{0.69, 1.34, 1.70, 1.79, 1.79, 1.87, 2.01, 2.04, 2.08, 2.08, 2.14, 2.14, 2.20, 2.20, 2.20, 2.30, 2.30, 2.40, 2.48, 2.64, 2.64, 2.71, 2.83, 2.89, 3.00, 3.00, 3.09, 3.18, 3.26, 3.37, 3.66, 4.22}	
Geometric Mean	$\bar{x}_G = \exp\left(\left(\frac{1}{n}\right) \cdot \sum_i \ln x_i\right) = \left(\prod_i x_i\right)^{\frac{1}{n}}$	$\bar{x}_G = 11.5 = \exp(2.45)$
Geometric Standard Deviation	$\sigma_G = \exp\left(\sqrt{\left(\frac{1}{n}\right) \cdot \sum_i (\ln x_i - \ln\bar{x}_G)^2}\right)$	$\sigma_G = 2.02 = \exp(0.702)$
95.5% Range	$\left[(\bar{x}_G \div \sigma_G^2) \leq x \leq (\bar{x}_G \times \sigma_G^2)\right]$	$[2.9\ mg \cdot kg^{-1} \leq x \leq 46.0\ mg \cdot kg^{-1}]$

Source: Ahrens, 1954.

arithmetic mean and *arithmetic standard deviation* are central tendency and variation statistics of the original data set, while the transformed data set employs the *geometric mean* and *geometric standard deviation*. See Appendix 1F for another example from Ahrens (1954).

Figure 1.15 plots the Pb abundance distribution on a linear concentration scale and the log-normal distribution using the data sets in Table 1.2. Although the arithmetic and geometric means do not differ much from one another, the geometric mean is clearly a better estimate of the *average* Pb concentration than the arithmetic mean, regardless of which concentration scale is used for plotting. The major difference, a difference crucial for establishing whether the most probable values of two different sample populations are significantly different from one another, is the range estimate.

Range estimates are designed to cover a specified interval for the variable in question. Normal distributions are *symmetric about the arithmetic mean*, so

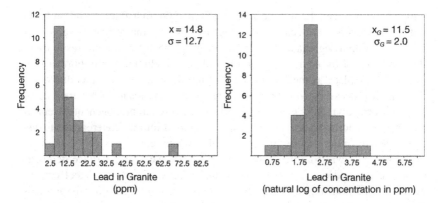

FIGURE 1.15 Frequency distributions of Pb in Canadian granite (Ahrens, 1954): original distribution (left) and logarithmic-transformed distribution (right). The appropriate central tendency concentration and variation appear in Table 1.2.

the range is symmetric about the mean. For example, a range extending from one standard deviation below the mean to one standard deviation above the mean covers 68.3% of the population, while a range spanning two standard deviations from the mean covers 95.5% (see Table 1.2). Log-normal distributions are asymmetric about the geometric mean. The lower limit of a range covering 68.3% of the population is the *geometric mean divided by the geometric standard deviation* while the upper limit is the *geometric mean multiplied by the geometric standard deviation*. Table 1.2 lists the 95.5% range for our logarithmic-transformed data set. The skewed nature of the Pb concentrations (see Figure 1.15, left) clearly demands the skewed range estimate provided by log-normal distribution statistics rather than the symmetric range estimate based on normal distribution statistics.

1.13. SUMMARY

The evolution of the universe and the Solar System, ultimately, determined the composition of planet Earth. Gravitational collapse of primordial gas clouds formed first-generation stars, triggering exothermic fusion reactions that generated sufficient heat to forestall complete collapse and producing a series of heavier elements culminating in the isotope $^{56}_{26}Fe$. The sequential *burning* of lighter elements produces heavier elements with less efficiency in each subsequent stage. Eventually the first-generation stars depleted their nuclear fuel, either collapsing into obscurity or exploding into supernovae that seeded the interstellar medium with heavy elements.

Later stellar generations repeat the process, producing still heavier elements by neutron capture and beta decay either slowly throughout the lifetime of the star (*s-process*) or in the paroxysm of a supernova (*r-process*). *R-process* neutron capture generates a host of radioactive isotopes, some of which have survived the formation of the Solar System and are found on Earth today, and others that decayed to lighter isotopes by the present time.

Solar System segregation occurred during planetary accretion, resulting in the formation of the four inner terrestrial planets, including Earth. Further segregation took place as the Earth separated into layers: core, mantle, and crust. Separation processes continue to the present day, the result of the rock cycle and soil formation. Segregation transformed the original homogeneous element distribution inherited from the collapsing gas cloud into the heterogeneous distributions we find in the Earth's terrestrial environment.

Samples drawn from the environment—rocks, weathered surface deposits, soils, and groundwater—reflect this heterogeneity. Efforts to detect enrichment or depletion require statistical analysis of sample populations of sufficient size to capture natural variability. Chemical analysis of environmental samples typically yields data sets with skewed frequency distributions. Estimates of the mean concentration of an element and the concentration variation of the sample population usually rely on the logarithmic transformation of the data set, a practice justified because log-normal distributions are generally good representations of natural abundance distributions.

APPENDIX 1A. FACTORS GOVERNING NUCLEAR STABILITY AND ISOTOPE ABUNDANCE

1A.1. The Table of Isotopes and Nuclear Magic Numbers

The Table of Isotopes (Figure 1A.1) organizes all known isotopes and elements into columns and rows, listing important information about each isotope: atomic number Z, neutron number N, mass number A, isotopic rest mass, binding energy, stability, half-life, and abundance.

Stable isotopes (black squares in Figure 1A.1) occupy a narrow band passing through the center of the known isotopes in the Table of Isotopes. Bordering this narrow *peninsula of stability* is a zone of radioactive (i.e., unstable) isotopes that are increasingly less stable (i.e., shorter half-lives) as the isotopes become increasingly proton-rich or neutron-rich. Lighter shades of gray in Figure 1A.1 represent increasingly less stable isotopes. Physicists and nuclear chemists call the several processes that form elements and isotopes *nucleosynthesis*.

There are parallels between the Periodic Table of the Elements and the Table of Isotopes. Protons and neutrons in the nucleus are also organized into shells; isotopes exhibit significantly greater stability when a *nuclear shell* is completely filled by either protons or neutrons or when there is an even number of protons or neutrons.

1A.2. Nuclear Magic Numbers

The Periodic Table of the Elements owes its periodicity to the filling of electronic shells. It takes 2, 8, 10, and 14 electrons to complete electronic shells

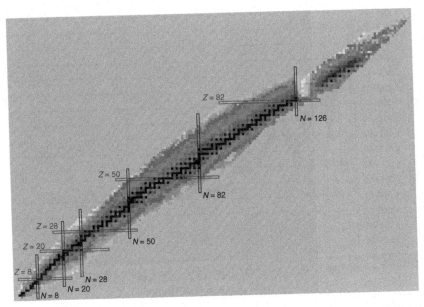

FIGURE 1A.1 The Table of Isotopes, with each square representing a known isotope and the shade representing stability. *Source: Reproduced with permission from NDDC, 2009. Interactive Chart of Nuclides. Retrieved August 26, 2009, from http://www.nndc.bnl.gov/chart/.*

designated s, p, d, and f. Chemists noticed that when elements react, they tend to lose, gain, or share a certain number of electrons. The *Octet Rule* associates unusual chemical stability with the sharing or transfer of electrons that result in completely filled s and p electronic shells. The preferred oxidation states of an element or charge of an ion can usually be explained by stability conferred by a filled electron shell. A filled electron shell and covalent chemical bonds increment by two spin-paired electrons.

Completely filled *nuclear shells* contain a *magic number* (2, 8, 20, 28, 50, 82, and 126) of nucleons. Nucleons are either protons or neutrons. The filling of proton nuclear shells is independent of the filling of neutron nuclear shells. Isotopes with *completely filled nuclear shells* have significantly greater stability—and greater abundance—than isotopes with partially filled nuclear shells. Isotopes with half-filled nuclear shells or shells with an even number of spin-paired nucleons represent stable nuclear configurations and explain the even-odd effect seen so clearly in the composition of the Solar System (Figure 1A.2) and, to a lesser degree in the Earth's crust and soil (Figure 1A.3), and in the number of stable isotopes for each element.

Let's make this very clear: atomic number Z or proton count determines the identity of each element, while neutron count N identifies each isotope. A magic atomic number Z or a half-filled nuclear proton shell (an atomic number Z of 1, 4, 10, 14, etc.) or an even number of protons confers added

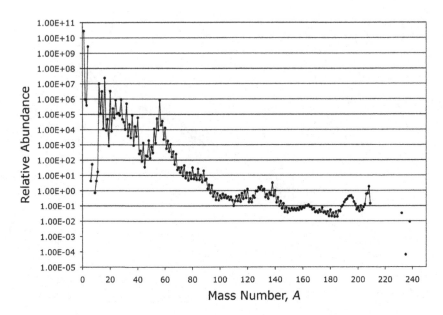

FIGURE 1A.2 Isotope abundance decreases with increasing mass number A. The abundance is the isotope mole fraction normalized by $^{28}_{14}Si$ (Anders and Grevesse, 1989).

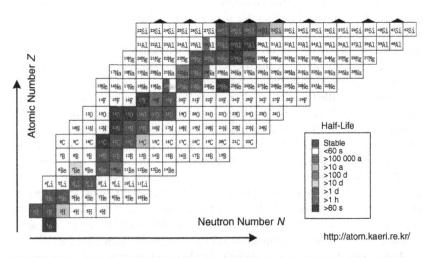

FIGURE 1A.3 A portion of the Table of Isotopes: the darkest shade of gray identifies stable isotopes, while lighter shades of gray indicate radioactive isotopes with progressively shorter half-lives. *Source: Reproduced with permission from NDDC, 2009. Interactive Chart of Nuclides. Retrieved August 26, 2009, from http://www.nndc.bnl.gov/chart/.*

stability to the element that translates into greater abundance than elements lacking those characteristics. A magic neutron number N or a half-filled nuclear neutron shell (a neutron number N of 1, 4, 10, 14, etc.) or an even number of neutrons confers added stability to the isotope, which translates into greater stability than isotopes lacking those characteristics.

A portion of the Table of Isotopes appears in Figure 1A.3. The elements, represented by their atomic number Z, appear as columns in this version, while the isotopes, represented by the number of neutrons in the nucleus N, appear as rows. The columns and rows are not completely filled because only known isotopes are shown. Stable isotopes appear in the darkest shade of gray, while unstable isotopes are shaded in lighter shades of gray and white, depending on the half-life of each radioactive isotope. The half-life of a radioactive isotope is the time it takes for half of the isotope to undergo radioactive decay into another isotope. Figure 1A.3 helps us understand how isotope stability contributes to element abundance.

Hydrogen isotope $_{1}^{3}H$—$N = 2$, a magic neutron number—is the only recognized unstable hydrogen isotope. Helium isotopes span a range from $_{2}^{3}He$—$N = 1$, a half-filled nuclear neutron shell—to $_{2}^{10}He$—$N = 8$, a magic neutron number. Oxygen isotopes span a range from $_{8}^{12}O$—$N = 4$, a half-filled nuclear neutron shell and an even number—to $_{8}^{24}O$—$N = 16$, an even number—with the most abundant isotope being $_{8}^{16}O$ with a magic atomic number Z and a magic neutron number N.

APPENDIX 1B. NUCLEOSYNTHESIS

1B.1. Nuclear Reactions

The particles that make up each element—protons, neutrons, and electrons—condensed from more fundamental particles in the early instants of the universe following the Big Bang. In those early moments, none of the elements that now fill the Periodic Table of the Elements existed other than the precursors of hydrogen: protons, neutrons, and electrons. All of the elements that now exist in the universe result from the nuclear reactions listed in Table 1B.1.

The last three nuclear reactions, along with fission reactions, result from the spontaneous decay of unstable or radioactive isotopes. Radioactive decay results in the emission of alpha particles ($_{2}^{4}He$ nuclei), electrons (beta-minus particles: $_{-1}^{0}e$), positrons (beta-plus particles: $_{+1}^{0}e$), or gamma rays (γ).

Fusion reactions produce proton-rich isotopes that, if they are unstable, decay by positron emission (beta plus decay)—a process that effectively converts a proton to a neutron in the nucleus coupled with the emission of a positron $_{+1}^{0}e$ and an electron neutrino v_e to preserve charge and nuclear spin. Neutron-capture reactions produce neutron-rich isotopes that are often unstable, leading to spontaneous beta decay emitting an electron $_{-1}^{0}e$ and an electron antineutrino $\bar{v}_e$ while converting a neutron to a proton in the nucleus.

TABLE 1B.1 Nucleosynthesis Reactions Yielding Elements and Isotopes

Nuclear Reaction	Example
Fusion	$^2_1H + ^1_1H \xrightarrow{\text{nuclear fusion}} \, ^3_2He + \gamma$
	$^1_1H + ^1_1H \xrightarrow{\text{nuclear fusion}} \, ^2_1H + ^0_{+1}e + \nu_e$
	$^4_2He + ^{13}_6C \xrightarrow{\text{nuclear fusion}} \, ^{16}_8O + ^1_0n$
Electron Capture	$^7_4Be + ^0_{-1}e \xrightarrow{\text{electron capture}} \, ^7_3Li + \nu_e$
Neutron Capture	$^{59}_{26}Fe + ^1_0n \xrightarrow{\text{neutron capture}} \, ^{60}_{26}Fe + \gamma$
Spallation	$^1_1H + ^{14}_7N \xrightarrow{\text{cosmic spallation}} \, ^9_4Be + ^4_2He + 2^1_1H$
Fission	$^{238}_{92}U \xrightarrow{\text{spontaneous fission}} \, ^{143}_{55}Cs + ^{90}_{37}Rb + 3^1_0n + \gamma$
Positron Emission	$^{13}_7N \xrightarrow{\text{positron emission}} \, ^{13}_6C + ^0_{+1}e + \nu_e$
Beta Decay	$^{60}_{26}Fe \xrightarrow{\text{beta decay}} \, ^{60}_{27}Co + ^0_{-1}e + \bar{\nu}_e$
Alpha Decay	$^{210}_{84}Po \xrightarrow{\text{alpha decay}} \, ^{206}_{82}Pb + ^4_2He$

Thus, fusion is often coupled with positron emission, while neutron capture commonly leads to beta decay.

1B.2. Nuclear Fusion

The early universe was very hot and dense—ideal conditions for nucleosynthesis but short-lived because the universe was also rapidly expanding. Rapid expansion cooled the early universe to temperatures that could no longer sustain nuclear fusion within a few minutes after the Big Bang. The composition of the universe was limited to the handful of elements that formed during the early minutes of its existence until about 100 million years later (Larson and Bromm, 2001), when the first stars formed and the second stage of stellar nucleosynthesis began.

Primordial nucleosynthesis began a fraction of a second after the Big Bang, when the temperature of the universe had dropped to 10^9 K, low enough for protons to capture neutrons, forming hydrogen isotopes and heavier elements through nuclear fusion. The absence of stable isotopes with $A = 5$ and $A = 8$ (black boxes on diagonal, Figure 1A.3) created a bottleneck for primordial nucleosynthesis beyond ^7_3Li and ^7_4Be.

Free protons and electrons are stable, but neutrons can survive only 10.3 m in an unbound state. The neutron-capture rate drops as the temperature of the expanding universe cools below 10^9 K, resulting in the decay of unbound neutrons to protons, electrons, and electron antineutrinos $\bar{v}_e$, bringing primordial nucleosynthesis to a close.

$$\,_0^1 n \xrightarrow{t_{1/2}=10.3\,m} \,_1^1 H + \,_{-1}^0 e + \bar{v}_e$$

About 100–250 million years after the Big Bang, gravitational collapse of gas clouds led to the first generation of stars (Larson and Bromm, 2001). The core temperature in collapsing gas clouds rose until, once again, they reached about 10^9 K.

A collapsing gas cloud emits infrared radiation as its temperature rises. Radiation emission is a cooling process that controls the eventual size of the star. Population III (first-generation) stars did not cool as efficiently as later-generation stars because they lacked heavy elements. As a result, Population III stars with masses of 100 solar masses or higher were much more common (Larson and Bromm, 2001). Massive Population III stars meant that supernovae were more common—a star must have a mass greater than 9 solar masses to produce a core-collapse or Type II supernova.

The Sun derives 98 to 99% of its energy output by the so-called *proton-proton I cycle*, depicted in the following reactions:

$$\,_1^1 H + \,_1^1 H \xrightarrow{\substack{nuclear \\ fusion}} \,_1^2 H + \,_{+1}^0 e + v_e$$

$$\,_1^2 H + \,_1^1 H \xrightarrow{\substack{nuclear \\ fusion}} \,_2^3 He + \gamma$$

$$\,_2^3 He + \,_2^3 He \xrightarrow{\substack{nuclear \\ fusion}} \,_2^4 He + 2\,_1^1 H$$

The intermediate isotopes in these reactions—$\,_1^2 H$ and $\,_2^3 He$—do not accumulate. The net product of the *proton-proton I* process is production of alpha particles:

$$4\,_1^1 H \xrightarrow{net} \,_2^4 He + 2\,_{+1}^0 e + 2 v_e + \underset{26.7\,MeV}{\gamma}$$

Regardless of the specific process, the temperatures in the stellar core produced by hydrogen burning are insufficient to fuse alpha particles into heavier elements. Hydrogen burning ceases once nearly all of the hydrogen in the stellar core is gone and gravity causes the core to collapse. This second

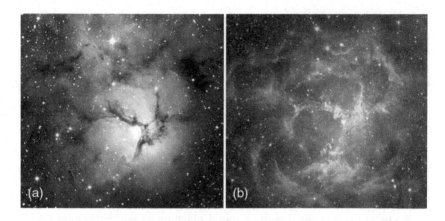

FIGURE 1B.1 False color images at two wavelengths of the Trifid Nebula.

collapse increases core temperatures above 10^9 K, sufficient for a second stage of stellar nucleosynthesis: helium burning.

A sequence of nuclear fusion reactions unfolds as heavier elements form and burn at ever-increasing temperatures as the stellar core continues to collapse (see Appendixes 1A and 1C). Each collapse and core temperature increase makes it possible to fuse heavier nuclei that resist fusion at lower temperatures. The sequential fusion of alpha particles with heavier nuclei tends to produce isotopes with even atomic numbers Z and mass numbers A of 32, 36, 40, 48, 52, and 56. About 97% of the stars in our galaxy lack sufficient mass to sustain fusion reactions beyond the *triple-alpha process* (see Appendix 1C). These stars evolve into white dwarfs and do not eject heavy elements into interstellar space. Stars exceeding 9 solar masses undergo a series of collapses, leading to higher temperatures and densities capable of sustaining carbon, oxygen, and silicon burning. Each stage is less efficient, producing less energy and fewer nuclei. The fusion reactions in this sequence are all exothermic up to $^{56}_{26}Fe$. The final collapse of these massive stars triggers a *core-collapse* or *Type II* supernova (Figure 1B.1), seeding the interstellar medium with dust containing heavy elements that eventually collapse to form second-generation *Population II* stars.

1B.3. Neutron Capture

Not only is the fusion of nuclei heavier than $^{56}_{26}Fe$ endothermic, consuming the energy released by the fusion of lighter nuclei, but also stellar core temperatures are too low to overcome Coulomb repulsions that resist the fusion of heavy elements. The nucleosynthesis of heavier elements must follow a different pathway that avoids Coulomb repulsion: neutron capture.

Neutron capture occurs simultaneously with exothermic nuclear fusion in the stellar core but under low flux conditions know as the *slow-* or *s-process* because the timescale for neutron capture τ_n under low neutron flux conditions in the stellar core is on the order of 100–1000 years. Unstable neutron-rich isotopes undergo *beta decay* to form more stable isotopes with fewer neutrons. Only those beta decay reactions with timescales $\tau_\beta \leq \tau_n$ can sustain s-process nucleosynthesis. The following *neutron capture* and *beta decay* reaction sequence illustrate s-process nuclear reactions:

$$^{59}_{26}Fe + {}^{1}_{0}n \xrightarrow[\text{capture}]{\text{neutron}} {}^{60}_{26}Fe + \gamma$$

$$^{60}_{26}Fe \xrightarrow[\text{decay}]{\text{beta}} {}^{60}_{27}Co + {}^{0}_{-1}e + \bar{\nu}_e$$

$$^{60}_{27}Co + {}^{1}_{0}n \xrightarrow[\text{capture}]{\text{neutron}} {}^{61}_{27}Co + \gamma$$

$$^{61}_{27}Co \xrightarrow[\text{decay}]{\text{beta}} {}^{61}_{28}Ni + {}^{0}_{-1}e + \bar{\nu}_e$$

S-process nucleosynthesis is very sensitive to unstable, short-lived isotopes, confining it to reaction pathways involving relatively stable isotopes. The s-process terminates with the production of $^{209}_{83}Bi$, unable to produce elements with atomic masses between Po and Ac ($84 \leq Z \leq 89$) because all of the elements in this range are extremely unstable and very short-lived. The most stable radium isotope—$^{226}_{88}Ra$—has a half-life of 1600 years, while the most stable francium isotope—$^{223}_{87}Fr$—has a half-life of a mere 21.8 minutes.

Not all of the elements present before the final collapse survive a supernova. Astronomers believe photodisintegration in the superheated conditions of the supernova destroys much of the $^{56}_{26}Fe$ formed in first-generation stars. In reality, supernova photodisintegration reactions affect all heavy nuclei, not just $^{56}_{26}Fe$, and reach equilibrium during the brief supernova episode:

$$^{56}_{26}Fe + \gamma \xleftrightarrow[\text{neutron capture}]{\text{photodisintegration}} {}^{55}_{26}Fe + {}^{1}_{0}n$$

The brief but intense (on the order of 10^{22} neutrons $cm^{-2}\,s^{-1}$) supernova neutron flux triggers *rapid-process* (*r-process*) neutron capture that occurs at a much faster rate than the decay rate of unstable isotopes ($\tau_\beta \ll \tau_n$), producing elements with atomic numbers $Z > 82$ and isotopes with mass numbers A as high as 270. Super-heavy elements are so unstable that their nuclei fragment through nuclear fission reactions as rapidly as *r-process* neutron capture produces them.

The presence of thorium and uranium on Earth is proof that the Solar System is a third-generation *Population I* stellar system because uranium could only form during the supernova of a massive second-generation *Population II* star.

1B.4. Cosmic Ray Spallation

The interstellar medium consists of gas and dust left over from the Big Bang and subsequent star formation eras bathed in cosmic rays, mostly protons and alpha particles traveling at extremely high velocities ($\leq 10^{-1}$ the speed of light). When cosmic rays strike heavy elements in the interstellar medium, the collision causes the fragmentation or spallation of the heavier nuclei.

Primordial and stellar nucleosynthesis is unable to account for the abundance of three light elements—*Li, Be,* and *B*—in the interstellar medium and Solar System. These three elements are unable to withstand the high temperatures and densities that sustain nuclear fusion reactions; the quantities formed during the initial few minutes following the Big Bang would have been destroyed in first-generation stars. Cosmic ray spallation, however, occurs at low enough temperatures and densities to allow *Li, Be,* and *B* isotopes to survive once they have formed. The following reactions are representative of cosmic ray spallation and account for the higher interstellar abundance of *Li, Be,* and *B* compared to their Solar System abundances:

$$\,^4_2He + \,^4_2He \xrightarrow{\text{cosmic spallation}} \,^6_3Li + \,^1_1H + \,^1_0n$$

$$\,^1_1H + \,^{12}_6C \xrightarrow{\text{cosmic spallation}} \,^{11}_5B + 2\,^1_1H$$

$$\,^1_1H + \,^{14}_7N \xrightarrow{\text{cosmic spallation}} \,^9_4Be + \,^4_2He + 2\,^1_1H$$

$$\,^4_2He + \,^{12}_6C \xrightarrow{\text{cosmic spallation}} \,^7_3Li + 2\,^4_2He + \,^1_1H$$

$$\,^1_1H + \,^{14}_7N \xrightarrow{\text{cosmic spallation}} \,^{10}_5B + 3\,^1_1H + 2\,^1_0n$$

1B.5. Transuranium Elements

Although elements heavier than uranium ($Z > 92$) either do not exist or are extremely rare on Earth, they are not *synthetic elements,* as they are often called. All transuranic elements comprise isotopes that are short-lived compared to the age of the Solar System $(4 \cdot 10^6 a)$. The transuranium elements and several other elements lacking stable isotopes are extinct, the quantities present in the primordial Solar System having decayed long ago.

Technetium, element 43, has no stable isotopes; the longest-lived— $^{98}_{43}Tc$—has a half-life $t_{1/2}$ on the order of 10^6 years. Polonium, $Z = 84$, is another element with no stable isotopes; all radioactive Po isotopes have half-lives less than 102 years. The most long-lived isotopes of the next three heavier elements—astatine ($Z = 85$), radon ($Z = 86$), and francium ($Z = 87$)—have half-lives ranging from 22 minutes to 3.8 days. None of the elements heavier than Pb form stable isotopes, and all are radioactive. Though unstable, thorium ($Z = 90$), and uranium ($Z = 92$) have isotopes with half-lives greater than the age of the Solar System (Table 1B.2).

How can we explain the presence of radon and other unstable elements in the environment with half-lives less than the age of the Solar System? The trace level of elements lacking stable isotopes results from the radioactive decay of $^{238}_{92}U$, a portion of which appears in Table 1B.3, and $^{232}_{90}Th$.

Trace amounts of transuranium isotopes are also found in nature, believed to be the products of the spontaneous or stimulated fission of long-lived uranium isotopes. Rather than emitting a simple alpha or beta particle, the nuclei of heavy uranium and transuranium isotopes disintegrate into two large fragments containing many neutrons and protons (Figure 1B.2). Nuclear fission may be spontaneous, requiring no external excitation:

$$^{236}_{92}U \xrightarrow{\substack{spontaneous \\ fission}} {}^{143}_{55}Cs + {}^{90}_{37}Rb + 3\,{}^{1}_{0}n + \gamma$$

TABLE 1B.2 The Long-Lived Isotopes $^{232}_{91}Th$ and $^{238}_{92}U$

Element	Isotope	Half-Life, $t_{1/2}$
Thorium	$^{232}_{90}Th$	$1.405 \cdot 10^{10}\,a$
Uranium	$^{238}_{92}U$	$4.468 \cdot 10^{10}\,a$

TABLE 1B.3 A Portion of the Radioactive Decay Series for $^{238}_{93}U$

Reaction	Half-Life, $t_{1/2}$	Reaction	Half-Life, $t_{1/2}$
$^{238}_{92}U \xrightarrow{\alpha} {}^{234}_{90}Th$	$4.468 \cdot 10^9\,a$	$^{234}_{90}Th \xrightarrow{\alpha} {}^{226}_{88}Ra$	$8 \cdot 10^4\,a$
$^{234}_{90}Th \xrightarrow{\beta} {}^{234}_{91}Pa$	$24\,d$	$^{226}_{88}Ra \xrightarrow{\alpha} {}^{222}_{86}Rn$	$1.6 \cdot 10^3\,a$
$^{234}_{91}Pa \xrightarrow{\beta} {}^{234}_{92}U$	$1\,m$	$^{222}_{86}Rn \xrightarrow{\alpha} {}^{218}_{84}Po$	$3.8\,d$
$^{234}_{92}U \xrightarrow{\alpha} {}^{230}_{90}Th$	$2.4 \cdot 10^5\,a$	$^{218}_{84}Po \xrightarrow{\beta} {}^{218}_{85}At$	$3.1\,m$

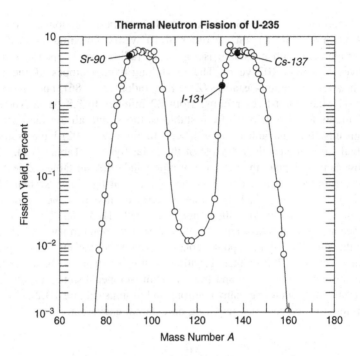

FIGURE 1B.2   Fission yield of $^{235}_{92}U$.

Certain uranium isotopes undergo fission when stimulated by neutron capture:

$$^{235}_{92}U + {}^1_0n \xrightarrow{\substack{\text{thermal neutron}\\\text{fission}}} {}^{137}_{56}Ba + {}^{95}_{36}Kr + 3\,{}^1_0n + \gamma$$

$$^{235}_{92}U + {}^1_0n \xrightarrow{\substack{\text{thermal neutron}\\\text{fission}}} {}^{139}_{54}Xe + {}^{95}_{38}Sr + 2\,{}^1_0n + \gamma$$

Nuclear reactors rely on the *stimulated nuclear fission* of radioactive uranium and transuranium isotopes to generate high neutron fluxes that mimic the *r-process*. Nuclear reactors produce a variety of neutron-rich isotopes by neutron capture. Some fission reactions release more than one neutron, a necessary condition for fission chain reactions that rely on neutrons to stimulate fission.

APPENDIX 1C. THERMONUCLEAR FUSION CYCLES

1C.1. The CNO Cycle

About 1% of the Sun's energy output and nearly all of the output of massive Population III stars is by the *CNO cycle* depicted below. Some of the products

($^{13}_{7}N$ and $^{15}_{8}O$) are unstable because they have too many neutrons and decay by positron emission:

$$^{1}_{1}H + ^{12}_{6}C \xrightarrow{\text{nuclear fusion}} ^{13}_{7}N + \underset{1.95\ MeV}{\gamma}$$

$$^{13}_{7}N \xrightarrow{\text{positron emission}} ^{13}_{6}C + ^{0}_{+1}e + \nu_e$$

$$^{1}_{1}H + ^{13}_{6}C \xrightarrow{\text{nuclear fusion}} ^{14}_{7}N + \underset{7.54\ MeV}{\gamma}$$

$$^{1}_{1}H + ^{14}_{7}N \xrightarrow{\text{nuclear fusion}} ^{15}_{8}O + \underset{7.35\ MeV}{\gamma}$$

$$^{15}_{8}O \xrightarrow{\text{positron emission}} ^{15}_{7}N + ^{0}_{+1}e + \nu_e$$

$$^{1}_{1}H + ^{15}_{7}N \xrightarrow{\text{nuclear fusion}} ^{12}_{6}C + ^{4}_{2}He$$

The carbon, nitrogen, and oxygen isotopes in these reactions also do not accumulate. The net product of the *CNO cycle* is nearly identical to the *proton-proton I* process—production of alpha particles—the only difference being the net energy output:

$$4^{1}_{1}H \xrightarrow{\text{net}} ^{4}_{2}He + 2^{0}_{+1}e + 2\nu_e + \underset{16.84\ MeV}{\gamma}$$

1C.2. The Triple-Alpha Process

Helium burning (*triple-alpha process*) follows the core collapse once hydrogen is depleted. Unlike hydrogen burning (e.g., *proton-proton I* or *CNO cycles*) the *triple-alpha process* faces a bottleneck. The fusion of two alpha particles produces a very unstable isotope $^{8}_{4}Be$ with a lifetime of about 10^{-16} s. If $^{8}_{4}Be$ fuses with another alpha particle before it decays it will form a stable $^{12}_{6}C$ that can sustain further fusion reactions. The two fusion reactions must be nearly simultaneous:

$$^{4}_{2}He + ^{4}_{2}He \xrightarrow[10^{-16}\ s]{\text{nuclear fusion}} [^{8}_{4}Be] + \gamma$$

$$^{4}_{2}He + [^{8}_{4}Be] \xrightarrow[10^{-16}\ s]{\text{nuclear fusion}} ^{12}_{6}C + \gamma$$

The ^8_4Be bottleneck means helium burning is much slower than hydrogen burning.

1C.3. Carbon Burning

Carbon burning follows the core collapse once helium is depleted. Fusion of two carbon nuclei must overcome significant Coulombic repulsions, requiring higher temperatures ($6\ 10^9\ K$) and densities that are needed to sustain helium burning:

$$^{12}_6C + ^{12}_6C \xrightarrow{\substack{nuclear \\ fusion}} ^{20}_{10}Ne + ^4_2He$$
$$4.617\ MeV$$

$$^{12}_6C + ^{12}_6C \xrightarrow{\substack{nuclear \\ fusion}} ^{23}_{11}Na + ^1_1H$$
$$2.241\ MeV$$

$$^{12}_6C + ^{12}_6C \xrightarrow{\substack{nuclear \\ fusion}} ^{23}_{12}Mg + ^1_0n$$
$$2.599\ MeV$$

These fusion reactions are all exothermic, the heat released as the kinetic energy of particles emitted from the nucleus (^4_2He, ^1_1H, and ^1_0n).

APPENDIX 1D. NEUTRON-EMITTING REACTIONS THAT SUSTAIN THE *S-PROCESS*

Numerous nuclear reactions in the stellar core, represented by the reaction sequences below, create a low neutron flux on the order of 10^5 to 10^{11} neutrons $\cdot cm^{-2} \cdot s^{-1}$ that supports *s-process* neutron capture in the stellar core. These neutron-producing nuclear reactions occurred in first-generation stars but in the absence of the heavy elements found in second and subsequent-generation stars:

$$^{12}_6C + ^1_1H \xrightarrow{\substack{nuclear \\ fusion}} ^{13}_7N + \gamma$$

$$^{13}_7N \xrightarrow{\substack{positron \\ emission}} ^{13}_6C + ^0_{+1}e + \nu_e$$

$$^4_2He + ^{13}_6C \xrightarrow{\substack{nuclear \\ fusion}} ^{16}_8O + ^1_0n$$

$$^4_2He + ^{14}_7N \xrightarrow{\substack{nuclear \\ fusion}} ^{18}_9F + \gamma$$

$$\mathrm{^{18}_{9}F} \xrightarrow{\substack{\textit{positron} \\ \textit{emission}}} \mathrm{^{18}_{8}O} + \mathrm{^{0}_{+1}}e + v_e$$

$$\mathrm{^{4}_{2}He} + \mathrm{^{18}_{8}O} \xrightarrow{\substack{\textit{nuclear} \\ \textit{fusion}}} \mathrm{^{22}_{10}Ne} + \gamma$$

$$\mathrm{^{4}_{2}He} + \mathrm{^{22}_{10}Ne} \xrightarrow{\substack{\textit{nuclear} \\ \textit{fusion}}} \mathrm{^{25}_{12}Mg} + \mathrm{^{1}_{0}}n$$

APPENDIX 1E. RANDOM SEQUENTIAL DILUTIONS AND THE LAW OF PROPORTIONATE EFFECT

Ott (1990) published a simple *random sequential dilution* model to illustrate the type of environmental process that would yield a log-normal concentration distribution. The random sequential dilution model is actually a special case of a more general process known as the *law of proportionate effect* first described by the Dutch astronomer and mathematician K. C. Kapteyn (Kapteyn, 1903).

Ott described two simple random processes that, in the limit of a large number of transformations, consistently yield log-normal concentration distribution. The first example imagines a *sequential random dilution*, in which the dilution factor $(0 < d_i \le 1)$ at each step is a random value.

The sequential random dilution (Figure 1E.1) begins with an initial concentration C_0 and transfers a random volume v_i in each step. The concentration at each stage C_i is a function of the transferred volume v_i, the concentration of the previous stage C_{i-1}, and a fixed volume V:

$$C_i = \left(\frac{v_i}{V}\right) \cdot C_{i-1} = d_i \cdot C_{i-1}$$

The concentration C_n after n random volume transfers is a function of the initial concentration C_0 and the random transfer volumes v_i in all previous steps in the dilution sequence:

$$d_1 = \frac{v_1}{V} \quad d_2 = \frac{v_2}{V} \quad d_3 = \frac{v_3}{V} \quad d_4 = \frac{v_4}{V}$$

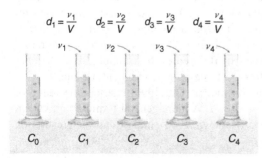

$C_0 \quad C_1 \quad C_2 \quad C_3 \quad C_4$

FIGURE 1E.1 A four-step sequential random dilution series begins with an initial concentration C_0 and uses a fixed volume V. The volume transferred in each step v_i is a random value, as is the resulting dilution factor d_i.

$$C_1 = \left(\frac{v_1}{V}\right) \cdot C_0 = d_1 \cdot C_0$$

$$C_2 = d_2 \cdot C_1 = d_2 \cdot (d_1 \cdot C_0)$$

$$\vdots$$

$$C_n = d_n \cdot C_{n-1} = d_n \cdot (d_{n-1} \cdot \cdots \cdot (d_2 \cdot (d_1 \cdot C_0)))$$

A logarithmic transformation of the product of these random dilutions yields a sum of logarithmic term. Since each dilution factor d_i is a random number, the logarithm of the dilution factor $\ln(d_i)$ is itself a random number:

$$\ln(C_n) = \ln(d_n) + \ln(d_{n-1}) + \ldots + \ln(d_2) + \ln(d_1) + \ln(C_0)$$

$$\ln(C_n) = \ln(C_0) + \sum_{i=1}^{n} \ln(d_i)$$

The *Central Limit Theorem* states that a sample population in which each value is a sum of random numbers yields a normal distribution. You can verify this by using a spreadsheet computer application to generate a large population of numbers (say, 100) in which each number is the sum of random numbers (say, 100 each). A plot of the frequency distribution of the numbers in this population of sums has the characteristics of a normal distribution.

Conversely, if you use the spreadsheet application to generate a large population of numbers (say, 100, as before) in which each number is the product of random numbers (say, 100 each), the plot of the frequency distribution of numbers generated by this population of products has the characteristics of a log-normal distribution. A logarithmic transformation of each product in the population generates a sum of random numbers, entirely analogous to the sums yielding a normal distribution, except the random numbers in each sum are actually the logarithm of a random number. Of course, it makes no difference whether the sum is based on random numbers or the logarithm of a random number.

Ott (1990) described a second example: a fluid-filled chamber (gas or liquid) of volume V and a chemical substance in the chamber at an initial concentration C_0. The chamber has two openings: an entrance and an exit. Through the entrance a random volume of pure fluid v_1 enters, forcing the expulsion of an equal volume and a quantity of the chemical substance $(v_1 \cdot C_0)$ through the exit. A random sequential dilution of the chemical substance from the initial concentration C_0 to some final concentration C_n results when n random pulses of fluid enter the chamber. The chamber could be a pore in or an arbitrary volume of soil, sediment, or aquifer.

The sequential random dilution model (Ott, 1990) restricts the range of each random dilution factor: $0 < d_i \leq 1$. A more general formulation, known as the *law of proportionate effect* (Kapteyn, 1903), imposes no such restriction on the random scaling factors f_i:

$$C_1 = f_1 \cdot C_0$$
$$C_2 = f_2 \cdot C_1 = f_2 \cdot (f_1 \cdot C_0)$$
$$\vdots$$
$$C_n = f_n \cdot C_{n-1} = f_n \cdot (f_{n-1} \cdot \ \cdots \ \cdot (f_2 \cdot (f_1 \cdot C_0)))$$
$$\ln(C_n) = \ln(C_0) + \sum_{i=1}^{n} \ln(f_i)$$

APPENDIX 1F. THE ESTIMATE OF CENTRAL TENDENCY AND VARIATION OF A LOG-NORMAL DISTRIBUTION

The following data set consists of chromium concentrations in 27 samples of Canadian granite (Ahrens, 1954).

Sample	Cr, ppm	Sample	Cr, ppm
G-1	22	KB-14	35
KB-1	40	KB-15	28
KB-2	30	KB-16	95
KB-3	120	KB-17	59
KB-4	25	KB-18	38
KB-5	3.5	KB-19	17
KB-6	5.3	48-63	19
KB-7	5.8	48-115	7
KB-8	17	48-158	3
KB-9	22	48-118	6
KB-10	410	48-485	5.5
KB-11	2	48-489	27
KB-12	24	48-490	2
KB-13	15		

The arithmetic mean $\bar{x}$ and arithmetic standard deviation σ are easily computed using standard functions in any spreadsheet program:

$$\bar{x} = \left(\frac{1}{n}\right) \cdot \sum_{i=1}^{n} x_i = 40.1 \text{ ppm}$$

$$\sigma = \sqrt{\left(\frac{1}{n}\right) \cdot \sum_{i=1}^{n}(x_i - \bar{x})^2}$$

$$\sigma = \sqrt{\left(\frac{1}{n}\right) \cdot \sum_{i=1}^{n}(x_i - 40.1)^2} = 78.9 \text{ ppm}$$

The geometric mean $\bar{x}_G$ is simply the exponential transform of an arithmetic mean computed from the (natural) logarithm transforms of the concentrations:

$$\bar{x}_G = \exp\left(\left(\frac{1}{n}\right) \cdot \sum_{i=1}^{n} \ln(x_i)\right) = \exp(2.85) = 17.2 \text{ ppm}$$

The geometric standard deviation σ_G is computed by performing a natural logarithm transformation of each chromium concentration, subtracting the natural logarithm of the geometric mean $\ln(\bar{x}_G)$, squaring the difference, finding the average of the differences, and taking the exponential of the square root:

$$\sigma_G = \exp\left(\sqrt{\left(\frac{1}{n}\right) \cdot \sum_{i=1}^{n} (\ln(x_i) - \ln(\bar{x}_G))^2}\right)$$

$$\sigma_G = \exp\left(\sqrt{\left(\frac{1}{n}\right) \cdot \sum_{i=1}^{n} (\ln(x_i) - 2.85)^2}\right) = 3.56 \text{ ppm}$$

Problems

1. Calculate the binding energy in *MeV* for the following isotopes: ^4_2He, ^6_3Li, $^{12}_6C$, and $^{13}_6C$. Proton mass m_{proton} is 1.007276 *u*, and the neutron mass $m_{neutron}$ is 1.008865 *u*. Convert the mass difference to a binding energy using 931.494 *MeV* u^{-1}. You can obtain a complete list of isotope masses from a website maintained by the National Institute of Standards and Technology: http://www.physics.nist.gov/PhysRefData/Compositions/index.html.

   Solution
   The isotope mass for ^6_3Li is 6.015122 *u* (6.015122 *g mol^{-1}*) found at the NIST website, typing in the atomic number *Z* for *Li*: 3. The page also lists the isotope mass (relative atomic mass) of the two stable *Li* isotopes and their relative abundance in percent.
   The nucleus of lithium-6 contains 3 protons and 3 neutrons. Following the example given in the chapter, the rest mass of the protons and neutrons in the nucleus is:
   The isotope mass of ^6_3Li is 6.015122 *u*—0.032701 *u* less than the mass found by adding the rest mass of 3 protons and 3 neutrons. The mass difference, multiplied by 931.494 *MeV* u^{-1} and divided by 6 nucleons, is the *binding energy per nucleon* for ^6_3Li: 5.077 *MeV*. This value can be checked against the value plotted in Figure 1.5.

2. Complete the following nuclear reactions:
   a. neutron capture: $^{58}_{26}Fe + ^1_0n \rightarrow ^{59}_{26}Fe + \_\_$
   b. beta (minus) decay: $^{59}_{26}Fe \rightarrow ^{59}_{27}Co + ^0_{-1}e + \bar{v}_e$
   c. nuclear fusion: $^{18}_8O + ^4_2He \rightarrow ^{22}_{10}Ne + \gamma$
   d. positron emission: $^{18}_9F \rightarrow \_\_ + ^0_{+1}e + v_e$

3. List the number of stable isotopes for the period 3 elements (*Na* through *Ar*). Explain why some elements have more stable isotopes than others.

   Solution
   Sodium has only one stable isotope with an odd number of protons: $^{23}_{11}Na$. Elemental abundances are listed on the National Physical Laboratory

website: http://www.kayelaby.npl.co.uk/chemistry/3_1/3_1_3.html. Sodium abundance in the Solar System is 2.0 (10^{-6}) = 5.7 (10^{4}) atoms-Na ÷ 2.8 (10^{10}) atoms-H.

Magnesium has three stable isotopes: Mg-24 and Mg-26 have an even number of protons and neutrons, while Mg-25 has an unpaired (odd) neutron. Magnesium abundance in the Solar System is 3.9 (10^{-5}) = (1.1 (10^{6}) atoms-Mg/2.8 (10^{10}) atoms-H).

Partially filled nuclear shells and an unpaired proton in the nucleus make sodium less stable than magnesium and Mg-25 less stable than Mg-24 and Mg-26. Na is less abundant than Mg, and Mg-25 is the least abundant Mg isotope because of isotope stability.

4. List the half-life of the longest-lived isotope for each element from Po to Th ($84 < Z \leq 90$). A complete Table of the Isotopes is available on the Abundances of the Elements (Kaye and Laby Tables of Physical and Chemical Constants, National Physics Laboratory): http://www.kayelaby.npl.co.uk/atomic_and_nuclear_physics/4_6/4_6_1.html.

5. The following data are the vanadium content of samples of Canadian granite (Ahrens, 1954). Determine the geometric mean and geometric standard deviation from this data set.

Sample	V, ppm
G-1	21
KB-1	75
KB-2	43
KB-3	200
KB-4	50
KB-5	7.5
KB-6	30
KB-7	33
KB-8	27
KB-9	34
KB-10	42
KB-11	10
KB-12	52
KB-13	51
KB-14	182
KB-15	630
KB-16	144
KB-17	94
KB-18	94
KB-19	29
48-63	8
48-115	11
48-158	5.5
48-118	23
48-485	19
48-489	7
48-490	5.6

Soil Moisture and Hydrology

2.1. INTRODUCTION

Water is the primary medium that transports nutrients and contaminants through terrestrial environments and a major environmental factor determining biological availability. This chapter examines the properties and behaviors of freshwater, both aboveground in lakes and streams and belowground, where it is confined to pores in soil and rocks. The selection of topics reflects our primary interest: chemical hydrology.

Chemical hydrology focuses on the changes in water chemistry caused by exposure to soil and aquifers. Residence time, regardless of aquifer chemical properties, influences pore water chemistry because mineral dissolution and precipitation reactions are often slow relative to residence time. Water residence time estimates require an understanding of both the water volume in connected hydrologic environments and the water flux between those environments.

Groundwater flow paths control the migration pathway of dissolved and suspended substances. Flow paths above and at the water table are quite different from flow paths below the water table. Changes in aquifer texture alter water flow rates and, potentially, the direction of flow. Dissolved and suspended substances typically migrate more slowly than pore water itself because interactions with the aquifer retard migration.

2.2. WATER RESOURCES AND THE HYDROLOGIC CYCLE

Freshwater accounts for about 0.26% of the total water on Earth's surface (Table 2.1); 70% of the freshwater is frozen in glaciers and polar icecaps. The remaining 30% of the fresh terrestrial water is either stored aboveground in lakes, rivers, and wetlands or belowground confined in the pores of soil and aquifers (water-bearing rock formations). The hydrologic cycle describes the exchanges of water between these reservoirs through precipitation, infiltration, evaporation, transpiration (water vapor loss from vegetation), surface runoff, and groundwater discharge.

TABLE 2.1 World Water Resources from Maidment

	Volume, km^3	Percent of Total[‡]	Percent of Freshwater[†]
Terrestrial Water	47,971,700	0.36%	
Fresh Lakes[†]	91,000	0.0007%	0.26%
Saline Lakes	85,400	0.0006%	
Rivers[†]	2,100	0.00002%	0.01%
Wetlands[†]	11,500	0.00009%	0.03%
Soils[†]	16,500	0.00012%	0.05%
Fresh Groundwater[†]	10,530,000	0.079%	30.07%
Saline Groundwater	12,870,000	0.097%	
Biomass Water[†]	1,100	0.000008%	0.003%
Glaciers & Icecaps[†]	24,364,100	0.183%	69.58%
Oceans	13,238,000,000	99.28%	
Atmosphere[†]	12,900	0.00010%	0.04%
Total[‡]	13,333,956,300	100.00%	

Source: Data reproduced with permission from Maidment, D.R., 1993. Handbook of Hydrology. McGraw-Hill, New York, NY, pp. 1.1–1.15.

2.3. WATER BUDGETS

Hydrologists use water budgets to quantify water exchange in the hydrologic cycle. The conservation of mass ensures the total volume does not change at steady state. The example of a lake surrounded by a catchment (watershed) serves to illustrate how a water budget is used. At steady state the lake volume does not change, provided we neglect daily and seasonal fluctuations:

$$\frac{\Delta V}{\Delta t} = 0 \tag{2.1}$$

Steady state does not mean that the lake and its catchment are static. Water enters the lake through mean annual precipitation $\overline{P}[m^3 \cdot y^{-1}]$, groundwater inflow $\overline{Q}_{g,in}[m^3 \cdot y^{-1}]$, and surface inflow (runoff) $\overline{Q}_{s,in}[m^3 \cdot y^{-1}]$ and is lost through evaporation and transpiration $\overline{ET}[m^3 \cdot y^{-1}]$, groundwater outflow $\overline{Q}_{g,out}[m^3 \cdot y^{-1}]$, and surface outflow $\overline{Q}_{s,out}[m^3 \cdot y^{-1}]$. The annual steady-state water budget for the lake is given by Eq. (2.2):

$$\frac{\Delta V_{lake}}{\Delta t} = 0 = \overline{P} + \left(\overline{Q}_{g,in} + \overline{Q}_{s,in}\right) - \overline{ET} - \left(\overline{Q}_{g,out} + \overline{Q}_{s,out}\right) \tag{2.2}$$

The rate of groundwater inflow and outflow is usually negligible when compared to the other exchange processes, leading to a simplified annual lake budget where surface inflow (surface runoff and stream flow) and outflow are combined in the mean annual *surface discharge* $\overline{D}_s[m^3 \cdot y^{-1}]$:

$$\overline{P} = \overline{ET} + \left(\overline{Q}_{s,out} - \overline{Q}_{s,in}\right) = \overline{ET} + \overline{D}_s \tag{2.3}$$

You can imagine an alternative water budget for catchments where water is stored in the soil rather than aboveground in a lake or wetland. The catchment water budget might cover a growing season measured in months, or it may represent an annual average. Water budget adds a term for soil moisture storage $\overline{S}_{storage}[m^3 \cdot yr^{-1}]$. Later, we will learn how to determine the plant-available water storage capacity. Details on the calculation of soil moisture recharge appear in Appendix 2A. Appendix 2B provides details on the calculation of soil water storage capacity:

$$\overline{P} = \overline{ET} + \overline{D}_s + \overline{S}_{storage} \tag{2.4}$$

2.4. RESIDENCE TIME AND RUNOFF RATIOS

Hydrologists use water budgets to estimate two important parameters of water systems: runoff ratio $\overline{D}_s/\overline{P}$ and residence time T_R for water in a particular reservoir. These parameters can be defined on many scales, ranging from the smallest catchment to the global scale. Table 2.2 lists the runoff ratios for the continents.

TABLE 2.2 Annual Continental Water Budgets and Runoff Ratios $\overline{D}_s/\overline{P}$

Continent	$\overline{P}$	$\overline{D}_s$	$\overline{ET}$	$\overline{D}_s/\overline{P}$
Africa	690 mm	140 mm	550 mm	0.20
Asia	720 mm	290 mm	430 mm	0.40
Australia	740 mm	230 mm	510 mm	0.31
Europe	730 mm	320 mm	410 mm	0.44
North America	670 mm	290 mm	380 mm	0.43
South America	1650 mm	590 mm	1060 mm	0.36

Source: Data reproduced with permission from L'vovich, M.I., 1979. World Water Resources. American Geophysical Union, Washington, DC.

The average global runoff ratio $\overline{D}_s/\overline{P}$ is 0.39, with evaporation-transpiration losses of 0.61.

Residence time T_R is the average time a particular reservoir in a water system retains water:

$$T_R = \frac{V}{Q_{in}} = \frac{V}{Q_{out}} \tag{2.5}$$

Using the reservoir volumes listed in Table 2.1 and the flow rates in Table 2.3, we can estimate the residence time for water in the atmosphere (0.02 year or 7 days), oceans (2650 years), subsurface aquifers (20,000 years), and surface waters (4 years).

TABLE 2.3 Global Hydrologic Cycle Flow Rates

Process	Flow Rate Q, $km^3 \cdot yr^{-1}$
Terrestrial Precipitation	119,000
Terrestrial ET	72,600
Oceanic Precipitation	458,200
Oceanic ET	504,600
Surface Discharge	452,000
Groundwater Discharge	1200

Source: Data reproduced with permission from Maidment, D.R., 1993. Handbook of Hydrology. McGraw-Hill, New York, NY, pp. 1.1–1.15.

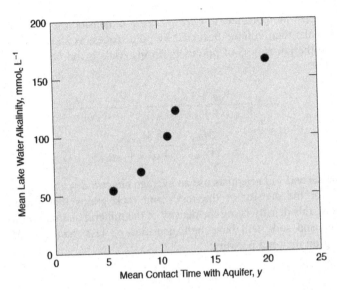

FIGURE 2.1   Lake water alkalinity increases the longer groundwater discharging into the lake remains in contact with aquifer formations. *Source: Reproduced with permission from Wolock, D.M., Hornberger, G.M., et al., 1989. The relationship of catchment topography and soil hydraulic characteristics to lake alkalinity in the northeastern United States. Water Resour. Res. 25: 829–837.*

Suppose we want to estimate the residence time for soil moisture, defined as subsurface water within the plant root zone. If we assume terrestrial evaporation-transpiration water flow comes primarily from soil (see Table 2.3) and use the global soil moisture volume listed in Table 2.1, the residence time of soil moisture is about 0.2 years, or 70 days.

Figure 2.1 illustrates the effect that groundwater residence time has on lake alkalinity. A portion of the inflow that sustains lake volume comes from surface runoff, and a portion comes from groundwater discharge, the latter accounting for lake water alkalinity. As the groundwater residence time increases, the alkalinity of the groundwater discharge and, consequently, lake water alkalinity increases. Alkalinity arises from the dissolution of carbonate minerals—principally calcite $CaCO_3$—into groundwater permeating the aquifer. The time scale on Figure 2.1 illustrates the time scale required for mineral dissolution reactions. Notice a 25-year residence time is still insufficient for alkalinity to reach a steady state.

2.5. GROUNDWATER HYDROLOGY

2.5.1. Water in the Porosphere

We can define the *porosphere* as that portion of the Earth surface where we find water-filled pores. This section considers a number of parameters used to characterize the porosphere. *Porosity* ϕ—a dimensionless number—is the fraction of the total volume occupied by pore spaces, as shown in Eq. (2.6). Related to the porosity ϕ of porous Earth materials is the *bulk density* ρ_B in Eq. (2.7).

$$\phi \equiv \frac{V_{air} + V_{water}}{V_{solids} + V_{air} + V_{water}} = \frac{V_{air} + V_{water}}{V_{total}} \tag{2.6}$$

$$\rho_B \equiv \frac{M_{dry}}{V_{total}} = \frac{M_{solids}}{(V_{air} + V_{solids})} \tag{2.7}$$

Hydrologists and soil scientists use an average *mineral density* $\overline{\rho}_{mineral}$ of 2.65 $Mg \cdot m^{-3}$ for the density of the rocks and rock grains that make up the porosphere, this density being the density of the mineral quartz. Porous rocks, sediments, and soils will have bulk densities ρ_B less than 2.65 $Mg \cdot m^{-3}$, depending on the total porosity:

$$\phi = 1 - \frac{\rho_B}{\overline{\rho}_{mineral}} \tag{2.8}$$

The method for measuring bulk density suggests a simple method for measuring *gravimetric water content* θ_M, which is simply mass loss caused by oven drying:

$$\theta_M = \frac{M_{water}}{M_{solids}} = \frac{M_{moist} - M_{dry}}{M_{dry}} \tag{2.9}$$

In many cases, soil scientists and hydrologists need to know the volume of water in a subsurface material, requiring conversion from gravimetric to *volumetric water content* θ_V:

$$\theta_V = \frac{V_{water}}{V_{solids} + V_{air} + V_{water}} \tag{2.10}$$

Box 2.1 Measuring Bulk Density ρ_B

If the sample is unconsolidated, the volume is simply the volume of a canister driven into *in situ* soil or sediment to extract the sample. A dry bulk measurement requires the drying of the sample to eliminate pore water, which is accomplished by drying the sample in an oven at about 100°C until mass loss with further drying is negligible. The bulk density is simply the dry mass of the sample divided by the sample volume. The exterior of both consolidated and unconsolidated samples can be sealed after drying to prevent water entry, permitting volume measurement by immersion.

Although it may not seem obvious at first, conversion from gravimetric water content θ_M to volumetric water content θ_V requires the bulk density ρ_B of the Earth material and the density of water ρ_W. Dividing a mass of a substance by its density yields the volume, as illustrated by the dimensional analysis:

$$\theta_V \left[\frac{m^3_{water}}{m^3_{dry\ soil}}\right] = \theta_M \left[\frac{Mg_{water}}{Mg_{dry\ soil}}\right] \times \frac{\rho_B}{\rho_W} \left[\frac{Mg_{dry\ soil}/m^3_{dry\ soil}}{Mg_{water}/m^3_{water}}\right] \quad (2.11)$$

Often soil scientists and hydrologists wish to determine water content relative to the water-holding capacity, a proportion called the *degree of saturation S:*

$$S \equiv \frac{V_{water}}{V_{air} + V_{water}} = \frac{V_{water}/V_{total}}{(V_{air} + V_{water})/V_{total}} = \frac{\theta}{\phi} \quad (2.12)$$

Appendix 2B illustrates the conversion of gravimetric to volumetric water content to estimate the water-holding capacity of soil profiles.

2.5.2. Hydrologic Units

Hydrologists distinguish geologic formations in the saturated (or phreatic) zone, whose water-holding capacity and hydraulic conductivities are relevant to groundwater flow (Figure 2.2). Formations that are effectively nonporous,

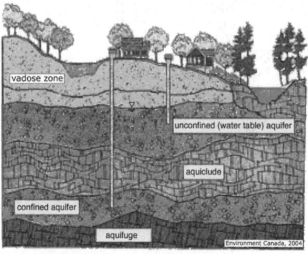

Environment Canada, 2004

FIGURE 2.2   Different aquifer formations: aquiclude, aquifuge, unconfined aquifer, and confined aquifer. The critical zone extends from the land surface to the lower boundary of the unconfined (or water table) aquifer. *Source: Reproduced with permission from Environment Canada, 2008. Nitrate Levels in the Abbotsford Aquifer an indicator of groundwater contamination in the Lower Fraser Valley. Retrieved November 2, 2010 from http://www.ecoinfo.org/env_ind/ region/nitrate/nitrate_e.cfm.*

holding negligible amounts of water and unable to conduct water, are called *aquifuges*. Certain formations, usually containing significant amounts of clay, may hold substantial amounts of water, but their low permeability impedes water movement. These formations are called *aquicludes*. Many hydrologists denote any formation with low permeability, regardless of porosity, as *aquitards*. An *aquifer* is any geologic formation, either consolidated rock or unconsolidated sediments, with substantial water-holding capacity and relatively high permeability. An unconfined, or *water table*, aquifer refers to any aquifer that is saturated in its lower depths but terminates at an unsaturated or *vadose zone* in its upper reaches, the two zones being separated by a water table. A confined, or artesian, aquifer is overlain by an aquitard. The hydrostatic pressure in a confined aquifer can be substantial, causing water in wells drilled into the aquifer to rise above the top of the aquifer.

2.5.3. Darcy's Law

Figure 2.3 shows the experimental apparatus used by Henry Darcy in 1856 to measure water flow rates Q_D through sand and other unconsolidated porous Earth materials. Darcy discovered that the water flow rate Q_D (Eq. (2.13))

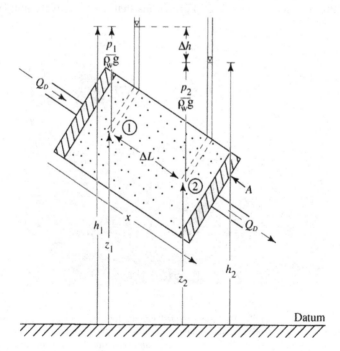

FIGURE 2.3   The experimental apparatus Darcy used to measure water flow under a hydraulic gradient through porous material. *Source: Reproduced with permission from Dingman, S.L., 1994. Physical Hydrology. Macmillan, New York.*

is directly proportional to both the cross-sectional area A of the cylinder containing the porous material and the difference in the hydraulic head Δh between the two ends of the cylinder and inversely proportional to the length L of the cylinder. The proportionality coefficient K_h (Eq. (2.14)) is called the *hydraulic conductivity* of the porous medium. Darcy could rotate the cylinder to vary the difference in the hydraulic head Δh, read by measuring the water level in two standpipes at either end of the cylinder relative to a *datum* (see Figure 2.3).

$$Q_D \propto \frac{A \cdot (h_1 - h_2)}{L} \tag{2.13}$$

$$Q_D \left[m^3 \cdot s^{-1}\right] = K_h \left[m \cdot s^{-1}\right] \cdot \left(\frac{A[m^2] \cdot (h_1 - h_2)[m]}{L[m]}\right) \tag{2.14}$$

2.5.4. Hydrostatic Heads and Hydrostatic Gradients

The hydraulic head h is central to Darcy's Law and the dynamics of water flow in the saturated aquifers. Hydrostatic pressure p is a component of the hydraulic head. Pressure is force per unit area and increases with depth d in a fluid. The downward-acting force in a column of fluid is simply the mass of the fluid in the column times the acceleration due to gravity g. The mass of the fluid at any given depth d is simply the density of the liquid ρ times the depth d, which yields the hydrostatic pressure p:

$$p_{hydrostatic}\left[N \cdot m^2\right] = \frac{F[N]}{A[m^2]} = \frac{m_{fluid}[Mg] \cdot g[m \cdot s^{-2}]}{A[m^2]}$$

$$p_{hydrostatic} = \left(\frac{m_{fluid}}{A \cdot d}\right) \cdot \left(\frac{d \cdot g}{1}\right) = \left(\rho_{fluid}[Mg \cdot m^{-3}] \cdot g[m \cdot s^{-2}]\right) \cdot d[m]$$

$$d_{hydrostatic} = \frac{p_{hydrostatic}}{\rho_{fluid} \cdot g} \tag{2.15}$$

Elementary thermodynamics from general chemistry tells us that the compression or expansion of a gas involves a type of work called *pressure-volume work*. Pressure-volume work involving a gas is usually illustrated by a change in volume at constant pressure, but it can also occur when there is a change in pressure at constant volume. Fluid flow associated with a change in hydrostatic pressure involves pressure-volume work:

$$work = p \cdot \Delta V = V \cdot \Delta p \tag{2.16}$$

Any change in elevation z represents a change in potential energy. The *Bernoulli equation* for fluid dynamics includes two primary contributions to the total hydraulic head h in groundwater systems: the pressure head that accounts for pressure-volume work during water flow $p/(\rho_W \cdot g)$ and the elevation

Soil and Environmental Chemistry

(gravitational) head z that accounts for changes in potential energy during water flow:

$$h_{total} = h_{hydrostatic} + h_{elevation} \equiv \frac{p_{hydrostatic}}{\rho_W \cdot g} + z \qquad (2.17)$$

Hydrologists and soil scientists measure pressure relative to atmospheric pressure. The hydrostatic head $h_{hydrostatic}$ is zero at the surface of free water (Figure 2.4, points A and B; Figure 2.4, points R and S). The water table in an unconfined aquifer (Figure 2.4, point R) is, by definition, the point where the hydrostatic head $h_{hydrostatic}$ is zero.

Figure 2.4 also illustrates the difference between hydrostatic head $h_{hydrostatic}$ and elevation head $h_{elevation}$. A pressure measurement taken at the water surface above a dam (Figure 2.4, point A) and at the water surface below a dam (Figure 2.4, point B) both record atmospheric pressure, which on our pressure gauge reads as zero ($p_{gauge} = 0$):

$$h^A_{pressure} = h^B_{pressure} \equiv \frac{p_{gage}}{\rho_W \cdot g} \equiv 0$$

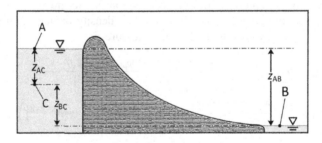

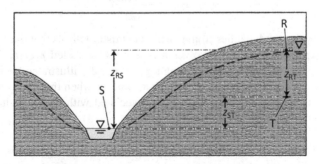

FIGURE 2.4 Impounded water (upper) and an unconfined aquifer (lower) illustrating reference points and components of the total head (hydrostatic and elevation). The elevation of a water level is denoted z_A and z_R, while the difference in the elevation of two water levels is denoted as z_{AB} and z_{RS}.

The water at point A above the dam is at a higher potential energy due to the difference in elevation z_{AB} between point A and point B, so the elevation head $h_{elevation}$ at point A is higher than point B:

$$h_{elevation}^{A} = h_{elevation}^{B} + z_{AB}$$

The hydrostatic pressure at point C is greater than zero and equal to the gravitation force (i.e., weight) of the overlying water column:

$$p_{hydrostatic}^{C} = \rho_W \cdot g \cdot z_{AC}$$

$$h_{hydrostatic}^{C} \equiv \frac{p_{hydrostatic}^{C}}{\rho_W \cdot g} = z_{AC}$$

The total hydraulic head h_{total} at point C relative to point B in Figure 2.4 includes contributions from both hydrostatic head $h_{hydrostatic}$ and elevation head $h_{elevation}$:

$$h_{total}^{C} - h_{total}^{B} = \left(h_{hydrostatic}^{C} + h_{elevation}^{C} \right) - \left(h_{hydrostatic}^{B} + h_{elevation}^{B} \right)$$

$$h_{total}^{C} - h_{total}^{B} = (z_{AC} + z_{BC}) - (0 + 0) = z_{AB}$$

As the depth z_{AC} below the water surface increases, the hydrostatic head $h_{hydrostatic}$ becomes more positive at exactly the same rate as the elevation head $h_{elevation}$ decreases. Consequently, the total hydraulic head h_{total} remains unchanged.

Now, consider the saturated zone at and below the water table in Figure 2.4 (points R and T). Pressure measurements taken at the water table (Figure 2.4, point R) and at the water surface of a stream in the valley bottom (Figure 2.4, point S) both record atmospheric pressure:

$$h_{hydrostatic}^{R} = h_{hydrostatic}^{S} \equiv \frac{p_{gage}}{\rho_W \cdot g} \equiv 0$$

The water at point R at the water table up slope is at a higher potential energy due to the difference in elevation z_{RS} between point R and point S, so the elevation head $h_{elevation}$ at point R is higher than point S:

$$h_{elevation}^{R} = h_{elevation}^{S} + z_{RS}$$

The hydrostatic pressure at point T is greater than zero and equal to the gravitation force (i.e., weight) of the overlying pore water. Unlike the case shown in Figure 2.4 (point C), the hydrostatic pressure $p_{hydrostatic}$ (and therefore the hydrostatic head $h_{hydrostatic}$) at point T depends on the porosity of the aquifer and, therefore, must be measured rather than inferred from the depth z_{RT} below the water table:

$$p^T_{hydrostatic} \neq \rho_W \cdot g \cdot z_{RT}$$

$$h^T_{hydrostatic} \equiv \frac{p^T_{hydrostatic}}{\rho_W \cdot g}$$

The total hydraulic head h_{total} at point T relative to point R in Figure 2.4 includes contributions from both hydrostatic head $h_{hydrostatic}$ and elevation head $h_{elevation}$:

$$h^T_{total} - h^R_{total} = \left(h^T_{hydrostatic} - h^R_{hydrostatic} \right) + \left(h^T_{elevation} - h^R_{elevation} \right)$$

$$h^T_{total} - h^R_{total} = \left(\frac{p^T_{hydrostatic}}{\rho_W \cdot g} - 0 \right) + (z_{RS} - z_{RT}) = \left(\frac{p^T_{hydrostatic}}{\rho_W \cdot g} \right) + z_{ST}$$

As the depth z_{RT} below the water table increases, the hydrostatic head $h_{hydrostatic}$ becomes more positive *but not at the same rate* as the elevation head $h_{elevation}$ decreases.

Darcy's Law can be applied directly to the example in Figure 2.5. As you compare Figures 2.4 and 2.5, note that the only contribution to the hydraulic gradient Δh between points A and B in Figure 2.5 is the difference in elevation between the two points $(z_A - z_B)$. Following the convention adopted in Figure 2.2, the elevations of points A and B are measured relative to a datum. The hydrostatic head is zero at both points because A is located at the water table and B is located at the water surface in the valley bottom. The discharge q_D between points A and B depends on the distance L between the two points

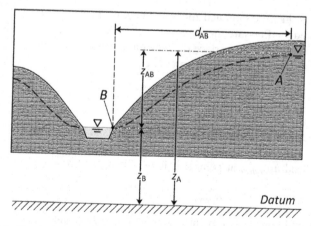

FIGURE 2.5 The hydrostatic gradient $\Delta h/\Delta L = z_{AB}/d_{AB}$ between two points at the water table of an unconfined aquifer is typically defined relative to an elevation *datum*. The elevations of water levels relative to a datum are denoted z_A and z_B, while the difference in the elevation of two water levels is denoted z_{AB}.

$\sqrt{d_{AB}^2 + z_{AB}^2}$, the hydraulic conductivity of the formation K_h, and the total hydraulic gradient Δh:

$$q_D\left[m\cdot s^{-1}\right] = \frac{Q_D\left[m^3\cdot s^{-1}\right]}{A\left[m^2\right]} = \frac{K_h\left[m\cdot s^{-1}\right]\cdot(h_A - h_B)}{\sqrt{d_{AB}^2 + z_{AB}^2}} = K_h\left[m\cdot s^{-1}\right]\cdot\left(\frac{\Delta h}{\Delta L}\right) \quad (2.18)$$

The effect of porosity requires the direct measurement of the hydrostatic head using either stand tubes (see Figure 2.3), which was how Darcy made the measurement, or *piezometers* (Figure 2.6). Piezometers are basically stand tubes that are inserted into the saturated zone at or below the water table. A piezometer has a screened opening at its lower end that allows water to enter the tube. The water level in the piezometer—measured from the screen—is the hydrostatic head $h_{hydrostatic}$ at the depth of the screened opening.

Figure 2.7 illustrates the measurement of the hydraulic gradient when both points are below the water table. The hydraulic gradient Δh simply reduces to the difference in the water levels of the two piezometers at

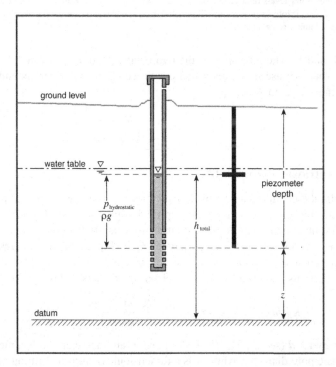

FIGURE 2.6   The components and placement of a piezometer used to measure hydrostatic gradient $\Delta h/\Delta L$. On the left are the screen and the water level in the piezometer tube, and on the right the screen depth is denoted by the heavy line and the water level denoted by a heavy crossbar. *Source: Reproduced with permission from Environment Canada, 2008. Nitrate Levels in the Abbotsford Aquifer an indicator of groundwater contamination in the Lower Fraser Valley. Retrieved on November 2, 2010 from http://www.ecoinfo.gc.ca/env_ind/region/nitrate/nitrate_ref_e.cfm.*

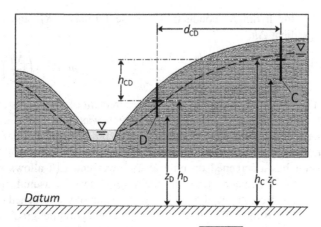

FIGURE 2.7 The hydrostatic gradient $\Delta h / \Delta L = h_{CD} / \sqrt{d_{CD}^2 + h_{CD}^2}$ between two points C and D below the water table in an unconfined aquifer requires piezometer measurements. The elevations of piezometer water levels relative to a datum are denoted h_C and h_D, while the difference in the elevation of two piezometer water levels is denoted h_{CD}. The elevations of the piezometer screens relative to a datum are denoted z_C and z_D.

points C and D. The relevance of the hydrostatic head $h_{hydrostatic}$ will become clearer when we use it to sketch hydrostatic equipotential contours and flow nets in the saturated zone:

$$h_{total}^C - h_{total}^D = \left(\frac{p_{hydrostatic}^C}{\rho_W \cdot g} \right) - \left(\frac{p_{hydrostatic}^D}{\rho_W \cdot g} \right) = h_{CD} \qquad (2.19)$$

2.5.5. Intrinsic Permeability

The hydraulic conductivity K_h in Darcy's Law is simply an empirical parameter. Hydrologists have derived an empirical expression (see Eq. (2.20)) for the hydraulic conductivity K_h that attempts to separate the contributions arising from the porous medium—parameter k_i in Eq. (2.20) and represented by terms enclosed by the second set of brackets in Eq. (2.21)—and the fluid—the terms enclosed by the second set of brackets in Eq. (2.21):

$$K_h \left[m \cdot s^{-1} \right] = k_i \left[m^2 \right] \cdot \left(\frac{\rho_w [Mg \cdot m^{-3}] \cdot g[m \cdot s^{-2}]}{\mu [Mg \cdot m^{-1} \cdot s^{-1}]} \right) \qquad (2.20)$$

The parameter d (see Eq. (2.21)) is the mean grain diameter and is a measure of intrinsic pore diameter, while N is a dimensionless empirical fitting parameter. The viscosity μ and density of the fluid ρ appear in the second set of terms. The empirical parameter k_i is called the *intrinsic permeability* of the porous medium and has units of area (m^2).

Freeze and Cherry (1979) measured the intrinsic permeability of a variety of rock types and unconsolidated materials. The median intrinsic permeability

Box 2.2 Determining Groundwater Flow Using Water Table Elevations.

The illustration below shows the elevation of the water table in meters (relative to mean sea level). Indicate the direction of groundwater flow on this illustration.

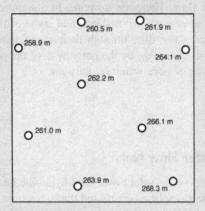

To determine the direction of groundwater flow at the water table, make a rough sketch of water table elevations.

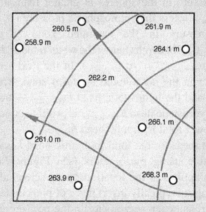

Water flow at the water table is "downhill" at right angles to the water table contours.

of gravel, sand, silt, and clay ranges 9 orders of magnitude from 10^{-9} m^2, 10^{-11} m^2, 10^{-14} m^2, and 10^{-18} m^2:

$$q_D = (N \cdot d^2)\left(\frac{\rho \cdot g}{\mu}\right)\frac{\Delta h}{\Delta L} = k \cdot \left(\frac{\rho \cdot g}{\mu}\right)\frac{\Delta h}{\Delta L} \qquad (2.21)$$

Before we leave this section on Darcy's Law, we need to discuss one final parameter that is relevant to chemical transport in the saturated zone. The specific discharge q_D (Eqs. (2.18) and (2.21)) has the dimensions of distance per unit time (m s^{-1}), equivalent to a velocity, and represents the fluid velocity averaged over the entire cross-sectional area of the measurement. The specific discharge q_D, however, is not a good representation of pore water velocity because a significant fraction of the cross-sectional area does not contribute to measurable water flow. There are no pores in a significant fraction of the cross-sectional area of any porous medium, and some of the pores have such a small diameter that water flow through their combined area is negligible. Scaling the specific discharge q_D by the porosity ϕ of the porous medium provides a better estimate of pore water velocity q_P:

$$q_P = \frac{q_D}{\phi} \tag{2.22}$$

2.5.6. Groundwater Flow Nets

Water flow in aquifers is toward a lower total hydraulic head h_{total}. By way of analogy, imagine two gas spheres connected through a tube fitted with a valve. Gas pressure in sphere A is higher than gas pressure in sphere B. Gas will flow from sphere A to sphere B when the valve is opened. This occurs regardless of whether the gas pressure in the spheres is greater than atmospheric pressure ($p_{gauge} > p_{atmospheric}$) or less than atmospheric pressure ($p_{gauge} < p_{atmospheric}$). The hydrostatic pressure in the saturated zone is greater than zero ($p_{hydrostatic} > p_{atmospheric}$) at depths below the water table and zero, by definition, at the water table ($p_{hydrostatic} = p_{atmospheric}$). In the next section we will learn about water behavior in the unsaturated (vadose) zone, where the hydrostatic pressure $p_{hydrostatic}$—and the hydrostatic head $h_{hydrostatic}$—are negative. Regardless of whether the hydrostatic head $h_{hydrostatic}$ is positive or negative, water flows in the direction of lower total hydraulic head h_{total}.

Hydrologists can measure the total hydraulic head h_{total} at any point at or below the water table using piezometers (see Figure 2.6). The measured hydraulic head h_{total} applies to the location of the screen at the bottom of the piezometer. Hydrologists typically insert nests of piezometers at selected locations on the land surface, each piezometer in the nest representing a different depth below the land surface. All measurements are relative to some elevation datum, whether it is mean sea level (MSL) or some convenient local datum. The total hydraulic head h_{total} of the piezometer (see Figure 2.6) is the height of the water level in the piezometer tube relative to the datum. The elevation head z of the piezometer is the position of the screen, referenced to the datum. The hydrostatic head $h_{hydrostatic}$ ($p_{hydrostatic}/\rho \cdot g$) at the depth of the screen is the elevation of the water level in the piezometer tube relative to the screen.

These components are also illustrated in Figure 2.7. The elevation of the datum is assigned a value of zero. Mean sea level is a very useful datum because land surface elevations are usually reported relative to MSL. The hydraulic gradient Δh is not an absolute value and, as such, is independent of the datum.

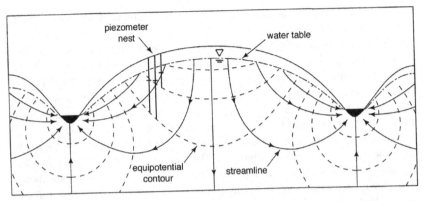

Adapted from Hubbert (1940)

FIGURE 2.8   Groundwater flow follows the flow net in an unconfined aquifer. The streamlines indicate the direction of groundwater flow. Vertical streamlines indicate groundwater divides. Note the relation between equipotential contours and piezometer water levels. *Source: Reproduced with permission from Hubbert, M.K., 1940. The theory of ground-water motion. J. Geol. 48 (8): 785–944.*

Figure 2.8 is an illustration adapted from Hubbert (1940) showing a land surface and the water table below it. A nest of three piezometers is located to the left of the center, and the water levels in the leftmost and center piezometers (indicated by the crossbar) are the same; both are lower than the rightmost piezometer. Water is flowing from the right piezometer toward the center and left piezometers because the hydraulic gradient Δh decreases in that direction. Although the water levels are the same in the center and left piezometers, the screens are not at the same depth. The dashed line running through the bottom of these two piezometers is an equipotential line connecting subsurface points with the same total hydraulic head h. The right piezometer indicates a higher hydraulic head and, therefore, lies on a different equipotential contour connecting points at a higher hydraulic head.

Several other dashed equipotential contours appear in Figure 2.8 radiating outward from the position where the water table elevation is highest (generally in the uplands) and converging where the water table elevation is lowest (generally in the lowlands). The illustration also plots a set of streamlines that intersect successive equipotential contours at right angles. The system of hydraulic equipotential contours and streamlines constitutes a groundwater flow net.

Streamlines diverge from groundwater recharge areas and converge at discharge areas. A *groundwater divide* occurs where streamlines are vertical, centered on recharge and discharge areas. In reality, Figure 2.8 illustrates two complete local flow cells, bordered by three groundwater divides, and portions of two more local flow cells on the left and right sides of the illustration. If the illustration is expanded vertically and horizontally to include regional trends in land surface elevation, then the features of regional ground-water flow would appear.

Local groundwater flow is dominated by water table topography and displays a complex pattern of local groundwater recharge and discharge areas. Regional groundwater flow is dominated by the orientation of geologic formations and large-scale changes in elevation. Regional groundwater flow is more strongly expressed in the lower depths of unconfined aquifers, while local groundwater flow is most evident in the vicinity of the water table.

The elevation of the water level in a piezometer h is, in general, not equal to the water table elevation (see Figure 2.6) because the water level is the hydraulic head h corresponding to conditions at the screen opening of the piezometer. The piezometer water level in Figure 2.6 is lower than the water table, indicating the hydraulic head at the bottom of the piezometer is lower than at the water table. This condition tells us water flow is downward at the piezometer screen and the piezometer is located in a groundwater recharge area.

In summary, piezometer measurements allow hydrologists to determine the hydraulic head anywhere in an aquifer. Groundwater flow nets are mapped using piezometer nests. Flow nets reveal seepage forces and the direction of groundwater flow. Environmental scientists working with hydrologists can predict the direction and rate of contaminant migration using flow nets.

2.6. VADOSE ZONE HYDROLOGY

2.6.1. Capillary Forces

Capillary forces—interactions between thin films of water and mineral surfaces—dominate water behavior in the unsaturated or vadose zone above the water table in unconfined aquifers and soils. The basic principles behind capillary forces are illustrated by a simple experiment in which the tip of a small-diameter glass tube is inserted into a dish of water (Figure 2.9).

As soon as the tip of the capillary tube touches the water surface, water is drawn into the tube and rises to a certain level before it stops. The adhesion of water molecules to the glass surface and the cohesion binding water molecules to one another accounts for the capillary forces that draw water into small-diameter glass tubes. Soil pores behave like capillary tubes, drawing in water and, like the experiment illustrated in Figure 2.9, holding the water in the pore against the force of gravity.

Our simplified description of capillary forces is cast in the same language used to describe forces in the saturated zone. The hydrostatic pressure at the water surface in the dish is zero—relative to atmospheric pressure—and increases with depth below the water surface, as we have already seen. The phenomenon we are interested in, however, is hydrostatic pressure within the capillary tube (or by analogy, within a small pore in the vadose zone). A clue emerges if we withdraw the capillary tube vertically from the water dish. The hydrostatic pressure at the bottom tip of the capillary tube (point B, Figure 2.9)

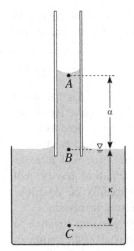

FIGURE 2.9   The hydrostatic pressures within and below a capillary tube illustrate capillary rise.

is zero—relative to atmospheric pressure—the same as the surface of the water outside the capillary tube (water table symbol indicates $p_{hydrostatic} = p_{atmospheric}$). If the hydrostatic pressure at a distance κ below the water surface (point C, Figure 2.9) is positive because of the gravitational force acting on the water column above, how can the hydrostatic pressure at the bottom of the water column in the capillary tube (point B, Figure 2.9) be zero?

The gravitational force per unit area in unconfined water at a depth κ *below* the water surface (point C, Figure 2.9) is the *compressive force* we call hydrostatic pressure p_C (23), reflecting the downward force of the water column at the depth κ. We are selecting a depth κ equal to the height α of water rise in the capillary tube:

$$p_C = \rho_W \cdot g \cdot \kappa > 0 \qquad (2.23)$$

Consider the change in hydrostatic pressure along the vertical axis of the capillary from point C at depth κ, through point B at the level of the water surface outside the capillary, upward to point A inside the capillary at a height α. Since the hydrostatic pressure at any depth is uniform, and the hydrostatic pressure at the unconfined water surface is atmospheric pressure, then the hydrostatic pressure at point B, p_B, must be 0—neglecting atmospheric pressure $p_{atmospheric}$:

$$p_B = \rho_W \cdot g \cdot 0 = 0 \qquad (2.24)$$

The hydrostatic pressure at the bottom tip of the capillary tube at point B is 0 regardless of whether the water is confined in the capillary tube or unconfined (outside the tube). The water in the capillary tube is *suspended* in the tube by capillary forces, and those forces are balanced at the bottom tip. The force suspending the column of water in the capillary tube against the force of gravity is a *tensional force*. This means the hydrostatic pressure at the meniscus height α in the capillary tube (p_A at point A, Figure 2.9) must be equal in magnitude to the hydrostatic pressure at an equivalent distance κ below the surface of the water dish (p_C at point C, Figure 2.9) but *negative in sign:*

$$p_A = -(\rho_W \cdot g \cdot \alpha) \tag{2.25}$$

A negative hydrostatic pressure may seem counterintuitive, but the force acting on the water is a tension force that, were it not for the cohesive forces that bind water molecules together, would tend to pull water molecules apart. Positive hydrostatic pressure is a compressive force that tends to push water molecules closer together. The compressive hydrostatic force will increase the density of water, while tensional force (caused by the pull of gravity) will decrease the density of water in the capillary tube. Appendix 2C describes how to predict capillary rise based on the balance of forces involved.

Henceforth we will refer to the negative hydrostatic pressure (relative to atmospheric pressure) in the vadose zone as $p_{tension}$ to emphasize the dominant role that tension forces play in the vadose zone. We reserve hydrostatic pressure $p_{hydrostatic}$ for the saturated zone where pore water hydrostatic pressure is positive relative to atmospheric pressure.

The tension force acting on water in a capillary tube (see Eq. (2.25)) depends on meniscus height α and capillary tube radius R (see Appendix 2C). The water tension $p_{tension}$ in an unsaturated porous medium is a complex function of the water content θ_V and texture of the medium as it influences porosity ϕ and the mean pore radius r_{pore} (see Appendix 2C).

Soil scientists often use the symbol $\psi[Pa]$ rather than $p[Pa]$ when referring to water under tension. The *International System of Units* (SI) uses *Pascal* as the preferred pressure unit $(1N \cdot m^{-2} = 1\,Pa)$. To avoid unnecessary confusion, we do not use the symbol $\psi[Pa]$ to represent water tension:

$$Vadose\ Zone : p_{tension}[Pa] < 0 \tag{2.26}$$

2.6.2. Soil Moisture Zones

Figure 2.10 illustrates the upper edge of an unconfined aquifer. An unsaturated or vadose zone lies between the land surface and the water table. Lying above the water table is a zone, called the *capillary fringe* zone, where the pores are mostly saturated but the water pressure is negative (i.e., under tension): $p_{tension} < p_{atmospheric}$. The capillary fringe is in direct contact with the water table where the hydrostatic pressure equals atmospheric pressure: $p_{hydrostatic} = p_{atmospheric}$. The soil-water zone occupies the upper edge of the vadose zone and, by definition, is the vadose zone inhabited by plant roots. The soil-water zone undergoes significant changes in water content as plants absorb water through their roots and transpire it into the atmosphere through their leaf canopy and as water infiltrates into the soil following precipitation. An intermediate zone often lies between the saturated zone and the soil-water

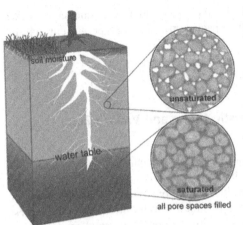

FIGURE 2.10 The unsaturated zone extends from the land surface to the water table. The zone below the water table is saturated. *Source: Reproduced with permission from Egger, A.E., 2003. The Hydrologic Cycle: Water's journey through time. Retrieved November 2, 2010 from http://visionlearning.com/library/module_viewer.php?mid=99&l=.*

zone—below the plant root zone and above the water table—where little change in water content takes place.

Soil scientists recognize three water contents with special significance to water behavior in the vadose zone: saturation, field-capacity FC, and the wilting point WP. Saturation is self-explanatory and usually does not persist long in upland sites but can persist at lowland sites where groundwater discharge occurs. Field capacity[1] is the water content of a soil after gravity has drained as much water from the soil as possible. Soil water at field capacity is held by capillary forces against the force of gravity and, by definition, represents a water tension of approximately $-10\ kPa$ (Eq. (2.27)) or tension head of $-1.02\ m$ (Eq. (2.28)):

$$p_{FC} \equiv -10\ kPa \qquad (2.27)$$

$$h_{FC} = \frac{p_{FC}}{\rho_W \cdot g} = \frac{\left(-10 \cdot 10^3\ kg \cdot m^{-1} \cdot s^{-2}\right)}{\left(1.0 \cdot 10^3\ kg \cdot m^{-3}\right) \cdot \left(9.807\ m \cdot s^{-2}\right)} = -1.02\ m \qquad (2.28)$$

The *wilting point* is also self-explanatory to a degree—it being the water content at which plants can no longer extract water from the soil. Needless to say, the WP of desert plants is different from plant species adapted to wetter climates, so the WP is defined as a water tension of $-1500\ kPa$ (Eq. (2.29)) or a tension head $-153.0\ m$ (Eq. (2.30)):

$$p_{WP} \equiv -1500\ kPa \qquad (2.29)$$

$$h_{WP} = \frac{p_{WP}}{\rho_W \cdot g} = \frac{\left(-1500 \cdot 10^3\ kg \cdot m^{-1} \cdot s^{-2}\right)}{\left(1.0 \cdot 10^3\ kg \cdot m^{-3}\right) \cdot \left(9.807\ m \cdot s^{-2}\right)} = -153.0\ m \qquad (2.30)$$

---

1. The United States Department of Agriculture Natural Resources Conservation Service defines *field moisture capacity* as the soil moisture content at a water tension of $-33\ kPa$ or $-10\ kPa$.

The plant-available water content AWC in a soil is the amount of water stored in a soil between field capacity θ_{FC} and the permanent wilting point θ_{WP}:

$$\theta_{AWC} = \theta_{FC} - \theta_{WP} \qquad (2.31)$$

2.6.3. The Water Characteristic Curve and Vadose Zone Hydraulic Conductivity

The soil tension $p_{tension}$ is a measure of how strongly capillary forces hold water in soil pores. Water rises higher in small-diameter capillaries, which shows that small-diameter pores hold water more strongly than large-diameter pores. As the water content of a soil decreases, water drains from the largest pores first, followed by the pores with progressively smaller diameters. Consequently, the soil tension $p_{tension}$ varies with the soil water content θ in a manner that is characteristic of the soil or porous medium. The *water characteristic curve* reflects the unique pore size distribution of each soil:

$$p_{tension} = p_{tension}(\theta) < 0 \qquad (2.32)$$

Likewise, the soil hydraulic conductivity $K_h(\theta)$ varies with the soil-water content θ because the mean pore velocity decreases with pore diameter, and the maximum pore diameter filled with water decreases with water content.

2.7. ELEMENTARY SOLUTE TRANSPORT MODELS

2.7.1. The Retardation Coefficient Model

Water percolating downward through the vadose zone or groundwater flow through the saturated zone carries dissolved chemical agents. Most agents do not travel at the same velocity as the water because interactions with the soil, saprolite, sediments, or aquifer retard their movement. Bouwer (1991) derived a simple expression for the retardation coefficient R_f that first appeared as an empirical expression for solute transport through an adsorbent (Vermeulen, 1952; Higgins, 1959). The following section, using the definitions in Appendix 2D, recapitulates the derivation by Bouwer and evaluates its limitations.

The retardation coefficient R_f expression derived by Bouwer is applicable to both saturated and unsaturated flow. Bouwer lists a single assumption: the timescale for water flow must be long relative to the timescale of the sorption reaction that binds the agent to the solid phase of the soil or aquifer. In this assumption the distribution coefficient K_D is an accurate representation of the sorption reaction.

The retardation coefficient R_f (Eq. (2.33)) is defined as groundwater velocity divided by agent velocity or, alternatively, the distance groundwater

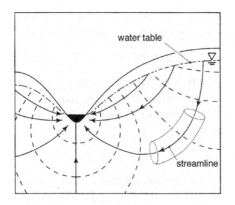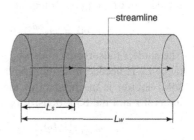

FIGURE 2.11   Flow paths (streamlines) tracing groundwater movement through an unconfined aquifer (left). Each streamline forms the axis of a cylindrical volume element. The retardation coefficient model represents interactions between the solute and aquifer delaying movement of a solute S, resulting in a shorter travel distance L_S than groundwater L_W during time interval t.

travels L_W (Eq. (2.34)) divided by the distance substance S travels L_S in time t (Figure 2.11, right):

$$R_f = \frac{L_W}{L_S} \tag{2.33}$$

$$L_W = q_D \cdot t \tag{2.34}$$

Consider a cylindrical volume of soil or aquifer centered on a streamline of water flow (see Figure 2.11). The base of the flow cylinder has an arbitrary area A. The total mass of agent S entering the volume during a time interval t equals the product of the concentration C_S of agent S dissolved in water and the volume swept out by water flowing along the streamline during time t. This volume is proportional to the distance water flows through the aquifer during time t. The volumetric water content θ_V makes this expression valid for both saturated and unsaturated flow:

$$m_S^{total}[Mg] = \left(L_W[m] \cdot A\left[m^2\right]\right) \cdot \theta_V \cdot C_S\left[Mg \cdot m^{-3}\right]$$

$$m_S^{total} = (L_W \cdot A) \cdot \theta_V \cdot C_S \tag{2.35}$$

The total dissolved mass of agent S in the flow cylinder is the product of concentration C_S and the volume swept out by solute flowing along the axis of the flow cylinder during time t. This volume is proportional to the distance L_S that agent S travels through the aquifer during time t:

$$m_S^m[Mg] = \left(L_S[m] \cdot A\left[m^2\right]\right) \cdot \theta_V \cdot C_S\left[Mg_S \cdot m^{-3}\right]$$

$$m_S^m = (L_S \cdot A) \cdot \theta_V \cdot C_S \tag{2.36}$$

The total mass of agent S retained by solid aquifer during time interval t is simply the difference between the total mass in the flow cylinder m_S^{total} and the dissolved mass m_S^m. Designating the dissolved agent as mobile m and the aquifer-bound agent as stationary s allows us to use a common set of symbols for the retardation coefficient and plate theory models of solute transport:

$$m_S^s = m_S^{total} - m_S^m = (L_W - L_S) \cdot A \cdot \theta_V \cdot C_S \qquad (2.37)$$

The aquifer (adsorbent) mass $m_{aquifer}$ in the flow cylinder encompassed by distance L_S is needed to determine the sorption coefficient K_D. The mass of the solid aquifer is the product of the volume accessible to agent S during time interval t and the bulk density ρ_B of the aquifer:

$$m_{aquifer}[Mg] = (L_S[m] \cdot A[m^2]) \cdot \rho_{bulk}[Mg \cdot m^{-3}] \qquad (2.38)$$

The total concentration of agent S immobilized through sorption by solid aquifer in the flow cylinder q_S is m_S^s divided by $m_{aquifer}$. The arbitrary area A cancels out:

$$q_S = \frac{m_S^s}{m_{aquifer}} = \left(\frac{A}{A}\right) \cdot \frac{(L_W - L_S) \cdot \theta_V \cdot C_S}{L_S \cdot \rho_{bulk}}$$

$$q_S = \frac{(L_W - L_S) \cdot \theta_V \cdot C_S}{L_S \cdot \rho_{bulk}} \qquad (2.39)$$

The sorption coefficient K_D is the ratio between sorbed and dissolved concentrations of agent S in the flow cylinder: q_S and C_S, respectively:

$$K_D = \frac{q_S}{C_S} = \left(\frac{1}{C_S}\right) \cdot \left(\frac{(L_W - L_S) \cdot \theta_V \cdot C_S}{L_S \cdot \rho_{bulk}}\right)$$

$$K_D = \frac{(L_W - L_S) \cdot \theta_V}{L_S \cdot \rho_{bulk}} \qquad (2.40)$$

Equation (2.40) is rearranged to yield the retardation factor R_f as defined by the ratio L_w/L_S:

$$L_W \cdot \theta_V - L_S \cdot \theta_V = K_D \cdot L_S \cdot \rho_{bulk}$$

$$L_W = L_S \cdot \left(1 + \frac{K_D \cdot \rho_{bulk}}{\theta_V}\right)$$

$$\frac{L_W}{L_S} = R_f = \left(1 + \frac{K_D \cdot \rho_{bulk}}{\theta_V}\right) \qquad (2.41)$$

The retardation coefficient R_f applies to both saturated and unsaturated flow, the former using porosity ϕ and the latter using the volumetric water content θ_V:

$$\phi\left[\frac{m^3}{m^3}\right] = \frac{V_w + V_a}{V_w + V_a + V_s} = \frac{V_{voids}}{V_{total}}$$

$$\theta_V\left[\frac{m^3}{m^3}\right] = \frac{V_w}{V_w + V_a + V_s} = \frac{V_w}{V_{total}}$$

$$\left(1 + \frac{K_D \cdot \rho_B}{\phi}\right) = \left(1 + \frac{K_D \cdot \rho_B}{\theta_V}\right) = R_f$$

Box 2.3 Estimating the Retardation Coefficient for a Soil Based on Its Properties

A soil from Illinois has a bulk density of $1.34\ Mg \cdot m^{-3}$ and a porosity of 0.49. Calculate the saturated-zone retardation coefficient R_f, for cyanazine, given a $K_D = 12\ m^3 \cdot Mg^{-1}$ for cyanazine adsorption by the soil.

$$R_f = \left(1 + \left(\frac{\rho_B \cdot K_D}{\phi}\right)\right) = \left(1 + \left(\frac{(1.34\ Mg \cdot m^{-3})(12\ m^3 \cdot Mg^{-1})}{0.49}\right)\right) = 33.8$$

Box 2.4 Estimating Contaminant Movement Based on Soil Properties and the Retardation Coefficient

Estimate the depth of cyanazine leaching from the soil surface after a 5 cm rainfall (assume no runoff or evaporation). The soil has a porosity of 0.49 and a volumetric water content of 0.25 at field capacity.

Since the water content is not given, we will assume that the water content of the soil is 0.25, the water-holding capacity at field capacity. A 5 cm rainfall will saturate the soil to a depth of $5.0/0.25 = 20$ cm. Since the depth of water movement is 20 cm, the depth of cyanazine leaching is *at least* that depth divided by the *retardation coefficient*.

$$R_f = \frac{L_{water}}{L_{cyanazine}} = 65.3$$

$$L_{cyanazine} = \frac{L_{water}}{65.3} = \frac{20\ cm}{65.3} = 0.30\ cm$$

The retardation coefficient model assumes the dissolved agent concentration is uniform everywhere behind the solute front and zero in advance of the solute front (see Figure 2.11). The plate theory model uses multiple sequential partitioning to represent dispersion at the solute front.

2.7.2. Plate Theory: Multiple Sequential Partitioning

Another model for solute transport comes from the field of separation chemistry, where it is known as the *plate theory* model. Plate theory is used to model affinity chromatography that relies on the partitioning of compounds dissolved in a mobile liquid phase onto a stationary phase; the mobile phase flows through a column filled with the porous stationary phase. The migration of the compound is determined by its relative affinity for the mobile and stationary phases. Plate theory uses a discrete representation of the flow cylinder (Figure 2.12)—dividing the cylinder into *n* cylindrical sections or plates of equal length—with water

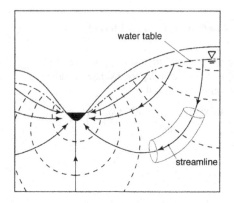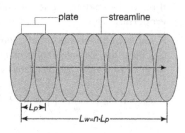

FIGURE 2.12 Flow paths (streamlines) tracing groundwater movement through an unconfined aquifer (left). Each streamline forms the axis of cylindrical volume element. Plate theory divides the flow cylinder into a series of plates of thickness L_p. Groundwater travel distance L_W during time interval t is equivalent to $L_W = n \cdot L_p$. Solute S undergoes partitioning between water and aquifer in each plate before water movement carries a fraction of dissolved solute to the next plate.

being the mobile phase and soil or aquifer material being the stationary phase. This model assumes that equilibrium partitioning is established in each plate before the fluid carries dissolved agent to the next plate.

Figure 2.13 illustrates the partitioning that occurs in plate 1 when the dissolved agent enters the first of a sequence of plates. A separatory funnel illustrates the agent partitioning that occurs in each plate; the upper solution is the mobile phase—the solution transferred to the next separatory funnel in each step—and the lower solution is the stationary phase.

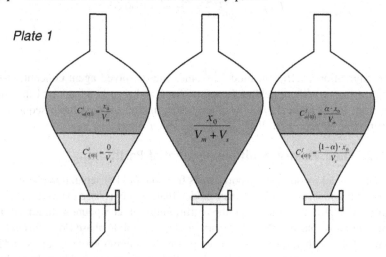

FIGURE 2.13 Partitioning of a solute in plate 1 (represented by a separatory funnel): initial state before partitioning (left), mixing of two phases (middle), and final state after partitioning (right). The volume transferred is $n = 0$.

The initial state before partitioning (Figure 2.13, left) consists of the agent in the mobile (upper) phase m; the stationary (lower) phase s does not contain any agent. The initial mass of agent added is m_0. The index $(n|p) = (0|1)$ indicates that the agent has not advanced beyond the first plate (volume transfer index $n = 0$), and the concentrations apply to the first plate $p = 1$.

$$m_{S(0|1)}^m\big|_{initial} = m_0$$

$$m_{S(0|1)}^s\big|_{initial} = 0$$

Partitioning during mixing of the two phases (Figure 2.13, middle), where the total volume accessible to the agent is the sum of the stationary and mobile phases, is $v_{plate} = v_m + v_s$.

$$\frac{m_0}{v_m + v_s}$$

The final state after partitioning (Figure 2.13, right) consists of a fraction $\alpha \cdot m_0$ of the agent in the mobile phase m and the remainder $(1 - \alpha) \cdot m_0$ in the stationary phase s.

Figure 2.14 illustrates the transfer of the mobile phase from the first separatory funnel (plate 1) to the second separatory funnel (plate 2). The transfer illustrated in Figure 2.14 is equivalent to water movement to the next in the sequence of plates, carrying agent to the next plate.

Rather than tracking concentrations, as in Figure 2.13, Figure 2.14 displays the agent fraction in each phase. Fresh solvent containing no dissolved agent replaces the mobile phase volume in the first funnel (Figure 2.14, plate 1, upper left), while the stationary phase in the second funnel initially contains no agent (Figure 2.14, plate 2, lower left). The stationary phase in funnel 1 is not transferred; consequently the initial stationary phase agent concentration (Figure 2.14, plate 1, upper left) is the same as final stationary phase concentration in Figure 2.13 (right). Initial concentrations for the mobile and stationary phases in funnel 1 appear below and in Figure 2.14 (plate 1, upper left): $(n|p) = (1|1)$:

$$m_{S(1|1)}^m\big|_{initial} = 0$$

$$m_{S(1|1)}^s\big|_{initial} = (1 - \alpha) \cdot m_0$$

Final concentrations for the mobile and stationary phases in funnel 1 appear following and in Figure 2.14 (plate 1, upper right): $(n|p) = (1|1)$:

$$m_{S(1|1)}^m\big|_{final} = \alpha \cdot (1 - \alpha) \cdot m_0$$

$$m_{S(1|1)}^s\big|_{final} = (1 - \alpha) \cdot (1 - \alpha) \cdot m_0 = (1 - \alpha)^2 \cdot m_0$$

As before, partitioning occurs during mixing (Figure 2.14, upper middle and lower middle), with the resulting agent concentrations in funnel 1 (Figure 2.14, plate 1, upper right) and funnel 2 (Figure 2.14, plate 2, lower right). Notice the relative partitioning is the same as in Figure 2.13.

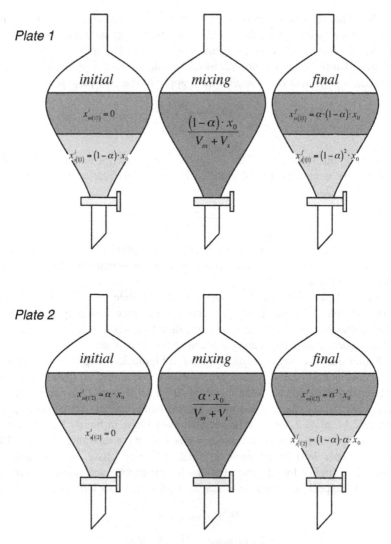

FIGURE 2.14 Partitioning of a solute in funnel 1 (plate 1) and funnel 2 (plate 2): initial state before partitioning (left), mixing of two phases (middle), and final state after partitioning (right). The volume transferred is $n = 1$.

One more transfer ($n = 1$) will suffice to establish the pattern. This requires adding a third funnel: adding fresh mobile phase solvent to funnel 1, transferring the final mobile phase from funnel 1 (Figure 2.14, plate 1, upper right) to funnel 2, and transferring the mobile phase from funnel 2 (Figure 2.14, plate 2, lower right) to funnel 3. The stationary phases in funnel 1 (Figure 2.14, plate 1, upper right) and funnel 2 (Figure 2.14, plate 1, lower right) remain in place, which means the stationary phase in funnel 3 initially contains no agent. After partitioning, the mobile and stationary phases appear in Table 2.4.

TABLE 2.4 Initial (before Partitioning) and Final (after Partitioning) Distribution of Agent between Mobile and Stationary Phases after Volume Transfer $n = 2$ for Plate 1, Index(2|1); Plate 2, Index (2|2); and Plate 3, Index (2|3)

Initial	Final
$m_{S(2\|1)}^m\big\|_{initial} = \alpha \cdot \left(0 + m_{S(1\|1)}^s\right)$	$m_{S(2\|1)}^m\big\|_{final} = \alpha \cdot \left((1-\alpha)^2 \cdot m_0\right) = \alpha \cdot (1-\alpha)^2 \cdot m_0$
$m_{S(2\|1)}^s\big\|_{initial} = (1-\alpha) \cdot \left(0 + m_{S(1\|1)}^s\right)$	$m_{S(2\|1)}^s\big\|_{final} = (1-\alpha) \cdot \left((1-\alpha)^2 \cdot m_0\right) = (1-\alpha)^3 \cdot m_0$
$m_{S(2\|2)}^m\big\|_{initial} = \alpha \cdot \left(m_{S(1\|1)}^m + m_{S(1\|2)}^s\right)$	$m_{S(2\|2)}^m\big\|_{final} = \alpha \cdot (2 \cdot \alpha \cdot (1-\alpha)) \cdot m_0 = 2 \cdot \alpha^2 \cdot (1-\alpha) \cdot m_0$
$m_{S(2\|2)}^s\big\|_{initial} = (1-\alpha) \cdot \left(m_{S(1\|1)}^m + m_{S(1\|2)}^s\right)$	$m_{S(2\|2)}^s\big\|_{final} = (1-\alpha) \cdot (2 \cdot \alpha \cdot (1-\alpha) \cdot m_0) = 2 \cdot \alpha \cdot (1-\alpha)^2 \cdot m_0$
$m_{S(2\|3)}^m\big\|_{initial} = \alpha \cdot \left(m_{S(1\|2)}^m + 0\right)$	$m_{S(2\|3)}^m\big\|_{final} = \alpha \cdot (\alpha^2 \cdot m_0) = \alpha^3 \cdot m_0$
$m_{S(2\|3)}^s\big\|_{initial} = \alpha \cdot \left(m_{S(1\|2)}^m + 0\right)$	$m_{S(2\|3)}^s\big\|_{final} = (1-\alpha) \cdot (\alpha^2 \cdot m_0) = \alpha^2 \cdot (1-\alpha) \cdot m_0$

If you were to proceed to volume transfer $n = 3$, you would discover that the amount of agent in each plate follows a binomial expansion of the distribution of α and $(1 - \alpha)$, as shown in Table 2.5. This was the result reported by Martin and Synge (1941) in their landmark paper on the theory of chromatography:

$$B(n,k) = \left(\frac{n! \cdot (1-\alpha)^{n-k} \cdot (\alpha)^k}{k! \cdot (n-k)!}\right) = \underbrace{\left(\frac{n!}{k! \cdot (n-k)!}\right)}_{binomial\ coefficient} \cdot (1-\alpha)^{n-k} \cdot (\alpha)^k \quad (2.42)$$

TABLE 2.5 Total Agent Mass after Volume Transfer $n = 3$ for Plate 1, Index (3|1); Plate 2, Index (3|2); Plate 3, Index (3|3), and Plate 4, Index (3|4)

Final
$m_{S(3\|1)}^m + m_{S(3\|1)}^s = \left(\frac{3! \cdot (1-\alpha)^{3-0} \cdot (\alpha)^0}{0! \cdot (3-0)!}\right) \cdot m_0 = (1-\alpha)^3 \cdot m_0$
$m_{S(3\|2)}^m + m_{S(3\|2)}^s = \left(\frac{3! \cdot (1-\alpha)^{3-1} \cdot (\alpha)^1}{1! \cdot (3-1)!}\right) \cdot m_0 = 3 \cdot (1-\alpha)^2 \cdot \alpha \cdot m_0$
$m_{S(3\|3)}^m + m_{S(3\|3)}^s = \left(\frac{3! \cdot (1-\alpha)^{3-2} \cdot (\alpha)^2}{2! \cdot (3-2)!}\right) \cdot m_0 = 3 \cdot (1-\alpha) \cdot \alpha^2 \cdot m_0$
$m_{S(3\|4)}^m + m_{S(3\|4)}^s = \left(\frac{3! \cdot (1-\alpha)^{3-0} \cdot (\alpha)^0}{0! \cdot (3-0)!}\right) \cdot m_0 = (1-\alpha)^3 \cdot m_0$

The index n represents the number of volume transfers that advance the agent through a series of plates (funnels in this illustration), p is the plate index ($p = k + 1; 0 \leq k \leq n$), and α represents the dimensionless fraction of total agent in each plate that partitions into the mobile phase (see Appendix 2E).

The binomial expansion is based on a discrete model of solute flow through a flow cylinder (see Figure 2.12). The flow cylinder is divided into a series of discrete increments representing water (and solute movement) as discrete advances indicated by the volume transfer index n. The binomial distribution probability function defines the probability of the agent appearing in some plate $p \leq n$, where n is the number of plates through which the mobile phase has advanced, equivalent to L_W in the retardation coefficient model (see Figure 2.12).

Figure 2.15 plots agent distributions predicted by plate theory for three different distances L_W. Plate theory—using a distribution coefficient α equivalent to the inverse of the retardation coefficient $\alpha = 1/R_f$ (Appendix 2E)—yields an agent distribution with the highest concentration being in the plate corresponding to $L_S = \alpha \cdot (n \cdot L_{plate})$.

The retardation coefficient model would set $L_W = n \cdot L_{plate}$ and, given the retardation coefficient R_f used in Figure 2.14, $L_S = (n \cdot L_{plate})/R_f$. Plate theory, using a distribution coefficient α equivalent to the retardation coefficient R_f used in Figure 2.14, yields a probability distribution with a most probable value corresponding to $L_S = \alpha \cdot (n \cdot L_{plate})$. Plate theory offers a more accurate representation of solute movement and dispersion as it advances along the flow cylinder.

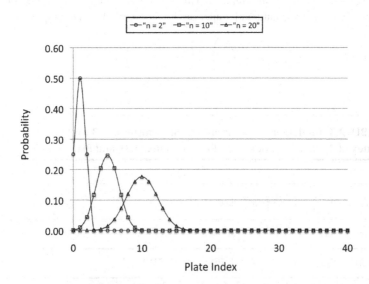

FIGURE 2.15 Solute probability distributions for three different water advancement distances: $2 \cdot L_{plate}, 10 \cdot L_{plate}, 20 \cdot L_{plate}$ for a distribution coefficient $\alpha = R_f^{-1}$.

2.8. SUMMARY

Residence time, water-holding capacity, groundwater flow paths and rates, and transport models all contribute to our capacity to predict the impact of aquifer properties on water chemistry. This chapter introduced water budgets as a means to estimate groundwater residence time, a key factor in determining groundwater chemistry. Changes in hydraulic head within the saturated zone determine groundwater flow paths that vary at several spatial scales, although environmental chemists are most interested in local flow paths. Variation in water content in the vadose zone influences water retention by capillary forces and unsaturated hydraulic conductivity (Appendix 2F). Darcy's Law provides a means to estimate pore water velocity that, in turn, influences solute transport rates. Finally, this chapter introduced two elementary transport models that predict the retardation of solute migration along flow paths with both the vadose and saturated zones caused by chemical interactions with the aquifer.

APPENDIX 2A. SOIL MOISTURE RECHARGE AND LOSS

Example 1-A

Vegetation at the North Carolina Agricultural Experiment Station in Oxford used an average 4.2 mm of water per day during May 2008. The soils near Oxford are about 60 cm deep and store 0.437 cm of water per cm of soil depth. At the beginning of May, the soil in the watershed held an average of 0.412 mm of water per mm of soil depth. Predict the net soil storage at Oxford at the end of May after a monthly total precipitation of 66.9 mm.

Precipitation: $\overline{P} = 66.9\ mm$

Evapotranspiration: $\overline{ET} = (30\ d) \cdot (4.2\ mm \cdot d^{-1}) = 126\ mm$

Total soil storage: $\overline{S_{storage}} = (600\ mm_{depth}) \cdot \left(0.437\ mm \cdot mm_{depth}^{-1}\right)$
$$= 262\ mm$$

Net soil storage: $\overline{S_{storage}} = (600\ mm_{depth}) \cdot \left(0.025\ mm \cdot mm_{depth}^{-1}\right)$
$$= 15\ mm$$

Surface discharge: $\overline{D_s} = \overline{P} - \left(\overline{S_{storage}} + \overline{ET}\right) = 66.9\ mm - 141\ mm$
$$= -74\ mm$$

The water budget is negative, indicating a loss of soil moisture.

$$\frac{74\ mm}{600\ mm_{depth}} = 0.123 \left(\frac{mm}{mm_{depth}}\right)$$

The average soil moisture is drawn down by soil moisture loss through evapotranspiration.

$$(0.412 - 0.124)\ mm \cdot mm_{depth}^{-1} = 0.289\ mm \cdot mm_{depth}^{-1}$$

The North Carolina Agricultural Station at Oxford reported an average soil moisture content of 0.350 $mm \cdot mm_{depth}^{-1}$ on May 30, 2008.

Example 2-A

The vegetation covering the 200 km^2 Gordon Creek watershed in Dane County, Wisconsin uses an average of 3.3 mm of water per day during a typical June. The average soil depth is 50 cm and can store 0.4 mm of water per mm of soil depth. At the beginning of June, the soil in the watershed holds an average of 0.2 mm of water per mm of soil depth. Predict surface discharge into Gordon Creek during a month when 306 mm of precipitation falls.

Precipitation: $\overline{P} = 306\ mm$

Evapotranspiration: $\overline{ET} = (30\ d) \cdot (3.3\ mm \cdot d^{-1}) = 99\ mm$

Total soil storage: $\overline{S_{storage}} = (500\ mm_{depth}) \cdot \left(0.4\ mm \cdot mm_{depth}^{-1}\right)$
$$= 200\ mm$$

Net soil storage: $\overline{S_{storage}} = (500\ mm_{depth}) \cdot \left(0.2\ mm \cdot mm_{depth}^{-1}\right)$
$$= 100\ mm$$

Surface discharge: $\overline{D_s} = \overline{P} - \left(\overline{S_{storage}} + \overline{ET}\right) = 306\ mm - 199\ mm$
$$= 107\ mm$$

Roughly one-third of the rainfall during this particular June appears as runoff into Gordon Creek.

APPENDIX 2B. THE WATER-HOLDING CAPACITY OF A SOIL PROFILE

Example 1-B

Table 2B.1 lists the physical and water retention data for a soil.

TABLE 2B.1 Gravimetric Water Contents θ_m as a Function of Tension (kPa) and Bulk Density ρ_B of a Soil Profile

Horizon	Depth, cm	Bulk density, $\rho_B\ [Mg_{drysoil} \cdot m^3]$	$\theta_m = \left(\frac{Mg_{water}}{Mg_{drysoil}}\right)$		
			−10 kPa	−100 kPa	−1500 kPa
A	0–30	1.28	0.28	0.20	0.08
B	30–70	1.40	0.30	0.25	0.15
C	70–120	1.95	0.20	0.15	0.05

To estimate the total available water-holding capacity (AWC) of the soil profile, begin by converting the gravimetric water of each horizon (see Table 2B.1) into the plant-available mass water content for each horizon (Table 2B.2). Plant-available water is the water content at field capacity ($p_{FC} = -10\ kPa$) minus the water content at the wilting point ($p_{PWP} = -1500\ kPa$).

TABLE 2B.2 Derived Gravimetric Water Contents (Field Capacity, Permanent Wilting Point, Available Water) of a Soil Profile

Depth, cm	$\theta_{m,FC}$	$\theta_{m,PWP}$	$\theta_{m,AWC}$
0–30	0.28	0.08	0.20
30–70	0.30	0.15	0.15
70–120	0.20	0.05	0.15

The available water content is not the same for each horizon, and, furthermore, each horizon differs in thickness. The *AWC* for the profile is the cumulative *AWC* accounting for the contribution from each horizon.

Horizon A: The *A*-horizon bulk density ρ_B^A is $1.28\ Mg_{drysoil} \cdot m^3$. Conversion from gravimetric to volumetric units requires Eq. (2.11): multiplying the gravimetric water content by the soil bulk density yields the volumetric water content.

$$\theta_V^A \left[\frac{m_{water}^3}{m_{dry\ soil}^3} \right] = 0.20 \left[\frac{Mg_{water}}{Mg_{dry\ soil}} \right] \times \left(\frac{1.28}{1.00} \right) \left[\frac{Mg_{dry\ soil}/m_{dry\ soil}^3}{Mg_{water}/m_{water}^3} \right]$$

$$\theta_V^A = 0.256\ m_{water}^3 \cdot m_{dry\ soil}^{-3}$$

Imagine a volume of soil with an area of $1\ m^2$ and a depth equal to the *A*-horizon: $30\ cm = 0.3\ m$. Multiplying the volumetric water content θ_V^A times the volume of the *A*-horizon yields the available water volume V for the *A*-horizon—a volume equivalent to a water depth of 0.077 meter for every square meter of soil (Table 2B.3).

TABLE 2B.3 Available Water Content of a Soil Profile

Depth, cm	$\rho_B\ [Mg_{drysoil} \cdot m^3]$	$\theta_{V,AWC}\ \left[m_{water}^3 \cdot m_{drysoil}^{-3} \right]$	$AWC[cm_{water}]$
0–30	1.28	0.256	7.7
30–70	1.40	0.210	8.4
70–120	1.95	0.293	14.6
Total Profile			30.7

$$d_{water}^A = 0.256 \left[\frac{m_{water}}{m_{dry\ soil}} \right] \times 0.3\ [m_{dry\ soil}]$$

$$d_{water}^A = 0.077\ m_{water}$$

In fact, the plant-available water content of the A-horizon of this soil is equivalent to a water depth of 0.077, or 7.7 cm, regardless of area. There is practical value in expressing water content this way. A 7.7 cm rainfall would saturate the A-horizon AWC of this soil because the rainfall is expressed as a depth regardless of area.

Example 2-B

Table 2B.4 lists the physical and water retention data for the Otter soil (Dane County, Wisconsin). Estimate the total available water-holding capacity (AWC) in centimeters of water for the entire soil profile.

TABLE 2B.4 Gravimetric Available Water Holding Capacity and Bulk Density ρ_B of Each Horizon (A: 0–50 cm, B: 50–135 cm, A: 135–150 cm) in a Soil

$d_{soil}[cm_{soil}]$	$\rho_B \left[Mg_{dry\ soil} \cdot m^3_{total} \right]$	$\theta_{m,AWC} \left[Mg^3_{water} \cdot Mg^{-3}_{dry\ soil} \right]$
0–50	1.25	0.19
50–135	1.45	0.15
135–150	1.55	0.13

The available water content and thickness of each horizon vary. The AWC for the profile is the cumulative AWC accounting for the contribution from each horizon.

Horizon A: The A-horizon bulk density ρ_B^A is $1.25\ Mg_{drysoil} \cdot m^{-3}$. Conversion from gravimetric to volumetric units requires Eq. (2.11): multiplying the gravimetric water content by the soil bulk density yields the volumetric water content.

$$\theta_V^A \left[\frac{m^3_{water}}{m^3_{dry\ soil}} \right] = 0.19 \left[\frac{Mg_{water}}{Mg_{dry\ soil}} \right] \times \left(\frac{1.25}{1.00} \right) \left[\frac{Mg_{dry\ soil}/m^3_{dry\ soil}}{Mg_{water}/m^3_{water}} \right]$$

$$\theta_V^A = 0.24\ m^3_{water} \cdot m^{-3}_{dry\ soil}$$

Multiplying the volumetric water content of each horizon $\theta_V^{horizon}$ times the volume of each horizon (depth $d_{soil}^{horizon}$ per square meter) yields the available water depth $d_{water}^{horizon}$. The volumetric water content of 0.24 $m^3_{water} \cdot m^{-3}_{dry\ soil}$ is equivalent to water depth of 0.24 meter for every meter of A-horizon (or 0.12 meter of water for an A-horizon 0.5 meter in depth).

$$d_{water}^A = 0.24 \left[\frac{m_{water}}{m_{dry\ soil}} \right] \times 0.5 \left[m_{dry\ soil} \right]$$

$$d_{water}^A = 0.12\ m_{water}$$

The AWC of the remaining horizons and the total profile appear in Table 2B.5.

TABLE 2B.5 Available Water Content of the Entire Soil Profile

Depth, cm	$\theta_{V,AWC}\left[m^3_{water} \cdot m^{-3}_{dry\ soil}\right]$	$AWC[cm_{water}]$
0–50	0.24	12.0
50–135	0.22	18.5
135–150	0.20	3.0
0–150		33.5

A 12.0 cm rainfall would saturate the A-horizon AWC of this soil because the rainfall is expressed as a depth regardless of area.

APPENDIX 2C. PREDICTING CAPILLARY RISE

Liquid in a capillary tube (Figure 2C.1) with radius r rises until the forces acting on the liquid in the tube are balanced at height l. The downward force is simply the gravitational force on the column of liquid: liquid mass times gravitational acceleration.

$$\left|\mathbf{F}_{gravitational}\right| = m_{liquid} \cdot g = (\rho \cdot V)_{liquid} \cdot g = \left(\rho \cdot \pi r^2 \cdot l\right)_{liquid} \cdot g$$

The upward force results from the surface tension of the liquid γ at the air-liquid interface. The surface tension γ is a force that acts at an angle relative to the vertical tube known as the contact angle θ and is distributed along the circumference of the capillary tube $2\pi r$. The surface tension γ is a property of the liquid, while the contact angle θ depends on the adhesion between the liquid and the material from which the tube is composed.

$$\left|\mathbf{F}_{capillary}\right| = \gamma_{liquid} \cdot 2\pi r \cdot \cos\theta$$

At equilibrium, the two forces are equal and opposite, enabling us to solve for the capillary rise l in a round tube as a function of the radius of the tube r, the density of the liquid ρ, the surface tension γ, and the contact angle θ.

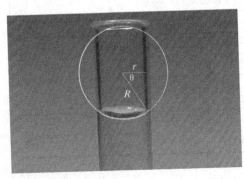

FIGURE 2C.1 The contact angle θ of water in a capillary tube is related to the meniscus radius R and diameter of the capillary tube r.

$$\left|\mathbf{F}_{gravitational}\right| = \left|\mathbf{F}_{capillary}\right|$$

$$\rho_{liquid} \cdot \pi r^2 \cdot l \cdot g = \gamma_{liquid} \cdot 2\pi r \cdot \cos\theta$$

The contact angle θ for water and glass is approximately 0 and $\cos\theta \approx 1$.

$$l = \frac{2 \cdot \gamma_{water}}{\rho_{water} \cdot g \cdot r}$$

Figure 2C.1 shows that the meniscus has a radius R that is greater than or equal to the radius of the capillary tube r. Simple geometry relates the radius of the meniscus R and the capillary tube radius r to the contact angle θ.

APPENDIX 2D. SYMBOLS AND UNITS IN THE DERIVATION OF THE RETARDATION COEFFICIENT MODEL OF SOLUTE TRANSPORT

Derivation of the Retardation Coefficient Model relies on the following expressions. The bulk density ρ_B is a function of the porosity ϕ, which typically assumes a mineral density $\overline{\rho}_{mineral}$ equal to the density of quartz: 2.65 $Mg \cdot m^{-3}$.

$$\rho_B = (1 - \phi) \cdot \overline{\rho}_{mineral}$$

Agent S distributes itself between the mobile water phase (dissolved state) and the stationary solid phase (sorbed state) according to the following distribution coefficient K_D expression:

$$K_D = \frac{q_S}{C_S}$$

TABLE 2D.1 Symbols and Units Used to Derive the Retardation Coefficient Model

Symbol	Units	Definition
C_S	$Mg \cdot Mg^{-1}$	Dissolved concentration of agent S
L_w	m	Distance water travels in time t
L_S	m	Distance dissolved agent S travels in time t
ϕ	$m^3 \cdot m^{-3}$	Porosity
θ_V	$m^3 \cdot m^{-3}$	Volumetric water content of porous medium ($\theta_V = \phi$ in the saturated zone)
$\overline{\rho}_{mineral}$	$Mg \cdot m^{-3}$	Mineral density (2.65 $Mg \cdot m^{-3}$)
ρ_B	$Mg \cdot m^{-3}$	Bulk density

Neutral organic agents associate with a specific component of the solid phase: the organic matter coating the mineral grains (cf. *Natural Organic Matter & Humic Colloids* and *Adsorption & Surface Chemistry*). This specific interaction allows us to define the distribution coefficient K_D in terms of the organic carbon content of the soil or aquifer f_{OC} and a universal distribution coefficient for the sorption of neutral organic compounds by natural organic matter in the soil or aquifer K_{OC}.

$$K_D = f_{OC} \cdot K_{OC}$$

APPENDIX 2E. SYMBOLS AND UNITS IN THE DERIVATION OF THE PLATE THEORY MODEL OF SOLUTE TRANSPORT

Derivation of the plate theory model relies on the following expressions. Agent S distributes itself between the mobile water phase (dissolved state) and the immobile solid phase (sorbed state) according to the same distribution coefficient K_D expression used in the retardation *coefficient* model.

$$K_D = \frac{q_S}{C_S}$$

Plate theory divides the flow cylinder into a series of identical cylindrical plates with a cross-sectional area A and length L_{plate}.

$$v_{plate} = A \cdot L_{plate}$$

The aquifer mass $m_{aquifer}$ (stationary phase mass) in each plate is the product of the plate volume v_{plate} and the bulk density of the aquifer ρ_B.

$$m_{aquifer} \equiv m_{plate} = \left(A \cdot L_{plate} \right) \cdot \rho_B$$

The mobile phase volume v_m in each plate along the flow cylinder is the product of the plate volume v_{plate} and the volumetric water content of the aquifer θ_V.

$$v_m = \left(A \cdot L_{plate} \right) \cdot \theta_V$$

The following two expressions define the mass concentration of agent bound to the stationary-phase aquifer q_s and the volume concentration of agent dissolved in mobile-phase water C_m. Plate theory adds two indices: the plate index (p) and the volume-transfer index (n). The latter tracks the advancement of agent dissolved in the mobile phase through the flow cylinder: $0 \leq n \leq p$:

$$q_{S(n|p)} = \frac{m^s_{S(n|p)}}{m_{aquifer}}$$

$$C_{S(n|p)} = \frac{m^m_{S(n|p)}}{v_m}$$

Combining the distribution coefficient expression with those defining the mass and volume concentrations of the agent, we arrive at the following expression that relates the mass of agent S bound to the stationary phase $m^s_{S(n|p)}$ to the agent mass dissolved in water $m^m_{S(n|p)}$ using the distribution coefficient K_D:

$$m^s_{S(n|p)} = K_D \cdot \left(\frac{m_{aquifer}}{v_m} \right) \cdot m^m_{S(n|p)}$$

Plate theory partitions the chemical agent between mobile phase (water) and stationary phase (aquifer) in each plate using the dimensionless coefficient α.

$$\alpha = \frac{m^m_{S(n|p)}}{m^m_{S(n|p)} + m^s_{S(n|p)}}$$

Plate theory coefficient α is related to the distribution coefficient commonly encountered in environmental chemistry by

$$m^m_{S(n|p)} = \alpha \cdot \left(m^m_{S(n|p)} + \left(\frac{K_D \cdot m_{aquifer}}{v_m} \right) \cdot m^m_{S(n|p)} \right)$$

$$\alpha = \left(1 + \left(\frac{K_D \cdot m_{aquifer}}{v_m} \right) \right)^{-1}$$

The dimensionless plate theory distribution coefficient α is the inverse of the dimensionless retardation coefficient R_f (Eq. (2.42)). Recognizing that plate volume is arbitrary leads us to replace $m_{aquifer}$ with ρ_B and v_m with θ_V with the result: $R_f = \alpha^{-1}$

TABLE 2E.1 Symbols and Units Used to Derive the Plate Theory Model

Symbol	Units	Definition
p		Plate index $1 \leq p \leq n+1$
n		Volume transfer or agent advancement index
$n \cdot L_{plate}$	m^3	Distance water and agent S travels after advancing through n plates
v_{plate}	m^3	Plate volume
v_m	m^3	Water (mobile phase) volume per plate
$m_{aquifer}$	Mg	Aquifer (stationary phase) mass per plate
ρ_B	$Mg \cdot m^{-3}$	Bulk density of aquifer
$m^m_{S(n|p)}$	Mg	Mass of mobile (dissolved) agent in plate p following the water advancement through n plates

Continued

TABLE 2E.1   Symbols and Units Used to Derive the Plate Theory Model—Cont'd

Symbol	Units	Definition
$m^s_{S(n|p)}$	Mg	Mass of stationary (sorbed) agent in plate p following the water advancement through n plates
α		Dimensionless fractional distribution coefficient
K_D	$m^3 \cdot Mg^{-1}$	Distribution coefficient

APPENDIX 2F. EMPIRICAL WATER CHARACTERISTIC FUNCTION AND UNSATURATED HYDRAULIC CONDUCTIVITY

Clapp and Hornberger (1978) published empirical functions that estimate hydraulic properties in the vadose zone. The parameters are for 11 different soil texture classes. The key variable is the degree of saturation S.

$$S \equiv \left(\frac{\theta_V}{\phi} \right)$$

The two empirical functions estimate the absolute value of tension head $|h_{tension}(S)|$ and the unsaturated hydraulic conductivity $K_h(S)$.

$$|h_{tension}(S)| = |h_S| \cdot S^{-b} \qquad K_h(S) = K_{h(sat)} \cdot S^c; \ c \approx 2b + 3$$

TABLE 2F.1 Hydraulic Parameters Based on Soil (<2 mm) Texture

Texture	ϕ	$K_{h(sat)} (cm \cdot s^{-1})$	$|h_S|(cm)$	b
Sand	0.395	$1.76 \cdot 10^{-2}$	12.1	4.05
Loamy Sand	0.410	$1.56 \cdot 10^{-2}$	9.0	4.38
Sandy Loam	0.435	$3.47 \cdot 10^{-3}$	21.8	4.90
Silty Loam	0.485	$7.20 \cdot 10^{-4}$	78.6	5.30
Loam	0.451	$6.95 \cdot 10^{-4}$	47.8	5.39
Sandy Clay Loam	0.420	$6.30 \cdot 10^{-4}$	29.9	7.12
Silty Clay Loam	0.477	$1.70 \cdot 10^{-4}$	35.6	7.75
Clay Loam	0.476	$2.45 \cdot 10^{-4}$	63.0	8.52

Continued

TABLE 2F.1 Hydraulic Parameters Based on Soil (<2 mm) Texture—Cont'd

Texture	ϕ	$K_{h(sat)}\,(cm \cdot s^{-1})$	$|h_S|\,(cm)$	b
Sandy Clay	0.426	$2.17 \cdot 10^{-4}$	15.3	10.4
Silty Clay	0.492	$1.03 \cdot 10^{-4}$	49.0	10.4
Clay	0.482	$1.28 \cdot 10^{-4}$	40.5	11.4

Source: Data reproduced with permission from Clapp, R.B., and G.M. Hornberger, 1978. Empirical equations for some soil hydraulic properties. Wat. Resour. Res. 14 (4), 601–604.

Problems

1. A cylindrical core ($L = 20$ cm, $r = 3$ cm) of the soil from problem 1 was recovered from the field site in Madison County, Iowa, and placed in a metal can (225 g empty) with a tight-fitting lid. The can of soil is weighed (1209.0 g field moist) and dried with the lid removed in an oven until it ceases to lose weight. The weight of the dried can with soil (including the lid) was 1005.0 g. Calculate the gravimetric water content θ_M of the field moist soil.

Solution

The gravimetric water content θ_M of the field moist soil is the water loss on drying divided by the mass of the dry soil. The mass of the dry soil must account for the mass of the container.

$$\theta_M = \frac{(1209 - 1005)_{moisture}}{(1005 - 225)_{dry\ soil}} = \frac{204\ g_{moisture}}{780\ g_{dry\ soil}} = 0.26\ g_{moisture} \cdot g_{dry\ soil}^{-1}$$

Calculate the bulk density ρ_B of the oven-dried soil.

Solution

The bulk density ρ_B of the oven-dried soil is the mass of dry soil divided by the soil volume: the volume of the cylindrical sample container.

$$\rho_B = \frac{780\ g_{dry\ soil}}{(9 \cdot 20 \cdot \pi)\ cm^3} = \frac{780\ g_{dry\ soil}}{565\ cm^3} = 1.38\ g_{dry\ soil} \cdot cm^{-3} = 1.38\ Mg_{dry\ soil} \cdot m^{-3}$$

Calculate the volumetric water content θ_V of the field moist soil.

Solution

Conversion from gravimetric to volumetric water content requires multiplying the former by the ratio of bulk density to water density.

$$\theta_V = \left(\frac{\rho_B}{\rho_w}\right) \cdot \theta_m = \left(\frac{1.38\ Mg_{dry\ soil} \cdot m_{dry\ soil}^{-3}}{1.00\ Mg_{water} \cdot m_{water}^{-3}}\right) \cdot \left(\frac{0.26\ Mg_{water}}{Mg_{dry\ soi}}\right) = 0.36\ m_{water}^3 m_{dry\ soil}^{-3}$$

2. The following table lists the physical and water retention data for a soil from Madison County, Iowa. The soil is located on a broad, nearly level, upland

divide. The soil has moderately low permeability and is considered poorly drained. The water table is located at a depth of 75 cm.

		Mass Water Content θ_m at Different Water Tensions $p_{tension}$.		
Horizon	$\rho_B \left[Mg_{dry\ soil} \cdot m^{-3}\right]$	−10 kPa	−100 kPa	−1500 kPa
A, 0–48 cm	1.49	0.29	0.22	0.14
B, 48–145 cm	1.38	0.31	0.25	0.19
C, 145–225 cm	1.33	0.32	0.25	0.16

Determine the plant-available water-holding capacity of the vadose zone of this soil, reporting the water-holding capacity as centimeters of water per unit area of soil.

Solution
The available water-holding capacity AWC is the difference between the water content at field capacity FC and the water content at the wilting point WP.

$$\theta_{m,AWC} = \theta_{m,FC} - \theta_{m,WP}$$

$$\theta_{m,AWC} = \theta_{m,-10\ kPa} - \theta_{m,-1500\ kPa}$$

Conversion from gravimetric to volumetric water content requires multiplying the former by the ratio of bulk density to water density. Since the water table is in the B-horizon at a depth of 75 cm, there is no need to calculate the AWC for the C-horizon.

$$\theta^A_{V,AWC} = \frac{(1.49\ Mg \cdot m^{-3}) \cdot (0.29 - 0.14)}{(1.00\ Mg \cdot m^{-3})} = \frac{0.22\ m^3_{water}}{m^3_{soil}} = \frac{0.22\ cm_{water}}{cm_{soil}}$$

$$\theta^B_V = \frac{(1.38\ Mg \cdot m^{-3}) \cdot (0.31 - 0.19)}{(1.00\ Mg \cdot m^{-3})} = \frac{0.17\ m^3_{water}}{m^3_{soil}} = \frac{0.17\ cm_{water}}{cm_{soil}}$$

The total AWC for the vadose zone accounts for the thickness of each horizon above the water table.

$$\theta_{V,AWC} = \left(48\ cm_{soil} \cdot \theta^A_{V,AWC}\right) + \left((75\ cm_{soil} - 48\ cm_{soil}) \cdot \theta^B_{V,AWC}\right)$$

$$\theta_{V,AWC} = \left(48\ cm_{soil} \cdot \frac{0.22\ cm_{water}}{cm_{soil}}\right) + \left(27\ cm_{soil} \cdot \frac{0.17\ cm_{water}}{cm_{soil}}\right)$$

$$\theta_{V,AWC} = 15\ cm_{water}$$

3. Two tensiometers were inserted into the soil on the University of Wisconsin–Madison campus (elevation 281.9 m above MSL) to a depth of 15 cm and 30 cm, and measurements of the tension head were recorded three times. Which way is the water flowing—upward from the 30 cm depth to the 15 cm depth or downward from the 15 cm depth to the 30 cm depth—at each measurement? Instead of using MSL as the elevation datum, set the datum at the 30 cm depth: $z_{elevation(30)} = 0$.

Depth	15 cm	30 cm
Elevation	15 cm	0 cm
Recording	$h_{tension(15)}$	$h_{tension(30)}$
1	−13 cm	−17 cm
2	−34 cm	−18 cm
3	−5 cm	−20 cm

Solution

Calculate the total head at the three individual recordings by adding the tension and elevation heads.

$$h_{total} = h_{tension} + z_{elevation}$$

Recording	$h_{tension(15)}$	$h_{tension} + z_{elevation}$	$h_{tension(30)}$	$h_{tension} + z_{elevation}$
1	−13 cm	2 cm	−17 cm	−17 cm
2	−34 cm	−19 cm	−18 cm	−18 cm
3	−5 cm	10 cm	−20 cm	−20 cm

Water flows in a direction that lowers the total head. When recordings 1 and 3 are made, water flows downward from the 15 cm depth to the 30 cm depth because the total head is lower at the 30 cm depth at those two times.

Recording	Relative Total Head	Water Flow between Sampling Depths
1	$h_{total(15)} > h_{total(30)}$	downward
2	$h_{total(15)} < h_{total(30)}$	upward
3	$h_{total(15)} > h_{total(30)}$	downward

When recording 2 is made, however, the total head at the 30 cm depth is higher than the 15 cm depth—because the soil is much drier at that depth and time—and water moves upward.

4a. A soil from Johnson County, Illinois has a bulk density ρ_B of 1.34 $Mg \cdot m^{-3}$ and a cyanazine sorption coefficient K_D of 12 $m^3 \cdot Mg^{-1}$. Cyanazine is a herbicide. Calculate the porosity ϕ and cyanazine retardation coefficient R_f for this soil.

Solution

The soil has a porosity of 0.49.

$$\phi = \left(1 - \frac{\rho_B}{\rho_S}\right)$$

$$\phi = \left(1 - \frac{(1.34\ Mg \cdot m^{-3})}{(2.65\ Mg \cdot m^{-3})}\right) = 0.494$$

$$R_f = \left(1 + \frac{K_D \cdot \rho_B}{\phi}\right)$$

$$R_f = \left(1 + \frac{(12\ m^3 \cdot Mg^{-1}) \cdot (1.34\ Mg \cdot m^{-3})}{0.49}\right) = 33.5$$

4b. Calculate how deep cyanazine applied at the surface of this soil will migrate in this soil after a 5 *cm* rainfall (assume no runoff or evaporation).

Solution

Since the water content is not given, assume that the water-holding capacity of the soil is equal to the porosity ϕ. This means 5 *cm* of water will penetrate $5/0.49 = 10.1$ *cm*.

$$L_{cyanazine} = \frac{L_{water}}{R_f}$$

$$L_{cyanazine} = \frac{L_{water}}{33.5} = \frac{10.1 \ cm}{33.5} = 0.30 \ cm$$

5a. The volumetric moisture content of a fine-sand soil is $\theta_{V,1} = 0.25 \ m^3_{water} \cdot m^{-3}_{dry \ soil}$ at an elevation $z_1 = 3.0 \ m$ above the water table and $\theta_{V,2} = 0.15 \ m^3_{water} \cdot m^{-3}_{dry \ soil}$ at an elevation $z_2 = 3.5 \ m$ above the water table. Use the empirical water characteristic function in Appendix 2F to determine the tension head $h_{tension}(S)$ at both points.

$$S \equiv \left(\frac{\theta_V}{\phi} \right)$$

$$|h_{tension}(S)| = |h_S| \cdot S^{-b}$$

Solution

Determine the degree of saturation S at each point using the empirical parameters for fine-sand texture.

$$S_1 = \left(\frac{\theta_{V,1}}{\phi} \right) = \left(\frac{0.25}{0.395} \right) = 0.63$$

$$S_2 = \left(\frac{\theta_{V,2}}{\phi} \right) = \left(\frac{0.15}{0.395} \right) = 0.38$$

Use the empirical water characteristic function to estimate the tension head $h_{tension}(S)$ at each point.

$$|h_{tension,1}| = |h_S| \cdot S_1^{-b} = (12.1) \cdot (0.63)^{-4.05}$$
$$h_{tension,1} = -77.2 \ cm = -0.772 \ m$$

$$|h_{tension,2}| = |h_S| \cdot S_2^{-b} = (12.1) \cdot (0.38)^{-4.05}$$
$$h_{tension,2} = -611 \ cm = -6.11 \ m$$

5b. Use the empirical hydraulic conductivity function $K_h(S)$ of Clapp and Hornberger (1978) to determine the hydraulic conductivity for water movement between these two points.

$$K_h(S) = K_{h(sat)} \cdot S^{(2b+3)}$$

Solution

The soil water contents is different at points 1 and 2. We can either compute the conductivity at the two points, using the water content at each point, and take the mean, or use a mean water content to estimate the unsaturated hydraulic conductivity. Using the latter: $\theta_{V,\,mean} = 0.20\ m^3_{water} \cdot m^{-3}_{dry\ soil}$.

$$K_h(S) = K_h\left(\frac{0.20}{0.395}\right) = K_h(0.51)$$

$$K_h(0.51) = (1.76 \cdot 10^{-2}) \cdot (0.51)^{(2 \cdot (4.05)+3)}$$

$$K_h(0.51) = (1.76 \cdot 10^{-2}) \cdot (0.51)^{(11.1)}$$

$$K_h(0.51) \approx 10^{-5}\ cm \cdot s^{-1}$$

5c. Applying Darcy's Law, estimate the direction of flow and magnitude of the specific discharge q_D.

Solution
Position 1 has an elevation 3.0 m above the water table and position 2 has an elevation 3.5 m above the water table. The elevation heads are $z_{datum} = z_1 = 0.0m$ and $z_2 = 0.5m$, respectively.

$$\frac{\Delta h}{\Delta z} = \frac{(0.5 + h_{tension,\,2}) - (0.0 + h_{tension,\,1})}{(0.5 - 0.0)}$$

$$\frac{\Delta h}{\Delta z} = \frac{(0.5 - 6.1) - (0.0 - 0.8)}{0.5} = \frac{-4.8}{0.5}$$

$$\frac{\Delta h}{\Delta z} \approx -9.6$$

The hydraulic gradient is higher at position 1 (–0.8 cm) than at position 2 (–5.6 cm), so the water will flow upward from position 1 to position 2. Darcy's Law yields a specific discharge q_D for upward water flow.

$$q_D \approx -K_h(0.51) \cdot \frac{\Delta h}{\Delta z} = -(10^{-5}) \cdot (-9.6) \approx 10^{-4}\ cm \cdot s^{-1}$$

Clay Mineralogy and Clay Chemistry

3.1. INTRODUCTION

Clay mineralogy is devoted to a group of silicate minerals with unique chemical and physical properties. The *clay minerals* connotation implies both specific mineralogy and particle size. Clays differ from other minerals found in most rocks because they are chemical alteration products formed at Earth's surface, where the conditions favor the formation of the hydrous, fine-grain minerals. Clays, by virtue of their extremely small particle size and high surface-to-volume ratio, are one of the most chemically active components of soils and sediments.

The chemical transformation—*weathering*—of rock produces clays and clay minerals. Our first task is to identify the geological materials where clay minerals occur and where clay minerals appear in the geochemical weathering sequence. Understanding mineral structures, while a worthy and interesting subject in and of itself, is generally not essential for environmental chemistry at the level of this book. Clay minerals, however, display physical and chemical behavior that simply cannot be appreciated without a grasp of their crystal structure. This is our second task. With these basics in hand, we are prepared for our final task: a description of clay mineral physical behavior and chemistry.

3.2. MINERAL WEATHERING

3.2.1. Mineralogy

Table 3.1 lists the major mineral classes. Silicate, phosphate, and sulfide minerals occur in all types of rock—*igneous, sedimentary,* and *metamorphic*—and weathering products. Oxyhydroxide, hydroxide, and certain hydrous silicate minerals are stable only at temperatures where liquid water can exist. Carbonate and sulfate minerals are found only in chemically weathered materials such as saprolite, sediments, soils, and sedimentary rock formations.

The Jackson Weathering Sequence

Physical weathering and abrasion generate fine-grained particles that retain the mineralogy of the rocks from which they form: sand (diameter range: 0.05–2.0 *mm*) particle-size class. Soil is defined as any material that passes a 2.0 *mm* sieve. Chemical weathering transforms primary (rock-forming) minerals into secondary minerals that are stable under the moist, low-temperature conditions existing at Earth's surface. The clay particle-size class (diameter range: <0.002 *mm*) is, with few exceptions, comprised of secondary minerals that include the clay minerals. Mineralogy of the silt particle-size class (diameter range: 0.002–0.05 *mm*) often contains a mixture of primary and secondary minerals.

M. L. Jackson (Jackson, Tyler et al., 1948) outlined a geochemical weathering sequence based on the mineralogy of the fine (≤ 5 μm) fraction in soils and sediments (Table 3.2). The minerals that are least resistant to chemical weathering, minerals in stages 1–7, are absent from the fine clay particle-size fraction (≤ 0.2 μm) and are confined to the coarse clay (0.2–2 μm) and fine silt

---

TABLE 3.1 Major Mineral Classes

Mineral Class	Representative Mineral Formulae
Silicates and aluminosilicates	$MgSiO_4$, $CaMgSi_2O_6$, $Al_2Si_2O_5(OH)_4$, $NaAlSi_3O_8$
Oxides, oxyhydroxides, and hydroxides	Al_2O_3, FeO, $MnOOH$, TiO_2
Sulfides	FeS_2, PbS, HgS, Cu_2S
Carbonates	$CaCO_3$, $MgCa(CO_3)_2$, $FeCO_3$
Sulfates	$CaSO_4$, $KFe_3(SO_4)_2(OH)_6$
Phosphates	$Ca_5(PO_4)_3(OH)$, $AlPO_4$, $2H_2O$
Halides	$NaCl$, CaF_2
Elements	Cu, Ag, Au, S

TABLE 3.2 Jackson Clay Mineral Weathering Stages

Weathering Stage	Clay Fraction[1] Mineralogy
1	Gypsum, halite
2	Calcite, dolomite
3	Olivine, pyroxene, amphibole
4	Biotite, chlorite
5	Feldspar (plagioclase and orthoclase)
6	Quartz
7	Muscovite, illite
8	Vermiculite
9	Smectite
10	Kaolinite
11	Aluminum hydrous oxides—gibbsite
12	Iron oxides and oxyhydroxides—hematite, goethite
13	Titanium oxides—rutile, anatase

[1]Particle-size diameter $\leq 5\ \mu m$.
Source: Jackson, Tyler et al., 1948.

(2–5 μm) size fractions. The minerals that are most resistant to chemical weathering, stages 8–13, occur predominantly in the clay fraction.

Minerals in stages 1–2 and 8–12 are exclusively secondary minerals, while the minerals in stages 3–7 and 13 are exclusively primary minerals. Making sense of the mineral weathering sequence requires some explanation. The stages represent chemical weathering as seen from a particular perspective.

Surficial deposits—soil, sediment, or saprolite—whose clay size fractions contain stage 1 or 2 minerals, may have undergone considerable chemical weathering during their past history, but the presence of relatively soluble chloride, sulfate, and carbonate minerals indicates that the recent chemical weathering history has not been sufficiently intense to dissolve these particular minerals. The absence of gypsum and halite in the fine-silt-size fraction means that the recent chemical weathering history has progressed beyond stage 1. Similarly, the absence of calcite or dolomite in the fine-silt- and coarse-clay-size fractions means that the recent chemical weathering history has progressed beyond stage 2.

Deposits whose fine-silt- and coarse-clay-size fractions contain the primary minerals of stage 3 have undergone primarily physical weathering and little chemical weathering. Stage 3 minerals are typical of igneous rocks considered vulnerable to chemical weathering. The absence of stage 3 minerals in the fine-silt-size fraction means that the recent chemical weathering history has progressed beyond stage 3. The chemical weathering of stage 3 minerals produces a variety of secondary minerals, but the key to advancing from stage to stage is the elimination of certain indicator minerals from the fine-silt- to clay-size fractions.

Stage 4–6 minerals are found in igneous rocks but represent increasing resistance to chemical weathering. As chemical weathering dissolves and transforms minerals in the fine silt to clay-size fractions, we witness the loss of biotite and chlorite (stage 5), feldspar minerals (stage 6), and, finally, quartz (stage 7). Regardless of whether these minerals occur in the coarser (>5 μm) size fractions, the chemical weathering stage is dependent on their elimination from the fine (<5 μm) size fractions.

Muscovite, the last igneous silicate mineral to disappear from the fine-silt-size fraction, has been eliminated by stage 8, along with a closely related secondary mineral: illite. The indicator minerals of stages 8–10 occur exclusively in the clay-size fraction. Vermiculite, smectite, and kaolinite, commonly known as *clay minerals*, are the major topic of this chapter.

Chemical weathering of sufficient duration and intensity rarely results in the loss of highly insoluble oxide minerals: aluminum hydrous oxides and iron oxides. Tropical soils with stage 11 clay mineralogy are often exploited as aluminum ores, while stage 12 soils are ideal iron ores.

Chemical Weathering Reactions

Minerals such as gypsum, halite, and most carbonate minerals completely dissolve. Other minerals consist of both soluble and insoluble elements. Stages 3–10 minerals partially dissolve, leaving behind products that are either highly insoluble or sufficiently reactive to form secondary minerals that persist throughout intermediate and advanced weathering stages. Given sufficient time, virtually all of the soluble *Na*, *K*, *Mg*, and *Ca* elements are leached from the vadose zone along with moderately soluble silica, leaving a highly insoluble residuum composed of aluminum, iron, and titanium oxide minerals.

Appendix 3A describes a simplified method for calculating the formal oxidation numbers of elements in minerals. The formal oxidation numbers of iron in the minerals of the *Jackson chemical weathering* sequence clearly indicate that primary (igneous) minerals undergo oxidation during chemical weathering; iron typically occurs in a ferrous oxidation state in primary minerals but occurs in the ferric oxidation state in secondary minerals.

Oxidation has several consequences: maintenance of charge neutrality requires changes in chemical composition, and the reduction in the ionic radius of iron ($r_{Fe^{2+}} > r_{Fe^{3+}}$) induces changes in crystal structure. Taken together, these consequences promote the chemical transformation of primary minerals containing ferrous ions.

Inspection of the chemical composition of minerals in the Jackson chemical weathering sequence reveals a second, profound transformation: secondary minerals typically contain hydrogen ions that are absent from igneous and other high-temperature rock-forming minerals. These added hydrogen ions represent increased hydration of mineral compositions as a result of chemical weathering. Hydration takes two forms: water molecules in the mineral crystal structure and hydroxyl ions, the latter known as structural hydration. The heating of minerals containing hydroxyl ions leads to weight loss: one water molecule vaporized for every two hydroxyl ions, leaving behind an oxygen ion:

$$2 \cdot OH^- \xrightarrow{heat} O^{2-} + H_2O \tag{3.1}$$

Silicate minerals resist chemical weathering to different degrees, but, in general, resistance to chemical weathering is proportional to the amount of cross-linking of the silicate network. Basalt rocks contain easily weathered olivine, pyroxene, and amphibole minerals, while granite rocks contain a higher proportion of resistant minerals: muscovite, feldspar, and quartz. Cleavage of the silicate network involves the hydrolysis (addition of a water molecule) of the link between silicate ions:

$$Si - O - Si \xrightarrow{H_2O} Si - OH + HO - Si \tag{3.2}$$

Secondary minerals fall into two broad groups: layer silicates (stages 7–10) and oxide minerals (stages 11–13). The crystal structure of oxide minerals has little direct bearing on their environmental chemistry. Oxide mineral chemistry arises from its solubility—which tends to be very low—and its surface properties—which is extremely important because oxide clays have a very high surface area–to–volume ratio.

3.3. THE STRUCTURE OF LAYER SILICATES

The three most abundant elements (oxygen, silicon, and aluminum) account for 83.3% of the total mass of Earth's outer crust (Taylor, 1964). Silicate and aluminosilicate minerals differ from other mineral classes because silicon and aluminum combine with oxygen to form tetrahedral units that polymerize into complex networks. This tendency to polymerize, more than elemental abundance, accounts for silicate mineral diversity—roughly 27% of the 4000 or so known minerals.

The chemical structural diversity of silicate minerals parallels the diversity of carbon compounds and derives from similar chemical principles. Both carbon and silicon atoms have four valence electrons and tend to form four bonds with other atoms to satisfy the *octet rule*. Carbon and silicon also favor tetrahedral bond geometry, as rationalized by both the valence orbital hybridization and the *valence shell electron-pair repulsion* (VSPER; Gillespie and Nyholm, 1957) chemical bonding models. The tetrahedral bond geometry of carbon-based compounds rests on direct carbon-carbon bonds, while oxygen-rich silicate minerals base their tetrahedral geometry on the silicate tetrahedron SiO_4^{4-} and its tendency to polymerize.

General chemistry books discuss silicate minerals under a special category of *network* solids. Network solids favor crystal structures that are less dense than *close-packed* ionic solids because the dominant bonding in network solids has a highly directional, covalent nature. Silicate minerals are organized into classes, depending on the type of silicate network. As a general rule, the melting point, mineral hardness, and resistance to chemical weathering tends to increase as the dimensionality of the silicate network increases.

3.3.1. Coordination Polyhedra

Chemists usually illustrate molecular structure using *ball-and-stick* models: atoms represented by balls and bonds represented by sticks drawn between bonded atoms. Pauling (1929) introduced an alternative model that became very popular among mineralogists. Minerals are composed mainly of cations and oxygen anions organized into coordination polyhedra. Oxygen anions occupy the vertices of these coordination polyhedra; lines drawn between the vertices define the shape of the coordination polyhedron. Cations occupy the center of each coordination polyhedron and are usually not illustrated in this convention (Appendix 3B).

Figure 3.1 shows the equivalence between the two models. Ball-and-stick models of tetrahedral and octahedral coordination appear in the left column, and the same ball-and-stick models appear in the middle column, with lines defining the vertices of the two polyhedra. Idealized symbols of a tetrahedron and an octahedron appear in the right column, similar to the illustrations found in Pauling (1929) and Bragg (1929).

The idealized octahedron in the bottom right of Figure 3.1 is oriented with its threefold rotation axis perpendicular to the page—nearly but not exactly the orientation of the ball-and-stick octahedron in the lower left. Two alternative orientations of idealized octahedra appear in Figure 3.2, where the orientation on the left is the same as in Figure 3.1, while the one on the right has been rotated so the twofold rotation axis is perpendicular to the page. You will often see both orientations in illustrations of layer silicate mineral structures.

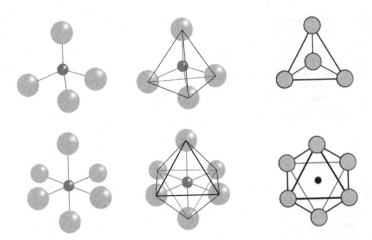

FIGURE 3.1 Ball-and-stick and polyhedral models for fourfold and sixfold coordination.

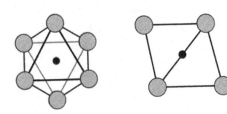

FIGURE 3.2 Two perspectives of an idealized octahedron: along sixfold screw symmetry axis (left) and along twofold rotation symmetry axis (right). The sixfold screw axis is perpendicular to the twofold rotation axis.

3.3.2. The Phyllosilicate Tetrahedral Sheet

Phyllosilicates—represented by the mica group minerals muscovite $KAl_2(AlSi_3O_{10})(OH)_2$ and biotite $K(Fe_xAl_{2-x})(AlSi_3O_{10})(OH)_2$—are common in granite and the coarse (sand and silt) fraction of moderately weathered saprolite. Aluminosilicate tetrahedra polymerize to form a two-dimensional network that serves as a scaffold for crystal layers 0.70–1.00 nm thick separated by cleavage planes (Pauling, 1930a, b).

To visualize the aluminosilicate tetrahedral network, imagine placing a single tetrahedron with one trigonal face resting on a flat surface as if it were a trigonal pyramid. The upper right image in Figure 3.1 is a vertical view of this tetrahedron: three oxygen atoms forming the base and one the apex. Several space-filling networks are possible from joining the basal vertices, but the network with the least bond strain consists of rings of six tetrahedra (Figure 3.3). The left image in Figure 3.3 is the aluminosilicate network found in phyllosilicates—all tetrahedra point in the same direction—while the right image in Figure 3.3 is the network found in tectosilicates such as quartz and feldspar—tetrahedra alternate pointing up and down around the sixfold ring. Figure 3.4 shows two unit cells for the two-dimensional aluminosilicate network.

FIGURE 3.3 Aluminosilicate networks in phyllosilicates (left) and tectosilicates (right). The phyllosilicate tetrahedra all point up in the illustrated sheet. The tecto-silicate tetrahedra alternate up and down in this sheet, with sheets jointed by tetrahedra pointing toward each other.

 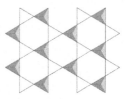

FIGURE 3.4 Alternative unit cell settings for the phyllosilicate tetrahedral sheet: a trigonal unit cell with composition $Si_2O_5^{2-}$ denoted by open points and an orthogonal unit cell with composition $Si_4O_{10}^{4-}$ denoted by filled points. The orthogonal unit cell is commonly used for phyllosilicates.

The phyllosilicate tetrahedral sheet is composed of three parallel atomic planes: the basal oxygen plane, the silicon-aluminum plane above that, and the apical oxygen plane. These atomic planes are crucial to understanding the phyllosilicate layer structure.

3.3.3. The Phyllosilicate Octahedral Sheet

The lower right image of Figure 3.1 (left image of Figure 3.2) is an octahedron viewed along the threefold rotation axis perpendicular to one of the eight triangular faces. Viewed from this perspective, the octahedron has a hexagonal profile. A planar arrangement of octahedra in this orientation fills space the same way hexagonal cells of a honeycomb fill space (Figure 3.5). Similar illustrations appear in the original papers describing the structure of phyllosilicates (Pauling 1930a, b).

The octahedra in Figure 3.5 share edges, while the tetrahedra in Figure 3.3 share corners (vertices). The symmetry and dimensions of these two sheets match one another (Pauling, 1930a, b). The apical oxygen atoms of the tetrahedral sheet align with vertices of the octahedral sheet to form the individual crystal layers of all phyllosilicates (Figures 3.6a and 3.6b).

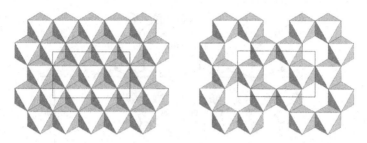

FIGURE 3.5 Octahedral sheets found in phyllosilicates: the "trioctahedral" $Mg_6(OH)_{12}$ sheet (left) and the "dioctahedral" $Al_4(OH)_{12}$ sheet (right).

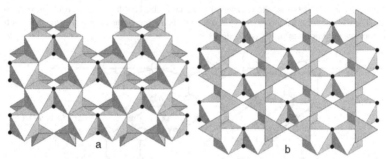

FIGURE 3.6 The apical oxygen atoms of an aluminosilicate tetrahedral sheet also coordinate trivalent ions of a "dioctahedral" $Al_4(OH)_{12}$ sheet. The illustrations offer two perspectives perpendicular to the sheets. The filled circles indicate OH^- ions that occupy the same atomic plane as the apical oxygen atoms of the tetrahedral sheet.

3.3.4. Kaolinite Layer Structure

A common clay mineral with the simplest layer structure is kaolinite $Al_4(OH)_8Si_4O_{10}$. Each kaolinite layer has the same composition and structure, consisting of one aluminosilicate tetrahedral sheet and one "dioctahedral" sheet (Figure 3.7), with a layer spacing of 0.70 *nm*. A less common mineral, lizardite $Mg_6(OH)_8Si_4O_{10}$, represents the "trioctahedral" end-member. Clay mineralogists classify kaolinite and lizardite as "1:1" phyllosilicates.

3.3.5. Talc Layer Structure

The mineral talc $Mg_6(OH)_4(Si_4O_{10})_2$ is a familiar phyllosilicate found in metamorphic rocks, and its structure is similar to the mineral lizardite. Each talc layer consists of two aluminosilicate tetrahedral sheets and one "triocta-hedral" sheet (Figure 3.8), with a layer spacing of 0.94 *nm*. The mineral pyrophyllite $Al_4(OH)_4(Si_4O_{10})_2$ represents the "dioctahedral" end-member. Clay mineralogists classify pyrophyllite and talc as "2:1" phyllosilicates.

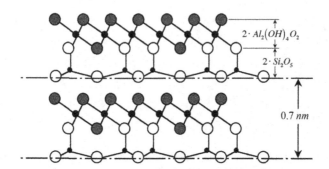

FIGURE 3.7 A ball-and-stick model illustrates the kaolinite layer structure (oxygen—large open circles, hydroxyl—large shaded circles, silicon—small filled circles, aluminum—medium filled circles). The compositions of the tetrahedral sheet and "dioctahedral" sheet appear next to each sheet. The overall composition $Al_4(OH)_8Si_4O_{10}$ avoids double counting apical oxygen atoms that also coordinate aluminum in the octahedral sheet.

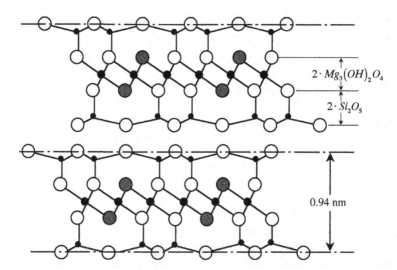

FIGURE 3.8 A ball-and-stick model illustrates the talc layer structure (oxygen—large open circles, hydroxyl—large shaded circles, silicon—small filled circles, aluminum—medium filled circles). The compositions of the tetrahedral sheet and "trioctahedral" sheet appear next to each sheet. The overall composition $Mg_6(OH)_4(Si_4O_{10})_2$ avoids double counting apical oxygen atoms that also coordinate magnesium in the octahedral sheet.

3.3.6. Mica-Illite Layer Structure

Two types of mica are common in granite: the "dioctahedral" variant called muscovite and the "trioctahedral" variant called biotite. The layer structure (Figure 3.9) and composition of muscovite $K_2 : Al_4(OH)_4(AlSi_3O_{10})_2$ and biotite $K_2 : Mg_{6-x}Fe_x(OH)_4(AlSi_3O_{10})_2$ closely resemble pyrophyllite and talc, respectively, except for two profound differences. Aluminum ions Al^{3+}

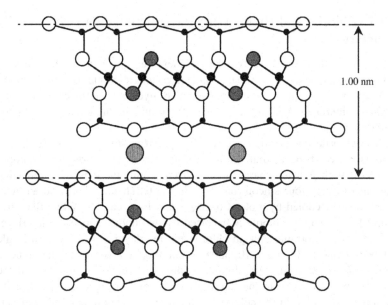

FIGURE 3.9   A ball-and-stick model illustrates the mica layer structure (oxygen—large open circles, hydroxyl—large shaded circles, silicon—small filled circles, aluminum—medium filled circles, potassium—large shaded circles between the layers).

replace one out of every four silicon ions Si^{4+} in the tetrahedral sheet, and one potassium ion K^+ occupies the interlayer (the space between the crystal layers) for every Al^{3+} substitution in the crystal layer. We know this because the molar content of potassium equals the molar aluminum content in mica (Pauling, 1930), and crystal structure refinements using x-ray diffraction locate potassium ions between the mica layers.

Cation substitution in micas—the replacement of Si^{4+} by Al^{3+} in the tetra-hedral sheet—results in negatively charged phyllosilicate layers whose charge is balanced by interlayer cations. The combined effect of strong electrostatic forces binding the mica layers together, the low enthalpy of hydration for K^+, and the snug fit of the bare K^+ cation into the sixfold rings of the aluminosili-cate network exclude water from the mica interlayer. The mica layer spacing is 1.00 nm, significantly larger than in talc and pyrophyllite because interlayer K^+ ions prop open the layers. The interlayer K^+ ions also align the stacking of mica layers by occupying every hexagonal ring in the tetrahedral sheet. The coordi-nation number of interlayer K^+ is 12—6 from each crystal layer.

Illite closely resembles mica but is recognized as a distinct mineral by clay mineralogists. The illite classification is based on layer charge, typically expressed using $O_{10}(OH)_2$ units. A mineral is classified as mica if its layer charge is greater than 0.8 per $O_{10}(OH)_2$ and illite if its layer charge is in the range of 0.7–0.8 per $O_{10}(OH)_2$. Layer charge in the illite range allows some of the interlayer K^+ to exchange with cations in solution.

3.3.7. Chlorite and Hydroxy-Interlayered Smectite Layer Structure

X-ray diffraction identifies two clay mineral groups with a characteristic 1.4 nm layer spacing: the primary mineral group called chlorite and a modified smectite variant—commonly known as hydroxy-interlayered smectite. Hydroxy-interlayered smectite is typically found in moderately weathered acidic soils.

The chlorite composition $Mg_4Al_2(OH)_{12} : Mg_{6-x}Fe_x(OH)_4(AlSi_3O_{10})_2$ is reminiscent of "trioctahedral" biotite (to the right of the colon in the chlorite formula), with a layer charge arising from aluminum ions Al^{3+} replacing one out of every four silicon ions Si^{4+} in the tetrahedral sheet and an octahedral layer occupied by a blend of Mg^{2+} and Fe^{2+} cations. While the tetrahedral layer charge in biotite is balanced by two nonexchangeable interlayer K^+ cations per formula unit, in chlorite the interlayer is filled with a single octahedral sheet (Figure 3.10, left) resembling a layer from the mineral brucite $Mg_6(OH)_{12}$ but with aluminum ions Al^{3+} replacing one-third of the magnesium ions Mg^{2+} in the interlayer octahedral sheet $Mg_4Al_2(OH)_{12}$ (to the left of the colon in the chlorite formula). Many clay mineralogists designate minerals of the chloride groups as 2:1:1 clay minerals (Figure 3.10, left) because a variant brucite octahedral layer lies between 2:1 biotite layers.

The structure and layer spacing of hydroxy-interlayered smectite (Figure 3.10, right) resemble chlorite, but there are significant compositional differences. First, the extent and location of cation substitution in the 2:1 primary layers are characteristic of smectite clay minerals (Table 3.3). Second,

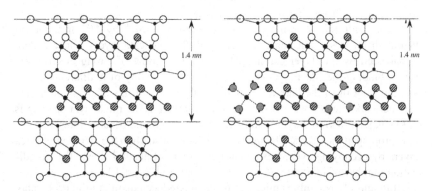

FIGURE 3.10 A ball-and-stick model illustrates the layer structures of chlorite (left) and a modified smectite common in moderately weathered acidic soils generally known as hydroxy-interlayered smectite (right). The symbols used are oxygen—large open circles, hydroxyl—large cross-hatched circles, silicon—small filled circles, aluminum or magnesium—medium filled circles. Exchangeable interlayer cations (right): small filled circles, water hydrating interlayer cations—large gray circles with small filled circles representing protons.

TABLE 3.3 Swelling Clay Minerals

Clay Mineral	Substitution Location	Cation Substitution	Substitutions per $O_{10}(OH)_2$	CEC, $cmol_c$ kg^{-1}
Montmorillonite	Octahedral	$Mg^{2+} \longrightarrow Al^{3+}$	0.25–0.6	60–140
Beidellite	Tetrahedral	$Al^{3+} \longrightarrow Si^{4+}$	0.25–0.6	60–140
Nontronite	Tetrahedral	$Al^{3+} \longrightarrow Si^{4+}$	0.25–0.6	60–160
Vermiculite	Tetrahedral	$Al^{3+} \longrightarrow Si^{4+}$	0.6–0.7	140–165

the interlayer octahedral sheet is not continuous; instead, the interlayer is occupied by a combination of hydrated exchangeable cations and *pillars,* whose structure and composition resemble gibbsite $Al_4(OH)_{12}$ instead of brucite. The gibbsitic interlayer of hydroxy-interlayered smectites results from the accumulation of aluminum ions Al^{3+} released by chemical weathering and their hydrolysis to form a product much like gibbsite between the smectite layers.

The interlayer octahedral sheet characteristic of both chlorite and hydroxy-interlayered smectites prevents a decrease in layer spacing as these minerals are dried or subjected to mild heating to drive off interlayer water ($<200°C$). Chlorites exhibit negligible weight loss upon mild heating, while hydroxy-interlayered smectites exhibit weight loss proportional to the exchangeable interlayer cations. The interlayer octahedral sheet also prevents swelling when these minerals are immersed in water.

3.3.8. Layer Structure of the Swelling Clay Minerals: Smectite and Vermiculite

Cation substitution also occurs in smectite and vermiculite. Cation substitution in vermiculite involves the replacement of Si^{4+} by Al^{3+} in the tetrahedral sheet, but in smectite minerals cation substitution also includes Mg^{2+} replacing Al^{3+} in the octahedral sheet. Table 3.3 lists some of the common smectite minerals found in nature. The layer structure and composition resemble the micas, the primary differences being a significantly lower layer charge and the presence of interlayer water. Lower layer charge means the electrostatic forces binding interlayer cations to the crystal layers are comparable in magnitude to the hydration energy of the interlayer cations. Strongly hydrating cations—Na^+, Ca^{2+}, Mg^{2+}—draw water molecules into the interlayer, forcing the crystal layers apart in a phenomena called *swelling* (Figure 3.11).

FIGURE 3.11 A ball-and-stick model illustrates interlayer water molecules and the layer structure of smectite. This illustration is based on a molecular dynamics simulation. The interlayer cations are not shown in this illustration. *Source: Reproduced with permission from Chavez-Paez, M., Van Workum, K., 2001. Monte Carlo simulations of Wyoming sodium montmorillonite hydrate. J. Phys. Chem. 114, 1405–1413.*

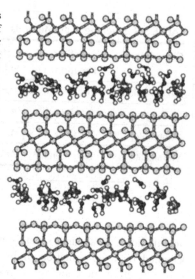

The Swelling of Smectite and Vermiculite Clay Minerals

A detailed picture of clay swelling began to emerge in the early 1950s, as illustrated by the x-ray diffraction results in Figures 3.12 and 3.13. (See Appendix 3C for an explanation of how the distance between crystal planes is measured using x-ray diffraction.) Increasing the relative humidity (Figure 3.12a) or decreasing the salt concentration in a clay suspension when salt concentrations are relatively high (Figure 3.12b) causes the layer spacing—as indicated by d_{001}—to increase in a stepwise manner as water enters the interlayer one layer of water molecules at a time. Crystalline clay swelling—the stepwise increases in layer spacing with the addition of 1, 2, and 3 water monolayers to the interlayer—leads to abrupt jumps in d_{001} equal to the diameter of a water molecule: almost 0.25 *nm*. Notice that Ca^{2+} saturated smectite jumps to a three-layer hydrate at a lower relative humidity than Na^+ saturated smectite (Figure 3.12a), demonstrating Ca^{2+} hydrates more strongly than Na^+, and draws water into the interlayer at a lower humidity.

Low humidity or a high osmotic potential in concentrated salt solutions limits the amount of water available to hydrate interlayer cations. Clay swelling in dilute salt solutions appears as a continuous increase in d_{001} as the salt concentration becomes increasingly more dilute (see Figure 3.13). Clay chemists refer to this continuous swelling regime as *osmotic clay swelling* (see Appendix 3D).

Structure of the Hydrated Clay Mineral Interlayer

Unfortunately, no currently available experimental technique can reveal the detailed structure of a hydrated smectite or vermiculite interlayer, leading some clay chemists to rely on molecular dynamic simulations. These

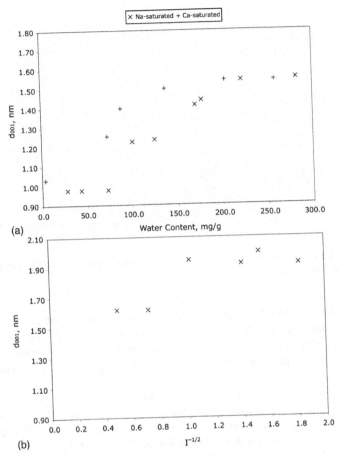

FIGURE 3.12 Crystalline swelling of smectite clays as influenced by (a) relative humidity (modified from Mooney, Keenan et al., 1952) and (b) ionic strength (modified from Norrish, 1954).

simulations generate a host of predictions about the most probable structure of the interlayer: the positions and orientation of water molecules, cation positions, interlayer spacing under specific conditions, and so on. To the extent simulation predictions match measurable parameters (e.g., interlayer spacing), we can cautiously accept interlayer structure predictions that are currently not verifiable by experiment.

Most simulations predict crystalline swelling behavior (see Figures 3.12 and 3.13) quite well. Figure 3.14 displays probability distributions of $Na^+, O_{water}, H_{water}$ in the interlayer normal to the crystal layer for three crystalline swelling states. Figure 3.15 displays probability distributions for $K^+, O_{water}, H_{water}$ for three crystalline swelling states. Notice that Na^+ ions

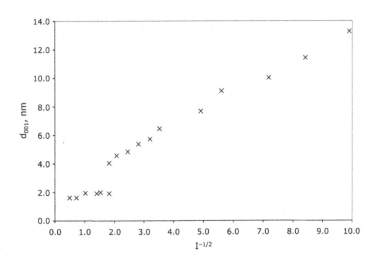

FIGURE 3.13 Crystalline and osmotic swelling of smectite clays as influenced by ionic strength (modified from Norrish, 1954).

(see Figure 3.14) tend to move from the clay surface in the one-layer hydrate to the center of the interlayer in the two-layer hydrate, while K^+ ions (see Figure 3.15) remain near the clay surface as water enters the interlayer. This behavior is consistent with the different swelling behavior of Na^+-saturated and K^+-saturated smectite.

Sodium ions promote continuous osmotic swelling of smectite (see Figure 3.13), but potassium ions resist osmotic swelling because they lack sufficient hydration energy to swell the interlayer. Simulations such as this (Boek, Coveney et al., 1995; Tambach, Bolhuis et al., 2006) and the hydration properties of these two ions are offered as an explanation for why potassium-saturated smectites resist swelling.

In the osmotic swelling regime, interlayer cations shift away from the midplane, returning to positions near the clay layers (Figure 3.16). This configuration reduces electrostatic repulsion between interlayer cations (by increasing the relative cation-cation distance) and increases the electrostatic attraction between the negatively charged clay layer and the interlayer cations (by decreasing the relative cation-layer distance).

The electrostatic force between the clay layer and the interlayer cation depends on the hydrated radius of the cation and the location of the cation-substitution site in the clay layer. The strongest force exists when cation substitution is in the tetrahedral sheet ($Al^{3+} \longrightarrow Si^{4+}$) and the cation hydration energy is relatively weak. In this configuration the cation comes into close contact (see Figure 3.15) with the negative charge in the tetrahedral sheet. The weakest force exists when cation substitution is in the octahedral sheet ($Mg^{2+} \longrightarrow Al^{3+}$) and the hydration energy is strong. In this configuration water

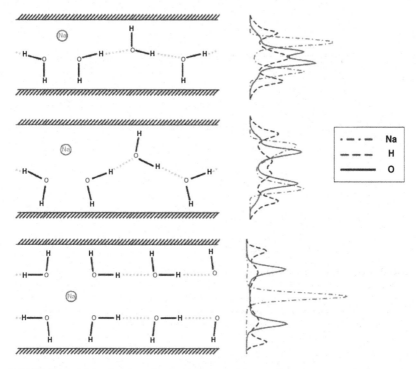

FIGURE 3.14 Molecular dynamic simulations of the crystalline swelling of Na⁺-saturated smectite clay representing $d_{001} = 1.25$ nm, one-layer hydrate (top); $d_{001} = 1.35$ nm, intermediate hydrate (middle); and $d_{001} = 1.45$ nm, two-layer hydrate (bottom). The probability distributions normal to the clay layer appear on the right. *Source: Reproduced with permission from Tambach, T.J., Bolhuis, P.G., et al., 2006. Hysteresis in clay swelling induced by hydrogen bonding: Accurate prediction of swelling states. Langmuir 22, 1223–1234.*

molecules prevent interlayer cations from coming into close contact with the clay layer (see Figure 3.14), and the negative layer charge is in the central atomic plane.

Table 3.4 lists estimates of the hydrated radius and absolute enthalpy of hydration $\Delta H^\circ_{absolute}$ for alkali and alkaline earth group cations. A decreasing hydrated radius correlates with a decreasing enthalpy of hydration in spite of the trend for increasing ionic radius within each period.

Smectite clay minerals exhibit crystalline swelling regardless of the type of interlayer cation; differences arise from either the number of hydration layers that form during crystalline swelling or the conditions (humidity or ionic strength) required to add a hydration layer. As a general rule, the greater the enthalpy of hydration, the lower the humidity or the higher the ionic strength at which a crystalline swelling jump occurs. Smectites saturated with divalent interlayer cations resist crystalline swelling relative to those saturated

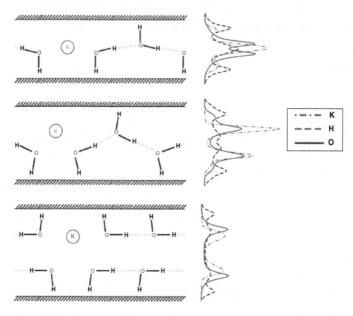

FIGURE 3.15 Molecular dynamic simulations of the crystalline swelling of K^+-saturated smectite clay representing $d_{001} = 1.25$ *nm*, one-layer hydrate (top); $d_{001} = 1.35$ *nm*, intermediate hydrate (middle); and $d_{001} = 1.45$ *nm*, two-layer hydrate (bottom). The probability distributions normal to the clay layer appear on the right. *Source: Reproduced with permission from Tambach, T.J., Bolhuis, P.G., et al., 2006. Hysteresis in clay swelling induced by hydrogen bonding: Accurate prediction of swelling states. Langmuir 22, 1223–1234.*

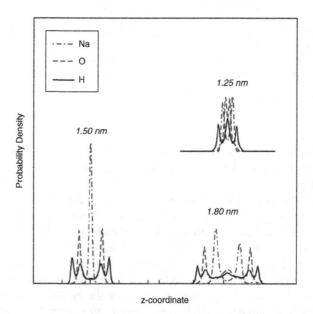

FIGURE 3.16 Molecular dynamic simulations of the crystalline swelling of Na^+-saturated smectite clay representing $d_{001} = 1.25$ *nm*, one-layer hydrate (upper right); $d_{001} = 1.50$ *nm*, two-layer hydrate (lower left); and $d_{001} = 1.80$ *nm*, three-layer hydrate (lower right). *Source: Reproduced with permission from Tambach, T.J., Hensen, E.J.M., et al., 2004. Molecular simulations of swelling clay minerals. J. Phys. Chem. B 108, 7586–7596.*

TABLE 3.4 Hydrated Radius and Absolute Enthalpy of Hydration $\Delta H^{\circ}_{absolute}$ for Alkali and Alkaline earth Cations

Ion	Hydrated Radius[1], pm	$\Delta H^{\circ}_{absolute}$[2], kJ/mol
Li^+	405	−531
Na^+	348	−416
K^+	317	−334
Rb^+	300	−308
Cs^+	270	−283
Mg^{2+}	480	−1949
Ca^{2+}	440	−1602
Sr^{2+}	435	−1470
Ba^{2+}	430	−1332

[1]Abbas, Gunnarsson et al., 2002.
[2]Marcus, 1987.

with univalent interlayer cations. Of the ions listed in Table 3.4, only Li^+ and Na^+ have the capacity to trigger osmotic swelling (see Figure 3.12) when they saturate the interlayer of smectite clay minerals; the remaining cations resist osmotic swelling.

Clay Colloids

Colloid chemistry is a special branch of chemistry devoted to the study of systems where surface properties have a measurable effect on physical and chemical behavior. Clay colloidal systems occur in the environment as a dispersion of ultra-fine clay particles in water—either a clay soil dispersed in the water column of a river or lake, or a gel at low water contents. Particles qualify as colloidal if at least one dimension is in the 1–1000 nm range. The upper limit of the clay particle-size fraction (<5000 nm) would seem to disqualify clays as true colloids, but smectite and vermiculite both easily qualify given the 1.00 nm thickness of the clay layers.

Under osmotic swelling conditions—low ionic strength and Na^+ as interlayer cation—smectite clay in the environment become a true colloidal dispersion. Even under crystalline swelling conditions, the extremely high surface area of smectite and vermiculite clay minerals affects the chemistry of the vadose zone, sediments, and bodies of water carrying a suspended load of clay minerals. The phenomenon of ion exchange, discussed in the following chapter, is a chemical reaction that can be reliably measured only if the particles are in the clay particle-size range.

APPENDIX 3A. FORMAL OXIDATION NUMBERS

Most general chemistry books introduce a scheme for calculating the formal oxidation number of elements in compounds. Following is an abbreviated scheme adapted for mineralogy.

1. The oxidation number of a free element is always 0.
2. The oxidation number of hydrogen is usually +1.
3. The oxidation number of oxygen in minerals is usually −2.
4. The oxidation number of a Group IA (alkali) element in minerals is +1.
5. The oxidation number of a Group IIA (alkaline earth) element in minerals is +2.
6. The oxidation number of a Group IIIA element in minerals is +3.
7. The oxidation number of a Group IVA element in minerals is usually +4.
8. The oxidation number of a Group VIIA element in minerals is −1.
9. The sum of oxidation numbers for all the atoms in the formula unit of a mineral is 0.
10. The sum of oxidation numbers in a polyatomic ion is equal to the charge of the ion.

Olivine

The formula notation places Mg and Fe in parentheses with a subscript of 2: $(Mg, Fe)_2 SiO_4$. This notation is equivalent to $Mg_{2-x} Fe_x SiO_4$ and signifies that Mg and Fe are present in varying rather than fixed proportions. The formal oxidation number for oxygen is −2. The formal oxidation numbers of Mg (Group IIA) and Si (Group IVA) are +2 and +4, respectively. The condition of charge neutrality requires assignment of formal oxidation number +2 to Fe.

$$(Mg, Fe)_2 SiO_4$$

Formula Charge	0
Formal Oxidation Numbers	
O	−2
Mg	+2
Si	+4
Fe	+2

Hedenbergite (Pyroxene)

The formal oxidation number for oxygen is −2. The formal oxidation numbers of Ca (Group IIA) and Si (Group IVA) are +2 and +4, respectively.

The condition of charge neutrality requires assignment of formal oxidation number $+2$ to Fe.

$$FeCaSi_2O_6$$

Formula Charge	0
Formal Oxidation Numbers	
O	-2
Ca	$+2$
Si	$+4$
Fe	$+2$

Goethite

The formal oxidation number for oxygen is -2 and hydrogen is $+1$. The condition of charge neutrality requires assignment of formal oxidation number $+3$ to Fe.

$$FeOOH$$

Formula Charge	0
Formal Oxidation Numbers	
O	-2
H	$+1$
Fe	$+3$

APPENDIX 3B. THE GEOMETRY OF PAULING'S RADIUS RATIO RULE

Linus Pauling (1929) proposed several rules governing mineral structures; here we will consider only the *Radius Ratio* rule:

Radius Ratio rule: *A polyhedron of anions surrounds each cation. The cation-anion distance is the sum of their respective radii, and the coordination number of the cation is determined by the radius ratio of cation to anion.*

The Radius Ratio rule uses geometry to define the minimum radius ratio for stable coordination, listed in Table 3B.1. Pauling (1929) based this rule on the notion that small cations fit into vacancies created by the packing together of larger oxygen ions. He believed that cations preferentially occupy those vacancies where the cation ionic radius just matches the space available or when it is somewhat too large, forcing the oxygen ions apart. A cation rarely occupies vacancies where its ionic radius is smaller than the space available.

Geometry defining the minimum radius of a small sphere—the cation—that can occupy the vacancy created by larger spheres is worked out in this Appendix for three of the most common coordination types encountered in

TABLE 3B.1 Minimum Radius Ratios for Stable Coordination In Minerals

Coordination Number	Minimum Radius Ratio	Formula
3	0.155	$\left(\frac{(2\sqrt{3}-3)}{3}\right)$
4	0.225	$\left(\frac{(\sqrt{6}-2)}{2}\right)$
6	0.414	$(\sqrt{2}-1)$
8	0.732	$(\sqrt{3}-1)$
12	1.000	1

minerals: trigonal, tetrahedral, and octahedral. In each case we will assume a unit radius for the larger sphere and use geometry to determine the radius of the smaller sphere.

Trigonal

Imagine three large spheres lying on a plane (Figure 3B.1). The large spheres define an equilateral triangle whose sides are each 2 times their radii. Assume that the large spheres have a radius of 1. The distance from the center of the equilateral triangle to each of the vertices l is a function of the cosine of 30°; the distance l is simply the radius of the small sphere plus the radius of the large sphere: $1 + r_c$.

FIGURE 3B.1   Trigonal close-packed spheres.

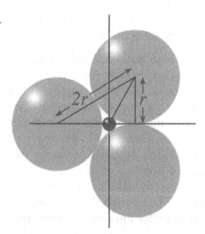

$$\cos(30°) = \frac{adjacent}{hypotenuse}$$

$$\frac{\sqrt{3}}{2} = \frac{1}{l} = \frac{1}{(r_c + 1)}$$

$$r_c + 1 = \frac{2}{\sqrt{3}} = \frac{2 \cdot \sqrt{3}}{3}$$

$$r_c = \left(\frac{2 \cdot \sqrt{3} - 3}{3}\right) \approx 0.155$$

Tetrahedron

This geometry is best understood by visualizing a tetrahedron inside of a cube (Figure 3B.2). The diagonal distance across the face of the cube is the length of the edge of the embedded tetrahedron. If the cube-face diagonal equals 2, then edge of the cube is $\sqrt{2}$

$$(e^2 + e^2) = 2^2$$
$$2 \cdot e^2 = 4$$
$$e = \sqrt{2}$$

The center of the cube is also the center of the tetrahedron. The distance from any vertex of the cube to the center is the sum of the radius of the anion and the radius of the cation: $l = 1 + r_c$. The distance from any vertex of the cube to the center is the hypotenuse of a right triangle whose base is half of the diagonal distance across the face of the cube and whose height is half the edge length of the cube.

$$l^2 = \left(\left(\frac{e}{2}\right)^2 + 1^2\right) = \left(\frac{2}{4} + 1\right)$$

$$l = (r_c + 1) = \sqrt{\frac{6}{4}} = \frac{\sqrt{6}}{2}$$

$$r_c = \left(\left(\frac{\sqrt{6}}{2}\right) - 1\right) = \left(\frac{\sqrt{6} - 2}{2}\right) \approx 0.2247$$

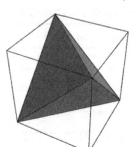

FIGURE 3B.2   Tetrahedron in a cube showing 2 times the anion radius along cube face diagonal.

Octahedron

Octahedral coordination is equivalent to four large spheres lying on a plane with two more spheres of the same radius placed above and below on the vertical axis. The geometry is completely defined by the four large spheres in the plane (Figure 3B.3). The spheres define a square whose sides are each 2 times the radii. Assume that each large sphere has a radius of 1. The distance from the center of the square to each of the vertices l is a function of the sine of 45°; the distance l is simply the radius of the small sphere plus the radius of the large sphere: $1 + r_c$.

FIGURE 3B.3 Square close-packed spheres for fourfold and sixfold coordination.

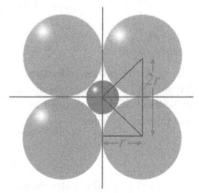

$$\sin(45°) = \frac{opposite}{hypotenuse}$$

$$\frac{\sqrt{2}}{2} = \frac{1}{l} = \frac{1}{(r_c + 1)}$$

$$r_c + 1 = \frac{2}{\sqrt{2}} = \frac{2 \cdot \sqrt{2}}{2}$$

$$r_c = (\sqrt{2} - 1) \approx 0.414$$

The Radius Ratio rule allows us to estimate cation-anion bond lengths (Table 3B.2) for each coordination polyhedra in minerals given the radius of oxygen: 0.140 *nm*.

TABLE 3B.2 Ideal Cation-Oxygen Distances Based on Coordination Number

Coordination Number	Cation-Oxygen Distance, nm
3	0.162
4	0.172
6	0.198
8	0.242
12	0.280

APPENDIX 3C. BRAGG'S LAW AND X-RAY DIFFRACTION IN LAYER SILICATES

The constructive interference of x-rays reflected by parallel atomic plane within a crystal occurs when the difference in the distance traveled by two reflected x-rays is an integral number times the x-ray wavelength: $n \cdot \lambda$.

Suppose x-ray beams reflect off parallel atomic planes—separated by a distance d_{hkl}—at an angle θ. The difference in the distance traveled by two reflected x-rays is an integral number times the x-ray wavelength $n \cdot \lambda$, provided the reflection angle and distance between atomic planes satisfies the following condition—known as *Bragg's Law* (Figure 3C.1).

$$n \cdot \lambda = 2 \cdot d_{hkl} \cdot \sin\theta$$

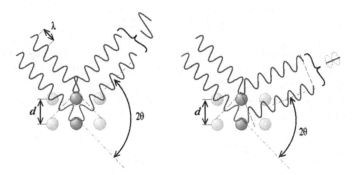

FIGURE 3C.1 Bragg's Law reflection of monochromatic x-rays from crystal planes. *(image source: Chan, 2004)*

Crystallographers identify x-ray reflections using *hkl* notation of specific atomic planes in crystals. The *001* x-ray reflection in phyllosilicate minerals results from Bragg diffraction by atomic planes parallel to the aluminosilicate tetrahedral sheet. The distance associated with each *001* x-ray reflection in phyllosilicates is denoted d_{001}. The first-order *001* x-ray reflection ($n = 1$) in a phyllosilicate is the repeat distance perpendicular to the crystal layers: $d_{001} = 0.70$ *nm* for kaolinite, $d_{001} = 0.94$ *nm* for talc, $d_{001} = 1.00$ *nm* for mica, and $d_{001} > 12$ *nm* for smectite.

APPENDIX 3D. OSMOTIC MODEL OF INTERLAYER SWELLING PRESSURE

Imagine a solution confined by a membrane that is permeable to water molecules but impermeable to solutes (e.g., the cytoplasm of a bacterium or a plant cell surrounded by a rigid cell wall). The equilibrium internal pressure acting against the confining membrane is given by the following expression for osmotic pressure $\prod$ (3D.1): R is the gas constant, T is the absolute temperature, n_{solute} is the number of moles of solute, $n_{solvent}$ is the number of moles of solvent, V^* is the molar volume of the solvent. Hydrologists use osmotic *potential* interchangeably with osmotic *pressure*.

$$\Pi[N \cdot m^2] \approx n_{solute} \cdot \left(\frac{R \cdot T}{n_{solvent} \cdot V^*}\right) = p_{osmotic}[Pa] \qquad (3D.1)$$

Smectite clay minerals swell when interlayer cations hydrate, forcing the layers to expand. The osmotic swelling pressure is given by Eq. (3D.2). Keep in mind that ionic solutes dissociate, so a 1 M $NaCl$ solution has the osmotic potential $p_{osmotic}$ of a 2 M sucrose solution. The smectite interlayer behaves like a solution, with the exception that one solute is the interlayer cation and the other is the negatively charged site (which is also hydrated); dilution occurs by water entering the interlayer and increasing the volume $n_{H_2O} \cdot V^*$ as the layers are forced apart (3D.2).

$$\Pi_{swelling} \approx \frac{2 \cdot n_{Na^+} \cdot R \cdot T}{n_{H_2O} \cdot V^*} \qquad (3D.2)$$

Using a surface area of 750,000 $m^2 \cdot kg^{-1}$, which counts the interlayer area of swelling clay minerals, we can calculate the number of moles of interlayer water per kilogram of clay (3D.3) based on the interlayer spacing $d_{001}(nm)$, clay interlayer area ($m^2 \cdot kg^{-1}$), and the molar volume V^* of water at 25°C ($1.807 \cdot 10^{-5} \, m^3 \cdot mol^{-1}$).

$$n_{H_2O} = \frac{d_{001} \cdot A}{V^*} = \frac{(d_{001}[nm] \cdot 10^{-9}[m \cdot nm^{-1}]) \cdot (7.5 \cdot 10^5[m^2 \cdot kg^{-1}])}{(1.807 \cdot 10^{-5}[m^3 \cdot mol^{-1}])} \qquad (3D.3)$$

The osmotic swelling pressure $\Pi_{swelling}$ becomes a simple function of the degree of cation substitution in the clay—typically n_{Na^+} reported using units of $cmol_c \cdot kg^{-1}$—and the interlayer spacing d_{001} in nanometers (3D.4).

$$\Pi_{swelling} \approx \left(\frac{2 \cdot n_{Na^+}}{d_{001}}\right) \cdot \left(\frac{R \cdot T}{A}\right)$$

$$\Pi_{swelling} \approx \left(\frac{2 \cdot (n_{Na^+}[cmol_c \cdot kg^{-1}])}{(d_{001}[nm])}\right) \cdot (33.05[kPa \cdot cmol_c^{-1} \cdot kg \cdot nm])$$

$$\qquad (3D.4)$$

The *osmotic* model of clay swelling imagines the interlayer as a type of solution where the interlayer cation concentration remains fixed, anions are excluded, and water is free to enter or leave depending on the relative osmotic potential $\Pi_{swelling}$ of the interlayer and the aqueous solution $\Pi_{solution}$ bathing the clay particles (3D.5): i is the Van't Hoff factor for the solute, equal to the number of moles of solute dissolved in solution per mole of solid added, and M is the concentration of the solute.

$$\Pi_{solution} \approx i \cdot C \cdot R \cdot T$$

$$\Pi_{solution} \approx i \cdot (mol \cdot L^{-1}) \cdot (2.4791 0^3 \, kPa \cdot L \cdot mol^{-1}) \qquad (3D.5)$$

The simplified osmotic model used by chemists does not precisely predict the experimental water vapor pressure above real aqueous solutions because it assumes that the aqueous solution is ideal and the vapor pressure is a simple function of the mole fraction of the solvent. A more detailed model is required to accurately predict experimental vapor pressures, one that accounts for the nonideal behaviors of both the solute and the solvent. The similar shortcomings apply to the osmotic model of clay swelling; still, the simple osmotic model used here predicts the swelling tendency observed experimentally.

Norrish (1954) used a clay product called *Volclay* (Volclay, American Colloid Co.) containing 89 $cmol_c \cdot kg^{-1}$ interlayer cations. Norrish washed the clay with concentrated *NaCl* to replace the interlayer cations in the naturally occurring clay to yield smectite containing 89 $cmol_c \cdot kg^{-1}$ interlayer Na^+ for the swelling experiment. Osmotic swelling begins with a d_{001} of 4.0 *nm* (see Figure 3.12) at an ionic strength of: $I = (1/16) = 0.0625 \ mol \cdot L^{-1}$:

$$\Pi_{swelling} \approx \frac{2 \cdot (89 \ cmol_c \cdot kg^{-1}) \cdot (33.05 \ kPa \cdot cmol_c^{-1} \cdot kg \cdot nm)}{(4 \ nm)} \approx 1455 \ kPa$$

(3D.6)

At equilibrium the osmotic swelling pressure of the interlayer water $\Pi_{swelling}$ (Eq. (3D.6)) equals the osmotic pressure of the aqueous solution $\Pi_{solution}$ (Eq. 3D.5)). Solving for C_{NaCl} provides a surprisingly good estimate of the actual salt concentration in the clay suspension—about fivefold larger than measured experimentally by Norrish (1954):

$$2 \cdot C_{NaCl} \cdot (2.479 10^3 \ kPa \cdot L \cdot mol^{-1}) = 1455 \ kPa$$

(3D.7)

$$C_{NaCl} = \frac{1455 \ kPa}{2 \cdot (2.479 10^3 \ kPa \cdot L \cdot mol^{-1})} = 0.294 \ mol \cdot L^{-1}$$

(3D.8)

APPENDIX 3E. EXPERIMENTAL ESTIMATES OF INTERLAYER SWELLING PRESSURE

Imagine two chambers, one containing pure water and the other containing a clay gel, separated by a semipermeable membrane that allows water but not clay to pass. The glass columns allow each chamber to adjust its volume to remain at atmospheric pressure. The height of each column indicates the osmotic pressure (a milliliter of water contains 0.056 mole of H_2O).

Energy units $[J \cdot mol^{-1}]$ can be converted into pressure units because energy per unit volume $[J \cdot m^{-3} = (N \cdot m) \cdot m^{-3}]$ has the same dimensions as pressure $[N \cdot m^{-2} = Pa]$.

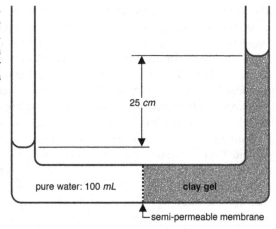

FIGURE 3E.1 Chambers containing pure water and a clay gel connected by a rigid semi-permeable membrane. An increase in the clay gel water content is denoted by changes in the water column height.

Question: What is the swelling pressure of the clay suspension (in atmosphere units) when the clay gel column attains a level 25 *cm* higher than the water column?

Solution: A pressure of 1 *atm* is equivalent to 1034 *cm* $H_2O(l)$ at 16°C. If we assume the density of the clay gel is equal to water (i.e., the mass of the clay is negligible relative to the mass of water), 25 *cm* represents a pressure differential of about $2.42 \cdot 10^{-2}$ *atm* or $p/p_0 = 2.42 \cdot 10^{-2}$

$$\Delta G_{swelling} = R \cdot T \cdot \ln \frac{P}{P_0}$$

$$\Delta G_{swelling} = R \cdot T \cdot \ln \left(\frac{25}{1034} \right)$$

$$\Delta G_{swelling} = R \cdot T \cdot \ln(0.0242)$$
$$\Delta G_{swelling} = R \cdot T \cdot (-3.72)$$

$$\Delta G_{swelling} = (8.314\ J \cdot mol^{-1} \cdot K^{-1}) \cdot (289K) \cdot (-3.72)$$
$$\Delta G_{swelling} = -8949\ J \cdot mol^{-1}$$
$$\Delta G_{swelling} = -8949\ N \cdot m \cdot mol^{-1}$$

Converting energy units to pressure units, using 0.056 mole H_2O per cubic centimeter:

$$\Pi_{swelling} = \left| \Delta G_{swelling} \right| \cdot V^*$$
$$\Pi_{swelling} = (8949\ N \cdot m \cdot mol^{-1}) \cdot (0.056 \cdot 10^6\ mol \cdot m^{-3})$$
$$\Pi_{swelling} = 4.962 \cdot 10^8\ N \cdot m^{-2}$$
$$\Pi_{swelling} = 4898\ atm$$

Since we do not know the water content of the clay gel and already assume that the density is unaffected by the gel, we will take this as the actual swelling pressure.

Question: What would happen if 15 g of $NaCl$ were added to the water (left side)? The osmotic pressure π is related to the electrolyte valence Z and concentration n by the following expression:

$$\Pi \approx \frac{nRT}{Z}$$

Solution: Adding $NaCl$ to the pure water will lower the activity of water in that cell, drawing water from the clay gel and increasing the height of the water column relative to the clay gel column. Using the osmotic pressure equation:

$$\Pi \approx \frac{n_B}{Z}RT$$

$$\Pi \approx \left\{ \frac{15}{10^{-1}} \left[\frac{g_{NaCl}}{L} \right] \times 10^3 \left[\frac{L}{m^3} \right] \times \frac{2}{58.44} \left[\frac{mole_{ions}}{g_{NaCl}} \right] \right\} RT = \left\{ 5130 \left[\frac{mole_{ions}}{m^3} \right] \right\} RT$$

$$\Pi \approx \left\{ 5130 \left[\frac{mole_{ions}}{m^3} \right] \right\} \cdot (8.314) \cdot (298) = 1.23 \times 10^7 \left[\frac{N \cdot m}{m^3} \right]$$

$$\Pi \approx 1.23 \times 10^7 \left[\frac{N}{m^2} \right] \times \frac{1}{1.013 \times 10^5} \left[\frac{atm}{N/m^2} \right] = 122[atm]$$

$$\Pi \approx 122[atm] \times \frac{1}{1034} \left[\frac{cm_{H_2O}}{atm} \right] = 0.12[cm_{H_2O}]$$

Problems

1. Discuss the physical processes and chemical reactions contributing to mineral weathering.
2. Apply Pauling's Radius Ratio rule (see Appendix 3B) to determine the preferred coordination of cation listed following using the ionic radius of oxygen.

Abundance Rank	Element	Ionic Radius, nm
1	O^{2-}	0.140
2	Si^{4+}	0.034
3	Al^{3+}	0.053
4	Fe^{2+}	0.077
4	Fe^{3+}	0.065
5	Ca^{2+}	0.100
5	K^+	0.138
6	Na^+	0.102
7	Mg^{2+}	0.072
8	Ti^{4+}	0.069

3. The following illustration shows the layer structure of four silicate minerals. Associate a specific mineral name with each structure, indicate which chemical weathering stage is dominated by the named mineral, and specify whether the mineral is capable of crystalline swelling.

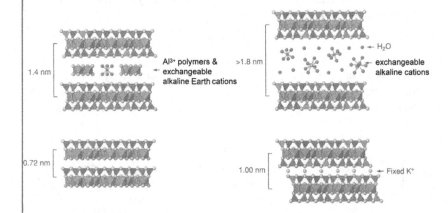

4. The surface area of a typical fine clay ($\leq 0.2\ \mu m$) fraction of kaolinite is $10\ m^2\ g^{-1}$, while smectite has a surface area of over $700\ m^2\ g^{-1}$. What accounts for this dramatic difference in surface area?

5. Consider a landscape underlain by granite that weathers *in situ* to saprolite. The x-ray diffraction pattern from the silt-size fraction of the saprolite collected near the contact between saprolite and the granite bedrock clearly shows the presence of fine-grained muscovite. The x-ray diffraction pattern from the clay-size fraction of the saprolite collected from the soil profile at the land surface lacks the characteristic diffraction lines of muscovite; instead, the dominant mineral is kaolinite. Explain the significance of these findings.

6. Based on the Jackson chemical weathering sequence, what minerals would occur in association with the following clay minerals: (a) kaolinite, (b) smectite, and (c) vermiculite?

7. Explain why feldspar mineral grains tend to be larger than kaolinite mineral particles and kaolinite particles tend to be larger than montmorillonite.

8. What type of minerals—primary or secondary—dominates the sand- and silt-size fractions of soil or sediments?

9. Layer charge of clay minerals influences a variety of physical properties: swelling behavior, surface area, and the exchangeability of interlayer ions. Explain the relationship between layer charge and these physical properties.

10. Layer charge in clay minerals arises from ion substitution in the layer structure. What determines whether a given ion can substitute for another in the layer structure of a clay mineral?

11. A soil contains 30% clay by weight: 20% montmorillonite, 5% vermiculite, and 5% iron oxide. What is its approximate ion exchange capacity?

12. Estimate the swelling pressure $\Pi_{swelling}[kPa]$ of a sodium-saturated smectite (SAz-1, Cheto, Arizona) with a cation exchange capacity of 97 $cmol_c \cdot kg^{-1}$ and a layer spacing (d_{001}) of 5.0 nm. Assume the surface area of this smectite is 750 $m^2 \cdot kg^{-1}$. The molar volume of water at 25°C (298.13 K) is $1.807 \cdot 10^{-5} \ m^3 \cdot mol^{-1}$.

13. Following are three illustrations of layer silicate structures. The repeat distance is shown for the 1:1 mineral kaolinite (0.70 nm) and the 2:1:1 mineral chlorite (1.40 nm). Based on these repeat distances and layer structure, estimate the repeat distance of 2:1 mica.

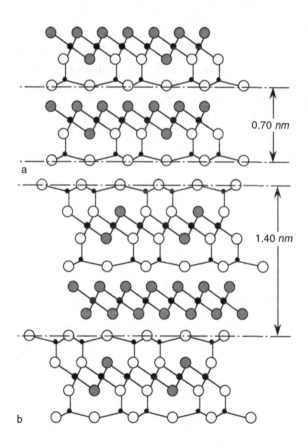

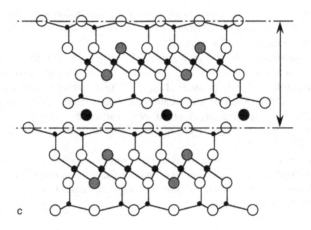

c

Solution

The 1.40 *nm* repeats after six planes of oxygen atoms, while the 0.70 *nm* clay mineral consists of three planes, so a clay mineral that repeats after four oxygen planes would have an approximate thickness of 93 *nm*. Since there are interlayer cations but no water, we can add a bit more to the repeat distance and conclude it would be more than 0.93 *nm*.

Ion Exchange

4.1. INTRODUCTION

While studying the chemistry of manure in soil, nineteenth-century agricultural chemists discovered ion exchange. This discovery was wholly unexpected and all the more puzzling because chemists at that time did not have an accurate understanding of electrolyte solutions. Regardless, a handful of scientists worked out the broad outlines of the ion exchange reaction within a remarkably short time. Ion exchange is now recognized as a very important chemical reaction with applications in nearly all chemistry fields.

This chapter begins with a brief account of the discovery and the implications of ion exchange for environmental chemistry. The next topic is a complete description of an ion exchange experiment itself and the data collected from such an experiment. Scientists studying ion exchange reactions have developed several ways of writing the equilibrium quotient, and, not surprisingly, the different formulations are equivalent to each other. The most important topic is a discussion of the physical parameters that determine ion exchange equilibrium.

4.2. THE DISCOVERY OF ION EXCHANGE

A chief concern of agriculturalists has been soil fertility—the capacity of soil to supply essential nutrients to crops. Nitrogen is one of the most important plant nutrients, and maintenance of soil nitrogen fertility is critical to crop yield. Manure is a rich and readily accessible nitrogen source, with the nitrogen in manure being released largely as ammonia through the microbial degradation process known generally as *mineralization*.

Much of the strong odor from animal manure comes from ammonia gas, and nineteenth-century agricultural chemists understood that ammonia loss from manure meant that there was less nitrogen for crop uptake. H. S. Thompson (1850) began studying methods to prevent ammonia loss from stored manure in 1845, including sulfuric acid additions that produced large amounts of ammonia sulfate. In a report to the Royal Agricultural Society, Thompson mentions an interesting property of soil: "the power of retaining ammonia." Thompson designed experiments to measure the "extent of this power and to ascertain whether it also extended to the [sulfate]" by mixing ammonium sulfate with soil, filling a glass column with the soil, washing the soil column with water to simulate rainfall, and analyzing the drainage solution. Thompson was certainly aware of an observation made by Mr. Huxtable when liquid manure drained through a soil bed: "it went in manure and came out water" (Way, 1850). Thompson found that the salt remaining after he evaporated the drainage solution "proved to be chiefly gypsum."

This was a complete surprise. The large portion of gypsum ... [in the drainage solution] ... showed that a considerable portion of the sulphate of ammonia mixed with the soil had been decomposed and that this process was in some way connected with the presence of lime in the soil, as the sulphuric acid was washed out in combination with lime.

Thompson performed an experiment to observe the result when "the absorptive powers of the soil were fully called into play by passing the filtered liquid repeatedly through the [soil column]." He found that "the whole of the ammonia was retained by the soil, whether applied in the form of sulphate or sesquicarbonate."

J. T. Way (1850) published a report in the same publication of the Royal Agricultural Society, discussing possible explanations for the "absorptive power" of soil and describing numerous experiments that extended those performed by Thompson. Way found that after adding ammonia sulfate solution to a soil column, the drainage solution was "entirely free of the pungent smell of ammonia." Way also noticed that the soil had a fixed capacity to retain ammonia; after continued washing of the soil column, "the ammonia would shortly have passed through, the soil being saturated with it."

Way (1850, 1852) traced the "absorbing power" of soil to the clay fraction, but how clays retain cations would remain a mystery until Linus Pauling (1930) solved the crystal structure of mica. The most abundant material in nature with significant ion exchange capacity is smectite. A number of other minerals have measurable ion exchange capacity—for example, oxide clays, natural organic matter, zeolite minerals—but the overwhelming number of ion exchange studies involved smectite.

4.3. ION EXCHANGE EXPERIMENTS

The surface of all solids has some capacity to retain cations and anions, but unless the surface-to-mass ratio is extremely high, it is impossible to reliably measure the ion exchange capacity. If the ion exchange capacity of the material is very low, ion exchange reactions will have a negligible effect on the composition of the solution. The coarse clay particle-size fraction—2.0 μm spherical diameter—has a surface area of about 2 $m^2 \cdot g^{-1}$. Particles much larger than this, with surface areas less than 1 $m^2 \cdot g^{-1}$, have negligible ion exchange capacity. The mineral kaolinite has a cation exchange capacity of 5 $cmol_c \cdot kg^{-1}$ and a surface area of about 10 $m^2 \cdot g^{-1}$, one of the lowest ion exchange capacities of any naturally occurring ion exchanger and among the lowest surface areas of any clay mineral.

4.3.1. Preparing Clay Saturated with a Single Cation

Naturally occurring ion exchanges found in the soils, sediments and aquifers are cation exchangers, with the notable exception of oxide clays under acidic conditions (more on this in Chapter 9). The first step when measuring cation exchange capacity is saturating the cation exchanger with a single cation, preferably a cation that is easy to chemically analyze. Suspend the clay in a centrifuge bottle or tube containing a concentrated (e.g., 1 M) solution of a highly soluble salt (Figure 4.1a). Agitate or shake for an hour or so to allow sufficient time for cation exchange (Figure 4.1b). Centrifuge until the supernatant solution contains no clay particles (Figure 4.1c). Discard the supernatant solution (Figure 4.1d) and repeat (suspend, agitate, centrifuge, discard) several times. Wash the clay repeatedly with pure water (suspend, agitate, centrifuge, discard). The clay is now saturated by a single cation and washed free of excess salts. All that remains is to suspend the clay in salt-free water and accurately determine the mass concentration of suspended clay. This is necessary because the ion exchange capacity is reported on a mass basis.

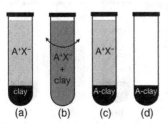

FIGURE 4.1   Preparing clay saturated by cation A^+. (a) Suspend clay in centrifuge tube containing a concentrated solution of a highly soluble salt A^+X^-. (b) Agitate to allow sufficient time for cation exchange. (c) Centrifuge until the supernatant solution contains no clay particles. (d) Discard the supernatant solution and repeat.

FIGURE 4.2 Measuring cation exchange capacity (CEC) by quantitatively replacing all of the exchangeable A^+ from A^+-saturated clay with a new cation B^+. (a) Suspend A^+-saturated clay in centrifuge tube containing a new salt B^+X^-. (b) Agitate. (c) Separate the supernatant solution from the clay, and (d) *reserve* the supernatant solution.

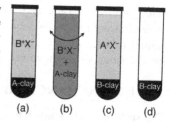

(a)　　(b)　　(c)　　(d)

4.3.2. Measuring Cation Exchange Capacity

Begin with a known volume of a well-mixed clay suspension whose mass concentration of clay has been measured. The next series of steps are a repetition of the cation saturation process just described, except this time using a new salt to displace the cation presently saturating the clay. Transfer clay suspension to a centrifuge bottle or tube containing a concentrated solution of the new salt (Figure 4.2a). Agitate or shake for an hour or so to allow sufficient time for cation exchange (Figure 4.2b). Centrifuge until the supernatant solution contains no clay particles (Figure 4.2c). Reserve the supernatant solution in another bottle (Figure 4.2d) and repeat (suspend, agitate, centrifuge, reserve) three or four times. The reserved supernatant solutions are combined in a volumetric flask, diluted to the volume mark on the flask, and analyzed for the concentration of the original saturating cation. The cation exchange capacity is the number of moles of original saturating cation divided by the dry mass of clay in the suspension aliquot; the units are centimoles of charge—$cmol_c$—per kilogram of clay.

4.3.3. Measuring the Cation Exchange Isotherm

The simplest experiment begins with a known volume of a well-mixed clay suspension containing negligible soluble electrolyte. The mass concentration of clay is known, and the clay is saturated with one or the other of the two cations undergoing exchange. In the example shown in Figure 4.3a, the clay is initially A^+-saturated. Solution aliquots from two stock solutions—A^+X^- and B^+X^-—are

FIGURE 4.3 Measuring the cation exchange isotherm by allowing a series of solutions containing varying concentrations (a) of two salts—A^+X^- and B^+X^-—to exchange with A^+-saturated clay, agitation (b), and separation of clay from solution (c).

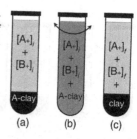

(a)　　(b)　　(c)

added to the clay suspension, yielding initial concentrations $[A^+]_{initial}$ and $[B^+]_{initial}$ (Figure 4.3a). It is important that the ionic strength of the initial solutions remains constant with only the concentration ratio of the two cations A^+ and B^+ being varied.

The exchange reaction is generally quite rapid at ionic strengths greater than 0.01 M, but it can take weeks in more dilute solutions. It is best to continually agitate the clay suspension until equilibrium is reached (Figure 4.3b). Analysis of the equilibrium solution to determine $[A^+]_{final}$ and $[B^+]_{final}$ usually requires centrifugation to remove clay from suspension (Figure 4.3c).

Analysis of the equilibrium solution becomes difficult under certain conditions: the clay particles are less than 0.2 µm, the exchanger is a smectite, and the saturating cation is either Li^+ or Na^+. Chapter 3 discusses both crystalline and osmotic swelling of clay minerals. Vermiculite exhibits crystalline swelling but not osmotic swelling, regardless of the saturating cation. Smectite exhibits both crystalline swelling and osmotic swelling; the latter with saturating cation is either Li^+ or Na^+, and the ionic strength (or osmotic potential) is low. A low ionic strength suspension of either Li^+- or Na^+-saturated smectite may form a stable colloidal dispersion that resists sedimentation in the operating range of conventional centrifuges. Under these circumstances, clay chemists resort to dialysis to analyze the concentrations of soluble cations.

Table 4.1 illustrates the experimental design of a typical exchange isotherm (Vanselow, 1932). The exchanger that Vanselow used for this experiment was a montmorillonite with a cation exchange capacity of 104.1 $cmol_c \cdot kg^{-1}$. Half of the suspensions used Na^+-saturated montmorillonite, and the other half used K^+-saturated montmorillonite. Vanselow based the clay content on the volume concentration of the exchangeable cation—24.0 $g \cdot L^{-1}$ of K^+-saturated montmorillonite, yielding a K^+ concentration of 25.0 $mmol_c \cdot L^{-1}$ as exchangeable K^+. For his experiment, he used different amounts of montmorillonite: 24.0 $g \, L^{-1}$ (25.0 $mmol_c \cdot L^{-1}$) and 41.2 $g \, L^{-1}$ (42.9 $mmol_c \cdot L^{-1}$).

Table 4.2 lists the equilibrium composition of the montmorillonite suspension, and Figure 4.4 plots the exchange isotherm. Vanselow did not have to measure the amounts of Na^+ and K^+ bound to the montmorillonite because he could determine those quantities by mass balance. The total Na^+ concentration is the sum of the soluble concentration $[Na^+]$ and the amount of clay-bound Na^+ expressed in volume concentration units q_{Na^+}. The initial and final total Na^+ concentration remains unchanged; only the relative soluble and clay-bound concentrations change.

$$([Na^+] + q_{Na^+})_{initial} = ([Na^+] + q_{Na^+})_{final}$$

$$(q_{Na^+})_{final} = ([Na^+] + q_{Na^+})_{initial} - [Na^+]_{final}$$

A comparison of Tables 4.1 and 4.2 clearly shows that cation exchange has altered the solution concentration of both cations and the quantities of both cations bound to the clay.

TABLE 4.1 Initial Conditions of a Symmetric Cation Exchange Experiment

$[Na^+]$ $(mmol_c \cdot L^{-1})$	$[K^+]$ $(mmol_c \cdot L^{-1})$	q_{Na^+} $(mmol_c \cdot L^{-1})$	q_{K^+} $(mmol_c \cdot L^{-1})$
100.0	0.0	0.0	25.0
100.0	0.0	0.0	42.9
85.7	14.3	0.0	42.9
71.4	28.6	0.0	42.9
57.1	42.9	0.0	42.9
42.9	57.1	0.0	42.9
25.0	75.0	0.0	25.0
75.0	25.0	25.0	0.0
57.1	42.9	42.9	0.0
42.9	57.1	42.9	0.0
28.6	71.4	42.9	0.0
14.3	85.7	42.9	0.0
0.0	100.0	42.9	0.0
0.0	100.0	25.0	25.0

Source: Data reproduced with permission from Vanselow, A.P., 1932. Equilibria of the base-exchange reactions of bentonites, permutites, soil colloids, and zeolites. Soil Sci. 33, 95–113.

Row 5 in Table 4.2—where $[Na^+] \approx [K^+]$—also shows that the fraction of clay-bound Na^+ is significantly less than the Na^+ fraction in solution.

4.3.4. Selectivity Coefficients and the Exchange Isotherm

Table 4.3 lists the most frequently encountered ion exchange expressions, commonly known as selectivity coefficients.[1] All of the expressions are based on specific exchange reactions, both symmetric and asymmetric. Some examples involving abundant ions are presented following.

---

1. Usage of *selectivity coefficient* can be confusing because it may refer to both the coefficient (e.g., K_{GT}) and the quotient. The superscript "c" (Table 4.3) indicates a *conditional* coefficient (see the discussion of *activity* in Chapter 5). Appendix 4A lists thermodynamic *selectivity coefficient* expressions corresponding to those listed in Table 4.3.

TABLE 4.2 Equilibrium Conditions of a Symmetric Cation Exchange Experiment

$[Na^+]$ $(mmol_c \cdot L^{-1})$	$[K^+]$ $(mmol_c \cdot L^{-1})$	q_{Na^+} $(mmol_c \cdot L^{-1})$	q_{K^+} $(mmol_c \cdot L^{-1})$
86.3	14.9	13.70	10.10
81.1	20.6	18.90	22.30
71.9	29.3	13.80	27.90
61.8	39.5	9.60	32.00
50.5	50.6	6.60	35.20
38.7	62.2	4.20	37.80
23.7	76.9	1.30	23.10
85.3	15.6	14.70	9.40
79.8	22.3	20.20	20.60
70.4	31.5	15.40	25.60
60.0	42.1	11.50	29.30
48.4	53.1	8.80	32.60
37.0	64.7	5.90	35.30
22.7	78.3	2.30	46.70

Source: Data reproduced with permission from Vanselow, A.P., 1932. Equilibria of the base-exchange reactions of bentonites, permutites, soil colloids, and zeolites. Soil Sci. 33, 95–113.

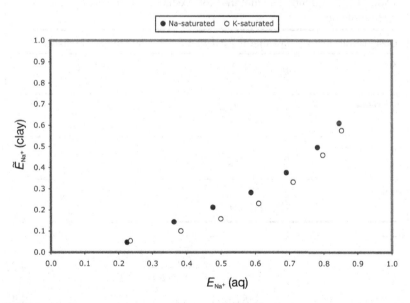

FIGURE 4.4 The *cation exchange isotherm* for (Na^+, K^+) exchange on a montmorillonite (Vanselow, 1932).

Common Symmetric Exchange Reactions

$$\underset{solution}{Na^+(aq)} + \underset{exchanger}{K^+} \leftrightarrow \underset{exchanger}{Na^+} + \underset{solution}{K^+(aq)}$$

$$\underset{solution}{Ca^{2+}(aq)} + \underset{exchanger}{Mg^{2+}} \leftrightarrow \underset{exchanger}{Ca^{2+}} + \underset{solution}{Mg^{2+}(aq)}$$

Common Asymmetric Exchange Reactions

$$2 \cdot \underset{solution}{Na^+(aq)} + \underset{exchanger}{Ca^{2+}} \leftrightarrow 2 \cdot \underset{exchanger}{Na^+} + \underset{solution}{Ca^{2+}(aq)}$$

$$2 \cdot \underset{solution}{K^+(aq)} + \underset{exchanger}{Mg^{2+}} \leftrightarrow 2 \cdot \underset{exchanger}{K^+} + \underset{solution}{Mg^{2+}(aq)}$$

The Kerr selectivity quotient (Kerr, 1928) employs solution concentrations $(mol \cdot L^{-1})$ and solid concentrations $(mol \cdot kg^{-1})$. The *Vanselow* selectivity quotient (Vanselow, 1932) expresses *exchanger-bound* ions using mole fractions $\tilde{N}_{A^{u+}}$ $(mol \cdot mol^{-1})$. The Vanselow formalism treats the exchanger-bound ions (commonly referred to as the *exchange complex*) as a mixture:

$$\tilde{N}_{A^{u+}} = \frac{\tilde{m}_{A^{u+}}}{\tilde{m}_{A^{u+}} + \tilde{m}_{B^{v+}}} \tag{4.1}$$

$$\tilde{N}_{A^{u+}} = 1 - \tilde{N}_{B^{v+}} \tag{4.2}$$

The *Gaines-Thomas* selectivity quotient (Gaines and Thomas, 1953) expresses exchanger-bound ions using equivalent fractions $\tilde{E}_{A^{u+}}$ $(mol_c \cdot mol_c^{-1})$ to quantify the concentration of ion A^{u+} bound to the ion exchanger:

TABLE 4.3 Frequently Encountered Ion Exchange Selectivity Coefficients and Quotients

Designation	Symmetric Exchange	Asymmetric Exchange
Kerr	$K_K^c = \dfrac{[K^+] \cdot q_{Na^+}}{q_{K^+} \cdot [Na^+]}$	$K_K^c = \dfrac{[Ca^{2+}] \cdot q_{Na^+}^2}{q_{Ca^{2+}} \cdot [Na^+]^2}$
Vanselow	$K_V^c = \dfrac{[K^+] \cdot \tilde{N}_{Na^+}}{\tilde{N}_{K^+} \cdot [Na^+]}$	$K_V^c = \dfrac{[Ca^{2+}] \cdot \tilde{N}_{Na^+}^2}{\tilde{N}_{Ca^{2+}} \cdot [Na^+]^2}$
Gaines-Thomas	$K_{GT}^c = \dfrac{[K^+] \cdot \tilde{E}_{Na^+}}{\tilde{E}_{K^+} \cdot [Na^+]} = K_V^c$	$K_{GT}^c = \dfrac{[Ca^{2+}] \cdot \tilde{E}_{Na^+}^2}{\tilde{E}_{Ca^{2+}} \cdot [Na^+]^2}$
Gapon	$K_G^c = \dfrac{[K^+] \cdot \tilde{E}_{Na^+}}{\tilde{E}_{K^+} \cdot [Na^+]} = K_V^c$	$K_G^c = \dfrac{[Ca^{2+}]^{1/2} \cdot \tilde{E}_{Na^+}}{\tilde{E}_{Ca^{2+}} \cdot [Na^+]}$

$$\tilde{E}_{A^{u+}} = \frac{u \cdot \tilde{m}_{A^{u+}}}{u \cdot \tilde{m}_{A^{u+}} + v \cdot \tilde{m}_{B^{v+}}} \tag{4.3}$$

$$\tilde{E}_{A^{u+}} = 1 - \tilde{E}_{B^{v+}} \tag{4.4}$$

The *Gapon* selectivity quotient (Gapon, 1933) employs a much different convention than the other expressions (cf. Appendix 4A) and is widely used by scientists studying salinity and sodicity problems (see Chapter 7). The two contenders for a thermodynamic formulation—the Vanselow selectivity quotient and coefficient and the Gaines-Thomas selectivity quotient and coefficient—ultimately yield the same Gibbs free energy for an ion exchange reaction.

Argersinger and colleagues (1950) developed an integral method for determining the thermodynamic equilibrium constant based on the Vanselow selectivity quotient. Their result, though derived from a selectivity quotient using mole fraction $\tilde{N}_{A^{u+}}$ to quantify the concentration of ions bound to the ion exchanger, yields an expression in which the equivalent fraction $\tilde{E}_{A^{s+}}$ became the master variable. An equivalent integral expression can be derived using the Gaines-Thomas selectivity quotient; the equivalent fraction $\tilde{E}_{A^{s+}}$ remains the master variable:

$$\tilde{N}_{A^{u+}} = \frac{\tilde{m}_{A^{u+}}}{\tilde{m}_{A^{u+}} + \tilde{m}_{B^{v+}}}$$

$$\tilde{m}_{A^{u+}} = \left(\frac{\tilde{N}_{A^{u+}}}{1 - \tilde{N}_{A^{u+}}}\right) \cdot \tilde{m}_{B^{v+}}$$

$$\tilde{E}_{A^{u+}} = \frac{u \cdot \tilde{m}_{A^{u+}}}{u \cdot \tilde{m}_{A^{u+}} + v \cdot \tilde{m}_{B^{v+}}} = \frac{u \cdot \left(\left(\frac{\tilde{N}_{A^{u+}}}{1 - \tilde{N}_{A^{u+}}}\right) \cdot \tilde{m}_{B^{v+}}\right)}{u \cdot \left(\left(\frac{\tilde{N}_{A^{u+}}}{1 - \tilde{N}_{A^{u+}}}\right) \cdot \tilde{m}_{B^{v+}}\right) + v \cdot \tilde{m}_{B^{v+}}}$$

$$\tilde{E}_{A^{u+}} = \frac{u \cdot \tilde{N}_{A^{u+}}}{\left(u \cdot \tilde{N}_{A^{u+}}\right) + v \cdot \left(1 - \tilde{N}_{A^{u+}}\right)} = \frac{u \cdot \tilde{N}_{A^{u+}}}{v + \left(u \cdot \tilde{N}_{A^{u+}}\right) - \left(v \cdot \tilde{N}_{A^{u+}}\right)}$$

$$\tilde{E}_{A^{u+}} = \frac{u \cdot \tilde{N}_{A^{u+}}}{v + (u - v) \cdot \tilde{N}_{A^{u+}}} \tag{4.5}$$

The exchange isotherm (Figure 4.4) plots the equivalent fraction of one cation, Na^+ in this case, bound to the exchanger $\tilde{E}_{Na^+}$ as a function of the equivalent fraction E_{Na^+} of the same cation in solution. This cation exchange experiment (Table 4.2 and Figure 4.4) clearly demonstrates the montmorillonite selectively binds K^+ from solution, yielding a higher clay-bound equivalent fraction: $\tilde{E}_{K^+} > E_{K^+}$.

4.4. INTERPRETING THE ION EXCHANGE ISOTHERM

The example in Table 4.2 and Figure 4.4 is from a classic paper by Vanselow on ion exchange phenomena. The purpose of that and other studies was an attempt to understand the quantitative details of the ion exchange reaction and the selective adsorption of certain ions, resulting in a different proportion of ions bound to the exchanger than the proportion dissolved in solution. Scientists studying ion exchange anticipated an equilibrium expression similar to other chemical reactions in which the equilibrium quotient equals a constant determined by the Gibbs free energy of the ion exchange reaction. Much of the scientific research devoted to ion exchange seeks to properly formulate the equilibrium quotient for the ion exchange reaction. The historical developments in ion exchange research are beyond the scope of this book, but a summary of the most prominent ion exchange expressions is in order.

4.4.1. The Ion Exchange Isotherm for Symmetric Exchange

Chemists commonly plot experimental data using an exchange isotherm rather than plotting the selectivity quotient. The exchange isotherm for symmetric exchange is derived from the selectivity quotient as follows:

$$\textit{General Symmetric Exchange Reaction}$$
$$\underset{solution}{A^{u+}(aq)} + \underset{exchanger}{B^{u+}} \;\leftrightarrow\; \underset{exchanger}{A^{u+}} + \underset{solution}{B^{u+}(aq)}$$

$$K_{GT}^c = \frac{[B^{u+}] \cdot \tilde{E}_{A^{u+}}}{[A^{u+}] \cdot \tilde{E}_{B^{u+}}} \tag{4.6}$$

$$K_{GT}^c = \frac{\left(\dfrac{[B^{u+}]}{[A^{u+}] + [B^{u+}]}\right) \cdot \tilde{E}_{A^{u+}}}{\left(\dfrac{[A^{u+}]}{[A^{u+}] + [B^{u+}]}\right) \cdot \tilde{E}_{B^{u+}}} = \frac{E_{B^{u+}} \cdot \tilde{E}_{A^{u+}}}{E_{A^{u+}} \cdot \tilde{E}_{B^{u+}}}$$

$$K_{GT}^c = \frac{(1 - E_{A^{u+}}) \cdot \tilde{E}_{A^{u+}}}{E_{A^{u+}} \cdot (1 - \tilde{E}_{A^{u+}})} \tag{4.7}$$

The conditional Gaines-Thomas selectivity coefficient K_{GT}^c includes activity coefficients for the exchanger equivalent fractions (Table 4A.1). Equation (4.7), which contains one parameter and two variables is readily rearranged to yield the symmetric exchange isotherm:

$$\textit{Symmetric Exchange Isotherm}$$
$$\tilde{E}_{A^{u+}} = \frac{K_{GT}^c \cdot E_{A^{u+}}}{1 - E_{A^{u+}} + K_{GT}^c \cdot E_{A^{u+}}} \tag{4.8}$$

Example 4.1 Estimate the conditional selectivity coefficient K_{GT}^c for the symmetric (Na^+, K^+) exchange results in Table 4.2.

Figure 4.4 is a plot of the exchange isotherm derived from the data listed in Table 4.2. An estimate of the selectivity coefficient K_{GT}^c can be made using any data point plotted in Figure 4.4, but the most reliable values will be those where the equivalent fraction of either cation in solution is about 0.5.

Taking the experimental results from row 5 of Table 4.2, first determine each term in the symmetric Gaines-Thomas selectivity coefficient expression (Eq. (4.6)). The equivalent fractions of Na^+ and K^+ on the exchanger (Eq. (4.3)) are determined following:

$$\tilde{E}_{Na^+} = \frac{\tilde{m}_{Na^+}}{\tilde{m}_{Na^+} + \tilde{m}_{K^+}}$$

$$\tilde{E}_{Na^+} = \frac{6.60}{6.60 + 35.20} = 0.16$$

$$\tilde{E}_{K^+} = 1 - \tilde{E}_{Na^+} = 1 - 0.16 = 0.84$$

The corresponding equilibrium solution composition (Table 4.2, row 5) is $[Na^+] = 0.0505 \, mol \cdot L^{-1}$ and $[K^+] = 0.0506 \, mol \cdot L^{-1}$. Entering these values in Eq. (4.6) yields a single-point estimate of Gaines-Thomas selectivity coefficient for the following ion exchange reaction:

$$\underset{solution}{Na^+(aq)} + \underset{exchanger}{K^+} \leftrightarrow \underset{exchanger}{Na^+} + \underset{solution}{K^+(aq)}$$

$$K^c_{GT} = \frac{[K^+] \cdot \tilde{E}_{Na^+}}{\tilde{E}_{K^+} \cdot [Na^+]}$$

$$K^c_{GT} = \frac{(0.0506 \, mol \cdot L^{-1}) \cdot 0.16}{0.84 \cdot (0.0505 \, mol \cdot L^{-1})}$$

$$K^c_{GT} \approx \frac{0.16}{0.84} = 0.19$$

This selectivity coefficient $\left(K^c_{GT} \approx 0.19 \right)$ is for the ion exchange reaction as it is written above. Notice that the solution concentrations of the two cations at equilibrium are essentially equal for the selected data (Table 4.2, row 5). If the exchange were nonselective, then the equivalent fractions of the two cations would be equal $\left(K^c_{GT} \approx 1 \right)$. The exchange results appearing in Table 4.2 and Figure 4.4, however, clearly indicate an enrichment of K^+ in the exchange complex $\left(\tilde{E}_{K^+} = 0.84 \right)$ and a depletion of Na^+ in the exchange complex $\left(\tilde{E}_{Na^+} = 0.16 \right)$.

4.4.2. The Ion Exchange Isotherm for Asymmetric Exchange

The exchange isotherm for asymmetric exchange is more complex than the symmetric exchange isotherm because the stoichiometric coefficients do not cancel. Rather than tackle the general asymmetric exchange isotherm, the following derivation is for the specific case of asymmetric $\left(Na^+, Ca^{2+} \right)$ exchange:

$$\underset{solution}{2 \cdot Na^+(aq)} + \underset{exchanger}{Ca^{2+}} \leftrightarrow \underset{exchanger}{2 \cdot Na^+} + \underset{solution}{Ca^{2+}(aq)}$$

$$K^c_{GT} = \frac{[Ca^{2+}] \cdot \tilde{E}^2_{Na^+}}{\tilde{E}_{Ca^{2+}} \cdot [Na^+]^2} \tag{4.9}$$

Replacing solution activities with solution equivalent fractions leads to a new solution parameter C_0 absent from the symmetric exchange isotherm.

Solution parameter C_0 is the solution charge concentration summed over all ions involved in the exchange $\left(C_0 \equiv [Na^+] + 2 \cdot [Ca^{2+}] \quad Units: mol_c \cdot L^{-1}\right)$. The solution charge concentration C_0 allows us to convert solution concentrations to solution equivalent fractions, but we must recognize a factor of 2 in the numerator for the equivalent fraction of the divalent ion:

$$Define : C_0 \equiv [Na^+] + 2 \cdot [Ca^{2+}]$$

$$E_{Ca^{2+}} = \frac{2 \cdot [Ca^{2+}]}{[Na^+] + 2 \cdot [Ca^{2+}]} = \frac{2 \cdot [Ca^{2+}]}{C_0}$$

$$K_{GT}^c = \frac{\left(\dfrac{C_0 \cdot E_{Ca^{2+}}}{2}\right) \cdot \tilde{E}_{Na^+}^2}{(C_0 \cdot E_{Na^+})^2 \cdot \tilde{E}_{Ca^{2+}}}$$

$$K_{GT}^c = \left(\frac{1}{2 \cdot C_0}\right) \cdot \left(\frac{(1 - E_{Na^+}) \cdot \tilde{E}_{Na^+}^2}{E_{Na^+}^2 \cdot (1 - \tilde{E}_{Na^+})}\right) \tag{4.10}$$

Equation (4.10) reduces the univalent-divalent asymmetric exchange isotherm to a quadratic equation (Eq. (4.11)) whose solution is given by Equation (4.12), where the term β is defined in Eq. (4.13):

$$(2 \cdot C_0) \cdot K_{GT}^c = \frac{(1 - E_{Na^+}) \cdot \tilde{E}_{Na^+}^2}{E_{Na^+}^2 \cdot (1 - \tilde{E}_{Na^+})}$$

$$(2 \cdot C_0 \cdot K_{GT}^c) \cdot E_{Na^+}^2 \cdot (1 - \tilde{E}_{Na^+}) = (1 - E_{Na^+}) \cdot \tilde{E}_{Na^+}^2$$

$$\tilde{E}_{Na^+}^2 + \left(\frac{2 \cdot C_0 \cdot K_{GT}^c \cdot E_{Na^+}^2}{1 - E_{Na^+}}\right) \cdot \tilde{E}_{Na^+} - \left(\frac{2 \cdot C_0 \cdot K_{GT}^c \cdot E_{Na^+}^2}{1 - E_{Na^+}}\right) = 0 \tag{4.11}$$

$$\tilde{E}_{Na^+} = \frac{-\beta \pm \sqrt{\beta^2 + 4 \cdot \beta}}{2} \tag{4.12}$$

$$\beta \equiv \frac{2 \cdot C_0 \cdot K_{GT}^c \cdot E_{Na^+}^2}{1 - E_{Na^+}} \tag{4.13}$$

Example 4.2 Estimate the conditional selectivity coefficient K_{GT}^c for the asymmetric (K^+, Ca^{2+}) exchange reaction on a smectite from Crook County, Wyoming.

The Twotop soil series from Crook County, Wyoming, whose clay fraction is predominantly smectite, has a CEC of 76.4 $cmol_c \cdot kg^{-1}$. The asymmetric

(K^+, Ca^{2+}) exchange experiment reports the following results: $[K^+] = 10.17$ mM, $[Ca^+] = 0.565$ mM, exchangeable K^+ equals 38.2 $cmol_c \cdot kg^{-1}$, and exchangeable Ca^{2+} equals 38.2 $cmol_c \cdot kg^{-1}$. What is the value of the Gaines-Thomas conditional selectivity coefficient K^c_{GT} for (K^+, Ca^{2+}) exchange in this soil?

The asymmetric ion exchange reaction and corresponding Gaines-Thomas selectivity quotient (Eq. (4.9)) follow:

$$2 \cdot K^+(ex) + Ca^{2+}(aq) \leftrightarrow 2 \cdot K^+(aq) + Ca^{2+}(ex)$$

$$K^c_{GT} = \frac{[K^+]^2 \cdot \tilde{E}_{Ca^{2+}}}{\tilde{E}^2_{K^+} \cdot [Ca^{2+}]}$$

As in Example 4.1, determine each term in the asymmetric Gaines-Thomas selectivity coefficient expression. The equivalent fraction of the univalent cation K^+ on the exchanger (Eq. (4.3)) is the simplest term because the denominator is the cation exchange capacity CEC:

$$\tilde{E}_{K^+} = \frac{\tilde{m}_{K^+}}{\tilde{m}_{K^+} + 2 \cdot \tilde{m}_{Ca^{2+}}} = \frac{\tilde{m}_{K^+}}{CEC}$$

$$\tilde{E}_{K^+} = \frac{38.2}{76.4} = 0.50$$

$$\tilde{E}_{Ca^{2+}} = 1 - \tilde{E}_{K^+} = 0.50$$

Entering the exchange equivalent fractions and cation concentrations in Eq. (4.9) yields a single-point estimate of Gaines-Thomas selectivity coefficient for the asymmetric exchange reaction:

$$K^c_{GT} = \frac{(0.50) \cdot (1.017 \cdot 10^{-2})^2}{(0.50)^2 \cdot (5.65 \cdot 10^{-4})}$$

$$K^c_{GT} = \frac{5.17 \cdot 10^{-5}}{1.41 \cdot 10^{-4}} = 0.37$$

4.4.3. Effect of Ionic Strength on the Ion Exchange Isotherm

The symmetric exchange isotherm does not contain any terms explicitly influenced by ionic strength. The exchange isotherm contains a conditional selectivity coefficient K^c_{GT}, but that term adds ionic strength dependence indirectly (see Chapter 5). Figure 4.5 is a plot of exchange isotherms for a symmetric cation exchange at two different ionic strengths. The effect of ionic strength is clearly negligible relative to experimental error.

The asymmetric exchange isotherm (Eq. (4.12)) does contain a term explicitly influenced by solution composition: solution parameter C_0. Figure 4.6 is a plot of experimental exchange isotherms for an asymmetric cation exchange at ionic strength 0.01 M and 0.001 M. Figure 4.7 plots asymmetric A^+-B^{2+} exchange isotherms in solutions, where the ionic strengths are 0.1 M, 0.01 M, and 0.001 M using the Gaines-Thomas exchange isotherm (Eq. (4.12)).

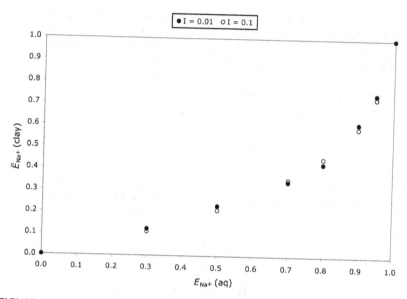

FIGURE 4.5 Experimental cation exchange isotherms for symmetric (K^+, Na^+) exchange (Jensen and Babcock, 1973), where the ionic strengths are 0.1 M (open circles) and 0.01 M (filled circles).

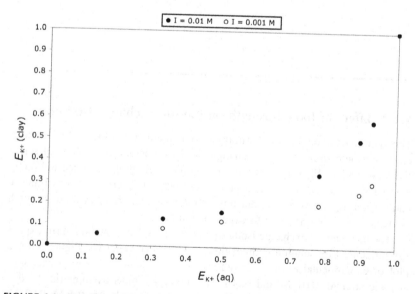

FIGURE 4.6 The cation exchange isotherm for asymmetric (K^+, Ca^{2+}) exchange (Jensen and Babcock, 1973) at two ionic strengths: 0.1 M (open circles) and 0.01 M (filled circles).

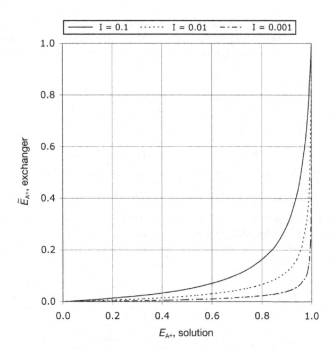

FIGURE 4.7 The asymmetric (A^+, B^{2+}) cation exchange isotherms in three solutions, where the ionic strengths are 0.1 M, 0.01 M, and 0.001 M. The isotherm plots are for a Gaines-Thomas selectivity coefficient $K_{GT} = 1$

4.4.4. Effect of Ion Selectivity on the Ion Exchange Isotherm

Figures 4.4 and 4.5 illustrate the second major influence on the exchange isotherm: ion selectivity. In both of these experiments, the exchanger, montmorillonite in the first case and a soil containing 18% clay dominated by montmorillonite and vermiculite in the second, exhibits exchange selectivity favoring K^+ over Na^+. Quantifying this selectivity and interpreting the basis for ion selectivity are the topics of this section. The approach presented here differs from the conventional approach described in the published ion exchange literature. The approach used here is based on a least sum-square regression of the exchange isotherm (Appendix 4B).

Example 4.3 Determine which cation—K^+ or Ca^{2+}—is selectively enriched during the ion exchange reaction described in Example 4.2.

The Gaines-Thomas conditional selectivity coefficient for the exchange reaction in Example 4.2 is $K^c_{GT} = 0.37$. Selectivity in an asymmetric exchange reaction can be determined only by comparing the experimental equivalent fraction—for example, $\tilde{E}_{Ca^{2+}}$—with the equivalent fraction of that same ion with the selectivity coefficient K_{GT} set equal to 1. (NOTE: The solution composition in both cases must be identical; only the equivalent fractions on the exchanger will differ.)

$$Define: \quad K_{GT} = \frac{\tilde{E}_{Ca^{2+}} \cdot [K^+]^2}{\tilde{E}_{K^+}^2 \cdot [Ca^{2+}]} = 1$$

$$\left(\tilde{E}_{K^+}^2 \cdot [Ca^{2+}]\right) \cdot 1 = \left(\tilde{E}_{Ca^{2+}} \cdot [K^+]^2\right)$$

$$\left(1 - \tilde{E}_{Ca^{2+}}\right)^2 \cdot \left(5.65 \cdot 10^{-4}\right) = \tilde{E}_{Ca^{2+}} \cdot \left(1.017 \cdot 10^{-2}\right)^2$$

$$\left(5.65 \cdot 10^{-4}\right) \cdot \left(1 - 2 \cdot \tilde{E}_{Ca^{2+}} + \tilde{E}_{Ca^{2+}}^2\right) = \left(1.034 \cdot 10^{-4}\right) \cdot \tilde{E}_{Ca^{2+}}$$

$$\left(5.65 \cdot 10^{-4}\right) \cdot \tilde{E}_{Ca^{2+}}^2 + \left(-2 \cdot \left(5.65 \cdot 10^{-4}\right) - \left(1.034 \cdot 10^{-4}\right)\right) \cdot \tilde{E}_{Ca^{2+}} + \left(5.65 \cdot 10^{-4}\right) = 0$$

The *exchanger-bound* equivalent fraction $\tilde{E}_{Ca^{2+}}$ for *nonselective* asymmetric exchange is found by solving the preceding quadratic equation.

$$\tilde{E}_{Ca^{2+}} = x_- = \frac{-b - \sqrt{b^2 - 4 \cdot a \cdot c}}{2 \cdot a}$$

$$\tilde{E}_{Ca^{2+}} = \frac{-\left(-1.233 \cdot 10^{-3}\right) - \sqrt{\left(-1.233 \cdot 10^{-3}\right)^2 - 4 \cdot \left(5.650 \cdot 10^{-4}\right) \cdot \left(5.650 \cdot 10^{-4}\right)}}{2 \cdot \left(5.65 \cdot 10^{-4}\right)}$$

$$\tilde{E}_{Ca^{2+}} = \frac{\left(1.233 \cdot 10^{-3}\right) - \sqrt{2.444 \cdot 10^{-7}}}{\left(1.130 \cdot 10^{-3}\right)}$$

$$\tilde{E}_{Ca^{2+}} = \frac{\left(1.233 \cdot 10^{-3}\right) - \left(4.944 \cdot 10^{-4}\right)}{\left(1.130 \cdot 10^{-3}\right)}$$

$$\tilde{E}_{Ca^{2+}} = \frac{7.390 \cdot 10^{-4}}{1.130 \cdot 10^{-3}} = 0.65$$

The exchanger-bound equivalent fraction $\tilde{E}_{Ca^{2+}}$ for nonselective asymmetric exchange is predicted to be 0.65 for the solution composition as specified in Example 4.2—$[K^+] = 10.17\ mM$ and $[Ca^+] = 0.565\ mM$. The experimental equivalent fraction $\tilde{E}_{Ca^{2+}}$ is 0.50, clearly demonstrating that selective exchange depletes Ca^+ (and enriches K^+).

A least sum-square regression of symmetric $\left(Mg^{2+}, Ca^{2+}\right)$ exchange from Jensen and Babcock (1973) appears in Figure 4.8. The symmetric exchange isotherm model (Eq. (4.8)) estimates of the Gaines-Thomas selectivity coefficients K_{GT} for these data are 0.616 ($I = 0.001$ M) and 0.602 ($I = 0.01$ M). A single-value Gaines-Thomas selectivity coefficient K_{GT} appears to be an acceptable model for the data in Figure 4.8.

Figure 4.9 is a symmetric exchange isotherm with a nonselective isotherm (i.e., $K_{GT} = 1$) extending from the lower left corner to the upper right corner. If data are plotted on the nonselective isotherm, the ion ratios in solution and in the exchange complex are identical for all compositions $E_{A^+} = \tilde{E}_{A^+}$.

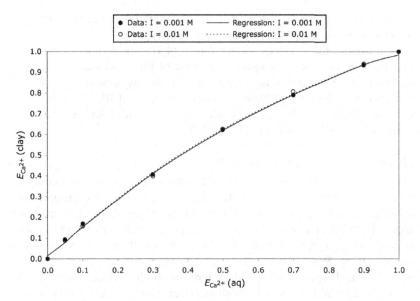

FIGURE 4.8 The experimental cation exchange isotherm for symmetric $\left(Mg^{2+}, Ca^{2+}\right)$ exchange (Jensen and Babcock, 1973) at two ionic strengths and sum-square regression estimates of single-value Gaines-Thomas selectivity coefficients K_{GT}.

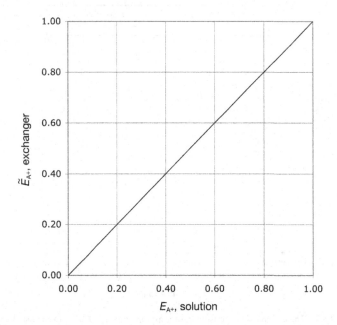

FIGURE 4.9 The exchange isotherm for nonselective symmetric (A^+, B^+) exchange (i.e., $K_{GT} = 1$) appears as a straight line.

Experimental symmetric exchange data rarely exhibit nonselective exchange; the results in Figures 4.4, 4.5, and 4.7 are more typical.

The asymmetric exchange isotherm (Eq. (4.9)) also permits estimates of K_{GT} through the least sum-square regression of the experimental exchange isotherm. A least sum-square regression of the asymmetric $\left(K^+, Ca^{2+}\right)$ exchange from Jensen and Babcock appears in Figure 4.10. The model estimates of the Gaines-Thomas selectivity coefficients K_{GT} for these data are 0.0463 ($I = 0.001$ M) and 0.0931 ($I = 0.01$ M).

Physical Basis for Ion Selectivity

Ion selectivity appears to arise from differences in the Coulomb interaction between ions of the exchange complex and charged sites of the exchanger. The hydration of ions in the exchange complex determines how close they can approach surface-charge sites, which determines the energy that binds the ions in the exchange complex. Ions that can approach the surface more closely—ions surrounded by fewer water molecules—are selectively adsorbed relative to ions whose hydrated radius keeps them further from the surface-charge site.

Argersinger and colleagues (1950) and Gaines and Thomas (1953) showed that the Gibbs free energy $\Delta G^{\circ}_{exchange}$ of ion exchange is related to the selectivity coefficient K_{GT}. In those cases where a single-value selectivity

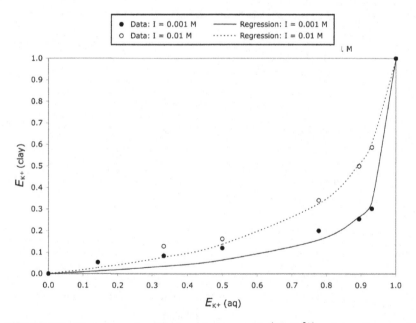

FIGURE 4.10 The cation exchange isotherm for asymmetric $\left(K^+, Ca^{2+}\right)$ exchange (Jensen and Babcock, 1973) at two ionic strengths and the sum-square regression estimates of single-value Gaines-Thomas selectivity coefficients.

coefficient K_{GT} yields a good fit of the exchange isotherm, the single-value selectivity coefficient K_{GT} is a good estimate of the thermodynamic Gibbs free energy of exchange constant $\Delta G^{\circ}_{exchange}$:

$$\Delta G^{\circ}_{exchange} \approx -RT \ln K_{GT} \qquad (4.13)$$

Gast (1969) proposed a Coulomb expression for the Gibbs free energy of exchange using the Debye-Hückel radius r_i of closed approach for ions A^+ and B^+ (e: charge of an electron; ε_0: permittivity of free space; ε: dielectric constant of water):

$$\underset{solution}{A^+(aq)} + \underset{exchanger}{B^+} \leftrightarrow \underset{exchanger}{A^+} + \underset{solution}{B^+(aq)}$$

$$\Delta G^{\circ}_{ion\ exchange} = \frac{e^2}{4\pi\varepsilon_0\varepsilon} \cdot \left[\frac{1}{r_{A^+}} - \frac{1}{r_{B^+}}\right] \qquad (4.14)$$

The radius r_i of the closed approach for the exchangeable interlayer cations K^+ and Na^+ is illustrated in Figure 4.11 based on molecular dynamic simulations of interlayer cations and water in a smectite (Tambach, Bolhuis et al.,

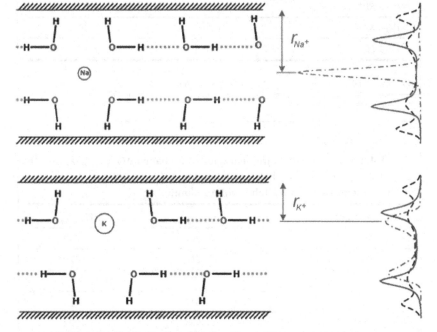

FIGURE 4.11   Molecular dynamic simulations of the crystalline swelling of K^+- (top) and Na^+-saturated (bottom) smectite clay representing the two-layer hydrate. The probability distributions normal to the clay layers appear on the right. *Source: Reproduced with permission from Tambach, T.J., Hensen, E.J.M., et al., 2004. Molecular simulations of swelling clay minerals. J. Phys. Chem. B 108, 7586–7596.*

2006). The position of interlayer K^+ is relatively closer to the clay layer (and the layer charge) than interlayer Na^+, meaning that K^+ cations are held by a stronger electrostatic force than Na^+ cations.

The Gibbs free energy $\Delta G^{\circ}_{exchange}$ values listed in Tables 4.4 and 4.5 are from Gast (1969) and Krishnamoorthy and Overstreet (1950). Recent research shows that the Debye-Hückel radii used by Gast lack a firm physical basis (Marcus, 1988; Ohtaki and Radnai, 1993; Abbas, Gunnarsson et al., 2002). A more reliable indication of ion hydration is the enthalpy of hydration (Marcus, 1987) listed in Table 4.4.

Figure 4.12 plots the standard Gibbs free energy of symmetric univalent exchange $\Delta G^{\circ}_{exchange}(kJ \cdot mol^{-1})$ as a function of the Debye-Hückel radius r_i

TABLE 4.4 Standard Gibbs free energies of exchange $\Delta G^{\circ}_{exchange}$ $(kJ\ mol^{-1})$ and absolute standard enthalpy of hydration $\Delta H^{\circ}_{absolute}$ $(kJ\ mol^{-1})$ for alkali metal cations on Wyoming montmorillonite.

Exchange Reaction	$\Delta G^{\circ}_{exchange}$	$\Delta H^{\circ}_{absolute}$, $kJ\ mol^{-1}$
$Na^+ \rightarrow Cs^+$	−4.52	−4.96
$Na^+ \rightarrow Rb^+$	−2.65	−2.85
$Na^+ \rightarrow K^+$	−1.28	−1.36
$Na^+ \rightarrow Li^+$	+0.20	+0.12

Source: Data reproduced with permission from Gast, R.G., 1969. Standard free energies of exchange for alkali metal cations on Wyoming bentonite. Soil Sci. 69, 41–55.; Marcus, Y., 1987. The thermodynamics of solvation of ions. Part 2. The enthalpy of hydration at 298.15 K. J. Chem. Soc., Faraday Trans. I 83. 339–349.

TABLE 4.5 Standard Gibbs free energies of exchange $\Delta G^{\circ}_{exchange}$ $(kJ\ mol^{-1})$ and absolute standard enthalpy of hydration $\Delta H^{\circ}_{absolute}$ $(kJ\ mol^{-1})$ for divalent metal cations on Utah montmorillonite.

Exchange Reaction	$\Delta G^{\circ}_{exchange}$	$\Delta H^{\circ}_{absolute}$
$Ca^{2+} \rightarrow Mg^{2+}$	+207	−1949
$Ca^{2+} \rightarrow Sr^{2+}$	−236	−1470
$Ca^{2+} \rightarrow Ba^{2+}$	−473	−1332
$Ca^{2+} \rightarrow Cu^{2+}$	−317	−2123
$Ca^{2+} \rightarrow Pb^{2+}$	−834	−1572

Source: Data reproduced with permission from Krishnamoorthy, C. and Overstreet, R., 1950. An experimental evaluation of ion-exchange relationships. Soil Sci. 69, 41–55; Marcus, Y., 1987. The thermodynamics of solvation of ions. Part 2. The enthalpy of hydration at 298.15 K. J. Chem. Soc., Faraday Trans. I 83. 339–349.

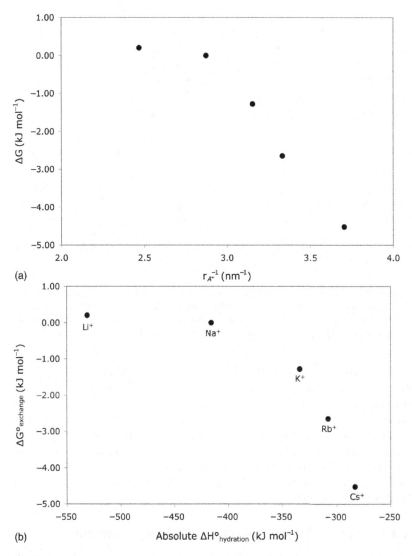

FIGURE 4.12 Ion exchange selectivity for Na^+ exchange with Li^+, K^+, Rb^+, and Cs^+ using data listed in Table 4.4 based on the correlation between the Gibbs free energy of exchange $\Delta G^{\circ}_{exchange}$ and (a) Debye-Hückel radii r_{A^+} (modified from (Gast (1969))) and (b) absolute standard enthalpy of hydration $\Delta H^{\circ}_{absolute}$ of the ions.

and the absolute standard enthalpy of hydration $\Delta H^{\circ}_{absolute}(kJ \cdot mol^{-1})$ for alkali cations. Figure 4.13 plots the standard Gibbs free energy of symmetric univalent exchange $\Delta G^{\circ}_{exchange}(kJ \cdot mol^{-1})$ as a function of the absolute standard enthalpy of hydration $\Delta H^{\circ}_{absolute}(kJ \cdot mol^{-1})$ for selected divalent cations using selectivity coefficients from Krishnamoorthy and Overstreet (1950).

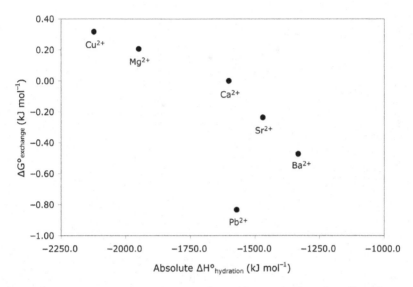

FIGURE 4.13 Ion exchange selectivity for Ca^{2+} exchange with Mg^{2+}, Sr^{2+}, Ba^{2+}, Cu^{2+}, and Pb^{2+} (data listed in Table 4.5) based on the correlation between the Gibbs free energy of exchange $\Delta G^{\circ}_{exchange}$ and the absolute standard enthalpy of hydration of the ions (modified from Krishnamoorthy and Overstreet (1950)).

The correlation between the Gibbs free energy of exchange $\Delta G^{\circ}_{exchange}$ and either the Debye-Hückel radii (Gast, 1969, Figure 4.12a) or the absolute standard enthalpy of hydration $\Delta H^{\circ}_{absolute}$ (Figures 4.12b and 4.13) is not without uncertainty. Gast noted the anomalous behavior of the (Na^+, Li^+) exchange.

Figures 4.12 and 4.13 demonstrate that ion selectivity is correlated with ion hydration, which affects the distance between ion and surface-charge site and, thus, the strength of the Coulomb interaction between ion and site. Similar results for symmetric alkaline earth cation exchange appear in the study by Wild and Keay (1964).

4.4.5. Other Influences on the Ion Exchange Isotherm

A single-value selectivity coefficient often yields a reasonable fit of the exchange isotherm, but there are numerous cases where experimental exchange isotherm data deviate from a single-value selectivity coefficient. One way to evaluate the relative importance of deviations from a single-value selectivity coefficient is to compare the deviation of data from a sum-square regression using a single-value selectivity coefficient to displacement of the selectivity isotherm caused by ion selectivity.

Two high-quality studies of symmetric exchange report sufficient data to plot the exchange isotherms: (Na^+, K^+) exchange (Jensen and Babcock, 1973) and (Ca^{2+}, Fe^{2+}) exchange (Saeki, Wada et al., 2004). The exchange isotherms appear in Figure 4.14. Displacement of the exchange isotherm in both cases indicates ion selectivity in the two exchange reactions (see Figure 4.9). Notice that deviations of the data from the sum-square regression on a single-value selectivity coefficient in the (Na^+, K^+) exchange example (Jensen and Babcock, 1973) are much smaller than the displacement caused

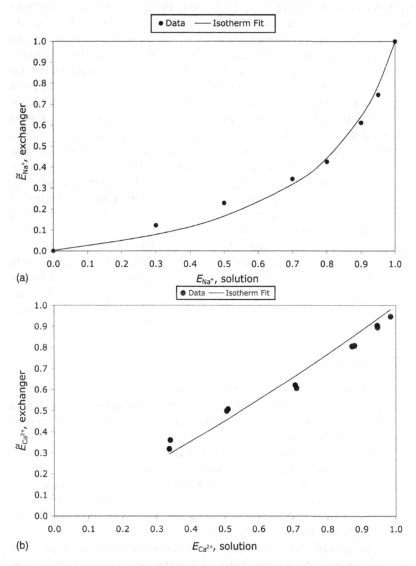

FIGURE 4.14 Ion exchange isotherm for symmetric exchange: (a) of (Na^+, K^+) (Jensen and Babcock, 1973) and (b) (Ca^{2+}, Fe^{2+}) (Saeki, Wada et al., 2004). The experimental data are plotted as filled circles and the sum-square regression estimate of a single-value Gaines-Thomas selectivity coefficients K_{GT} is plotted as a smooth line.

by ion selectivity. Deviations in the (Ca^{2+}, Fe^{2+}) exchange example (Saeki, Wada et al., 2004) are comparable to the displacement caused by ion selectivity because, in this case, the selectivity coefficient is rather small ($\approx$1.2).

Two high-quality studies of asymmetric exchange report sufficient data to plot the exchange isotherms at two different ionic strengths: (K^+, Ca^{2+}) exchange (Jensen and Babcock, 1973; Udo, 1978). The exchange isotherms appear in Figure 4.15. Displacement of the exchange isotherm in both cases indicates the effects of ionic strength (dashed line) and ion selectivity (solid line) on the exchange isotherm (see Figure 4.10). Notice that deviations of the data from the sum-square regression on a single-value selectivity coefficient in the (K^+, Ca^{2+}) exchange example (Jensen and Babcock, 1973; Udo, 1978) are much smaller than the displacement caused by both ionic strength and ion selectivity.

Deviations of the exchange isotherm data in all of these cases from a single-value selectivity coefficient are significant, but the magnitude of these deviations is small relative to the major influences we have identified: ionic strength (asymmetric exchange only) and ion selectivity.

Do the deviations plotted in Figures 4.14 and 4.15 have thermodynamic significance? Systematic deviations of the type shown in Figures 4.14 and 4.15 have little effect on the Gibbs free energy of exchange. Ruvarac and Vesely (1970) found that the value of the selectivity coefficient determined when the equivalent fraction of either ion is 0.5 is a good estimate of the selectivity coefficient determined by the integral method of Argersinger and colleagues (1950). A notable exception to the deviations from a single-value selectivity coefficient that is shown in Figures 4.4, 4.5, 4.6, 4.8, 4.10, 4.14, and 4.15 is the symmetric (Mg^{2+}, Ca^{2+}) cation exchange on the Libby vermiculite (Petersen, Rhoades et al., 1965; Rhoades, 1967), which is discussed in Appendix 4C.

4.5. SUMMARY

Ion exchange is an extremely important chemical reaction in soils, sediments, and aquifers, a process that has profound effects on the solution concentration of many ions. The ion exchange isotherm that describes the relation between solution composition and the ion exchange complex is influenced by two major factors: ionic strength and ion selectivity.

Ionic strength affects the exchange isotherm only for asymmetric exchange reactions, the effect being proportional to the equivalent concentration of the solution C_0. Ion selectivity influences symmetric and asymmetric exchange reactions and arises from Coulomb interactions between the ion and the exchange site (see Figure 4.11).

The strength of the Coulomb interaction between ion and site relate to ion hydration because the hydration shell determines how closely each ion can approach the exchange site. Weakly hydrated ions that are able to shed their hydration shell can bind more strongly to the exchange site than strongly hydrated ions. As a general rule, a single-value selectivity coefficient is a

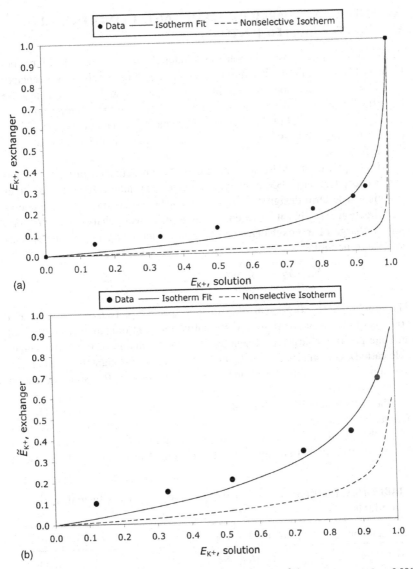

FIGURE 4.15 Ion exchange isotherm for asymmetric (K^+, Ca^{2+}) exchange: (a) $I = 0.001$ (Jensen and Babcock, 1973) and (b) $I = 0.012$ (Udo, 1978). The experimental data are plotted as filled circles and the sum-square regression estimate of a single-value *Gaines-Thomas* selectivity coefficients K_{GT} is plotted as a smooth line. The nonselectivity isotherm is plotted as a dashed line in both cases.

good model of the exchange isotherm. In certain cases (Appendix 4C) the selectivity coefficient adopts two limiting values representing preferred inter-layer hydration states and undergoes an abrupt transition between the two limiting values when the exchange complex reaches a critical value.

APPENDIX 4A. THERMODYNAMIC AND CONDITIONAL SELECTIVITY COEFFICIENTS

Table 4A.1 lists the most commonly encountered ion exchange selectivity quotients and coefficients. Each solution concentration term in Table 4.3 is replaced by the corresponding solution activity (see Chapter 5), and each exchangeable mole fraction or equivalent fraction is multiplied by the corresponding exchange activity coefficient. All of the expressions are based on specific exchange reactions, symmetric and asymmetric, and some examples involving abundant ions appear following.

The Gapon quotient for asymmetric exchange demands further explanation because Gapon (1933) employed a convention quite different from other scientists. The term designating an exchanger-bound cation A^{u+} is the equivalent fraction $\tilde{E}_{A^{u+}}$, but it represents a single negatively charged site *plus* one unit of charge of the exchanger cation. The Gapon exchanger-bound species are charge neutral:

$$\underset{solution}{Na^+(aq)} + \underset{exchanger}{S_{Ca^{2+}}} \quad \leftrightarrow \quad \underset{exchanger}{S_{Na^+}} + \underset{solution}{\tfrac{1}{2} \cdot Ca^{2+}(aq)}$$

The stoichiometric coefficient for exchange specie $S_{A^{u+}}$ is always 1 because it consists of one negative charge (representing one unit of charge from the site) and one positive charge (representing one unit of charge from the exchangeable cation). Gapon (1933) counts one unit of charge throughout the exchange reaction, meaning that the stoichiometric coefficient for solution cation $A^{u+}(aq)$ is $(1/u)$.

The Gapon activity coefficient $f''_{A^{u+}}$ is different from the Gaines-Thomas activity coefficient $f'_{A^{u+}}$ even though both use equivalent fraction $\tilde{E}_{A^{u+}}$ to quantify exchanger-bound species. The numerical value of $\tilde{E}_{A^{u+}}$ is identical regardless of convention; both count one unit of charge. The activity coefficients, on

---

TABLE 4A.1 Thermodynamic Ion Exchange Selectivity Coefficients and Quotients

Designation	Symmetric Exchange	Asymmetric Exchange
Vanselow	$K_V = \dfrac{a_{K^+} \cdot \left(f_{Na^+} \cdot \tilde{N}_{Na^+}\right)}{\left(f_{K^+} \cdot \tilde{N}_{K^+}\right) \cdot a_{Na^+}}$	$K_V = \dfrac{a_{Ca^{2+}} \cdot \left(f_{Na^+} \cdot \tilde{N}_{Na^+}\right)^2}{\left(f_{Ca^{2+}} \cdot \tilde{N}_{Ca^{2+}}\right) \cdot a^2_{Na^+}}$
Gaines-Thomas	$K_{GT} = \dfrac{a_{K^+} \cdot \left(f'_{Na^+} \cdot \tilde{E}_{Na^+}\right)}{\left(f'_{K^+} \cdot \tilde{E}_{K^+}\right) \cdot a_{Na^+}} = K_V$	$K_{GT} = \dfrac{a_{Ca^{2+}} \cdot \left(f'_{Na^+} \cdot \tilde{E}_{Na^+}\right)^2}{\left(f'_{Ca^{2+}} \cdot \tilde{E}_{Ca^{2+}}\right) \cdot a^2_{Na^+}}$
Gapon	$K_G = \dfrac{a_{K^+} \cdot \left(f''_{Na^+} \cdot \tilde{E}_{Na^+}\right)}{\left(f''_{K^+} \cdot \tilde{E}_{K^+}\right) \cdot a_{Na^+}} = K_V$	$K_G = \dfrac{a^{1/2}_{Ca^{2+}} \cdot \left(f''_{Na^+} \cdot \tilde{E}_{Na^+}\right)}{\left(f''_{Ca^{2+}} \cdot \tilde{E}_{Ca^{2+}}\right) \cdot a_{Na^+}}$

the other hand, are different because the two conventions apply different stoichiometric coefficients to the exchanging cations.

APPENDIX 4B. NONLINEAR LEAST SQUARE FITTING OF EXCHANGE ISOTHERMS

Symmetric Exchange: The symmetric exchange isotherm (Eq. (4.8)) is the basis for the nonlinear least sum-square regression estimate of selectivity coefficient K_{GT}. The simplest regression model uses a single value of the selectivity coefficient K_{GT}, which is tantamount to assuming that the exchange complex behaves as an ideal mixture. Section 4.4.5 discusses the relative importance of nonideal behavior of the exchange complex.

Consider the symmetric $Ca^{2+} - Mg^{2+}$ exchange reaction:

$$Ca^{2+}(aq) + \underset{exchanger}{Mg^{2+}} \leftrightarrow \underset{exchanger}{Ca^{2+}} + Mg^{2+}(aq)$$
$$\underset{solution}{} \qquad \qquad \qquad \qquad \underset{solution}{}$$

The Ca^{2+} equivalent fraction on the exchanger $\tilde{E}_{Ca^{2+}}$ varies as a function of the corresponding equivalent fraction in solution $E_{Ca^{2+}}$. The single-value selectivity coefficient model for this exchange reaction $\tilde{E}_{Ca^{2+}}$ is a function of a single parameter K_{GT} and a single solution composition variable—the Ca^{2+} equivalent fraction in solution $E_{Ca^{2+}}$:

$$\tilde{E}_{Ca^{2+}} = \frac{K_{GT} \cdot E_{Ca^{2+}}}{1 - E_{Ca^{2+}} + K_{GT} \cdot E_{Ca^{2+}}} \tag{4B.1}$$

The optimal single-value selectivity coefficient parameter K_{GT} is found by minimizing the sum-square error for all data points in the exchange isotherm. A least sum-square regression of symmetric $Ca^{2+} - Mg^{2+}$ exchange from Jensen and Babcock (1973) appears in Figure 4.8:

$$S = \sum_i \left(\tilde{E}_{Ca^{2+}} - \tilde{E}_{Ca^{2+}} \right)^2 \tag{4B.2}$$

Asymmetric Exchange: The asymmetric exchange isotherm (Eq. (4.9)) also permits estimates of K_{GT} through the least sum-square regression of the experimental exchange isotherm. The simplest model, once again, uses a single value of the selectivity coefficient K_{GT}.

Consider the asymmetric $K^+ - Ca^{2+}$ exchange reaction:

$$2 \cdot K^+(aq) + \underset{exchanger}{Ca^{2+}} \leftrightarrow \underset{exchanger}{2 \cdot K^+} + Ca^{2+}(aq)$$
$$\underset{solution}{} \qquad \qquad \qquad \qquad \underset{solution}{}$$

The K^+ equivalent fraction on the exchanger $\tilde{E}_{K^+}$ varies as a function of the corresponding equivalent fraction in solution E_{K^+}. The single-value selectivity coefficient model (Eq. (4B.3)) for this exchange reaction $\tilde{E}_{K^+}$ is a quadratic function of a single parameter K_{GT} and two solution composition variables—solution normality C_0 and the K^+ equivalent fraction in solution E_{K^+} (Eq. (4B.4)):

$$\tilde{E}_{K^+} = \frac{-\beta - \sqrt{\beta^2 + 4 \cdot \beta}}{2} \qquad (4B.3)$$

$$\beta \equiv \frac{2 \cdot C_0 \cdot K_{GT}^c \cdot E_{K^+}^2}{1 - E_{K^+}} \qquad (4B.4)$$

The optimal single-value selectivity coefficient parameter K_{GT} is found by minimizing the sum-square error for all data points in the exchange isotherm:

$$S = \sum_i \left(\tilde{E}_{K^+} - \tilde{E}_{K^+} \right)^2 \qquad (4B.5)$$

A least sum-square regression of asymmetric (K^+, Ca^{2+}) exchange from Jensen and Babcock (1973) appears in Figure 4.15.

APPENDIX 4C. EQUIVALENT FRACTION-DEPENDENT SELECTIVITY COEFFICIENT FOR (Mg^{2+}, Ca^{2+}) EXCHANGE ON THE LIBBY VERMICULITE

Petersen and colleagues (1965) and Rhoades (1967) reported the symmetric (Mg^{2+}, Ca^{2+}) exchange reaction on vermiculite from Libby, Montana. Similar results were found for (Mg^{2+}, Ba^{2+}) exchange reaction involving vermiculite (identified as the World vermiculite) from Transvaal, South Africa (Wild and Keay, 1964):

$$\underset{exchanger}{Ca^{2+}} + Mg^{2+}(aq) \leftrightarrow Ca^{2+}(aq) + \underset{exchanger}{Mg^{2+}}$$

$$K_{GT}^c = \frac{\tilde{E}_{Mg^{2+}} \cdot [Ca^{2+}]}{\tilde{E}_{Ca^{2+}} \cdot [Mg^{2+}]}$$

The exchange isotherm (Figure 4C.1) strongly deviates from a single-value selectivity coefficient, jumping from $K_{GT}^c \approx 0.6$ in the range $\tilde{E}_{Mg^{2+}} < 0.40$ to $K_{GT}^c \approx 10$ as $\tilde{E}_{Mg^{2+}} > 0.70$.

The explanation for this dramatic change in cation selectivity appears to be related to the effect of interlayer composition on the crystalline swelling state (see Chapter 3). Petersen and colleagues (1965) suggest the exceedingly structured (i.e., crystal-like) hydrated interlayer for Mg^{2+}-saturated vermiculite apparently accounts for $K_{GT}^c \approx 10$ when the exchange complex is dominated by Mg^{2+} (i.e., $\tilde{E}_{Mg^{2+}} > 0.60$). In contrast, the interlayer for Ca^{2+}-saturated

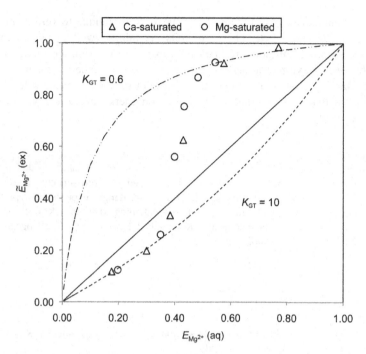

FIGURE 4C.1 The ion exchange isotherms for symmetric $\left(Mg^{2+}, Ca^{2+}\right)$ exchange on a vermiculite sample from Libby, Montana (Petersen, Rhoades et al., 1965), exhibit a dramatic change in selectivity.

vermiculite (i.e., $\tilde{E}_{Mg^{2+}} < 0.40$) may be less structured (i.e., fluid-like), allowing differences in cation hydration to determine the position of interlayer cations and, therefore, selectivity (see Figure 4.13).

Picture the evolution of the hydrated interlayer of the Libby vermiculite as the interlayer Mg^{2+} equivalent fraction $\tilde{E}_{Mg^{2+}}$ decreases from 1.0 (i.e., Mg^{2+}-saturated interlayer). Initially, the interlayer behaves as if it is a solid solution—the position of the interlayer cations fixed—negating any differences in the hydration shell surrounding the cations. In this state, selectivity favors Mg^{2+} relative to Ca^{2+}. When the interlayer equivalent fraction $\tilde{E}_{Mg^{2+}}$ decreases below 0.6, the interlayer begins to melt, allowing interlayer cations to adopt positions relative to the clay layer that are consistent with the dimensions of their respective hydration shells. Interlayer melting is complete once the equivalent fraction $\tilde{E}_{Mg^{2+}}$ decreases below 0.4. At this interlayer composition, the lower hydration energy of Ca^{2+} ions (see Figure 4.13) allows them to reside closer to the clay layer, and the higher hydration energy of Mg^{2+} ions forces them to reside further from the clay layer. As a consequence, exchange selectivity favors Ca^{2+} relative to Mg^{2+} for compositions where $\tilde{E}_{Mg^{2+}} < 0.40$.

The behavior seen in Figure 4C.1 appears to be unique to vermiculite, a layer silicate whose layer charge permits crystalline swelling but prevents osmotic swelling. Furthermore, the cation exchange reaction apparently *must involve* Mg^{2+}. Symmetric and asymmetric exchange isotherms for reactions where Mg^{2+} is not involved appear to conform to a single-value selectivity coefficient that is appropriate to the hydration energy of the cations undergoing exchange.

Problems

1. The CEC of a soil whose clay fraction is predominantly smectite is 18.6 $cmol_c \cdot kg^{-1}$. An asymmetric (K^+, Ca^{2+}) exchange experiment reports the following results for a solution containing 0.40 mM $K^+(aq)$ and 0.20 mM $Ca^{2+}(aq)$: exchangeable K^+ equals 2.2 $cmol_c \cdot kg^{-1}$, exchangeable Ca^{2+} equals 16.4 $cmol_c \cdot kg^{-1}$.

$$2 \cdot K^+(ex) + Ca^{2+}(aq) \leftrightarrow Ca^{2+}(ex) + 2 \cdot K^+(aq)$$

$$K_{GT} = \frac{\tilde{E}_{Ca^{2+}} \cdot [K^+]^2}{\tilde{E}_{K^+}^2 \cdot [Ca^{2+}]} \quad \tilde{E}_{Ca^{2+}} + \tilde{E}_{K^+} = 1$$

What is the value of the Gaines-Thomas selectivity coefficient K_{GT} for (K^+, Ca^{2+}) exchange in this soil?

Solution
Determine the equivalent fraction of each exchangeable cation before entering each value into the Gaines-Thomas selectivity expression.

$$\tilde{E}_{K^+} = \frac{m_{K^+}}{m_{K^+} + 2 \cdot m_{Ca^{2+}}} = \frac{m_{K^+}}{CEC}$$

$$\tilde{E}_{K^+} = \frac{2.2}{18.6} = 0.12$$

$$\tilde{E}_{Ca^{2+}} = 1 - \tilde{E}_{K^+} = 0.88$$

$$K_{GT} = \frac{(0.88) \cdot (0.40 \cdot 10^{-3})^2}{(0.12)^2 \cdot (0.20 \cdot 10^{-3})} = \frac{1.41 \cdot 10^{-7}}{2.80 \cdot 10^{-6}} = 5.04 \cdot 10^{-2}$$

2. Determine which cation—K^+ or Ca^{2+}—is selectively enriched during ion exchange reaction from Problem 1. This requires analysis of the exchange isotherm for this reaction. Please show results to support your answer. (HINT: You may wish to use the quadratic relation from the exchange isotherm for asymmetric univalent-divalent exchange.)

Solution
Selectivity in an asymmetric exchange reaction can be determined only by comparing the experimental equivalent fraction on the exchanger of an ion involved in the exchange—for example, $\tilde{E}_{Ca^{2+}}$—with the equivalent fraction of that same ion with the selectivity coefficient K_{GT} set equal to 1.

$$K_{GT} = \frac{\tilde{E}_{Ca^{2+}} \cdot [K^+]^2}{\tilde{E}_{K^+}^2 \cdot [Ca^{2+}]}$$

Define : $K_{GT} \equiv 1$

$$\left(\tilde{E}_{K^+}^2 \cdot [Ca^{2+}]\right) \cdot 1 = \left(\tilde{E}_{Ca^{2+}} \cdot [K^+]^2\right)$$

$$\left(1 - \tilde{E}_{Ca^{2+}}\right)^2 \cdot (0.2 \cdot 10^{-3}) = \tilde{E}_{Ca^{2+}} \cdot \left(0.4 \cdot 10^{-3}\right)^2$$

$$(2.0 \cdot 10^{-4}) \cdot \left(1 - 2 \cdot \tilde{E}_{Ca^{2+}} + \tilde{E}_{Ca^{2+}}^2\right) = (1.6 \cdot 10^{-7}) \cdot \tilde{E}_{Ca^{2+}}$$

$$(2.0 \cdot 10^{-4}) \cdot \tilde{E}_{Ca^{2+}}^2 + (-4.0 \cdot 10^{-4}) \cdot \tilde{E}_{Ca^{2+}} + (2.0 \cdot 10^{-4}) = 0$$

Solve the quadratic equation for the Ca^{2+} equivalent fraction on the exchanger.

$$\tilde{E}_{Ca^{2+}} = x_- = \frac{-b - \sqrt{b^2 - 4 \cdot a \cdot c}}{2 \cdot a}$$

$$\tilde{E}_{Ca^{2+}} = \frac{-(-4.0016 \cdot 10^{-4}) - \sqrt{(-4.0016 \cdot 10^{-4})^2 - 4 \cdot (2.0 \cdot 10^{-4}) \cdot (2.0 \cdot 10^{-4})}}{2 \cdot (2.0 \cdot 10^{-4})}$$

$$\tilde{E}_{Ca^{2+}} = \frac{(4.0016 \cdot 10^{-4}) - \sqrt{1.6013 \cdot 10^{-7} - 1.6000 \cdot 10^{-7}}}{(4.0000 \cdot 10^{-4})}$$

$$\tilde{E}_{Ca^{2+}} = \frac{(4.0016 \cdot 10^{-4}) - (1.13 \cdot 10^{-5})}{(4.0000 \cdot 10^{-4})}$$

$$\tilde{E}_{Ca^{2+}} = \frac{3.89 \cdot 10^{-4}}{4.00 \cdot 10^{-4}} = 0.97$$

The Ca^{2+} equivalent fraction on the exchanger for nonselective exchange would be 0.97, while the experimental equivalent fraction is 0.88. The soil clay favors K^+ over Ca^{2+}.

3. A soil whose clay fraction is predominantly smectite has a CEC of 26 $cmol_c \cdot kg^{-1}$ balanced by 0.117 $mol \cdot kg^{-1}$ of exchangeable Ca^{2+} and 0.026 $mol \cdot kg^{-1}$ of exchangeable Na^+. Measured concentrations of Ca^{2+} and Na^+ in the soil solution, which occupies 30% of the soil volume, are 0.025 M and 0.04 M, respectively. We will neglect activity coefficients for ions dissolved in solution.

$$2 \cdot Na^+(ex) + Ca^{2+}(aq) \leftrightarrow Ca^{2+}(ex) + 2 \cdot Na^+(aq)$$

$$K_{GT} = \frac{\tilde{E}_{Ca^{2+}} \cdot [Na^+]^2}{\tilde{E}_{Na^+}^2 \cdot [Ca^{2+}]} \qquad \tilde{E}_{Ca^{2+}} + \tilde{E}_{Na^+} = 1$$

What is the value of the Gaines-Thomas selectivity coefficient K_{GT} for (Na^+, Ca^{2+}) exchange in this soil?

Solution

Determine the equivalent fraction of each exchangeable cation before entering each value into the Gaines-Thomas selectivity expression.

$$\tilde{E}_{Na^+} = \frac{m_{Na^+}}{m_{Na^+} + 2 \cdot m_{Ca^{2+}}} = \frac{m_{Na^+}}{CEC}$$

$$\tilde{E}_{Na^+} = \frac{2.6}{26} = 0.10$$

$$\tilde{E}_{Ca^{2+}} = 1 - \tilde{E}_{Na^+} = 0.90$$

$$K_{GT} = \frac{(0.90) \cdot (0.04)^2}{(0.10)^2 \cdot (0.025)} = \frac{1.44 \cdot 10^{-3}}{2.50 \cdot 10^{-4}} = 5.76$$

4. Determine which cation—Ca^{2+} or Na^+—is selectively enriched during the ion exchange reaction described in Problem 3. This requires analysis of the exchange isotherm for this reaction. Please show results to support your answer. (HINT: You may wish to use the quadratic relation from the exchange isotherm for asymmetric univalent-divalent exchange.)

Solution 1
Selectivity in an asymmetric exchange reaction can be determined only by comparing the experimental equivalent fraction on the exchanger of an ion involved in the exchange—for example, $\tilde{E}_{Ca^{2+}}$—with the equivalent fraction of that same ion with the selectivity coefficient K_{GT} set equal to 1.

$$K_{GT} = \frac{\tilde{E}_{Ca^{2+}} \cdot [Na^+]^2}{\tilde{E}_{Na^+}^2 \cdot [Ca^{2+}]}$$

$$\text{Define}: K_{GT} \equiv 1$$

$$\left(\tilde{E}_{Na^+}^2 \cdot [Ca^{2+}] \right) \cdot 1 = \left(\tilde{E}_{Ca^{2+}} \cdot [Na^+]^2 \right)$$

$$\left(1 - \tilde{E}_{Ca^{2+}} \right)^2 \cdot (0.025) = \tilde{E}_{Ca^{2+}} \cdot (0.04)^2$$

$$(0.025) \cdot \left(1 - 2 \cdot \tilde{E}_{Ca^{2+}} + \tilde{E}_{Ca^{2+}}^2 \right) = (1.60 \cdot 10^{-3}) \cdot \tilde{E}_{Ca^{2+}}$$

$$(0.025) \cdot \tilde{E}_{Ca^{2+}}^2 + \left(-2 \cdot (0.025) - (1.60 \cdot 10^{-3}) \right) \cdot \tilde{E}_{Ca^{2+}} + (0.025) = 0$$

Solve the quadratic equation for the Ca^{2+} equivalent fraction on the exchanger.

$$\tilde{E}_{Ca^{2+}} = x_- = \frac{-b - \sqrt{b^2 - 4 \cdot a \cdot c}}{2 \cdot a}$$

$$\tilde{E}_{Ca^{2+}} = \frac{-(-5.16 \cdot 10^{-2}) - \sqrt{(-5.16 \cdot 10^{-2})^2 - 4 \cdot (2.50 \cdot 10^{-2}) \cdot (2.50 \cdot 10^{-2})}}{2 \cdot (2.50 \cdot 10^{-2})}$$

$$\tilde{E}_{Ca^{2+}} = \frac{(5.16 \cdot 10^{-2}) - \sqrt{1.63 \cdot 10^{-4}}}{(5.00 \cdot 10^{-2})}$$

$$\tilde{E}_{Ca^{2+}} = \frac{(5.16 \cdot 10^{-2}) - (1.28 \cdot 10^{-2})}{(5.00 \cdot 10^{-2})}$$

$$\tilde{E}_{Ca^{2+}} = \frac{3.88 \cdot 10^{-7}}{5.00 \cdot 10^{-2}} = 0.776$$

The Ca^{2+} equivalent fraction on the exchanger for nonselective exchange would be 0.776, while the experimental equivalent fraction is 0.90. The soil clay favors Ca^{2+} over Na^+.

Solution 2

An alternative solution adapts the expression for asymmetric (Na^+, Ca^{2+}) exchange isotherm.

$$2 \cdot \underset{solution}{Na^+(aq)} + \underset{exchanger}{Ca^{2+}} \leftrightarrow 2 \cdot \underset{exchanger}{Na^+} + \underset{solution}{Ca^{2+}(aq)}$$

$$K_{GT}^c = \frac{[Ca^{2+}] \cdot \tilde{E}_{Na^+}^2}{\tilde{E}_{Ca^{2+}} \cdot [Na^+]^2}$$

The univalent-divalent asymmetric exchange isotherm as a function of the univalent cation equivalent fractions follows:

$$\tilde{E}_{Na^+}^2 + \left(\frac{2 \cdot C_0 \cdot K_{GT}^c \cdot E_{Na^+}^2}{1 - E_{Na^+}} \right) \cdot \tilde{E}_{Na^+} - \left(\frac{2 \cdot C_0 \cdot K_{GT}^c \cdot E_{Na^+}^2}{1 - E_{Na^+}} \right) = 0$$

$$\tilde{E}_{Na^+} = \frac{-\beta \pm \sqrt{\beta^2 + 4 \cdot \beta}}{2}$$

$$\beta \equiv \frac{2 \cdot C_0 \cdot K_{GT}^c \cdot E_{Na^+}^2}{1 - E_{Na^+}}$$

Selectivity in an asymmetric exchange reaction can be determined only by comparing the experimental equivalent fraction on the exchanger of an ion involved in the exchange—for example, $\tilde{E}_{Ca^{2+}}$—with the equivalent fraction of that same ion with the selectivity coefficient K_{GT} set equal to 1.

$$C_0 \equiv [Na^+] + 2 \cdot [Ca^{2+}] = 4.0 \cdot 10^{-2} + 2 \cdot (2.5 \cdot 10^{-2})$$

$$E_{Na^+} = \frac{4.0 \cdot 10^{-2}}{9.0 \cdot 10^{-2}} = 0.44$$

$$\beta \equiv \frac{2 \cdot C_0 \cdot K_{GT} \cdot E_{Na^+}^2}{1 - E_{Na^+}} = \frac{2 \cdot (9.0 \cdot 10^{-2}) \cdot (1) \cdot (0.44)^2}{1 - (0.44)} = 6.4 \cdot 10^{-2}$$

$$\tilde{E}_{Na^+} = \frac{(-6.4 \cdot 10^{-2}) + \sqrt{(6.4 \cdot 10^{-2})^2 + 4 \cdot (6.4 \cdot 10^{-2})}}{2}$$

$$\tilde{E}_{Na^+} = 0.22$$

The Na^+ equivalent fraction on the exchanger for nonselective exchange would be 0.22, while the experimental equivalent fraction is 0.10. The soil clay favors Ca^{2+} over Na^+.

5. What determines the relative selectivity of two ions in an ion exchange reaction?

Water Chemistry

5.1. THE EQUILIBRIUM CONSTANT

5.1.1. Thermodynamic Functions for Chemical Reactions

You first encountered thermodynamics and the equilibrium concept in general chemistry. This initial encounter covered several important ideas while passing over other ideas that, in the context of environmental chemistry, we must confront. Every chemical reaction involves the conversion of energy: capturing radiant energy, kinetic energy, or heat in chemical bonds or transforming the energy stored in chemical bonds into heat, kinetic energy, and radiant energy. This energy conversion is quantified in various ways: the enthalpy of reaction ΔH_{rxn}, the entropy of reaction ΔS_{rxn}, and the *Gibbs energy* of reaction ΔG_{rxn}. Each of these energy variables changes with absolute temperature T and pressure P. The symbols v_i represent the stoichiometric coefficients for each educt $(A, B, \ldots)$ and product $(M, N, \ldots)$ involved in the reaction in the hypothetical chemical reaction:

$$\underset{educts}{v_A \cdot A + v_B \cdot B + \ldots} \leftrightarrow \underset{products}{v_M \cdot M + v_N \cdot N + \ldots}$$

$$\Delta G_{rxn} = \Delta H_{rxn} - T \cdot \Delta S_{rxn} \tag{5.1}$$

A chemical reaction exists for every *compound*, quantifying the energy conversion upon formation of the compound *from its constituent elements*: enthalpy of formation ΔH_f, entropy of formation ΔS_f, and Gibbs energy of formation ΔG_f. The formation of one mole of the mineral calcite $CaCO_3(s)$ from its constituent elements follows:

$$\underset{elements}{Ca(s) + C(s) + \tfrac{3}{2} \cdot O_2(g)} \leftrightarrow \underset{compound}{CaCO_3(s)}$$

$$\Delta G_{f,CaCO_3(s)} = \Delta H_{f,CaCO_3(s)} - T \cdot \Delta S_{f,CaCO_3(s)}$$

Chemists have adopted a *standard state* convention—the chemical state of every element and compound at 298.15 K (25°C) and 1 atmosphere. By this convention, the enthalpy, entropy, and Gibbs energy of formation of any compound from its constituent elements are the standard enthalpy ΔH_f°, standard entropy ΔS_f°, and standard Gibbs energy ΔG_f° of formation, respectively. The standard Gibbs energy of any reaction ΔG_{rxn}° is a function of the standard Gibbs energy of formation ΔG_f° for the educts and products involved:

$$\Delta G_{rxn}^\circ = \left(\sum_i v_i \cdot \Delta G_{f,i}^\circ \right)_{products} - \left(\sum_j v_j \cdot \Delta G_{f,j}^\circ \right)_{reactants} \tag{5.2}$$

5.1.2. Gibbs Energy of Reaction and the Equilibrium Constant

The standard state, of course, is not representative of prevailing conditions in the environment. Remarkably enough, a simple function relates the Gibbs energy of reaction under prevailing conditions ΔG_{rxn} to the Gibbs energy of reaction under standard conditions ΔG_{rxn}°. The second term on the right-hand side of Eq. (5.3) is the product of a constant and two parameters: the universal gas constant R (8.314 472 $J \cdot K^{-1} \cdot mol^{-1}$), the absolute temperature T (K), and the natural logarithm of the reaction quotient Q (Eq. (5.4)). We will symbolize the concentration of component (i) as C_i rather than the more familiar square brackets:

$$\Delta G_{rxn} \cong \Delta G_{rxn}^\circ + RT \cdot \ln(Q) \tag{5.3}$$

$$Q \cong \frac{[M]^{v_M} \cdot [N]^{v_N} \ldots}{[A]^{v_A} \cdot [B]^{v_B} \ldots} = \frac{C_M^{v_M} \cdot C_N^{v_N} \ldots}{C_A^{v_A} \cdot C_B^{v_B} \ldots} \tag{5.4}$$

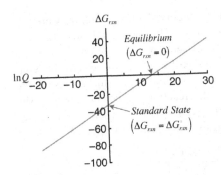

FIGURE 5.1 The transformation of educts into products—regardless of prevailing conditions (horizontal axis)—yields a Gibbs energy of reaction ΔG_{rxn} (vertical axis) dependent on the reaction quotient Q and absolute temperature T.

The reaction quotient Q (Eq. (5.4)) is written as a product of concentrations, just as it would appear in any general chemistry textbook. Equation (5.4) relating the Gibbs energy of reaction under prevailing conditions ΔG_{rxn} to the Gibbs energy of reaction under standard conditions ΔG°_{rxn} and the reaction quotient Q, however, is an *approximate equality* (more on that in the following section).

Figure 5.1 demonstrates graphically that the Gibbs energy of reaction ΔG_{rxn} under prevailing conditions is a linear function of the natural logarithm of Q with a temperature-dependent slope: $-RT$. Two significant intercepts occur in the linear relation between the Gibbs energy of reaction under prevailing conditions ΔG_{rxn} and the natural logarithm of Q (Figure 5.1, right): when $\ln(Q) = 0$ (standard conditions: $\Delta G_{rxn} = \Delta G^\circ_{rxn}$) and when $\ln(Q) = 1$ (equilibrium: $\Delta G_{rxn} = 0$ and $\Delta G^\circ_{rxn} = -RT \cdot \ln(K)$). At equilibrium the reaction quotient Q_{eq} attains an invariant value known as the thermodynamic equilibrium constant K:

$$\ln(K) \equiv -\frac{\Delta G^\circ_{rxn}}{RT} \tag{5.5A}$$

$$K = e^{-\Delta G^\circ_{rxn}/RT} \tag{5.5B}$$

5.2. ACTIVITY AND THE EQUILIBRIUM CONSTANT

5.2.1. Concentrations and Activity

If we write the reaction quotient Q or the equilibrium constant K using concentrations (Eq. (5.4))—the way it is written in most general chemistry books—then Eq. (5.3) must include an approximate equality. The need for this became apparent when component concentrations are used to calculate the equilibrium constant K under what appear to be equilibrium conditions.

It soon became evident that the apparent equilibrium coefficient for reactions usually encountered in environmental chemistry—aqueous *electrolyte* solutions—clearly varied as a function of the electrolyte concentration—even if some of the ions dissolved in the solution were not educts or products in the chemical reaction. This prompted two chemists, Peter Debye and Erich Hückel (1923), to seek an explanation. Debye and Hückel collected all of the variability caused by changes in electrolyte concentration into parameters—ion activity coefficients γ_i—that adjusted the concentration of each ion involved in the reaction—educt or product. The conditional equilibrium coefficient—the equilibrium coefficient that varies with electrolyte concentration—is typically assigned the symbol K^c:

$$K = \frac{(\gamma_M \cdot C_M)^{\nu_M} \cdot (\gamma_N \cdot C_N)^{\nu_N} \cdots}{(\gamma_A \cdot C_A)^{\nu_A} \cdot (\gamma_B \cdot C_B)^{\nu_B} \cdots}$$

$$K = \left(\frac{\gamma_M{}^{\nu_M} \cdot \gamma_N{}^{\nu_N} \cdots}{\gamma_A{}^{\nu_A} \cdot \gamma_B{}^{\nu_B} \cdots} \right) \cdot \left(\frac{C_M{}^{\nu_M} \cdot C_N{}^{\nu_N} \cdots}{C_A{}^{\nu_A} \cdot C_B{}^{\nu_B} \cdots} \right)$$

$$K = \left(\frac{\gamma_M{}^{\nu_M} \cdot \gamma_N{}^{\nu_N} \cdots}{\gamma_A{}^{\nu_A} \cdot \gamma_B{}^{\nu_B} \cdots} \right) \cdot K^c \tag{5.6}$$

Both the reaction quotient Q and the equilibrium constant K are the product of ion activities (Eq. (5.7)). Each ion activity a_i is the product of the concentration C_i and an ion activity coefficient γ_i that depends on electrolyte concentration (Eq. (5.8)):

$$K = \left(\frac{a_M{}^{\nu_M} \cdot a_N{}^{\nu_N} \cdots}{a_A{}^{\nu_A} \cdot a_B{}^{\nu_B} \cdots} \right) \tag{5.7}$$

$$a_i = \gamma_i \cdot C_i \tag{5.8}$$

5.2.2. Ionic Strength I

In their theory of electrolyte solutions, Debye and Hückel derived an expression—known as the *ionic strength* I—that was the natural measure of electrolyte concentration accounting for variability in the conditional equilibrium coefficient K^c (Eq. (5.6)). The ionic strength I is a sum over the concentration of each ionic solute C_i multiplied by the square of the ion charge z_i

$$I = \tfrac{1}{2} \cdot \sum_i \left(z_i^2 \cdot C_i \right) \tag{5.9}$$

Ions with a larger valence have a disproportionate influence on ionic interactions in electrolyte solutions.

5.2.3. Empirical Ion Activity Coefficient Expressions

Debye and Hückel derived a simple expression for ion activity coefficients based on fundamental physical principles. The Debye-Hückel *limiting law*

(Eq. (5.10)) is a single-parameter expression ($A = 0.5109$ at $25°C$) that is adequate for most freshwater environments where ionic strength I tends to be ≤ 0.01 M. Guntelberg (1926) developed an empirical single-parameter expression (Eq. (5.11)) based on the Debye and Hückel limiting law, extending the ionic strength I range to ≤ 0.1 M:

$$\log(\gamma_i) = -A \cdot z_i^2 \cdot \sqrt{I} \tag{5.10}$$

$$\log(\gamma_i) = -A \cdot z_i^2 \cdot \left(\frac{\sqrt{I}}{1 + \sqrt{I}} \right) \tag{5.11}$$

Davies (1938) developed a two-parameter empirical expression that extends the ionic strength range to ≤ 0.5 M. None of these expressions is adequate for seawater or extremely saline lake or groundwater:

$$\log(\gamma_i) = -A \cdot z_i^2 \cdot \left(\frac{\sqrt{I}}{1 + \sqrt{I}} - (0.2 \cdot I) \right) \tag{5.12}$$

How significant are ion activity coefficients under prevailing environmental conditions? Were Debye and Hückel, Guntelberg, and Davies engaged in minutiae, or would the use of concentration rather than activity yield significant error when applying equilibrium principles to environmental chemistry? Figure 5.2 illustrates the magnitude of the error.

In Chapter 7 we estimate the ionic strength I of calcareous water in contact with the atmosphere, an electrolyte solution whose major constituent is

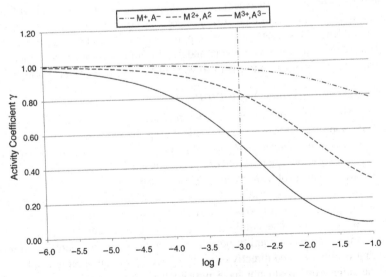

FIGURE 5.2 The effect of increasing electrolyte concentration C on the Davies activity coefficients for univalent, divalent, and trivalent symmetric electrolytes.

calcium carbonate dissolved from the mineral calcite $CaCO_3(s)$ and aqueous carbon dioxide from the atmosphere. The ionic strength I of this representative environmental electrolyte solution is about 1 millimolar.

This ionic strength I representative of natural waters is indicated in Figure 5.2 by a dashed line. The error is roughly 48% for trivalent ions and roughly 18% for divalent ions in 10^{-3} M solutions, significantly greater than experimental error that tends to be on the order of 1%. Based on a reasonable estimate of the ionic strength, we are likely to find that the error introduced from reliance on concentrations rather than activities in the environmental context is much too great to neglect.

5.3. MODELING WATER CHEMISTRY

5.3.1. Simple Equilibrium Systems

General chemistry demonstrates the properties of simple equilibrium systems such as the hydrolysis of weak monoprotic acids, the formation of soluble ion pairs, or the solubility of systems containing a single ionic compound. Somewhat more complicated equilibrium systems—requiring either simplifying assumptions or complex algebraic manipulations—include the hydrolysis of weak diprotic acids, the effect of ion-pair formation on the solubility of a single ionic compound, or the simultaneous solubility of two ionic compounds with a single common ion. None of these simple systems approach the complicated aqueous solutions commonly encountered by environmental chemists.

In this section we briefly consider simple equilibrium systems of the type you would have encountered in general chemistry, pausing to consider the simplifying assumptions used in general chemistry to solve what soon become very complicated problems. This review serves two purposes: it reminds us of familiar methods for solving equilibrium problems, and it will demonstrate the need for activity corrections and computer models to solve realistic water chemistry problems.

Hydrolysis of a Weak Monoprotic Acid

General chemistry usually examines two problems involving the hydrolysis of weak monoprotic acids (and bases): calculating the hydrolysis constant for a solution containing a weak acid (or base) when the pH is known or calculating the pH of a solution where the hydrolysis constant is known.

An important problem-solving method employs an *equilibrium table* listing the initial concentration of the weak acid (or base) and a variable x representing the moles of weak acid or base converted into conjugate base (or conjugate acid). The equilibrium table assigns concentration values to each component in the equilibrium expression, creating an algebraic expression that is either solved directly or is simplified before attempting a solution.

One thing you would not have noticed from the examples appearing in general chemistry is a tendency to use weak acid concentrations that are far more concentrated than those found in nature. A concentrated solution of a weak acid results in equilibrium conditions where simplifying assumptions are appropriate. Example 5.1 illustrates how an equilibrium table is used to solve a problem under conditions where the solution concentration is quite high, relative to the hydrolysis constant K_a. Example 5.2 involves the same weak acid but at a much lower concentration to illustrate the difficulty of using this method when the hydrolysis constant K_a is about the same magnitude as or larger than the solute concentration.

Example 5.1 Calculate the equilibrium hydrolysis constant K_a for a weak acid solution based on the measured pH.

Propionic acid is one of several weak monoprotic acids produced by soil bacteria as they ferment organic matter in the absence of oxygen. A 0.10 M aqueous solution of propionic acid has a pH of 2.94. What is the value of the K_a for propionic acid?

The equilibrium table used to assign values to each component in the equilibrium expression follows.

Stage	propionic acid(aq)	$\leftrightarrow$	propionate$^-$(aq)	+	H^+(aq)
Initial	0.10		0		0
Reaction	$-x$		$+x$		$+x$
Equilibrium	$0.10 - x$		x		x

The equilibrium expression using concentrations from the equilibrium table follows.

$$K_a = \frac{C_{A^-} \cdot C_{H^+}}{C_{HA}} = \frac{x \cdot x}{(0.10 - x)}$$

We know x from the pH: $x = C_{H^+} = 10^{-pH} = 10^{-2.94} = 1.15 \cdot 10^{-3}$. Since $x \ll 0.1$, then $(0.10 - x) \approx 0.10$.

$$K_a = \frac{10^{-2.94} \cdot 10^{-2.94}}{(0.10 - 1.15 \cdot 10^{-3})} \approx \frac{10^{-5.88}}{(0.10)}$$

$$K_a = 10^{-4.88} = 1.32 \cdot 10^{-5}$$

Example 5.2 Calculate the equilibrium hydrolysis constant K_a for a weak acid solution based on the measured pH.

Typical propionic acid concentrations in soil pore water rarely exceed 10^{-3} M. An aqueous solution containing 10^{-3} M propionic acid has a pH of 3.96. What is the value of the K_a for propionic acid?

An equilibrium table assigns values to each component in the equilibrium expression.

Stage	propionic acid(aq)	$\leftrightarrow$	propionate⁻(aq)	+	H⁺(aq)
Initial	10^{-3}		0		0
Reaction	$-x$		$+x$		$+x$
Equilibrium	$10^{-3} - x$		x		x

We know x from the pH: $x = C_{H^+} = 10^{-pH} = 10^{-3.96} \approx 1.0 \cdot 10^{-4}$. Since $x \approx 10^{-4}$, then $(10^{-3} - x) \approx 0.9 \; 10^{-3}$. The expression below results in an 85% error in the estimate of K_a, which equals $10^{-4.88}$.

$$K_a = \frac{10^{-4} \cdot 10^{-4}}{(10^{-3} - 10^{-4})} \approx 10^{-4.95}$$

A second simple equilibrium problem involves calculating solution pH when the hydrolysis constant is known. The equilibrium table method yields a quadratic equation when the solute is a weak monoprotic acid. A further simplifying assumption, based on the relative magnitude of x to the weak acid concentration, eliminates the need to solve the quadratic equation (Example 5.3).

Example 5.3 Calculate the pH of a weak acid solution when given the hydrolysis constant of the acid.

Acetic acid is a monoprotic acid found dissolved in soil pore water. Calculate the pH of a 20 mM aqueous solution of acetic acid ($K_a = 1.86 \cdot 10^{-5}$).

The solution uses an equilibrium table to assign values to each component in the equilibrium expression.

Stage	acetic acid(aq)	$\leftrightarrow$	acetate⁻(aq)	+	H⁺(aq)
Initial	$2.0 \cdot 10^{-2}$		0		0
Reaction	$-x$		$+x$		$+x$
Equilibrium	$2.0 \cdot 10^{-2} - x$		x		x

The equilibrium expression using concentrations from the preceding table follows.

$$K_a = \frac{C_{A^-} \cdot C_{H^+}}{C_{HA}}$$

$$K_a = \frac{x \cdot x}{(2.0 \cdot 10^{-2} - x)}$$

One simplifying assumption—which can be difficult to apply—begins with the notion that acetic acid undergoes minimal hydrolysis and concludes $x \ll 2.0 \cdot 10^{-2}$.

$$K_a \approx \frac{x^2}{2.0 \cdot 10^{-2}}$$

$$x \approx \sqrt{(2.0 \cdot 10^{-2}) \cdot K_a}$$

$$x \approx \sqrt{(2.0 \cdot 10^{-2}) \cdot (1.86 \cdot 10^{-5})} = \sqrt{3.72 \cdot 10^{-7}}$$

$$x = C_{H^+} \approx 6.10 \cdot 10^{-4}$$

$$pH \approx 3.21$$

The assumption made in Example 5.3—$x \ll 2.0 \cdot 10^{-2}$—is valid. The estimated value for x is almost two orders of magnitude smaller than the weak acid concentration. Some weak acids reach concentrations as high as 10 millimolar in the environment, but typical weak acid concentrations tend to be lower by one to four orders of magnitude. Concentrations encountered in the environment require solving the quadratic equation, as illustrated in Example 5.4.

Example 5.4 Calculate the pH of a weak acid solution when given the hydrolysis constant of the acid.

The typical acetic acid concentration in soil pore water rarely exceeds 150 μM. Calculate the pH of a 150 μM aqueous solution of acetic acid ($K_a = 1.86 \cdot 10^{-5}$).

The solution uses an equilibrium table to assign values to each component in the equilibrium expression.

Stage	acetic acid(aq)	$\leftrightarrow$	acetate$^-$(aq)	+	H^+(aq)
Initial	$1.50 \cdot 10^{-4}$		0		0
Reaction	$-x$		$+x$		$+x$
Equilibrium	$1.50 \cdot 10^{-4} - x$		x		x

The equilibrium expression using concentrations from the preceding table follows.

$$K_a = \frac{C_{A^-} \cdot C_{H^+}}{C_{HA}}$$

$$K_a = \frac{x \cdot x}{(1.50 \cdot 10^{-4} - x)}$$

In this example we solve the quadratic equation rather than using the simplifying assumption made in Example 5.3.

$$x_+ = \frac{-K_a + \sqrt{K_a^2 + 4 \cdot C_T \cdot K_a}}{2}$$

$$x_+ = \frac{-(1.86 \cdot 10^{-5}) + \sqrt{(1.86 \cdot 10^{-5})^2 + 4 \cdot (1.50 \cdot 10^{-4}) \cdot (1.86 \cdot 10^{-5})}}{2}$$

$$x_+ = 4.44 \cdot 10^{-5} = C_{H^+}$$
$$pH = 4.35$$

The simplifying assumption made in Example 5.3 would definitely not work in a more dilute acetic acid solution. Dilute solutions require application of the quadratic equation.

Aqueous Solubility of an Ionic Compound

The solubility of ionic solids is another simple chemical equilibrium process discussed in general chemistry. Solubility is a key water chemistry process involving both the dissolution and precipitation of sparingly soluble minerals during the chemical weathering stage of the rock cycle and soil formation. The chemistry of many inorganic pollutants, as it pertains to both biological availability to living organisms and remediation treatment, involves solubility reactions.

Consider the solubility reaction of the mineral gypsum $CaSO_4 \cdot 2H_2O(s)$ and the equilibrium solubility expression. The conditional solubility coefficient K_{s0}^c is the product of ion concentrations in solution, while the thermodynamic solubility constant K_{s0} is the product of ion activities in solution. The symbol K_{s0} indicates an equilibrium expression in which the solute ions are identical to the ions in the ionic solid:

$$CaSO_4 \cdot 2H_2O(s) \leftrightarrow Ca^{2+}(aq) + SO_4^{2-}(aq)$$
$$K_{s0}^c = C_{Ca^{2+}} \cdot C_{SO_4^{2-}}$$
$$K_{s0} = a_{Ca^{2+}} \cdot a_{SO_4^{2-}}$$

Example 5.5 demonstrates how solubility is determined when given the solubility constant K_{s0}.

Example 5.5 Calculate the solubility of a sparingly soluble ionic solid given the solubility constant K_{s0} of the solid.

Gypsum is a relatively soluble soil mineral found in arid and semiarid climates, where chemical weathering is slowed by the scarcity of rainfall. Calculate the solubility of gypsum in aqueous solution ($K_{s0} = 2.63 \cdot 10^{-5}$; $\log K_{s0} = -4.58$).

Use an equilibrium table to assign values to each component in the equilibrium expression. All of the Ca^{2+} and SO_4^{2-} dissolved in solution at equilibrium come from the mineral gypsum.

Stage	$CaSO_4 \cdot 2H_2O(s)$	$\leftrightarrow$	$Ca^{2+}(aq)$	$+$	$SO_4^{2-}(aq)$
Initial	1		0		0
Reaction	1		$+x$		$+x$
Equilibrium	1		x		x

The equilibrium expression using concentrations from the preceding table follows.

$$K_{s0} \approx C_{Ca^{2+}} \cdot C_{SO_4^{2-}}$$
$$K_{s0} \approx x \cdot x$$

Since the right side of the expression is a product of ion concentrations—rather than activities—it does not exactly equal the thermodynamic solubility constant.

$$x \approx \sqrt{K_{s0}} = \sqrt{2.63 \cdot 10^{-5}} = \left(10^{-4.58}\right)^{1/2}$$
$$x \approx 10^{-2.29} = 5.13 \cdot 10^{-3}$$
$$C_{Ca^{2+}} \approx 10^{-2.29} M$$
$$C_{SO_4^{2-}} \approx 10^{-2.29} M$$

Dividing $\log K_{s0}$ by 2 is equivalent to taking the square root. Aqueous solution is saturated by (i.e., in solubility equilibrium with) gypsum when $10^{-2.29}$ moles dissolve per liter.

Example 5.6 represents a more complicated solubility reaction because the dissolution of each mole of the aluminum hydroxide mineral gibbsite $Al(OH)_3(s)$ releases four moles of ions into solution: one mole of the ion Al^{3+} and three moles of the hydroxide ion OH^-. As in the solubility reaction of the mineral gypsum, the symbol K_{s0} indicates an equilibrium expression in which the solute ions are identical to the ions in the ionic solid:

$$Al(OH)_3(s) \leftrightarrow Al^{3+}(aq) + 3 \cdot OH^-(aq)$$
$$K_{s0}^c = C_{Al^{3+}} \cdot C_{OH^-}^3$$
$$K_{s0} = a_{Al^{3+}} \cdot a_{OH^-}^3$$

Example 5.6 Calculate the solubility of a sparingly soluble ionic solid given the solubility constant K_{s0} of the solid.

Gibbsite is a soil mineral found in humid climates, where chemical weathering is quite advanced. Calculate the solubility of gibbsite in aqueous solution ($K_{s0} = 1.29 \cdot 10^{-34}$; $\log K_{s0} = -33.89$).

Use an equilibrium table to assign values to each component in the equilibrium expression. We will assume no further reactions occur; all of the Al^{3+} and OH^- dissolved in solution at equilibrium come from the mineral gibbsite.

Stage	$Al(OH)_3(s)$	$\leftrightarrow$	$Al^{3+}(aq)$	+	$3 \cdot OH^-(aq)$
Initial	1		0		0
Reaction	1		$+x$		$+3 \cdot x$
Equilibrium	1		x		$3 \cdot x$

Three moles of OH^- dissolve for each mole of Al^{3+}, and the OH^- concentration at equilibrium is $3 \cdot x$. The exponent 3 results from the way the stoichiometric coefficient appears in the equilibrium expression (Eq. (5.4)).

$$K_{s0} \approx C_{Al^{3+}} \cdot C_{OH^-}^3$$
$$K_{s0} \approx (x) \cdot (3 \cdot x)^3 = 27 \cdot x^4$$

Since the right side of the expression is a product of ion concentrations—rather than activities—it does not exactly equal the thermodynamic solubility constant.

$$27 \cdot x^4 \approx K_{s0}$$

$$x \approx \left(\frac{K_{s0}}{27}\right)^{1/4} = \left(\frac{1.29 \cdot 10^{-34}}{27}\right)^{1/4}$$

$$x \approx 10^{-8.83} = 1.48 \cdot 10^{-9}$$
$$C_{Al^{3+}} = x \approx 1.48 \cdot 10^{-9} M$$
$$C_{OH^-} = 3 \cdot x \approx 4.44 \cdot 10^{-9} M$$

Aqueous solution is saturated by (i.e., in solubility equilibrium with) gibbsite when $10^{-8.83}$ moles dissolve per liter.

The solubility of gibbsite $Al(OH)_3(s)$ is more complicated than represented in Example 5.6 because the aluminum ion Al^{3+} forms a complex with water molecules, and this complex cation, the hydrated aluminum ion $Al(H_2O)_n^{3+}$, is a polyprotic acid that undergoes several hydrolysis steps. We will return to the gibbsite solubility problem in a later section, taking into full account the hydrolysis of the hydrated aluminum ion $Al(H_2O)_n^{3+}$.

Henry's Law: Aqueous Solubility of Gases

A system common in water chemistry is liquid water in contact with an atmosphere containing several gases. Each gas in the atmosphere will dissolve to some degree in water. *Henry's Law* states that the amount of a gas dissolved in a liquid $X(aq)$ is proportional to the partial pressure of the gas P_X in the atmosphere in contact with the liquid. The proportionality constant is the Henry's Law constant k_H and is characteristic of both the gas and the liquid:

$$X(g) \leftrightarrow X(aq)$$
$$C_{X(aq)} \overset{def}{=} k_H \cdot P_X$$

Henry's Law constants for several gases commonly dissolved in water in the environment appear in Table 5.1. Some of these gases—dioxygen, dinitrogen, and methane—do not react with water other than hydrating. Others—nitrogen monoxide, nitrogen dioxide, sulfur dioxide, sulfur trioxide, and carbon dioxide, to name a few—react with water to form acids. Hydrogen sulfide dissolved in water undergoes hydrolysis without combining with water.

The Henry's Law constant k_H is a good predictor of solubility for nonreactive gases. The solubility of reactive gases depends on acid-base reactions occurring in solution in addition to the partial pressure of the gas in the atmosphere in contact with the solution. This becomes apparent in the following section.

TABLE 5.1 Henry's Law Constants

Gas	Henry's Law Constant k_H $\left[mol \cdot L^{-1} \cdot Pa^{-1}\right]$
O_2	$1.3 \cdot 10^{-8}$
N_2	$6.0 \cdot 10^{-9}$
CH_4	$1.3 \cdot 10^{-8}$
NO	$1.9 \cdot 10^{-8}$
NO_2	$1.2 \cdot 10^{-7}$
SO_2	$1.2 \cdot 10^{-5}$
CO_2	$3.6 \cdot 10^{-7}$
H_2S	$8.6 \cdot 10^{-7}$

Hydrolysis of a Weak Diprotic Base

The hydrolysis of a weak acid produces a conjugate base:

$$propionic\ acid(aq) + H_2O(l) \leftrightarrow \underset{conjugate\ base}{propionate^-(aq)} + H_3O^+(aq)$$

$$\underset{weak\ acid}{benzoic\ acid(aq)} + H_2O(l) \leftrightarrow \underset{conjugate\ base}{benzoate^-\ (aq)} + H_3O^+(aq)$$

The weak acid hydrolysis constant K_a is related to the hydrolysis constant of the conjugate base K_b (Eq. (5.13)) through the water ionization constant K_w (Eq. (5.14)):

$$K_a \cdot K_b = K_w \tag{5.13}$$

$$K_w = 1.0 \cdot 10^{-14} = a_{H_3O^+(aq)} \cdot a_{HO^-(aq)} \tag{5.14}$$

Consider the weak acid propionic acid with acid dissociation constant $K_a = 1.32 \cdot 10^{-5}$. The base association constant for the conjugate base propionate is $K_b = 7.58 \cdot 10^{-10}$:

$$\underset{weak\ acid}{propionic\ acid(aq)} + H_2O(l) \xleftarrow{\quad K_a \quad} \underset{conjugate\ base}{propionate^-(aq)} + \underset{acid}{H_3O^+(aq)}$$

$$\underset{conjugate\ base}{propionate^-(aq)} + H_2O(l) \xleftarrow{\quad K_b \quad} \underset{weak\ acid}{propionic\ acid(aq)} + HO^-(aq)$$

$$K_b = K_w \cdot K_a^{-1}$$
$$K_b = \left(1.0 \cdot 10^{-14}\right) \cdot \left(1.32 \cdot 10^{-5}\right)^{-1}$$
$$K_b = 7.58 \cdot 10^{-10}$$

Sulfide $S^{2-}(aq)$ and hydrogen sulfide $HS^-(aq)$ function as conjugate bases of the weak acid $H_2S(aq)$ in water chemistry. The hydrolysis reactions for these two bases follow:

$$\underset{conjugate\ base}{S^{2-}(aq)} + H_2O(l) \rightarrow \underset{weak\ acid}{HS^-(aq)} + \underset{base}{OH^-(aq)} \quad K_{b1} = 7.94 \cdot 10^{-1}$$

$$\underset{conjugate\ base}{HS^-(aq)} + H_2O(l) \rightarrow \underset{weak\ acid}{H_2S(aq)} + \underset{base}{OH^-(aq)} \quad K_{b2} = 1.05 \cdot 10^{-7}$$

The sulfide ion $S^{2-}(aq)$ is an important solute in water chemistry. The most common source in soil pore water, surface water, and many unconfined aquifers is the anaerobic respiration of bacteria. In Chapter 8 we examine anaerobic respiration, but for now it is enough to know that *sulfate-reducing bacteria* (*SRB*) respire in the absence of molecular oxygen by reducing sulfate to sulfide. Sulfide production by SRB, because it is linked to respiration, requires a carbon source. Substantial dissolved sulfide in groundwater aquifers that contain negligible dissolved organic or organic carbon coating mineral surfaces probably results from sulfide leached from organic rich sediments and soil.

Examples 5.7 and 5.8 examine a sulfide solution produced by an SRB of genus *Clostridium* (Cunningham and Lundie, 1993). Cunningham and Lundie were studying the precipitation of greenockite $CdS(s)$ by sulfide evolved during the anaerobic respiration of the *Clostridium* bacteria. Cadmium is toxic to bacteria, and the precipitation of greenockite substantially lowers the solubility and, thus, the biological availability of Cd to the organism. If sulfide production were linked to Cd detoxification, Cunningham and Lundie reasoned, then this *Clostridium* bacteria would overproduce sulfide, possibly through the overproduction of cysteine and its subsequent transformation to hydrogen sulfide by the enzyme cysteine desulfhydrase to lower Cd biological availability. *Clostridium thernoaceticum* produces 0.38 *mM* of sulfide in the absence of dissolved Cd:

$$L - cysteine(aq) + H_2O(l)$$

$$\xrightarrow{cysteine\ desulfhydrase} HS^-(aq) + NH_4^+(aq) + pyruvic\ acid(aq)$$

Example 5.7 Calculate the pH of a polyprotic base solution given hydrolysis constants for the base. What is the pH of a solution containing 0.38 *mM* of total dissolved sulfide?

First approximating assumption: The second base association constant K_{b2} is much smaller than the first base association constant K_{b1}, so the hydroxide ion $OH^-(aq)$ concentration results almost entirely from the first ionization step. Assume, as a first approximation, that only the first step occurs.

Use an equilibrium table to assign values to each component in the equilibrium expression.

Stage	$S^{2-}(aq)$	$\leftrightarrow$	$HS^-(aq)$	$+$	$OH^-(aq)$
Initial	$3.8 \cdot 10^{-4}$		0		0
Reaction	$-x$		$+x$		$+x$
Equilibrium	$3.8 \cdot 10^{-4} - x$		x		x

The equilibrium expression using concentrations from the preceding table follows.

$$K_{b1} = 7.94 \cdot 10^{-1} \approx \frac{C_{HS^-} \cdot C_{OH^-}}{C_{S^{2-}}}$$

$$7.94 \cdot 10^{-1} \approx \frac{x \cdot x}{(C_T - x)}$$

$$K_{b1} \cdot (C_T - x) = x^2$$

$$x^2 + K_{b1} \cdot x - K_{b1} \cdot C_T = 0$$

The solution is found by solving the quadratic equation.

$$x_+ = \frac{-K_{b1} + \sqrt{K_{b1}^2 + 4 \cdot C_T \cdot K_{b1}}}{2}$$

$$x_+ = \frac{-(7.94 \cdot 10^{-1}) + \sqrt{(7.94 \cdot 10^{-1})^2 + 4 \cdot (3.8 \cdot 10^{-4}) \cdot (7.94 \cdot 10^{-1})}}{2}$$

$$x_+ = 3.78 \cdot 10^{-4} = C_{OH^-}$$

$$pOH = 3.42$$

$$pH = 14 - pOH = 10.58$$

We estimated the pH of a 0.38 mM sulfide solution by assuming that the second hydrolysis step, the hydrolysis reaction that produces dihydrogen sulfide from the hydrolysis of hydrogen sulfide, is negligible relative to the first hydrolysis step. The result confirms the validity of the simplifying assumption subject to a 10% error. Suppose we wish to estimate the concentration of dihydrogen sulfide in a 0.38 mM sulfide solution. This estimate can be made using a second simplifying assumption, illustrated in Example 5.8.

Example 5.8 Calculate the dihydrogen sulfide concentration in sulfide solution given the hydrolysis constants of dihydrogen sulfide and hydrogen sulfide.

Suppose we estimate the concentration of dihydrogen sulfide $H_2S(aq)$, the product of the second hydrolysis step, from Example 5.7. To do this we will construct an equilibrium table for the hydrolysis of hydrogen sulfide

$HS^-(aq)$ to assign values to each component in the equilibrium expression as before.

Stage	$HS^-(aq)$	$\leftrightarrow$	$H_2S(aq)$	$+$	$OH^-(aq)$
Initial	$3.78 \cdot 10^{-4}$		0		$3.78 \cdot 10^{-4}$
Reaction	$-y$		$+y$		$+y$
Equilibrium	$3.78 \cdot 10^{-4} - y$		y		$3.78 \cdot 10^{-4} + y$

The equilibrium expression using concentrations from the preceding table follows.

$$K_{b2} \approx \frac{C_{OH^-} \cdot C_{H_2S}}{C_{HS^-}}$$

$$K_{b2} \approx \frac{(3.78 \cdot 10^{-4} + y) \cdot y}{(3.78 \cdot 10^{-4} - y)}$$

$$1.05 \cdot 10^{-7} \approx \frac{(3.78 \cdot 10^{-4}) \cdot y}{(3.78 \cdot 10^{-4})}$$

$$y \approx 1.05 \cdot 10^{-7}$$

Because the second hydrolysis constant K_{b2} is relatively small, the second step occurs to a much smaller extent than the first. This means that the amount of $H_2S(aq)$ and $OH^-(aq)$ produced in the second step (y) is much smaller than $3.78 \cdot 10^{-4}$ M. It is therefore reasonable that both C_{HS^-} and C_{OH^-} are very close to $3.78 \cdot 10^{-4}$ M.

Simultaneous Equilibrium: Aqueous Solubility of Two Ionic Solids

Simultaneous equilibrium of numerous minerals is common in water chemistry problems. General chemistry introduces the concept of simultaneous equilibrium involving two ionic solids to illustrate the behavior of these systems. The example we will use involves two insoluble phosphate minerals: apatite $Ca_5(PO_4)_3OH$ and pyromorphite $Pb_5(PO_4)_3OH$.

These two minerals are believed to be the most stable (i.e., the most insoluble) calcium and lead phosphate minerals in the environment. Other, more soluble minerals may precipitate and persist for a time, but eventually these *metastable* calcium and lead phosphate minerals convert to apatite and pyromorphite. The composition of either mineral can, and most probably will, vary in nature. The fluoride and chloride variants of both apatite—$Ca_5(PO_4)_3F$ and $Ca_5(PO_4)_3Cl$—and pyromorphite occur in nature. The rationale for fluoridating water and using a fluoride dental wash is to convert the surface layer of dental mineral from hydroxy-apatite into the more insoluble—and harder—fluoro-apatite.

A favored remediation technology for lead(II) contaminated soil involves adding soluble phosphate salts, with the intention of converting all of the lead (II) compounds into pyromorphite. By virtue of its insolubility, pyromorphite is less biologically available to organisms than the more soluble lead(II) minerals that precipitate in the absence of sufficient phosphate.

Example 5.9 Estimate the soluble Pb^{2+} concentration in soil pore water when there is simultaneous equilibrium between aqueous solution and the two phosphate minerals apatite $Ca_5(PO_4)_3OH(s)$ and pyromorphite $Pb_5(PO_4)_3OH(s)$.

The solubility equilibrium expression for the minerals apatite $Ca_5(PO_4)_3OH(s)$ and pyromorphite $Pb_5(PO_4)_3OH(s)$ follows. The thermodynamic solubility constants K_{s0} are the product of ion activities in solution, while the conditional solubility coefficients K_{s0}^c are the product of ion concentrations and, as such, are subject to the ionic strength of the solution. For this example, since the ionic strength is not specified, we will neglect activity coefficients, recognizing $K_{s0}^c \underset{APATITE}{} \approx 10^{-30.2}$ and $K_{s0}^c \underset{PYROMORPHITE}{} \approx 10^{-48.8}$ are both approximate.

$$Ca_5(PO_4)_3OH(s) \leftrightarrow 5 \cdot Ca^{2+}(aq) + 3 \cdot PO_4^{3-}(aq) + OH^-(aq)$$
$$\underset{APATITE}{}$$

$$K_{s0}^c \underset{APATITE}{} = C_{Ca^{2+}}^5 \cdot C_{PO_4^{3-}}^3 \cdot C_{OH^-} = \frac{10^{-30.2}}{\left(\gamma_{Ca^{2+}}^5 \cdot \gamma_{PO_4^{3-}}^3 \cdot \gamma_{OH^-}\right)} \approx 10^{-30.2}$$

$$Pb_5(PO_4)_3OH(s) \leftrightarrow 5 \cdot Pb^{2+}(aq) + 3 \cdot PO_4^{3-}(aq) + OH^-(aq)$$
$$\underset{PYROMORPHITE}{}$$

$$K_{s0}^c \underset{PYROMORPHITE}{} = C_{Pb^{2+}}^5 \cdot C_{PO_4^{3-}}^3 \cdot C_{OH^-} = \frac{10^{-48.8}}{\left(\gamma_{Pb^{2+}}^5 \cdot \gamma_{PO_4^{3-}}^3 \cdot \gamma_{OH^-}\right)} \approx 10^{-48.8}$$

Combining the two solubility expressions, after inverting the apatite expression, yields the net conditional coefficient K_{net}^c for the simultaneous equilibrium. The greater insolubility of pyromorphite relative to apatite is revealed in the concentration ratio of the two ions.

$$Pb_5(PO_4)_3OH(s) + 5 \cdot Ca^{2+}(aq) \leftrightarrow Ca_5(PO_4)_3OH(s) + 5 \cdot Pb^{2+}(aq)$$
$$\underset{PYROMORPHITE}{} \qquad\qquad\qquad\qquad \underset{APATITE}{}$$

$$\frac{K_{s0}^c \underset{PYROMORPHITE}{}}{K_{s0}^c \underset{APATITE}{}} = \frac{C_{Pb^{2+}}^5 \cdot C_{PO_4^{3-}}^3 \cdot C_{OH^-}}{C_{Ca^{2+}}^5 \cdot C_{PO_4^{3-}}^3 \cdot C_{OH^-}} \approx \frac{10^{-48.8}}{10^{-30.2}}$$

$$K_{net}^c \approx 10^{-18.6} = \frac{C_{Pb^{2+}}^5}{C_{Ca^{2+}}^5}$$

$$\frac{C_{Pb^{2+}}}{C_{Ca^{2+}}} \approx 10^{-3.72}$$

A typical concentration of $Ca^{2+}(aq)$ in natural water is about $10^{-3.3}$ M. The preceding concentration ratio suggests a probable $Pb^{2+}(aq)$ concentration (not to be confused with a total soluble lead(II) concentration) when both apatite and pyromorphite are present of about $10^{-7} M$.

Simultaneous Equilibrium: The Effect of pH on the Solubility of a Mineral

Acidity is such an important issue in environmental chemistry that we devote an entire chapter to the topic. Since many processes ultimately determine water pH in the environment, a simple cause and effect are usually impossible to discern. For this reason, and because pH is so easy to measure, water pH is usually treated as an independent variable in water chemistry problems—a known quantity that controls the solubility of most minerals in the system.

The treatment of simultaneous equilibrium in general chemistry usually includes several examples of how pH influences the solubility of insoluble ionic solids. Example 5.10 examines the effect of pH on the solubility of the ferrous sulfide mineral troilite $FeS(s)$.

Example 5.10 Determine the effect of pH on the solubility of troilite $FeS(s)$. The pH effect involves the simultaneous equilibrium between sulfide ion hydrolysis in aqueous solution and solubility of the mineral troilite $FeS(s)$.

The solubility equilibrium expression for the mineral troilite $FeS(s)$ appears following. The thermodynamic solubility constant K_{s0} is the product of ion activities in solution, while the conditional solubility coefficient K_{s0}^c is the product of ion concentrations and, as such, is subject to the ionic strength of the solution. For this example we will neglect activity coefficients, recognizing that $K_{s0}^c \approx 10^{-19.15}$ is approximate.

$$\underset{\text{TROILITE}}{FeS(s)} \leftrightarrow Fe^{2+}(aq) + S^{2-}(aq)$$

$$\underset{\text{TROILITE}}{K_{s0}^c} = C_{Fe^{2+}} \cdot C_{S^{2-}} = \frac{10^{-19.15}}{(\gamma_{Fe^{2+}} \cdot \gamma_{S^{2-}})} \approx 10^{-19.15}$$

The sulfide ion $S^{2-}(aq)$ is a weak conjugate base that reacts with water to form the hydrogen sulfide ion $HS^-(aq)$.

$$HS^-(aq) \leftrightarrow S^{2-}(aq) + H^+(aq)$$

$$K_{a2}^c = \frac{C_{S^{2-}} \cdot C_{H^+}}{C_{HS^-}} \approx 10^{-13.90}$$

Multiplying the troilite $FeS(s)$ solubility expression by the inverse of the hydrogen sulfide hydrolysis expression yields a pH-dependent troilite $FeS(s)$ solubility expression.

$$\underset{\text{TROILITE}}{FeS(s)} + H^+(aq) \leftrightarrow Fe^{2+}(aq) + HS^-(aq)$$

$$\left(\frac{\underset{\text{TROILITE}}{K_{s0}^c}}{K_{a2}^c}\right) = (C_{Fe^{2+}} \cdot C_{S^{2-}}) \cdot \left(\frac{C_{HS^-}}{C_{S^{2-}} \cdot C_{H^+}}\right) \approx \frac{10^{-19.15}}{10^{-13.90}}$$

$$K_{net}^c = \frac{C_{Fe^{2+}} \cdot C_{HS^-}}{C_{H^+}} \approx 10^{-5.25}$$

To understand the effect of pH on troilite $FeS(s)$ solubility, we apply *Le Chatelier's Principle* to the preceding pH-dependent solubility expression. Increasing pH (i.e., lowering the $H^+(aq)$ concentration) causes the equilibrium to shift to the left, decreasing troilite $FeS(s)$ solubility.

Simultaneous Equilibrium: The Effect of Ion Complexes on the Solubility of a Mineral

Bacteria, fungi, and plants secrete a host of organic acids, some of which are by-products of respiration and other physiological processes, but some are excreted specifically to increase the solubility of essential nutrients. Lupine (*Lupinus* spp.) is a legume genus known for its ability to thrive in soils with low soluble phosphate, an essential nutrient. Lupines will excrete prodigious amounts of citrate in phosphate-limiting conditions. Example 5.11 examines the solubility of calcite $CaCO_3(s)$ in a solution containing 1 mM of citric acid.

Example 5.11 What is the solubility of calcite $CaCO_3(s)$ in a 1 mM citric acid solution?

The solubility equilibrium expression for the mineral calcite appears following.

$$\underset{CALCITE}{CaCO_3(s)} + H^+(aq) \leftrightarrow Ca^{2+}(aq) + HCO_3^-(aq) \qquad \underset{CALCITE}{K_s^c} \approx 10^{+1.85}$$

$$\underset{CALCITE}{K_s^c} = \frac{C_{Ca^{2+}} \cdot C_{HCO_3^-}}{C_{H^+}}$$

Notice that the $Ca(Citrate)^-(aq)$ formation constant K_f (below) is rather large, indicating that equilibrium favors the product $Ca(Citrate)^-(aq)$ over the educts.

$$Citrate^{3-}(aq) + Ca^{2+}(aq) \leftrightarrow Ca(Citrate)^-(aq) \quad K_f = 10^{+4.70}$$

Multiplying the calcite solubility expression by the $Ca(Citrate)^-(aq)$ formation expression yields a solubility expression showing the effect of complex formation.

$$\underset{CALCITE}{CaCO_3(s)} + Citrate^{3-}(aq) + H^+(aq) \leftrightarrow Ca(Citrate)^-(aq) + HCO_3^-(aq) + H_2O(l)$$

$$\underset{CALCITE}{K_s^c} \cdot K_f^c = \left(\frac{C_{Ca^{2+}} \cdot C_{HCO_3^-}}{C_{H^+}}\right) \cdot \left(\frac{C_{Ca(Citrate)^-}}{C_{Ca^{2+}} \cdot C_{Citrate^{3-}}}\right) \approx \left(10^{+1.85}\right) \cdot \left(10^{+4.70}\right)$$

$$K_{net}^c = \frac{C_{Ca(Citrate)^-} \cdot C_{HCO_3^-}}{C_{Citrate^{3-}} \cdot C_{H^+}} \approx 10^{+6.55}$$

In calcareous water, the pH is ≈ 10, and the $HCO_3^-(aq)$ concentration is $\approx 10^{-4}$ M. The ratio of $Ca(Citrate)^-(aq)$ to $Citrate^{3-}(aq)$ under these conditions is $10^{+0.6} \approx 4$. The citrate mass balance is solved following, assuming all of the dissolved citrate is $Citrate^{3-}(aq)$ at pH = 10.

$$C_{Citrate} = 10^{-3} = C_{Ca(Citrate)^-} + C_{Citrate^{3-}}$$
$$C_{Ca(Citrate)^-} \approx \left(10^{+0.6}\right) \cdot C_{Citrate^{3-}} \approx 4 \cdot C_{Citrate^{3-}}$$
$$10^{-3} = 4 \cdot C_{Citrate^{3-}} + C_{Citrate^{3-}}$$
$$C_{Citrate^{3-}} \approx 0.2 \cdot 10^{-3} \, M = 2.0 \cdot 10^{-4} \, M$$
$$C_{Ca(Citrate)^-} \approx 8.0 \cdot 10^{-4} \, M$$

Calcite solubility in the presence of 1 mM of citric acid is $\approx 8.0 \cdot 10^{-4}$ M compared to 10^{-4} M in its absence.

Another example of the effect that complexes have on mineral solubility appears in Appendix 5I. This example describes a very important chemical reaction in which a siderophore (a bioorganic compound) forms a complex with ferric iron to increase its solubility. Bacteria, fungi, and grass plants excrete siderophores to meet their iron nutritional requirements.

5.3.2. Water Chemistry Simulations

The Importance of Validating Water Chemistry Simulations

Examples 5.1 through 5.11 serve several purposes. First, they review the equilibrium principles and problem-solving methods from general chemistry. Second, several of the examples illustrate limitations in applying simplifying assumptions to aid problem solving. As a general rule, the simplifying assumptions introduced in general chemistry tend to fail when solute concentrations become more dilute, which is typical for solutes in most practical water chemistry problems. Third, these examples demonstrate the challenges a water chemist faces when simple equilibrium systems become more complex—involving numerous minerals, soluble complexes, and hydrolysis reactions where two or more protons or hydroxyl ions dissociate from weak acids and bases. Complicated water chemistry systems demand an entirely different numerical strategy to solve the numerous reactions in simultaneous equilibrium.

Environmental chemists can choose from several applications designed to simulate chemical equilibrium in water. They differ in the computational options, the output options, and the reactions and equilibrium constants included in their basic database; the fundamental numerical method used to simulate the equilibrium reactions, however, is essentially the same. The determining issue, besides ease of use, is whether a version of the model is available for the computer operating system that you prefer or at your disposal. This book uses ChemEQL (Müller, 2005) in all of the examples and problems. ChemEQL offers an opportunity to learn the basics of water chemistry modeling—setting up a problem for simulation and result validation (Table 5.2)—and illustrates important water chemistry principles.

Computer-based numerical models designed to solve practical water chemistry problems relieve you of an enormous computational burden, but does not mean you can check your knowledge of equilibrium chemistry at

> **TABLE 5.2** Validation Checklist for Computer-Based Water Chemistry Models
>
> 1. Validate charge balance.
> 2. Validate mass balance for each constrained component.
> 3. Validate the ionic strength estimate.
> 4. Validate ion activity coefficients.
> 5. Validate the ion activity products for important reactions match thermodynamic equilibrium constants listed in the model database.
> 6. Validate the thermodynamic equilibrium constants used in the model database, especially the most critical reactions in the simulation, against published values.

the door. Computer-based numerical models do not understand chemistry. Your task, besides generating the input data necessary to solve the problem and interpreting the results, is to use your chemistry knowledge to assess the validity of each simulation.

Validity assessment requires an understanding of equilibrium chemistry and a handful of simple calculations that reveal whether the numerical results are consistent with basic equilibrium principles. We will refer to the validation checklist in Table 5.2 repeatedly throughout the remainder of this chapter.

Modeling the Hydrolysis of Weak Monoprotic Acids or Bases

The first modeling exercise—Example 5.12—revisits the simple equilibrium problem discussed in Example 5.3, calculating solution pH given the hydrolysis constant and weak-acid concentration. The ChemEQL website offers an electronic user's guide, *ChemEQL Manual*, and you are encouraged to refer to it for detailed instructions. Example 5.12 is very similar to the first example on page 8 (*ChemEQL Manual*, version 3.0).

Example 5.12 Calculate the pH of a 20 *mM* acetic acid solution when given the hydrolysis constant $K_a = 1.86 \cdot 10^{-5}$.

The ChemEQL simulation has two components: 2.0E-2 M Acetic Acid (total) and 1.0E-7 M H+ (total). Solution pH is a dependent variable in this simulation; therefore, any input value is acceptable for component H+ as long as the **Total** radio button is selected. The results were copied from the data file generated by ChemEQL. Select **Activity** from the **Options** menu, and then select the Davies equation (p. 21, *ChemEQL Manual*, version 3.0), A = 0.5109 at 25°C.

Species	Log K	Conc. [mol/L]	Activity
Acetic Acid	0.00	1.94E-02	1.94E-02
Acetate-	−4.73	6.09E-04	5.92E-04
OH−	−14.00	1.69E-11	1.64E-11
H+	0.00	6.09E-04	6.09E-04

Validate by opening the data file using the spreadsheet application Excel (Appendix 5A).

Charge Balance and Ionic Strength. The following table lists the charge balance (1.00E-07 mol_c L^{-1}) and ionic strength I (6.09E-04 mol L^{-1}) calculated from the concentrations listed above. The charge balance is validated because it is much smaller than the most concentrated ion concentrations. The ionic strength I reported by ChemEQL is also validated because it matches the ionic strength calculated from the concentrations listed in the results table.

Species	Conc. [mol/L]	Charge	Conc. [mol_c/L]	Ionic Strength. [mol/L]
Acetic Acid	1.94E-02	0	0.00E+00	0.00E+00
Acetate-	6.09E-04	−1	−6.09E-04	3.05E-04
OH-	1.69E-11	−1	−1.69E-11	8.44E-12
H+	6.09E-04	1	6.09E-04	3.05E-04
		Sum	1.00E-07	6.09E-04

Mass Balances. The acetic acid/acetate mass balance is validated because the concentration, calculated from the preceding table, matches the input value 2.0E-2 M Acetic Acid (total).

Database Equilibrium Constants. The logarithm of thermodynamic equilibrium constant in the ChemEQL database is −4.73. This compares with −4.74 listed in a popular general chemistry book. The logarithm of the reaction quotient $\log(Q)$ for acetic acid hydrolysis, computed using activities listed in the preceding table, is LOG(1.862E-05) = −4.73, validating agreement between the database constant and solute activities generated by the simulation.

Most computer-based water chemistry models generate a file containing the results of each simulation. Transferring data from the results file to a spreadsheet application will greatly simplify the validation process. Appendix 5A addresses file format issues specific to ChemEQL. A detailed discussion of each assessment appears in Appendix 5B.

Since Example 5.12 uses the same weak acid and acid concentration as Example 5.3, we can evaluate the effect of simplifying assumptions and activity coefficients. Example 5.3 estimates that a 20 mM acetic acid solution will have $C_{H^+} \approx 6.10 \cdot 10^{-4}$ M ($pH \approx 3.21$) at equilibrium. The ChemEQL simulation in Example 5.12 yields identical results: $a_{H^+} = 6.09 \cdot 10^{-4}$ M ($pH \approx 3.21$). Under these conditions the simplifying assumptions had little effect on the accuracy of the results, and ignoring the activity coefficient corrections had a negligible effect on the estimated pH. Dividing the activities by the concentrations in the first table in Example 5.12 reveals that the Acetate-ion activity coefficient is 0.972.

To reinforce the importance of validation assessment, Example 5.12 applies the entire validation checklist (see Table 5.2). Complete validation assessments for examples 5.13 and 5.16 appear in Appendix 5C.

Modeling Mineral Solubility

Example 5.13 revisits the equilibrium problem discussed in Example 5.5, calculating the solubility of the mineral gypsum.

Example 5.13 Calculate the solubility of the mineral gypsum given the mineral's solubility constant $K_{s0} = 2.63 \cdot 10^{-5}$; $\log K_{s0} = -4.58$.

The ChemEQL simulation has three components: 0 M Ca++ (total), 0 M SO4−− (total), and 1.0E-7 M H+ (total). After compiling and saving the input matrix, select **Insert Solid Phase...** from the **Matrix** menu, following the directions in the *ChemEQL Manual*. You can replace either Ca++ or SO4−− with CaSO4:2H2O (gypsum).

Unlike Example 5.5, the ChemEQL matrix lists several ion species neglected in the typical general chemistry treatment of gypsum solubility: CaOH+, CaSO4(aq), and HSO4−.

Select **Activity** from the **Options** menu, and then select the Davies equation (p. 21, *ChemEQL Manual*, version 3.0), A = 0.5109 at 25°C. Solubility and solution pH are both dependent variables in this simulation; therefore, any input value is acceptable for component H+ as long as the **Total** radio button is selected. The results were copied from the data file generated by ChemEQL.

Species	Conc. [mol/L]	Activity	Activity Coefficient
Ca++	1.11E-02	5.12E-03	0.461
CaOH+	1.14E-08	9.41E-09	0.824
CaSO4(aq)	5.37E-03	5.37E-03	1.000
SO4−−	1.11E-02	5.12E-03	0.461
HSO4−	5.50E-08	4.53E-08	0.824
OH−	1.34E-07	1.11E-07	0.824
H+	9.05E-08	9.05E-08	1.000

Solubility. ChemEQL estimates a solubility of 1.648E-02 M ($10^{-1.78}$ M) compared to results from Example 5.5 when $10^{-2.29}$ moles dissolve per liter. See Appendix 5C for a validation assessment.

We can anticipate a significant deviation between concentration and activity, especially because the two dominant ions are both divalent and the concentration is relatively high (see Figure 5.2). Example 5.13 is very similar to the example of calcite $CaCO_3(s)$ solubility that appears on page 9 (*ChemEQL Manual*, version 3.0).

Example 5.5 underestimates gypsum solubility by 69%. This error has several sources: neglecting formation of the specie CaSO4(aq) represents 33% of the error, while the remainder arises from the lower activity of the two major cations Ca++ and SO4−−. Monovalent ions require activity coefficients of 0.824, while divalent ions require activity coefficients of 0.461.

ChemEQL Treatment of Hydrogen Ion Activity

The results in Examples 5.12 and 5.13 reveal an important attribute of Chem-EQL: the model treats the $H^+(aq)$ concentration C_{H^+} as an activity a_{H^+}: $C_{H^+} \stackrel{def}{=} a_{H^+}$. This is not as arbitrary as it might appear; $H^+(aq)$ concentration is generally measured using a pH glass electrode, which is a type of ion-selective electrode. It is possible to directly measure the activity of certain ions (cyanide, nitrate, nitrite, most halides, most alkaline cations, Ag^+, NH_4^+, and a few divalent cations) using ion-selective electrodes; otherwise, analytical results from all methods other than ion-specific electrodes should be considered concentrations.

Henry's Law: Aqueous Solubility of Gases

Dissolved gases have an important impact on mineral solubility in most water chemistry systems. Carbonate chemistry, a major topic in aqueous geochemistry and many environmental problems, is deferred to Chapter 7.

Example 5.14 Calculate the pH of a solution containing 0.38 mM total sulfide.

Cunningham and Lundie (1993) found that *C. thernoaceticum* produced a 0.38 mM sulfide solution under normal culture conditions.

A ChemEQL simulation uses two components in the input matrix: the total S−− concentration is 3.80E-04 M, while the H+ concentration is allowed to vary, select Total radio button. The results in the following table are copied from the output data file. The activity coefficients were computed using the Davies equation (Eq. (5.12)).

Species	Conc. [mol/L]	Activity	Activity Coefficient
H2S(g)	1.02E-06	1.02E-06	
H2S(aq)	1.04E-07	1.04E-07	1.000
HS−	3.79E-04	3.70E-04	0.978
S−−	1.90E-07	1.74E-07	0.915
OH−	3.81E-04	3.72E-04	0.978
H+	2.69E-11	2.69E-11	1.000

The approximate solution in Example 5.7 made estimates quite similar to this simulation. The greatest difference is found in the predicted S−− concentration: 1.05E-07 M and 1.90E-07 M, respectively. The simulation also estimates the partial pressure of H2S(g): 1.02E-06 atm.

Bacteria and other organisms produce a variety of gases during respiration. As a consequence, the partial pressure of these gases in soil pores is often orders of magnitude higher than in the aboveground atmosphere. Bacteria produce dihydrogen sulfide by the enzyme cysteine desulfhydrase. Some microbiologists believe the production of dihydrogen sulfide is essential for the optimal functioning of the respiratory system located in the cell envelope.

The simulation of equilibrium involving the mineral cinnabar $HgS(s)$ and dihydrogen sulfide gas with soluble mercury and sulfides appears in Appendix 5D. The results in Appendix 5D show the effect that dissolved dihydrogen sulfide from SRB has on mercury solubility under anaerobic conditions.

Modeling Simultaneous Equilibrium: Aqueous Solubility of Two Ionic Solids

Example 5.15 revisits the same problem discussed in Example 5.9: simultaneous equilibrium between aqueous solution and two minerals with a common ion in the mineral solubility expressions. The common ion in both Examples 5.9 and 5.15 is $PO_4^{3-}(aq)$.

Example 5.15 Estimate the soluble Pb^{2+} concentration in soil pore water when simultaneous equilibrium exists with the two phosphate minerals apatite and pyromorphite.

The results of the ChemEQL simulation appear in the table following. Activity coefficients were computed using the Davies equation (Eq. (5.12)) based on an estimated ionic strength I = 2.441E-05 M.

Species	Conc. [mol/l]	Activity	Activity Coefficient
Ca++	6.81E-06	6.65E-06	0.977
CaOH+	6.14E-10	6.10E-10	0.994
CaPO4−	1.85E-08	1.84E-08	0.994
CaHPO4(aq)	1.42E-08	1.42E-08	1.000
CaH2PO4+	1.87E-11	1.86E-11	0.994
Pb++	1.30E-09	1.27E-09	0.977
PbOH+	1.41E-08	1.40E-08	0.994
Pb(OH)2	3.08E-09	3.08E-09	1.000
Pb(OH)3−	1.71E-11	1.70E-11	0.994
Pb2OH+++	3.72E-16	3.53E-16	0.949
Pb3(OH)4++	2.45E-16	2.39E-16	0.977
Pb4(OH)4++++	3.32E-22	3.03E-22	0.912
Pb6(OH)8++++	9.93E-28	9.05E-28	0.912
PbHPO4	6.19E-12	6.19E-12	1.000
PbH2PO4+	4.48E-15	4.46E-15	0.994
PO4−−−	1.01E-09	9.57E-10	0.950
HPO4−−	3.97E-06	3.88E-06	0.977
H2PO4−	1.12E-07	1.11E-07	0.994
H3PO4	2.84E-14	2.84E-14	1.000
OH−	5.56E-06	5.53E-06	0.994
H+	1.81E-09	1.81E-09	1.000

The total calcium concentration in equilibrium with apatite is $10^{-5.2}$ M, and the total lead concentration in equilibrium with pyromorphite is $10^{-7.7}$ M.

The way ChemEQL handles this process is discussed in the *ChemEQL Manual* under the topic "Several Solid Phases" (p. 13, ChemEQL Manual version 3.0).

After compiling the input matrix, the user replaces solution components by selecting **Insert Solid Phase...** from the **Matrix** menu. Components are replaced in a specific order; each replacement removes a component and the ChemEQL database; equilibrium expressions of subsequent replacements must match the remaining components. The ChemEQL simulation in Example 5.15 has four components: 0 M Ca++ (total), 0 M Pb++ (total), 0 M PO4−−(total), and 1.0E-7 M H+ (total). After compiling the input matrix, select **Insert Solid Phase...** from the **Matrix** menu, replacing components Ca++ and PO4 − − − with Ca10(OH)2(PO4)6 (hydroxyapatite) and Pb5(PO4)3OH (hydroxypyromorphite). Replacement in that order eliminates Ca++ as a component first and PO4−−− as a component second, leaving Pb++ and H+ as the two remaining variables.

The solubility equilibrium expressions used in Example 5.15 are from the ChemEQL database:

$$10 \cdot Ca^{2+}(aq) + 6 \cdot PO_4^{3-}(aq) \leftrightarrow 2 \cdot H^+(aq) + Ca_{10}(PO_4)_6OH_2(s) \quad K$$
$$= 10^{+88.40}$$

$$5 \cdot Pb^{2+}(aq) + 3 \cdot PO_4^{3-}(aq) \leftrightarrow H^+(aq) + Pb_5(PO_4)_3OH(s) \quad K = 10^{+62.40}$$

Dividing the apatite stoichiometric coefficients by 2 (and taking the square root of the solubility constant) would make the two expressions easier to compare but has no effect on the simulation results.

Adding another component CO3−− generates a system more likely to occur in the environment because calcite $CaCO_3(s)$ is probably present wherever apatite is found (see Appendix 5E). Constraining phosphate solubility with apatite and carbonate solubility with calcite yields the following solution composition: Ca (total) = 1.22E-04 M, Pb (total) = 1.34E-05 M, and PO4−(total) = 2.70E-09 M, and pH = 9.86. The inclusion of calcite in the equilibrium system results in a 100-fold increase in lead solubility accompanied by a 100-fold decrease in phosphate solubility—an example of the common-ion effect in nature.

Appendix 5F illustrates another example of simultaneous equilibrium involving aqueous solution and minerals sharing a common component. The simultaneous equilibrium in Appendix 5F is representative of phosphate solubility in acidic soils, where the aluminum phosphate mineral variscite $AlPO_4 \cdot 2H_2O(s)$ controls phosphate solubility, and aluminum solubility must satisfy the coexistence of both variscite and gibbsite $Al(OH)_3(s)$.

Modeling Simultaneous Equilibrium: the Effect of pH on the Solubility of a Mineral

Example 5.6 calculates the solubility of the soil mineral gibbsite using a familiar general chemistry method: the equilibrium table. The example did not explicitly examine the pH effect, but the equilibrium solubility expression implies pH-dependent solubility:

$$Al(OH)_3(s) \leftrightarrow Al^{3+}(aq) + 3 \cdot OH^-(aq)$$
$$K_{s0} = 1.29 \cdot 10^{-34} = a_{Al^{3+}} \cdot a_{OH^-}^3$$

$$Al(OH)_3(s) + 3 \cdot H^+(aq) = Al^{3+}(aq) + 3 \cdot H_2O(l)$$
$$K_{net} = K_{s0} \cdot K_w^{-3} = 10^{+8.11} = \frac{a_{Al^{3+}}}{a_{H^+}^3}$$

Taking the logarithm of the equilibrium solubility expression containing a_{H^+} yields a linear expression between the logarithm of $Al^{3+}(aq)$ activity and solution pH.

$$\log(K_{net}) = 8.11 = \log(a_{Al^{3+}}) - 3 \cdot \log(a_{H^+})$$
$$8.11 = \log(a_{Al^{3+}}) + 3 \cdot pH$$
$$\log(a_{Al^{3+}}) = 3 \cdot pH - 8.11$$

Example 5.16 is a ChemEQL simulation illustrating the equilibrium principles discussed earlier in Examples 5.6 (mineral solubility), 5.10 (pH-dependent mineral solubility), and 5.11 (the effect of ion complexes on mineral solubility).

The solubility of gibbsite $Al(OH)_3(s)$ is more complicated than represented in Example 5.6 because the aluminum ion Al^{3+} forms a complex with water molecules, and this complex cation, the *hydrated* aluminum ion $Al(H_2O)_n^{3+}$, is a polyprotic acid that undergoes several hydrolysis steps. We will return to the gibbsite solubility problem in Chapter 7, taking into full account the hydrolysis of the hydrated aluminum ion $Al(H_2O)_n^{3+}$. The hydrolysis species should be considered ion complexes (see Example 5.11) when applying the equilibrium gibbsite solubility expression.

Example 5.16 Calculate the solubility of a sparingly soluble ionic solid given the solubility constant K_{s0} of the solid.

Gibbsite is an insoluble soil mineral found in humid climates, where chemical weathering is quite advanced. Calculate the solubility of gibbsite in aqueous solution ($K_{s0} = 1.29 \cdot 10^{-34}$; $\log K_{s0} = -33.89$).
The simulation results appear following.

Species	Conc. [mol/L]	Activity	Activity Coefficient
Al+++	6.52E-13	6.50E-13	0.996
AlOH++	3.80E-11	3.79E-11	0.998
Al(OH)2+	1.76E-09	1.76E-09	0.999
Al(OH)3(aq)	1.62E-09	1.62E-09	1.000
Al(OH)4-	1.50E-08	1.50E-08	1.000
Al2(OH)2++++	2.88E-19	2.86E-19	0.993
Al3(OH)4+++++	4.03E-24	3.99E-24	0.990
OH-	5.83E-08	5.83E-08	1.000
H+	1.72E-07	1.72E-07	1.000

Aqueous solution is saturated by (i.e., in solubility equilibrium with) gibbsite when $10^{-7.73}$ moles dissolve per liter. Validation assessment appears in Appendix 5C.

ChemEQL allows the user to vary a single component as an independent variable, which is discussed in the *ChemEQL Manual* under the topics "Array of Concentrations" and "Graphic Representation" (pp. 24–25, ChemEQL Manual version 3.0).

The logarithm of the equilibrium expression—a linear activity function—for gibbsite is plotted as the $Al^{3+}(aq)$ line in Figure 5.3, which is a complete representation of pH-dependent gibbsite solubility with minimum solubility $10^{-5.6} M$ occurring near pH 6.3.

Another example of pH-dependent solubility appears in Appendix 5G, where the mineral is the calcium phosphate mineral apatite $Ca_5(PO_4)_3OH(s)$. Appendices 5H and 5I give examples of the effect that soluble complexes have on mineral solubility. The example in Appendix 5H is another facet of apatite solubility, this time showing the effect of complexes formed when the citric acid is present. Certain plants secrete citric acid from their roots to increase phosphate solubility in the surrounding soil. Bacteria, fungi, and certain plants secrete compounds that have a highly specific capacity to complex and dissolve ferric iron from insoluble iron minerals. This process is discussed in Appendix 5I.

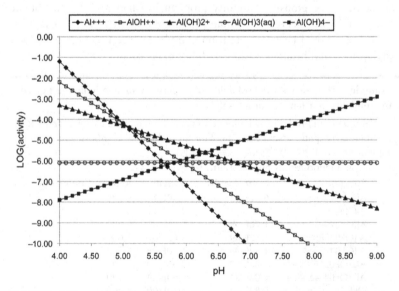

FIGURE 5.3 This graph plots the logarithm of ion activity for the five most abundant aluminum solution species as a function of pH as simulated using ChemEQL. The solubility of $Al^{3+}(aq)$ is constrained by gibbsite solubility as discussed in Example 5.16, except here $H^+(aq)$ is treated as an independent variable covering the range depicted in the graph.

5.3.3. Modeling the Chemistry of Environmental Samples: Groundwater, Soil Pore Water, and Surface Water

The Gibbs Phase Rule

Environmental chemists simulate the chemistry of water samples for many reasons, the most important being the identification of minerals that control the chemistry in water samples. The minerals that control solution concentrations are the minerals in solubility equilibrium with the solution. The control of solution chemistry implies a very specific and concrete concept: the activities of two or more solutes are constrained by equilibrium solubility and, therefore, are not independent of each other.

Some of the minerals in contact with water—minerals in the aquifer formation if the sample is groundwater, soil minerals if the sample is soil pore water, or minerals suspended in the water column or sediments if the sample is lake or river water—will not or cannot attain solubility equilibrium. Many primary rock-forming minerals that crystallize from magma (igneous rock minerals) or sinter at high temperatures (metamorphic rock minerals) are not stable in the presence of liquid water. Some sedimentary minerals precipitate from water whose composition is very different from prevailing conditions (May et al., 1986). Solubility equilibrium can exist only if both dissolution and precipitation reactions are possible.

Another important instance when a mineral cannot control solution chemistry occurs when the residence time of the water is less than the time necessary for the mineral to dissolve sufficiently to saturate the solution. This is illustrated by the effect of groundwater discharge on lake water alkalinity (Figure 2.1, *Soil Moisture and Hydrology*). Lake water alkalinity increases proportionately with the mean residence time of groundwater discharged into the lake (Wolock et al., 1989).

Environmental chemists rely on the principle of chemical equilibrium to identify minerals that control water chemistry. If the chemistry of a water sample is controlled by the equilibrium solubility of a suite of minerals, it should be possible to identify the mineral by calculating the ion activity product for mineral solubility (see Table 5.2 and Appendix 5B). The application of equilibrium methods requires water chemistry analyses that report solution pH and the concentration of one or more elements or compounds.

Each element or compound in the analysis, including solution pH, is considered a *component*. Chemical reactions distribute each solution component over a constellation of solution *species* (see Appendix 5B). The number of minerals that control water chemistry is subject to the *Gibbs Phase Rule*, a physical rule defining the degrees of freedom F in a system containing C components and P phases. The Gibbs Phase Rule for a system at fixed pressure and temperature is a very simple expression:

$$F = C - P \geq 0 \tag{5.15}$$

A more general expression for the Gibbs Phase Rule adds an additional two degrees of freedom if temperature and pressure are allowed to vary, but most environmental chemistry models assume constant temperature and pressure. The phase rule, as we will see, provides much needed guidance when modeling equilibrium in environmental water samples because the number of phases P must be less than or equal to the number of components C.

Phase Rule Examples

Suppose the system consists of two phases—aqueous solution and a gaseous phase—and two components: $H^+(aq)$[1] and carbon dioxide CO_2: $F = 2 - 2 = 0$. The $CO_2(g)$ partial pressure P_{CO_2} fixes the activity of dissolved carbon dioxide $CO_2(aq)$ and solution pH or, conversely, there are no degrees of freedom F. Solution pH is a direct consequence of P_{CO_2}.

Suppose the system consists of three components C: liquid water $H_2O(l)$, $Mg^{2+}(aq)$, and carbon dioxide $CO_2(aq)$. The number of phases can range from 1 to 3, and, consequently, the degrees of freedom F can range from 2 to 0: $F = 3 - P \geq 0$. A single-phase system ($F = 3 - 1 = 2$) would be an aqueous solution of $Mg^{2+}(aq)$ and carbon dioxide $CO_2(aq)$. These two components would react to form numerous solute species by hydrolysis and the formation of ion complexes. The two degrees of freedom allow two independent variables: $a_{Mg^{2+}}$ and $a_{CO_2(aq)}$ determine pH (or permutations thereof).

Two 2-phase systems ($F = 3 - 2 = 1$) are possible: an aqueous solution and gaseous carbon dioxide $CO_2(g)$, or an aqueous solution saturated by a single magnesium mineral (e.g., magnesite $MgCO_3(s)$ or brucite $Mg(OH)_2$ (s)). The single degrees of freedom allow one independent variable: $a_{Mg^{2+}}$ determines $a_{CO_2(aq)}$ and pH (or permutations thereof). The composition of the aqueous solution will determine which of the two magnesium minerals is the solid phase.

Two 3-phase systems ($F = 3 - 3 = 0$) are possible: an aqueous solution saturated by a single magnesium mineral (e.g., magnesite $MgCO_3(s)$ or brucite $Mg(OH)_2(s)$) and gaseous carbon dioxide $CO_2(g)$, or an aqueous solution simultaneously saturated by magnesite $MgCO_3(s)$ and brucite $Mg(OH)_2(s)$. Zero degrees of freedom allow no independent variables: $a_{Mg^{2+}}$, $a_{CO_2(aq)}$, and pH are fixed, depending on the nature of the three-phase system.

Saturated zone water encompasses systems where the only allowed phases are the aqueous solution and minerals. Since pH is always a component, the potential number of mineral phases P in equilibrium with the solution cannot exceed the number of components (elements or compounds other than pH). It is unlikely that some components (e.g., sodium, potassium, or other alkaline

---

1. Molecular water—a component and phase in all water chemistry problems—has no impact on the degrees of freedom F. A component in every water chemistry problem, however, is the hydrogen ion $H^+(aq)$. If we identify $H^+(aq)$ as a component, then $OH^-(aq)$ must be regarded a solution species, related to component $H^+(aq)$ through water hydrolysis equilibrium (Eq. (5.14)).

elements) will be constrained by mineral solubility. Some components are constrained by mineral solubility within a given pH range and by some other process not included in equilibrium water chemistry models (ion exchange or adsorption) in another pH range. The bottom line is that the number of mineral phases P that can saturate solution cannot exceed the number of components C. The following two sections describe methods for identifying the minerals that control water chemistry (i.e., solubility) in environmental water samples.

Reaction Quotients and Saturation Indices

The dissolution reaction for $Al(OH)_3(s)$ and its associated reaction quotient Q appear following. Remember that the reaction quotient Q is computed from the activities of specific solutes in the reaction expression and can assume any numerical value (see Figure 5.1):

$$Al(OH)_3(s) + 3 \cdot H^+(aq) \rightarrow Al^{3+}(aq) + 3 \cdot H_2O(l)$$
$$Q = a_{Al^{3+}} \cdot a_{H^+}^{-3}$$

When the solid phase $Al(OH)_3(s)$ is in solubility equilibrium with aqueous solution, the reaction quotient Q is numerically equal to the thermodynamic equilibrium constant K:

$$Al(OH)_3(s) + 3 \cdot H^+(aq) = Al^{3+}(aq) + 3 \cdot H_2O(l)$$
$$K = 10^{+8.11} = a_{Al^{3+}} \cdot a_{H^+}^{-3}$$

Solubility equilibrium may be approached as the mineral dissolves, rising toward saturation ($Q = K_s$) from a state of undersaturation ($Q < K_s$). A mineral may also approach solubility equilibrium by precipitating, drifting downward toward saturation ($Q = K_s$) from a state of oversaturation ($Q > K_s$). The saturation index SI reveals whether solubility equilibrium exists between a mineral and the surrounding aqueous solution. The saturation index SI (Eq. (5.16)) is defined as the ratio of the reaction quotient Q to the thermodynamic solubility constant K_s. Some environmental chemists use the expression *ion activity product IAP* rather than *reaction quotient Q*:

$$SI \equiv \frac{Q}{K_s} \tag{5.16}$$

In some instances the reaction quotient Q may have to exceed a threshold oversaturation ($SI > 1$) before a mineral will begin to precipitate. Under these circumstances the SI tends to decrease with time. A precipitate consisting of extremely small particles will also yield $SI > 1$. Precipitate *ripening* occurs when small particles dissolve at the expense of large particles, leading eventually to a saturated solution $SI \approx 1$. Variations in chemical composition, crystallinity, and particle size commonly alter the solubility of actual minerals in the environment (May et al., 1986).

Environmental chemists can identify minerals in solubility equilibrium with aqueous solution by performing a water chemistry simulation where precipitation is suppressed. This would mean not employing the **Insert Solid Phase...** command under the ChemEQL **Matrix** menu after compiling the input matrix, applying the **Activity** command under the ChemEQL **Option** menu, and running the simulation to generate an output data file listing species activities. If you open the output data file as an Excel document, you can generate SI values for likely mineral candidates. Your final list of minerals will be less than or equal to the number of components other than pH.

Example 5.17 illustrates the use of saturation indexes SI for the identification of minerals that control water chemistry. The water chemistry analyses come from a study of calcareous river and groundwater by Crandall and colleagues (1999).

ChemEQL provides another method for assessing whether the saturation index SI of a particular mineral indicates undersaturation, saturation (solubility equilibrium), or oversaturation. This is described under the heading "Check for Precipitation" (p. 12). First, identify a candidate mineral and use the **Insert Solid Phase...** command to replace one of the solution components with the candidate mineral. The mineral component is placed in solidPhase mode (the solid phase activity is 1.0) after performing the **Insert Solid Phase...** command. Change the mode from solidPhase to checkPrecip mode, following the instructions in the *ChemEQL Manual*. Finally, enter the total concentration of the original solution component below the **checkPrecip** mode listing (hit the tab key to write the concentration in the input matrix data file). Select **Activity** from the **Options** menu and run the simulation. The resulting output data file will indicate whether the chemical analysis concentration of the solution component is oversaturated or undersaturated by the results reported under **In or Out of System**.

Example 5.17

The following data are from water samples drawn from Wingate Sink, located 5.2 km west of O'Brien, Suwannee County, Florida (Crandall et al., 1999). The ChemEQL simulation does not replace any solution component with mineral phases. The reported pH values are much lower than expected for calcareous groundwater (pH 9.91). We can conclude that the water at Wingate Sink is in equilibrium with atmospheric carbon dioxide.

Date	pH	Ca, mol/L	Mg, mol/L	Na, mol/L	K, mol/L	Fe, mol/L
2/11/95	7.55	1.62E-03	2.35E-04	1.17E-04	6.39E-05	
4/8/95	7.31	1.50E-03	2.30E-04	1.17E-04	2.30E-05	1.43E-07
7/2/96	7.33	1.47E-03	2.43E-04	1.13E-04	1.28E-05	
11/4/96	7.39	9.98E-04	2.14E-04	1.44E-04	2.30E-05	2.69E-06

SO4, mol/L	HCO3, mol/L	SiO2(aq)	NO3, mol/L	Cl, mol/L
1.04E-04	3.11E-03	1.12E-04	1.21E-04	1.41E-04
1.04E-04	3.56E-03	1.05E-04	1.36E-04	1.55E-04
9.47E-05	3.08E-03	1.12E-04	1.07E-04	1.38E-04
8.43E-05	2.34E-03	1.05E-04	4.85E-05	1.66E-04

It is unlikely that the following components are controlled by mineral solubility: $Na+$, $K+$, $NO3-$, and $Cl-$. It is possible, however, the following components are controlled by mineral solubility: $Ca++$, $Mg++$, $Fe+++$, $SO4--$, $HCO3-$, and $SiO2(aq)$. A completely thorough analysis would calculate the saturation indexes SI for every mineral in the ChemEQL Solid Phase Library containing these components in their composition. The following table, however, lists the saturation indexes SI for the most likely candidate minerals: calcite, dolomite $CaMg(CO_3)_2(s)$, quartz $SiO_2(s)$, and amorphous $FeOOH(s)$.

Date	pH	Calcite SI	Dolomite SI	Quartz SI	FeOOH SI
2/11/95	7.55	1.5096	0.4485	1.0648	
4/8/95	7.31	0.8898	0.1648	0.9999	3.80E+5
7/2/96	7.33	0.8043	0.1451	1.0667	
11/4/96	7.39	0.5092	0.0753	0.9999	7.51E+6

The saturation indexes SI for every water sample clearly indicate that calcite and quartz control the solubility of two components: $Ca++$ and $SiO2(aq)$. The mineral dolomite is significantly undersaturated and probably does not control the solubility of $Mg++$. The solid phase amorphous $FeOOH(s)$ is significantly oversaturated, indicating that $Fe+++$ is probably actively precipitating. Oversaturation by ferric iron suggests that the groundwater feeding Wingate Sink likely contains high concentrations of ferrous iron that oxidize to ferric iron and precipitate when exposed to oxygen in the sink.

Logarithmic Activity Diagrams

Figure 5.3 plots solute activities using a logarithmic scale on both axes in what is commonly called an *activity diagram*. The use of activity diagrams is illustrated in Lindsay (1979) and Schwab (2004). The equilibrium solubility expression is converted into a linear expression by taking the logarithm of the thermodynamic equilibrium constant and the activity quotient in the matter demonstrated in the section "Modeling Simultaneous Equilibrium: Effect of pH on the Solubility of a Mineral" for the solubility expression for gibbsite $Al(OH)_3(s)$.

The log-log plot in Figure 5.3 shows the pH-dependent activity of several solutes, but the activity diagrams used to identify minerals controlling solute activities in environmental water samples typically plot one activity function (typically a straight line in a log-log plot) for each candidate solubility-controlling mineral.

Water chemistry models play a critical role in graphical assessment. The model simulates solution reactions (hydrolysis and ion-complex formation) to predict the activity of specific solutes used as variables in the activity diagram. If data points representing solute species in the water sample plot on or near the activity function representing a particular mineral, it is likely that this mineral is controlling solute activities in the water sample.

Figures 5.4 to 5.7 illustrate the use of activity diagrams to identify potential mineral phases constraining the chemistry of water samples. Figure 5.4 plots $Ca^{2+}(aq)$ activities from simulations of Wingate Sink water samples (Crandall et al., 1999), along with solubility lines for two systems: one that includes a gas phase ($P_{CO_2} = 3.47 \cdot 10^{-4}$ atm) and one that lacks a gas phase. The groundwater seeping into Wingate Sink would come from the latter type of system but potentially come into equilibrium with the atmosphere while in the Sink. The simulation suggests that the water is in equilibrium with calcite and shows little effect of dissolved carbon dioxide from the atmosphere.

Figure 5.5 plots $Mg^{2+}(aq)$ activities from simulations of Wingate Sink water samples (Crandall et al., 1999), along with the solubility line for a two-phase system: dolomite-controlling $Mg^{2+}(aq)$ activity and calcite-controlling $Ca^{2+}(aq)$ activity. The groundwater seeping into Wingate Sink is nearly 100-fold undersaturated relative to dolomite solubility and relatively insensitive to pH. The $Mg^{2+}(aq)$ activity is apparently unconstrained by mineral solubility in these water samples.

Figure 5.6 plots $H_4SiO_4^0(aq)$ activities from simulations of Wingate Sink water samples (Crandall et al., 1999), along with the solubility line for three

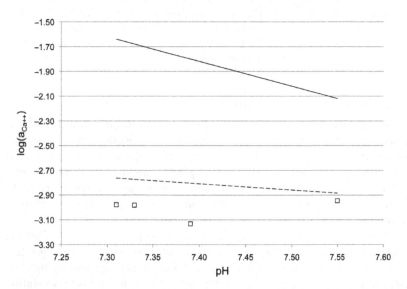

FIGURE 5.4 This graph plots the logarithm of $Ca^{2+}(aq)$ activity as a function of pH. The open square symbols represent Wingate Sink water samples (Crandall et al., 1999) as simulated using ChemEQL. The solid line plots $Ca^{2+}(aq)$ activity constrained by calcite solubility and atmospheric $CO_2(g)$. The dashed line plots $Ca^{2+}(aq)$ activity constrained by calcite solubility for a closed system without atmospheric $CO_2(g)$.

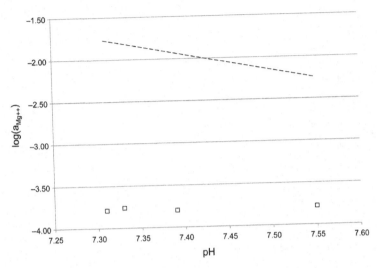

FIGURE 5.5 This graph plots the logarithm of $Mg^{2+}(aq)$ activity as a function of pH. The open square symbols represent Wingate Sink water samples (Crandall et al., 1999) as simulated using ChemEQL. The dashed line plots $Mg^{2+}(aq)$ activity constrained by simultaneous equilibrium of dolomite and calcite solubility (constraining $Ca^{2+}(aq)$ activity) in this pH range.

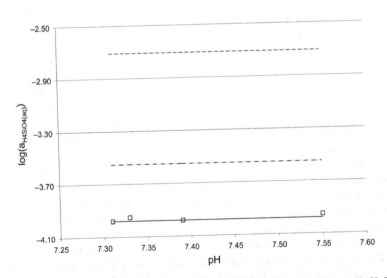

FIGURE 5.6 This graph plots the logarithm of $H_4SiO_4^0(aq)$ activity as a function of pH. The open square symbols represent Wingate Sink water samples (Crandall et al., 1999) as simulated using ChemEQL. The solid line plots $H_4SiO_4^0(aq)$ activity constrained by quartz solubility, the dot-dashed line indicates chalcedony solubility, and the dashed line indicates opal solubility.

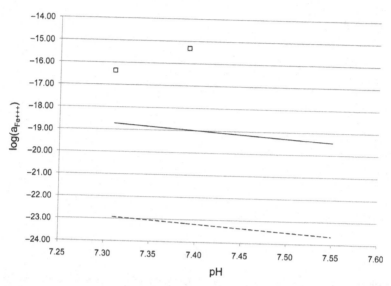

FIGURE 5.7   This graph plots the logarithm of $Fe^{3+}(aq)$ activity as a function of pH. The open square symbols represent Wingate Sink water samples (Crandall et al., 1999) as simulated using ChemEQL. The solid line plots $Fe^{3+}(aq)$ activity constrained by ferrihydrite $Fe(OH)_3(s)$ solubility, and the dashed line indicates goethite $FeOOH(s)$ solubility.

$SiO_2(s)$ minerals of increasing solubility and decreasing crystallinity: quartz, chalcedony and opal. The groundwater seeping into Wingate Sink is saturated relative to quartz, the most insoluble $SiO_2(s)$ mineral.

Figure 5.7 plots $Fe^{3+}(aq)$ activities from simulations of Wingate Sink water samples (Crandall et al., 1999), along with the solubility line for two ferric oxyhydroxide minerals of increasing solubility and decreasing crystallinity: goethite and ferrihydrite. The groundwater seeping into Wingate Sink is clearly oversaturated relative to ferrihydrite, the most soluble ferric oxyhydroxide mineral. This result is intriguing but easy to explain.

The groundwater before entering the Sink is probably anaerobic, characterized by high dissolved ferrous iron concentrations. Upon entering Wingate Sink, the water exposure to oxygen in the atmosphere causes the oxidation of ferrous iron to ferric iron. This oxidation process causes ferric oxyhydroxide minerals to precipitate. The chemistry clearly indicates that the precipitation of ferric oxyhydroxide minerals is underway. A more careful chemical analysis of the water that measures both total soluble iron and soluble ferrous iron would probably result in a significant reduction in the estimate of oversaturation.

Perhaps you may recall fresh well water with a distinct metallic taste. The metallic taste is almost entirely the result of dissolved ferrous iron. The ferrous iron originates from water percolating from a vadose zone whose porosity is nearly saturated with water, rich in organic matter, and relatively warm. These circumstances produce reducing conditions characterized by an active anaerobic bacteria community. The ferrous iron is a by-product of the anaerobic respiration of iron-reducing bacteria.

If you let a glass of this metallic-tasting water sit out, exposed to the air, after several hours the water will be cloudy and a yellow precipitate will cover the bottom of the glass. This yellow precipitate is the ferric mineral ferrihydrite. If you taste the water, you will find that it has lost its metallic taste because you can only taste dissolved iron. Much the same process is occurring in Wingate Sink.

5.4. SUMMARY

In this chapter we discussed the thermodynamic basis for chemical equilibrium and the equilibrium constant. In most water chemistry problems we need to adjust concentrations using activity coefficients; otherwise, the apparent equilibrium coefficient will depend on the concentration of dissolved ions in solution—the ionic strength I. Chemists have proposed several different equations for calculating ion activity coefficients (all of the examples used in the chapter are based on the Davies equation).

General chemistry introduced you to several methods for solving simple equilibrium problems. Few of these methods are practical when faced with actual water chemistry problems. The same basic principles apply, but computer-based numerical methods are required to solve the numerous simultaneous equilibrium relations. Your knowledge of chemistry and chemical equilibrium comes into play whenever you assess the validity of any water chemistry simulation. We covered the basic steps of a validity assessment, and you are strongly encouraged to develop the habit of validating any computer simulation.

The final topic we covered addresses a central issue in any simulation of environmental water samples: the identification of minerals that control water chemistry. A simple rule—the Gibbs Phase Rule—places an upper limit on the number of mineral phases that can actively control water chemistry. With this limit in hand, you can use water chemistry simulations to compute the saturation index SI for each candidate mineral, or you can plot data from your simulation in activity diagrams. Either method is a suitable strategy for identifying minerals controlling the chemistry of the water samples.

APPENDIX 5A. CHEMEQL RESULT DATA FILE FORMAT

An important attribute of ChemEQL is the format of the output data file. When the simulation is complete, a window appears displaying the results. You save the output data file by selecting **Save Data...** from the **File** menu (Figure 5A.1). The data file is saved in **text** format regardless of the default suffix (Figure 5A.2). Get into the habit of saving the file in an appropriate Excel **workbook** format: (Figure 5A.3). This is an important step because results validation becomes quite simple using spreadsheet functions.

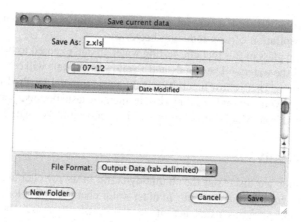

FIGURE 5A.1 Save the ChemEQL results data file by selecting **Save Data...** from the **File** menu. You will be prompted to name the file (default: z.xls). This naming convention allows you to open the file using the spreadsheet application Excel.

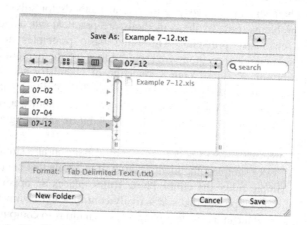

FIGURE 5A.2 The ChemEQL results data file default format is **text**.

APPENDIX 5B. VALIDATING WATER CHEMISTRY SIMULATIONS

Table 5.2 contains a checklist for validating computer-based water chemistry simulations. The list is not intended to be comprehensive. Consider it a framework that you can add to as your experience increases and you become more familiar with water chemistry modeling. Any user of computer-based models should develop the habit of routinely validating the data used to simulate the process being studied and specific simulation results.

Charge Balance Validation: This assessment evaluates the condition of charge neutrality in solution. It determines whether the charge sum over all

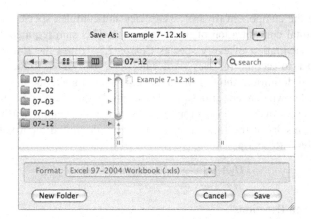

FIGURE 5A.3 Save the ChemEQL data files after opening in Excel by selecting an appropriate **workbook** format.

cation species equals the charge sum over all anion species in solution. Computing this sum requires multiplying the concentration of each ion $C_i[mol \cdot L^{-1}]$ times its valence z_i to convert molar concentration to moles of charge concentration (formerly called *normality*): $z_i \cdot C_i[mol_c \cdot L^{-1}]$. If this sum is performed using a spreadsheet, then sum over all cations (using moles of charge units) plus the sum over all anions (using moles of charge units) should equal zero to within rounding error:

$$\underbrace{\sum_i (z_i \cdot C_i)}_{cations} = \underbrace{\sum_j (z_j \cdot C_j)}_{anions}$$

$$\underbrace{\sum_i (z_i \cdot C_i)}_{cations} - \underbrace{\sum_j (z_j \cdot C_j)}_{anions} = 0 \qquad (5B.1)$$

The following table lists the species and concentration from a ChemEQL simulation. The third column lists the charge of each species, and the fourth the product of charge and concentration.

Species	Conc. [mol/L]	Charge	Charge Conc. [mol_c/L]
Acetic Acid	1.939E-02	0	0.000E+00
Acetate-	6.093E-04	-1	-6.093E-04
OH-	1.687E-11	-1	-1.687E-11
H+	6.094E-04	1	6.094E-04

Suppose the columns are labeled A–D and the rows are labeled 1–5 in the table. The formula in cell D2 is =B2*C2. The charge neutrality sum (Eq. 5B.1) is a sum over terms in column D: =SUM(D2:D5).

Mass Balance Validation: There is a separate mass balance sum for each component (e.g., Eq. (5B.2A), Eq. (5B.2B)). The sum is over all species containing the component (hydrolysis species and complex species) and must equal the total concentration of that component. This sum requires multiplying the concentration of every species $C_i[mol \cdot L^{-1}]$ times the stoichiometric coefficient v_i to account for every molecule: $v_i \cdot C_i[mol \cdot L^{-1}]$. Rather than rely on a general definition, the following example should make the mass balance explicit. If known concentrations of components M and A are added to a solution, then a sum over the concentrations of all species containing each component in the results of the simulation should equal the initial input values.

Hydrolysis Reactions

$$H_2A(aq) \leftrightarrow HA^-(aq) + H^+(aq)$$
$$HA^-(aq) \leftrightarrow A^{2-}(aq) + H^+(aq)$$

Complexation Reactions

$$M^{2+}(aq) + HA^-(aq) \leftrightarrow M(HA)^+(aq)$$
$$M^{2+}(aq) + 2 \cdot HA^-(aq) \leftrightarrow M(HA)_2^0(aq)$$
$$M^{2+}(aq) + 3 \cdot HA^-(aq) \leftrightarrow M(HA)_3^-(aq)$$
$$M^{2+}(aq) + A^{2-}(aq) \leftrightarrow M(A)^0(aq)$$
$$M^{2+}(aq) + 2 \cdot A^{2-}(aq) \leftrightarrow M(A)_2^{2-}(aq)$$

Mass Balance Expression: Component A

$$C_A = \sum_i \left(v_{A,i} \cdot C_{A,i} \right) \tag{5B.2A}$$

$$C_A = (C_{H_2A} + C_{HA^-} + C_{A^{2-}}) + \left(C_{M(HA)^+} + 2 \cdot C_{M(HA)_2^0} + 3 \cdot C_{M(HA)_3^-} \right)$$
$$+ \left(C_{M(A)^0} + 2 \cdot C_{M(A)_2^{2-}} \right)$$

Mass Balance Expression: Component B

$$C_M = \sum_j \left(v_{M,j} \cdot C_{M,j} \right) \tag{5B.2B}$$

$$C_M = \left(C_{M(HA)^+} + C_{M(HA)_2^0} + C_{M(HA)_3^-} \right) + \left(C_{M(A)^0} + C_{M(A)_2^{2-}} \right)$$

The following table lists the species and concentration from a ChemEQL simulation involving two components: Ca++ and SO4--. Suppose the columns are labeled A and B and the rows are labeled 1–8 in the following table. The mass balance formula for component Ca++ is =(B2+B3+B4). The mass balance formula for component SO4-- is =(B4+B5+B6).

Species	Conc. [mol/L]
Ca++	1.11E-02
CaOH+	1.14E-08
CaSO4(aq)	5.37E-03

SO4––	1.11E-02
HSO4–	5.50E-08
OH–	1.34E-07
H+	9.05E-08

Ionic Strength Validation: The simulation may or may not report the ionic strength I in the results, but ionic strength I, which also depends on an accurate charge balance, affects the estimate of each ion activity coefficient. If the ionic strength I sum is calculated using a spreadsheet, the sum is best performed using Eq. (5B.3) rather than Eq. (5.9):

$$I = \sum_i \left(\frac{z_i^2 \cdot C_i}{2} \right) \tag{5B.3}$$

The following table lists the species and concentration from a ChemEQL simulation involving two components: Ca++ and SO4–. Suppose the columns are labeled A–D and the rows are labeled 1–9 in the following table. The formula in cell D2 is =(C2^2)*(B2/2). The ionic strength I, computed using Eq. (5B.3), is the sum in cell D9: =SUM(D2:D8).

Species	Conc. [mol/L]	Charge	Ionic Strength [mol/L]
Ca++	1.11E-02	2	2.22E-02
CaOH+	1.14E-08	1	5.71E-09
CaSO4(aq)	5.37E-03	0	0.00E+00
SO4––	1.11E-02	–2	2.22E-02
HSO4–	5.50E-08	–1	2.75E-08
OH–	1.34E-07	–1	6.71E-08
H+	9.05E-08	1	4.52E-08
			4.44E-02

The ionic strength I from the model simulation is validated when the value reported in the results matches the value based on the preceding calculation.

Activity Coefficient Validation: Water chemistry models usually allow you to select the equation and parameters used to calculate activity coefficients or activities in the simulation results (see Eqs. (5.10), (5.11), and (5.12)). If the model reports both concentrations and activities, you can determine the activity coefficient for each ion by dividing the reported activity by the concentration (see Eq. (5.8)). Your validation of reported activities should use the concentrations in the output data file and the ionic strength I computed from those results. The validation results should agree with the values listed in the output data file.

The following table is from the same ChemEQL simulation used to illustrate the preceding ionic strength validation. Suppose the columns are labeled A–G and the rows are labeled 1–9 in the following table. Concentrations and activities from the ChemEQL simulation result data file appear in columns C and D. The activity coefficients in column E are computed by dividing the activity of each species (column D) by its concentration (column C). The activity coefficient in cell E2 is calculated by the terms =(D2/C2). The ionic strength terms in column F are identical to the terms in column D of the previous table.

The validated activity coefficients listed in column F are computed from activity coefficient expression used by the model and the ionic strength I computed from the simulation (cell F9). In the following example, all activity coefficients are calculated using the Davies equation (Eq. (5.12))—with temperature-dependent parameter $A = 0.5109$ for 25°C—and the ionic strength from the simulation in cell F9: 4.44E-02 [mol/L]. The formula in cell G2 is given following, where the ionic strength I is indicated by its absolute cell address F9.

$$= 10^{\wedge}((-0.5109) * (B2^{\wedge}2)$$
$$* ((SQRT(\$F\$9)/(1 + SQRT(\$F\$9))) - (0.2 * \$F\$9)))$$

Species	Charge	Conc. [mol/L]	Activity	Activity Coefficient	Ionic Strength [mol/L]	Activity Coefficient
Ca++	2	1.11E-02	5.12E-3	0.461	2.22E-02	0.461
CaOH+	1	1.14E-08	9.40E-9	0.824	5.71E-09	0.824
CaSO4(aq)	0	5.37E-03	5.37E-3	1.000	0.00E+00	1.000
SO4--	-2	1.11E-02	5.12E-3	0.461	2.22E-02	0.461
HSO4-	-1	5.50E-08	4.54E-8	0.824	2.75E-08	0.824
OH-	-1	1.34E-07	1.11E-7	0.824	6.71E-08	0.824
H+	1	9.05E-08	9.05E-8	1.000	4.52E-08	1.000
					4.44E-02	

If the explicitly calculated activity coefficients (column G) are identical to the simulated activity coefficients (column E), then the activity coefficients and activities are validated.

Activity Product Validation: The reaction quotient Q for a chemical reaction under any circumstances is a *product*. Earlier we represented the reaction quotient Q as an approximate equality (Eq. (5.4)), the product of educt and product concentrations (Eq. (5.4)). The reaction quotient Q, in reality, is equal to the product of educt and product activities (Eq. (5B.4)), where the stoichiometric quotient v_i is positive for all products and negative for educts.

$$Q = \left(\frac{a_M^{v_M} \cdot a_N^{v_N} \cdots}{a_A^{v_A} \cdot a_B^{v_B} \cdots}\right) = \prod_i a_i^{v_i} = \prod_i (\gamma_i \cdot C_i)^{v_i} \tag{5B.4}$$

The *activity product* is equal to the thermodynamic equilibrium constant K (Eq. (5B.4)) when the reaction is at equilibrium (Figure 5.1).

$$K = \left(\frac{a_M^{v_M} \cdot a_N^{v_N} \cdots}{a_A^{v_A} \cdot a_B^{v_B} \cdots}\right)_{equilibrium} = \prod_i a_i^{v_i} = \prod_i (\gamma_i \cdot C_i)^{v_i} \tag{5B.5}$$

ChemEQL lists the logarithm of the thermodynamic solubility constant for gypsum: −4.58. This is found in the Solid Phase Library Species under the **Library** menu. The activity product (Eq. (5B.5)) for this simulation is computed from the preceding table used to validate activity coefficients. The cell formula is = (D2*D5), equal to the value 2.626E-05. The base-10 logarithm of 2.626E-05 is =LOG(2.626E-05), which equals −4.58, validating the ion activity product IAP for gypsum. Stated another way, the *IAP* computed from the activities of the two species Ca++ and SO4−− equals the gypsum solubility product, meaning these activities are consistent with an aqueous solution saturated by the mineral.

Database Validation: Every water chemistry model uses an equilibrium reaction database listing the educts and products for each reaction, their stoichiometric coefficients, and the corresponding thermodynamic equilibrium constant. Validation of the ion activity products *IAP* involves selecting one or more equilibrium expressions from the database, computing the ion activity product using activities from the data file generated by the model and the appropriate stoichiometric coefficients, and then comparing the product with the thermodynamic equilibrium constant in the database. It is also wise to check the thermodynamic equilibrium constants in the database, especially for the reactions most critical for the chemistry you are modeling, against published values.

APPENDIX 5C. VALIDATION ASSESSMENT FOR EXAMPLES 5.13 AND 5.16

Example 5.13: Gypsum Solubility

Charge Balance and Ionic Strength: The following table lists the charge balance (−8.72E-08 mol_c L^{-1}) and ionic strength I (4.44E-02 mol L^{-1}) calculated from the concentrations generated by ChemEQL and listed in Example 5.13. The charge balance is validated because error is six orders of magnitude less than the most concentrated ion. The ionic strength *I* reported by ChemEQL is also validated because it matches the ionic strength calculated from the concentrations listed in the results table.

Species	Conc. [mol/L]	Charge	Conc. [mol$_c$/L]	Ionic Strength. [mol/L]
Ca++	1.11E-02	2	2.22E-02	2.22E-02
CaOH+	1.14E-08	1	1.14E-08	5.71E-09
CaSO4(aq)	5.37E-03	0	0.00E+00	0.00E+00
SO4−−	1.11E-02	−2	−2.22E-02	2.22E-02
HSO4−	5.50E-08	−1	−5.50E-08	2.75E-08
OH−	1.34E-07	−1	−1.34E-07	6.71E-08
H+	9.05E-08	1	9.05E-08	4.52E-08
			−8.72E-08	4.44E-02

Mass Balances: This example simulates the dissolution of a mineral into pure water. The mass balance for the two components Ca++ and SO4−− is validated against the solubility predicted by the simulation (1.648E-02 mol L^{-1}). The data file does not list Ca++ as a component (Ca++ was replaced by CaSO4:2H2O (gypsum) when executing **"Insert Solid..."** after compiling the input matrix and before running the simulation). The concentration sum over all species containing component Ca++ (Ca++, CaOH+, CaSO4(aq)) and the concentration sum over all species containing component SO4-- (SO4--, HSO4-, CaSO4(aq)) both equal the predicted solubility (1.648E-02 mol L^{-1}), validating the two mass balance equations.

Example 5.16: Gibbsite Solubility

Charge Balance and Ionic Strength: The following table lists the charge balance (1.000-07 mol_c L^{-1}) and ionic strength I (1.243E-07 mol L^{-1}) calculated from the concentrations generated by ChemEQL and listed in Example 5.14. The charge balance is invalid because it is the same order of magnitude as most concentrated ion. This example reveals a characteristic of the ChemEQL model: the condition of charge neutrality is not imposed on the solution. An additional 1.000-07 mol_c L^{-1} is required to achieve charge neutrality.

The ionic strength I reported by ChemEQL is also invalid because it fails to account for 1.000-07 mol_c L^{-1} of anion charge. The ionic strength calculated from the concentrations listed in the results table following, however, is identical to the ionic strength reported by the simulation (1.234E-07 mol L^{-1}). If we add the 1.000-07 mol_c L^{-1} required to achieve charge neutrality as a monovalent anion, the ionic strength becomes 1.734E-07 mol L^{-1} (or 2.234E-07 mol L^{-1} if added as a divalent anion). The ionic-strength error ranges from 29% to 45%.

Species	Conc. [mol/l]	Charge	Charge Conc. [mol_c/L]	Ionic Strength [mol/L]
Al+++	6.52E-13	3	1.957E-12	2.935E-12
AlOH++	3.80E-11	2	7.590E-11	7.590E-11
Al(OH)2+	1.76E-09	1	1.756E-09	8.780E-10
Al(OH)3(aq)	1.62E-09	0	0.000E+00	0.000E+00
Al(OH)4-	1.50E-08	−1	−1.499E-08	7.495E-09
Al2(OH)2++++	2.88E-19	4	1.153E-18	2.306E-18
Al3(OH)4+++++	4.03E-24	5	2.017E-23	5.043E-23
OH−	5.83E-08	−1	−5.833E-08	2.917E-08
H+	1.72E-07	1	1.715E-07	8.575E-08
			1.000E-07	1.234E-07

Mass Balances: This example simulates the dissolution of a mineral into pure water. The mass balance for component Al+++ is validated against the solubility predicted by the simulation (1.841E-08 mol L^{-1}). The data file does not list Al+++ as a component (Al+++ was replaced by Al(OH)3 (Gibbsite, crist) when **Insert Solid...** was executed after the input matrix was compiled and before the simulation was run). The solid phase specie Al(OH)3 (Gibbsite, crist) represents crystalline gibbsite, whose solubility differs from other solid phase species in the database for the same chemical composition: microcrystalline gibbsite, bayerite, and the amorphous solid $Al(OH)_3(s)$.

The concentration sum over all species containing the component Al+++ (first seven species in the preceding table) is equal to the predicted solubility (1.841E-08 mol L^{-1}), validating the mass balance equation.

Activities and Activity Product: The following table lists the results from the ChemEQL simulation of Al(OH)3 (Gibbsite, crist) dissolution in columns 2 and 3 and the activity coefficients generated by the simulation in column 4 (activity divided by concentration). The validated activity coefficients appear in column 5 calculated using the invalid ionic strength from the simulation.

Species	Conc. [mol/L]	Activity	Simulated Activity Coefficient	Validated Activity Coefficient
Al+++	6.52E-13	6.50E-13	0.996	0.996
AlOH++	3.80E-11	3.79E-11	0.998	0.998
Al(OH)2+	1.76E-09	1.76E-09	0.999	1.000
Al(OH)3(aq)	1.62E-09	1.62E-09	1.000	1.000
Al(OH)4-	1.50E-08	1.50E-08	1.000	1.000
Al2(OH)2++++	2.88E-19	2.86E-19	0.993	0.993
Al3(OH)4+++++	4.03E-24	3.99E-24	0.990	0.990
OH-	5.83E-08	5.83E-08	1.000	1.000
H+	1.72E-07	1.72E-07	1.000	1.000

If we calculate activity coefficients using ionic strength values that explicitly satisfy charge neutrality—1.734E-07 mol L^{-1} or 2.234E-07 mol L^{-1} (see ionic-strength errors above)—the error never exceeds 0.4%. This demonstrates that failing to impose charge neutrality has an insignificant effect on ChemEQL simulation results.

The logarithm of thermodynamic equilibrium constant for the solid phase specie Al(OH)3 (Gibbsite, crist) in the ChemEQL database is –8.11 for the following equilibrium expression. This product is identical to the *IAP* using the activities in the preceding table.

$$Al^{3+}(aq) \leftrightarrow 3 \cdot H^+(aq) + \underset{CRYSTALLINE\ GIBBSITE}{Al(OH)_3(s)}$$

$$K = 10^{-8.11} = \frac{a_{H^+}^3}{a_{Al^{3+}}}$$

$$7.76 \cdot 10^{-8} = \frac{\left(1.72 \cdot 10^{-7}\right)^3}{6.50 \cdot 10^{-13}}$$

The logarithm of thermodynamic equilibrium constant for the solution specie Al(OH)4- in the ChemEQL database is –22.70 for the following equilibrium expression. This product is identical to the *IAP* using the activities in the preceding table.

$$Al^{3+}(aq) + 4 \cdot H_2O(l) \leftrightarrow 4 \cdot H^+(aq) + Al(OH)_4^-(aq)$$

$$K = 10^{-22.70} = \frac{a_{H+}^4 \cdot a_{Al(OH)_4^-}}{a_{Al^{3+}}}$$

$$2.00 \cdot 10^{-23} = \frac{\left(1.72 \cdot 10^{-7}\right)^4 \cdot \left(1.50 \cdot 10^{-8}\right)}{6.50 \cdot 10^{-13}}$$

APPENDIX 5D. CINNABAR SOLUBILITY IN AN OPEN SYSTEM CONTAINING THE GAS DIHYDROGEN SULFIDE

The solubility equilibrium expression for the mineral cinnabar $HgS(s)$—as it appears in the ChemEQL library for Solid Phase Species—is appropriate for *closed systems* where equilibrium is confined to the solution and mineral phases.

$$Hg^{2+}(aq) + S^{2-}(aq) \leftrightarrow HgS(s)$$
$$\text{\tiny CINNABAR}$$
$$K = 10^{+52.00} = \frac{1}{a_{Hg^{2+}} \cdot a_{S^{2-}}}$$

An open system adds another phase: an atmosphere in contact with and containing gases that are free to dissolve into the solution. The preceding expression is not appropriate for a system whose components include, for example, dihydrogen sulfide gas $H_2S(g)$, the mineral phase cinnabar $HgS(s)$, and an aqueous solution containing $H_2S(aq)$, $Hg^{2+}(aq)$, $H^+(aq)$, and other species produced by hydrolysis and the formation of soluble complexes. The following solubility reaction and equilibrium expression are appropriate for an open, 3-phase system consisting of an atmosphere with dihydrogen sulfide gas $H_2S(g)$, the mineral cinnabar $HgS(s)$, and an aqueous solution. The numerical value for the following net solubility equilibrium expression depends on the pressure units. ChemEQL designates the partial pressure of gas using the pressure unit *atmosphere* (*atm*) rather than *Pascal* (*Pa*).

$$Hg^{2+}(aq) + H_2S(g) \leftrightarrow 2 \cdot H^+(aq) + HgS(s)$$
$$\text{\tiny CINNABAR}$$
$$K = 10^{+30.09} = \frac{a_{H+}^2}{a_{Hg^{2+}} \cdot P_{H_2S}}$$

Generating New Equilibrium Expressions for a Water Chemistry Model: Each water chemistry model has its own protocol for modifying its equilibrium database, but all permit the user to tailor the database to suit their specific modeling requirements by adding equilibrium expressions for components, solution species, and mineral phases that do not already appear in the database.

ChemEQL uses two databases, one for solution species and one for solid-phase (mineral) species. The *ChemEQL Manual* gives detailed instructions under the heading "Access the Libraries" (p. 15). The first step is to identify entries in the existing database that can be modified or combined to generate equilibrium expressions that are appropriate for an open system where $H_2S(g)$ is a component rather than a species. The standard ChemEQL solution species database (i.e., the Regular Library) and the mineral species database (i.e., the Solid Phase Library) do not contain $H_2S(g)$ as a component. Most of the solution and mineral species containing sulfide, and the associated equilibrium expressions and constants that define these sulfide-containing solution and mineral species are based on the component $S^{2-}(aq)$. Hydrogen sulfide $HS^-(aq)$ is a component in the standard Regular and Solid Phase libraries. A few solution and mineral species are defined by equilibrium expressions based on hydrogen sulfide $HS^-(aq)$.

The solution species database lists nine equilibrium expressions that define solution species that contain both sulfide and Hg(II). This list does not contain all of the solution species that contain sulfide involving other metallic elements. A comprehensive modification of the Regular Library would require many more expressions than the ones listed in the following table.

Equilibrium Expression Based on Component: S–	log(K)
S-- + 2.H+ = H2S(g)	+21.91
S-- + 2.H+ = H2S(aq)	+20.92
S-- + H+ = HS-	+13.90
Hg++ + S-- = HgS(aq)	+7.90
Hg++ + 2.S-- = HgS2--	+14.30
Hg++ + S-- = H+ + Hg(OH)S-	+4.50
Hg++ + 2.S-- + 2.H+ = Hg(HS)2(aq)	+65.51
Hg++ + 2.S-- + H+ = Hg(HS)S-	+59.32
Hg++ + S-- = HgS (Cinnabar)	+52.00

Next, we generate nine new equilibrium expressions and constants where $H_2S(g)$ is a component. The following examples illustrate how this is done. Invert the first expression, placing H2S(g) on the left and changing its role from product to educt.

$$H2S(g) = 2.H + + S - -; \log(K) = -21.91$$

Adding the second expression to the preceding one generates a second expression, defining the specie H2S(aq) in terms of the new component H2S(g).

$$H2S(g) = H2S(aq); \log(K) = -21.91 + 20.92 = -0.99$$

Adding the third expression to the inverse of the first expression generates a third expression, defining the specie HS- in terms of the new component H2S(g).

$$H2S(g) = H + + HS-; \log(K) = -21.91 + 13.90 = -8.01$$

The remaining expressions listed following were generated in the same way.

Equilibrium Expression Based on Component: H2S(g)	log(K)
Hg++ + H2S(g) = 2.H+ + HgS(aq)	−14.01
Hg++ + 2.H2S(g) = 4.H+ + HgS2−−	−29.52
Hg++ + H2S(g) = 3.H+ + Hg(OH)S−	−17.41
Hg++ + 2.H2S(g) = 2.H+ + Hg(HS)2(aq)	+21.69
Hg++ + 2.H2S(g) = 3.H+ + Hg(HS)S−	+15.50
Hg++ + H2S(g) = 2.H+ + HgS (Cinnabar)	+30.09

Since the standard Regular and Solid Phase libraries do not contain $H_2S(g)$ as a component, it must be added to both libraries before adding the new equilibrium expressions and constants. Each new expression is considered as adding a new species.

The ChemEQL simulation of cinnabar solubility has three components: 0 M Hg++ (total), 1.0E-6 atm H2S(g) (free), and 1.0E-7 M H+ (total). Designating H2S(g) as a "free" variable ensures that the H2S(g) partial pressure remains fixed at 1.0E-6 atm during the simulation. After compiling, select **Insert Solid Phase...** from the **Matrix** menu and replace component H2S (g) with HgS (Cinnabar).

Simulation results for cinnabar solubility in an open system containing 1.0E-6 atm H2S(g) partial pressure appear following, using Davies activity coefficients and I = 1.49E-7 M.

Species	Conc. [mol/L]	Activity
H2S(aq)	1.02E-07	1.02E-07
H2S(g)	1.00E-06	1.00E-06
Hg(HS)2(aq)	3.98E-15	3.98E-15
Hg(HS)S−	1.29E-14	1.29E-14
Hg(OH)S−	1.59E-41	1.59E-41
Hg++	3.23E-38	3.23E-38
HgOH+	6.45E-35	6.45E-35
Hg(OH)2	5.13E-31	5.13E-31
Hg(OH)3−	3.24E-39	3.24E-39
Hg2OH+++	2.63E-72	2.62E-72
Hg3(OH)3+++	1.70E-99	1.69E-99
HgS(aq)	7.94E-45	7.94E-45
HgS2−−	6.19E-53	6.18E-53
HS−	4.91E-08	4.90E-08
S−−	3.10E-15	3.10E-15
OH−	5.02E-08	5.02E-08
H+	1.99E-07	1.99E-07

The notable result from this simulation is the relative activity of Hg++ (3.23E-38 M) and S−− (3.10E-15). When cinnabar dissolves to saturate the aqueous solution in a closed system without an atmosphere and no other sulfide minerals, the activity of the two ions would be Hg++ (3.21E-26 M) and

S– – (3.12E-27). A 1.0E-6 atm H2S(g) partial pressure increases the S– – activity by 12 orders of magnitude, driving down the Hg++ activity by 12 orders of magnitude so as to maintain the *IAP* $= 10^{+52.00}$.

APPENDIX 5E. SIMULTANEOUS CALCITE-APATITE-PYROMORPHITE SOLUBILITY

This ChemEQL simulation has five components: 0 M Ca++ (total), 0 M Pb++ (total), 0 M PO4– – – (total), 0 M CO3– – (total), and 1.0E-7 M H+ (total). After compiling, select **Insert Solid Phase...** from the **Matrix** menu and replace components in the following order: (1) Pb++ with Pb5(PO4) 3OH (hydroxypyromorphite), (2) PO4– – – with Ca10(OH)2(PO4)6 (hydroxyapatite), and (3) CO3– – with Ca(CO3) (Calcite). This insert solid phase sequence removes Pb++ (remaining components: Ca++, PO4– – –, CO3– –, and H+), PO4– – – (remaining components: Ca++, CO3– –, and H+), and CO3– – (final components: Ca++ and H+) from the component list. Simulation results appear following using Davies activity coefficients.

Species	Conc. [mol/l]	Activity	Activity Coefficient
Ca++	1.17E-04	1.07E-04	0.914
CaOH+	1.30E-07	1.27E-07	0.978
CaCO3(aq)	5.50E-06	5.50E-06	1.000
CaHCO3+	1.30E-07	1.27E-07	0.978
CaPO4–	1.26E-09	1.23E-09	0.978
CaHPO4(aq)	7.32E-11	7.32E-11	1.000
CaH2PO4+	7.57E-15	7.41E-15	0.978
Pb++	2.22E-08	2.03E-08	0.914
PbOH+	2.97E-06	2.91E-06	0.978
Pb(OH)2	8.28E-06	8.28E-06	1.000
Pb(OH)3-	6.07E-07	5.93E-07	0.978
Pb2OH+++	1.44E-12	1.18E-12	0.816
Pb3(OH)4+	3.05E-08	2.78E-08	0.914
Pb4(OH)4++++	8.11E-13	5.66E-13	0.697
Pb6(OH)8++++	1.76E-11	1.23E-11	0.698
PbHCO3+	2.85E-09	2.79E-09	0.978
PbCO3	1.35E-06	1.35E-06	1.000
Pb(CO3)2– –	9.79E-08	8.95E-08	0.914
PbHPO4	3.20E-14	3.20E-14	1.000
PbH2PO4+	1.82E-18	1.78E-18	0.977
H2CO3	2.90E-08	2.90E-08	1.000
HCO3–	9.48E-05	9.27E-05	0.978
CO3– –	3.40E-05	3.10E-05	0.914
PO4– – –	4.89E-12	3.99E-12	0.817
HPO4– –	1.37E-09	1.25E-09	0.914
H2PO4–	2.83E-12	2.76E-12	0.978
H3PO4	5.45E-20	5.45E-20	1.000
OH–	7.32E-05	7.16E-05	0.978
H+	1.40E-10	1.40E-10	1.000

Soil and Environmental Chemistry

Final Solution Concentrations Ca (total) = 1.22E-04 M, Pb (total) = 1.34E-05 M, and PO4–––(total) = 2.70E-09 M, and *pH* = 9.86.

APPENDIX 5F. SIMULTANEOUS GIBBSITE-VARISCITE SOLUBILITY

This ChemEQL simulation has three components: 0 M Al+++ (total), 0 M PO4–––(total), and 1.0E-7 M H+ (total). After compiling, select **Insert Solid Phase...** from the **Matrix** menu and replace components in the following order: (1) PO4––– with AlPO4(H2O)2 (variscite) and (2) Al+++ with Al(OH)3 (amorphous). Simulation results appear following, using Davies activity coefficients (I = 3.14E-6 M).

Species	Conc. [mol/L]	Activity	Davies Coefficient
Al+++	6.05E-09	5.94E-09	0.982
AlOH++	1.32E-07	1.31E-07	0.992
Al(OH)2+	2.29E-06	2.28E-06	0.998
Al(OH)3(aq)	7.94E-07	7.94E-07	1.000
Al(OH)4–	2.77E-06	2.77E-06	0.998
Al2(OH)2++++	3.52E-12	3.40E-12	0.968
Al3(OH)4+++++	6.49E-14	6.16E-14	0.950
PO4–––	1.36E-14	1.34E-14	0.982
HPO4––	1.37E-08	1.36E-08	0.992
H2PO4–	9.84E-08	9.82E-08	0.998
H3PO4	6.31E-12	6.31E-12	1.000
OH–	2.20E-08	2.20E-08	0.998
H+	4.55E-07	4.55E-07	1.000

The *IAP* for both minerals—Al(OH)3 (amorphous) and AlPO4(H2O)2 (variscite)—match the LOG(K) values in the ChemEQL Solid Phase Library. At equilibrium, with the solution saturated by both mineral phases, a total of 5.88E-06 M of Al(OH)3 (amorphous) and 1.12E-07 M of AlPO4(H2O)2 (variscite) have dissolved into solution. The total soluble aluminum 5.99E-06 M is the sum of the solubility of both minerals.

APPENDIX 5G. APATITE SOLUBILITY AS A FUNCTION OF pH

This ChemEQL simulation has three components: 0 M Ca++ (total), 0 M PO4––– (total), and 0 M H+ (free). After compiling, select **Insert Solid Phase...** from the **Matrix** menu and replace component PO4––– with Ca10(OH)2(PO4)6 (hydroxyapatite). The simulation treats H+ as a "free" or independent variable that can be assigned values within a specified range (see p. 24 of *ChemEQL Manual*). Simulation results appear following, using Davies activity coefficients. The graph uses a logarithmic scale for the activity of each species (only the most abundant are plotted) and resembles the pH-dependent gibbsite solubility plot in Figure 5.3.

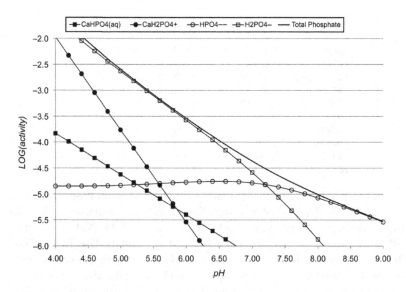

FIGURE 5G.1   The graph plots pH-dependent activities for the four most abundant phosphate species and the total phosphate concentration in a two-phase system: aqueous solution and the mineral apatite $Ca_{10}(PO_4)_6(OH)_2(s)$.

APPENDIX 5H. EFFECT OF THE CITRATE ON THE SOLUBILITY OF THE CALCIUM PHOSPHATE MINERAL APATITE

Phosphorus nutrition is problematic for plants in both acidic and alkaline soils. Low phosphorus availability in acid soil results from low strengite $FePO_4 \cdot 2H_2O(s)$ or variscite $AlPO_4 \cdot 2H_2O(s)$ solubility. Phosphorus deficiency in alkaline soil arises from low apatite $Ca_{10}(PO_4)_6(OH)_2(s)$ solubility. Legumes of the lupine genus (*Lupinus* spp.) secrete citrate and other weak organic acids to solubilize phosphate and enhance uptake. The present water chemistry simulation illustrates the effect of citrate complexes on the solubility of apatite.

The solubility equilibrium expression for the mineral apatite $Ca_{10}(PO_4)_6(OH)_2(s)$ as written in the ChemEQL database appears following.

$$10 \cdot Ca^{2+}(aq) + 6 \cdot PO_4^{3-}(aq) \leftrightarrow 2 \cdot H^+(aq) + \underset{\text{APATITE}}{Ca_{10}(PO_4)_6(OH)_2(s)}$$

$$K = 10^{+88.40} = \frac{a_{H^+}^2}{a_{Ca^{2+}}^{10} \cdot a_{PO_4^{3-}}^6}$$

The ChemEQL simulation of apatite solubility has three components: 0 M Ca++ (total), 0 M PO4– – – (total), and 1.0E-7 M H+ (total). After compiling, select **Insert Solid Phase...** from the **Matrix** menu and replace component PO4– – – with Ca10(OH)2(PO4)6 (hydroxyapatite). Simulation results for apatite solubility in the absence of citrate appear following, using Davies activity coefficients and I = 2.45E-5 M.

Soil and Environmental Chemistry

Species	Conc. [mol/l]	Activity	Davies Coefficient
Ca++	6.85E-06	6.69E-06	0.977
CaOH+	6.11E-10	6.07E-10	0.994
CaPO4–	1.85E-08	1.84E-08	0.995
CaHPO4(aq)	1.43E-08	1.43E-08	1.000
CaH2PO4+	1.91E-11	1.90E-11	0.994
PO4– – –	1.00E-09	9.50E-10	0.949
HPO4– –	3.98E-06	3.89E-06	0.977
H2PO4–	1.13E-07	1.13E-07	0.995
H3PO4	2.91E-14	2.91E-14	1.000
OH–	5.50E-06	5.47E-06	0.994
H+	1.83E-09	1.83E-09	1.000

Apatite solubility (in the absence of citrate) at saturation is 6.88E-7 moles of Ca10(OH)2(PO4)6 per liter. The ChemEQL simulation with added citrate has four components: 0 M Ca++ (total), 0 M PO4– – – (total), 1.0E-3 M Citrate– – – (total), and 1.0E-7 M H+ (total). Use **Insert Solid Phase…** to replace PO4– – – with Ca10(OH)2(PO4)6 (hydroxyapatite) as previously. Simulation results for apatite solubility in the absence of citrate, using Davies activity coefficients and I = 4.44E-3 M, appear in the following table.

The presence of 1.0E-3 M citrate increases ionic strength 100-fold and apatite solubility from 6.88E-7 moles of Ca10(OH)2(PO4)6 per liter in the absence of citrate to 2.96E-5 moles per liter. Although all calcium citrate species represent a mere 2.8% of the total added citrate, these species represent 95% of the total dissolved calcium.

Species	Conc. [mol/L]	Activity	Davies Coefficient
Ca++	1.37E-06	1.03E-06	0.750
CaOH+	4.18E-10	3.89E-10	0.931
CaPO4–	4.27E-08	3.98E-08	0.930
CaHPO4(aq)	7.46E-09	7.46E-09	1.000
CaH2PO4+	2.55E-12	2.37E-12	0.930
CaCitrate–	2.82E-05	2.62E-05	0.930
CaHCitrate	7.27E-10	7.27E-10	1.000
CaH2Citrate+	2.17E-16	2.02E-16	0.930
H2Citrate–	1.33E-11	1.24E-11	0.930
H3Citrate	8.23E-18	8.23E-18	1.000
PO4– – –	2.56E-08	1.34E-08	0.523
HPO4– –	1.76E-05	1.32E-05	0.750
H2PO4–	9.87E-08	9.18E-08	0.930
H3PO4	5.70E-15	5.70E-15	1.000
Citrate– – –	9.71E-04	5.08E-04	0.523
HCitrate– –	7.41E-07	5.56E-07	0.750
OH–	2.44E-05	2.27E-05	0.930
H+	4.40E-10	4.40E-10	1.000

Finally, compare the effect of added citrate on the activity coefficients (fourth column, both tables). The change in ionic strength induced by adding citrate has a substantial effect on activity corrections in this example.

APPENDIX 5I. EFFECT OF THE BACTERIAL SIDEROPHORE *DESFERRIOXAMINE B* ON THE SOLUBILITY OF THE IRON OXYHYDROXIDE GOETHITE

Iron nutrition is problematic for plants and soil microbes because ferric iron solubility is extremely low. Grass species, bacteria, and fungi secrete compounds called *siderophores* to dissolve ferric iron in order to meet their nutritional requirements. Siderophores are organic compounds that form highly specific, extremely stable complexes with the $Fe^{3+}(aq)$ ion in solution. The ferric-siderophore complex binds to a receptor on the cell surface of bacteria and fungi specifically designed to transport the complex across the cell membrane into the cytoplasm, where enzymes reduce Fe^{3+} to Fe^{2+}. The ferrous-siderphore complex, being relatively weak, releases Fe^{2+} following reduction to cytoplasmic *chaperone* complexes that specifically bind and transport ferrous iron. The following example illustrates the effect of siderophore complexes on the solubility of the ferric oxyhydroxide mineral goethite $FeOOH(s)$.

The solubility equilibrium expression for the mineral goethite $FeOOH(s)$ appears following. This solubility expression takes into account goethite pH-dependent solubility in much the same way as in Example 5.10.

$$\underset{GOETHITE}{FeOOH(s)} + 3 \cdot H^+(aq) \leftrightarrow Fe^{3+}(aq) + 2 \cdot H_2O(l)$$

$$\underset{GOETHITE}{K_s} = \frac{a_{Fe^{3+}}}{a_{H^+}^3} = 10^{-1.00}$$

The bacterial siderophore *ferrioxamine B* is trivalent when fully dissociated: $FOB^{3-}(aq)$. The pH-dependent formation expression follows (Smith and Martell, 2001) and was added to the ChemEQL database for this simulation (see Appendix 5D). Notice that the formation constant $K_f = 10^{+41.39}$ is very large, indicating equilibrium favors the product $Fe(FOB)^+(aq)$ over the educts.

$$Fe^{3+}(aq) + FOB^{3-}(aq) + H^+(aq) \leftrightarrow Fe(FOB)^+(aq)$$

$$K_f = \frac{a_{Fe(FOB)^+}}{a_{Fe^{3+}} \cdot a_{FOB^{3-}} \cdot a_{H^+}} = 10^{+41.39}$$

Multiplying the goethite solubility expression by the siderophore formation expression yields a solubility expression showing the effect of complex formation.

$$\underset{GOETHITE}{FeOOH(s)} + FOB^{3-}(aq) + 4 \cdot H^+(aq) \leftrightarrow Fe(FOB)^+(aq) + 2 \cdot H_2O(l)$$

$$\underset{GOETHITE}{K_s} \cdot K_f = \left(\frac{a_{Fe^{3+}}}{a_{H^+}^3} \right) \cdot \left(\frac{a_{Fe(FOB)^+}}{a_{Fe^{3+}} \cdot a_{FOB^{3-}} \cdot a_{H^+}} \right) = \left(10^{-1.00} \right) \cdot \left(10^{+41.39} \right)$$

$$K_{net} = \frac{a_{Fe(FOB)^+}}{a_{FOB^{3-}} \cdot a_{H^+}^4} = 10^{+40.39}$$

At neutral pH the ratio of $Fe(FOB)^+(aq)$ to $FOB^{3-}(aq)$ is $10^{+12.39}$. Adding 1 μmole of *ferrioxamine B* per liter will dissolve 10^{-6} moles of goethite.

A ChemEQL simulation of ferrihydrite $Fe(OH)_3$ solubility has three components: 0 M Fe+++ (total), 1.0E-4 M FOB– – – (total), and 0 M H+ (free). Ferrihydrite is a poorly crystalline hydrous ferric oxide mineral of indeterminate structure and composition; its mineralogy is defined by a characteristic x-ray diffraction pattern. Typical siderophore concentrations in soil pore water are on the order of 100 μM.

After compiling, select **Insert Solid Phase...** from the **Matrix** menu and replace component Fe+++ with Fe(OH)3 (s), making the initial Fe+++ irrelevant. The simulation treats H+ as a "free" or independent variable forced to assume values within a specified range (see p. 24 of the *ChemEQL Manual*). Simulation results for ferrihydrite solubility in a solution containing 100 μM *ferrioxamine B* appear graphically following. The graph uses a logarithmic scale for the activity of each solution species (only the most abundant are plotted), which resembles the pH-dependent gibbsite solubility plot in Figure 5I.1.

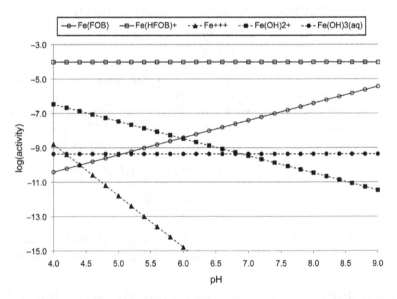

FIGURE 5I.1 The graph plots pH-dependent activities for the five most abundant ferric species in a two-phase system: aqueous solution containing 100 μM *ferrioxamine B* and the mineral ferrihydrite $Fe(OH)_3$. The activity and concentration of $Fe(FOB)^+(aq)$ are virtually independent of pH and represent ≥96% of the total dissolved ferric iron and *ferrioxamine B*.

Problems

1. The ChemEQL equilibrium database for minerals (Solid Phase Library) contains the following entry for silicon dioxide minerals quartz, chalcedony, and opal (amorphous silica).

$$H_4SiO_4(aq) \leftrightarrow \underset{QUARTZ}{SiO_2(s)} \quad K = 10^{+3.98}$$

$$H_4SiO_4(aq) \leftrightarrow \underset{CHALCEDONY}{SiO_2(s)} \quad K = 10^{+3.55}$$

$$H_4SiO_4(aq) \leftrightarrow \underset{OPAL}{SiO_2(s)} \quad K = 10^{+2.71}$$

These minerals are nominally silicon dioxide, but as crystallinity decreases, in the sequence just listed, water content increases. Figure 5.6 plots $H_4SiO_4(aq)$ activity as a line on a logarithmic activity diagram. Using the preceding equilibrium expressions and constants, give the equations for the three solubility lines plotted in Figure 5.6.

Solution
Follow the gibbsite solubility example discussed prior to Example 5.16. Write each equilibrium expression as an activity quotient, and then take the logarithm of both sides of the expression. The logarithmic activity expression for quartz follows.

$$\frac{1}{a_{H_4SiO_4}} = K = 10^{+3.98}$$

$$\log\left(\frac{1}{a_{H_4SiO_4}}\right) = \log(K) = \log(10^{+3.98})$$

$$QUARTZ: \quad \log(a_{H_4SiO_4}) = -3.98$$

The logarithmic activity expressions for chalcedony and opal are similar except for the values for $\log(K)$.

$$CHALCEDONY: \quad \log(a_{H_4SiO_4}) = -3.55$$
$$OPAL: \quad \log(a_{H_4SiO_4}) = -2.71$$

2. The ChemEQL equilibrium database for minerals (Solid Phase Library) contains the following entry for ferric oxyhydroxide minerals goethite and ferrihydrite (amorphous ferric hydroxide).

$$Fe^{3+}(aq) \leftrightarrow \underset{FERRIHYDRITE}{Fe(OH)_3(s)} +3 \cdot H^+(aq) \quad K = 10^{-3.20}$$

$$Fe^{3+}(aq) \leftrightarrow \underset{GOETHITE}{FeOOH(s)} +3 \cdot H^+(aq) \quad K = 10^{+1.00}$$

The chemical composition and crystal structure of goethite are known. Ferrihydrite is identified by its x-ray diffraction pattern, but its chemical composition and crystal structure remain undetermined. Figure 5.7 plots $Fe^{3+}(aq)$ activity as a line on a logarithmic activity diagram. Using the preceding

equilibrium expressions and constants, give the equations for the two solubility lines plotted in Figure 5.7.

Solution
Follow the gibbsite solubility example discussed prior to Example 5.16. Write each
equilibrium expression as an activity quotient, and then take the logarithm of both
sides of the expression. The logarithmic activity expression for goethite follows.

$$\frac{a_{H^+}^3}{a_{Fe^{3+}}} = K = 10^{+1.00}$$

$$\log\left(\frac{a_{H^+}^3}{a_{Fe^{3+}}}\right) = \log(K) = \log(10^{+1.00})$$

$$3 \cdot \log(a_{H^+}) - \log(a_{Fe^{3+}}) = 1.00$$

$$GOETHITE: \quad \log(a_{Fe^{3+}}) = -3 \cdot pH - 1.00$$

The logarithmic activity expression for ferrihydrite is similar except for the
$\log(K)$ value.

$$FERRIHYDRITE: \quad \log(a_{Fe^{3+}}) = -3 \cdot pH + 3.20$$

The $Fe^{3+}(aq)$ activity at pH 7.25 for solutions in equilibrium with these
two minerals are –22.75 and –18.55, respectively.

3. Install ChemEQL on a computer you have access to, and simulate the following two solutions. Remember: the weak acid is entered as "total," while the
   proton concentration is entered as "free" (meaning fixed). (NOTE: Review
   the example appearing on pages 8–9 of the *ChemEQL Manual*: 1 mM acetic
   acid solution at pH 4.)
   - 1 μM tartaric acid at pH 4.
   - 1 μM tartaric acid at pH 5.

   Please save your output file and perform a charge balance calculation to
   determine whether this restriction is satisfied.

4. Using your results from Problem 3, specifically the charge balance calculation, add a suitable "total" concentration of a strong-acid anion (chloride,
   nitrate, etc.) and repeat the calculation.

   Please save your output file and perform a charge balance calculation to
   determine whether this restriction is satisfied after adding sufficient strong-
   acid anion to satisfy the solution charge balance.

5. The standard free energies of formation ΔG_f° for each component in the struvite
   solubility reaction appear in the following table (Smith and Martell, 2001).

$$Mg(NH_4)(PO_4) \cdot 6H_2O(s) \leftrightarrow Mg^{2+}(aq) + NH_4^+(aq) + PO_4^{3-}(aq) + 6 \cdot H_2O(l)$$
$$\text{STRUVITE}$$

Component	$\Delta G_f^\circ [kJ \cdot mol^{-1}]$
$Mg^{2+}(aq)$	–456.40
$NH_4^+(aq)$	–79.51

$$PO_4^{3-}(aq) \qquad\qquad -1026.6$$
$$H_2O(l) \qquad\qquad\qquad -237.34$$
$$Mg(NH_4)(PO_4)\cdot 6H_2O(s) \qquad -3061.6$$
$$\underset{STRUVITE}{}$$

Calculate the standard Gibbs energy and the thermodynamic equilibrium constant for the solubility reaction at 298.15 K (25°C). The appropriate ideal gas constant R has units $[kJ\cdot mol^{-1}\cdot K^{-1}]$.

Solution
Apply Eqs. (5.2) and (5B.5) to calculate ΔG_{rxn}° and K, respectively.

$$\Delta G_{rxn}^{\circ} = ((-456.40)+(-79.51)+(-1026.6)+6\cdot(-237.34))_{products}$$
$$\qquad - (-3061.6)_{reactants}$$
$$\qquad = 75.05[kJ\cdot mol^{-1}]$$

$$K = \exp\left(\frac{-\Delta G_{rxn}^{\circ}}{R\cdot T}\right) = \exp\left(\frac{-\left(+75.05\ kJ\cdot mol^{-1}\right)}{\left(8.3145\cdot 10^{-3}\ kJ\cdot mol^{-1}\cdot K^{-1}\right)\cdot(298.15\ K)}\right)$$
$$K = 7.110\cdot 10^{-14}$$

6. Sewage treatment facilities rely on anaerobic digesters to eliminate pathogens and reduce waste volume. A typical two-stage anaerobic digester has an initial acid stage where anaerobic bacteria ferment organic solids into organic acids, followed by a second stage where methane-producing bacteria consume the organic acids from the first stage. Metabolic heating raises the temperature in the acid-stage to about 40°C, while the second stage is typically heated to about 50°C to promote the activity of thermophilic bacteria essential for rapid second-stage digestion. Not surprisingly, the pH of the sewage increases sharply as acid-stage sewage is pumped into the second-stage digesters.

   Coincidental with the second-stage digestion is the prodigious precipitation of the mineral struvite $Mg(NH_4)(PO_4)\cdot 6H_2O(s)$. Thick struvite deposits accumulate on the inner wall of the second-stage digester, the pipes carrying digested sewage from the second-stage digester, and the components of the dewatering unit that receive the digested sludge from the second digester. Use ChemEQL to estimate the struvite and hydroxyapatite saturation indexes for a water sample from a second-stage thermophilic digester at the Madison Metropolitan Sewerage District in Madison, Wisconsin (data provided by Prof. Phillip Barak, Soil Science Department, University of Wisconsin-Madison).

Component	Concentration [mol L^{-1}]
Ca++	5.0E-4
Mg++	3.0E-4
NH4+	6.09E-2
PO4---	6.3E-3
pH	7.53

7. The ChemEQL equilibrium database for minerals (Solid Phase Library) contains the following entry for ferric oxyhydroxide mineral goethite.

$$Fe^{3+}(aq) \leftrightarrow \underset{GOETHITE}{FeOOH(s)} + 3 \cdot H^+(aq) \quad K = 10^{+1.00}$$

Calculate the conditional equilibrium coefficient K^c for this reaction using Eq. (5.6) and one of the empirical activity coefficient expressions (5.10–5.12) when the ionic strength $I = 0.001$ M.

Solution

Apply Eq. (5.6) to the equilibrium solubility expression for goethite to yield an expression for the conditional equilibrium constant K^c as a function of the activity coefficients for $Fe^{3+}(aq)$ and $H^+(aq)$.

$$K = \frac{a_{H^+}^3}{a_{Fe^{3+}}} = 10^{+1.00}$$

$$\frac{a_{H^+}^3}{a_{Fe^{3+}}} = \frac{\gamma_{H^+}^3 \cdot C_{H^+}^3}{\gamma_{Fe^{3+}} \cdot C_{Fe^{3+}}} = \left(\frac{\gamma_{H^+}^3}{\gamma_{Fe^{3+}}}\right) \cdot \left(\frac{C_{H^+}^3}{C_{Fe^{3+}}}\right)$$

$$K = \left(\frac{\gamma_{H^+}^3}{\gamma_{Fe^{3+}}}\right) \cdot K^c$$

$$K^c = \left(\frac{\gamma_{Fe^{3+}}}{\gamma_{H^+}^3}\right) \cdot K$$

Calculate the activity coefficients $\gamma_{Fe^{3+}}$ and γ_{H^+}, using the Davies expression (5.12).

$$\log(\gamma_{H^+}) = -(0.5085) \cdot (+1)^2 \cdot \left(\frac{\sqrt{0.001}}{1 + \sqrt{0.001}} - (0.2 \cdot 0.001)\right)$$

$$\gamma_{H^+} = 10^{-0.015} = 0.965$$

$$\log(\gamma_{Fe^{3+}}) = -(0.5085) \cdot (+3)^2 \cdot \left(\frac{\sqrt{0.001}}{1 + \sqrt{0.001}} - (0.2 \cdot 0.001)\right)$$

$$\gamma_{H^+} = 10^{-0.139} = 0.725$$

The conditional equilibrium coefficient $K^c = 10^{+0.91}$ at an ionic strength $I = 0.001$.

Natural Organic Matter and Humic Colloids

6.1. INTRODUCTION

This chapter follows a trajectory that begins with the soil carbon cycle and ends with a discussion of naturally occurring organic colloids. This trajectory parallels the path taken by carbon fixed as biomass as it undergoes microbial decomposition. The path leading from biomass to the stable natural organic matter fraction takes hundreds to thousands of years, a path subject to considerable speculation but little verification. Our focus will be the chemical properties of natural organic matter that are, in turn, intimately linked to their colloidal properties. The chemically active components lie at the extreme ends of the carbon transformation trajectory: identifiable biomolecules in *dissolved organic carbon* and colloidal *humic substances*—the end product of microbial decomposition.

6.2. SOIL CARBON CYCLE

The global carbon cycle (Figure 6.1) encompasses both marine and terrestrial environments, but here we focus on a key component of the terrestrial

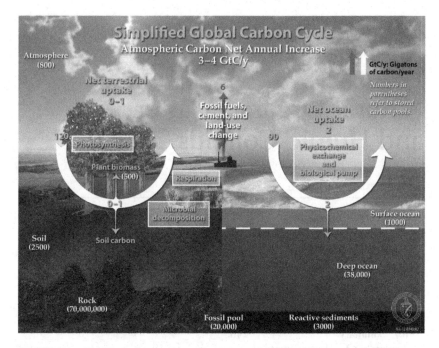

FIGURE 6.1   Simplified representation of the global carbon cycle (Genomics:GTL, 2005).

carbon cycle:[1] the soil (or *belowground*) carbon cycle. The cycle begins with biological carbon fixation: the conversion of carbon dioxide to biomass carbon by photolithotrophic (i.e., photosynthetic eukaryotes and prokaryots) and, to a much lesser degree, chemolithotrophic microorganisms. The cycle ends with the mineralization of organic carbon into carbon dioxide through a process called *oxidative metabolism*. Oxidative metabolism releases the chemical energy stored during carbon fixation.

6.2.1. Carbon Fixation

Photosynthetic carbon fixation converts light energy into chemical energy. Photosynthesis reduces $C(+4)$ in carbon dioxide to $C(+1)$ in the terminal carbon in glyceraldehyde-3-phosphate, the feedstock for simple sugars, amino acids, and lipids. The process involves four steps in the *Calvin cycle,* which is shown in Figures 6.2 to 6.5.

In the first step (Figure 6.2), a single CO_2 molecule attaches to carbon-2 of ribulose 1,5-bisphosphate. The product—2-carboxy-3-keto-arabinitol-1,5-bisphosphate—is an extremely unstable compound. The second step (Figure 6.3)

---

1. The freshwater and marine carbon cycles involve similar carbon transformations and yield organic matter products with chemical properties similar to soil organic matter.

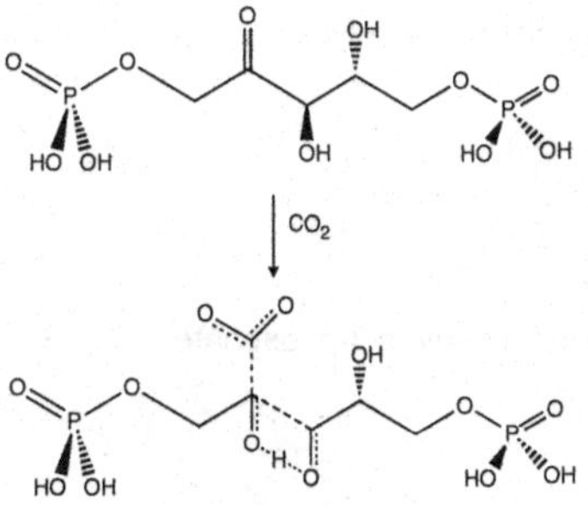

ribulose 1,5-bisphosphate

2-carboxy-3-keto-arabinitol-1,5-bisphosphate

FIGURE 6.2   Ribulose 1,5-bisphosphate and carbon dioxide combine to form 2-carboxy-3-keto-arabinitol-1,5-bisphosphate.

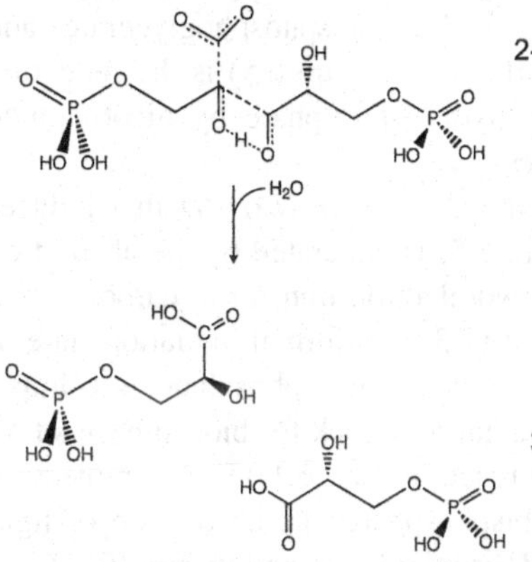

2-carboxy-3-keto-arabinitol-1,5-bisphosphate

3-phosphoglycerate

FIGURE 6.3   The unstable molecule 2-carboxy-3-keto-arabinitol-1,5-bisphosphate undergoes hydrolysis to yield two molecules of 3-phosphoglycerate.

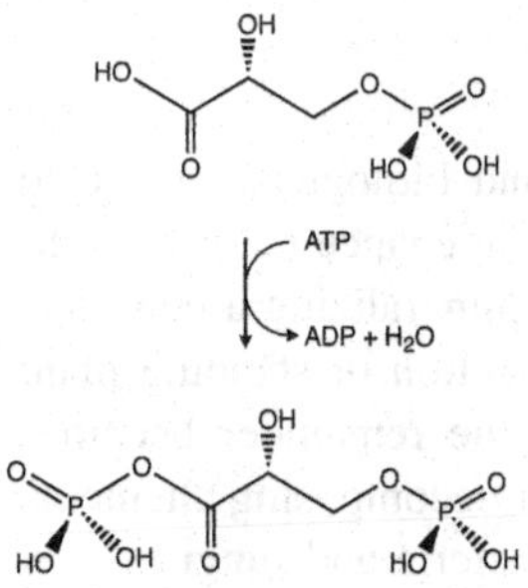

3-phosphoglycerate

FIGURE 6.4   *ATP* phosphorylates each 3-phosphoglycerate molecule, forming 1,3-bisphosphoglycerate.

1,3-bisphosphoglycerate

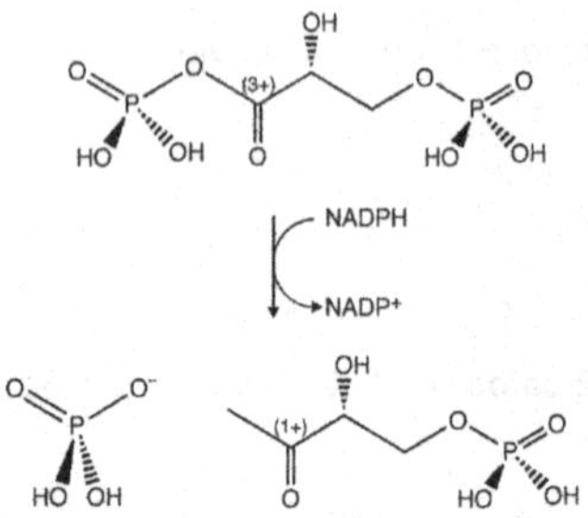

FIGURE 6.5 *NADPH* reduces 1,3-bisphosphoglycerate to glyceraldehyde 3-phosphate.

is the hydrolysis of the bond between carbons 2 and 3 in 2-carboxy-3-keto-arabinitol-1,5-bisphosphate, the product of the reaction in Figure 6.2, to yield two molecules of 3-phosphoglycerate.

The third step (Figure 6.4) is the phosphorylation of 3-phosphoglycerate by adenosine-5'-triphosphate *ATP* to yield 1,3-bisphosphoglycerate and adenosine-5'-diphosphate *ADP*. The fourth step (Figure 6.5) is the reduction of 1,3-bisphosphoglycerate to glyceraldehyde 3-phosphate by nicotinamide adenine dinucleotide phosophate *NADPH*.

The *nicotinamide adenine dinucleotide phosophate NADPH* that reduces 1,3–bisphosphoglycerate in step 4 (Figure 6.5) is generated by the absorption of light. The reduction (Figure 6.5) is a two-electron transfer to carbon 1 that reduces it from a formal oxidation state of $(+3)$ to a formal oxidation state of $(+1)$. Glyceraldehyde 3-phosphate, also known as triose phosphate, is a simple 3-carbon sugar that serves as the fundamental feedstock for biosynthesis of all bioorganic compounds. The net chemical reaction (Eq. (6.1)) fixes 6 molecules of CO_2 to 6 molecules of ribulose 1,5-bisphosphate; the absorption of light (Eq. (6.2)) generates the *ATP* and *NADPH* required by reaction (Eq. (6.1)).

$$6 \cdot CO_2 + 6 \cdot ribulose\ 1,5 - bisphosphate + 12 \cdot ATP + 12 \cdot NADPH \atop \rightarrow 12 \cdot glyceraldehyde\ 3\text{-}phosphate \quad (6.1)$$

$$12 \cdot H_2O + 12 \cdot NADP^+ + 18 \cdot ADP^{3-} + 18 \cdot H_2PO_4^- \atop \rightarrow 6 \cdot O_2 + 12 \cdot NADPH + 12 \cdot H^+ + 18 \cdot ATP^{4-} \quad (6.2)$$

6.2.2. Carbon Mineralization

The mineralization of reduced carbon in biomass and biological residue to CO_2 releases chemical energy for the growth and maintenance of soil organisms. The plants that fix virtually all carbon end up mineralizing a considerable fraction for their growth and maintenance. A fraction of standing plant biomass is consumed directly by herbivores, while the remainder becomes biological residue supporting a diverse community of decomposing chemoorganotrophic organisms ranging from arthropods to bacteria and fungi.

In Chapter 8 we examine mineralization in greater detail. *Catabolism* deconstructs large organic molecules into smaller molecules. The chemical energy released during catabolism is accumulated in the same compound found in Eq. (6.1)—nicotinamide adenine dinucleotide phosphate *NADPH*— now serving as the electron donor for the electron transport chain. Aerobic respiration is the reverse of the net reaction shown in Eq. (6.2).

An often overlooked characteristic of the carbon cycle is the chemical stability of biomass and biological residue in the presence of dioxygen O_2. The carbon in biomass and biological residue is thermodynamically unstable when in the presence of dioxygen in the aboveground and soil atmosphere. The persistence of biomass and biological residue has more to do with the properties of the dioxygen molecule than the chemical bonding in bioorganic compounds.

6.2.3. Oxidation of Organic Compounds by Dioxygen

You were introduced to *molecular orbital* theory in general chemistry. This theory represents the chemical bonds in any molecule as molecular orbitals ψ_i, each consisting of a sum of the atomic orbitals ϕ_j from the constituent atoms (Eq. (6.3)). The coefficients c_{ij} weight the relative contribution of each atomic orbital ϕ_j with the sign denoting the phase along the bonding axis. If two atomic orbitals are "in phase" along the bond axis, then orbital overlap increases electron density in the bond; if they are "out of phase," the overlap decreases electron density. The magnitude of the electron density along the bond axis determines the strength of each chemical bond.

$$\psi_i = \sum_j c_{ij} \cdot \phi_j \tag{6.3}$$

Figure 6.6 is the molecular orbital diagram for methane CH_4. The single electron (depicted as arrows) in each hydrogen atom pair, along with the four electrons from the carbon atom, form four bonding molecular orbitals $\psi_{\sigma g}$ in the methane CH_4 molecule (one electron spin in each pair is *spin-up*, and the other is *spin-down*). The hydrogen atomic orbitals in methane are grouped into four linear combinations that are determined by the tetrahedral symmetry of methane: combination (6.4) has the symmetry of the central carbon $2s$ orbital, while combinations (6.5), (6.6), and (6.7) have the symmetry of the central carbon $2p$ orbitals: $2p_x$, $2p_y$, and $2p_z$. All of the combinations appear in Figure 6.7.

$$\sigma_{1a_1} = \tfrac{1}{2} \cdot (\phi_A + \phi_B + \phi_C + \phi_D) \tag{6.4}$$

$$\sigma_{1t_2}(x) = \tfrac{1}{2} \cdot (\phi_A + \phi_B - \phi_C - \phi_D) \tag{6.5}$$

$$\sigma_{1t_2}(y) = \tfrac{1}{2} \cdot (-\phi_A + \phi_B + \phi_C - \phi_D) \tag{6.6}$$

$$\sigma_{1t_2}(z) = \tfrac{1}{2} \cdot (\phi_A - \phi_B + \phi_C - \phi_D) \tag{6.7}$$

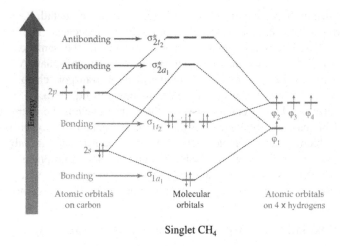

Singlet CH$_4$

FIGURE 6.6 The molecular orbital diagram for methane CH_4 illustrates how valence hydrogen and carbon atomic orbitals combine to form an equal number of molecular orbitals.

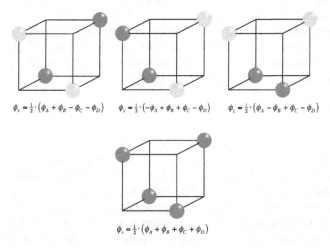

$\phi_x = \frac{1}{2} \cdot (\phi_A + \phi_B - \phi_C - \phi_D)$ $\phi_y = \frac{1}{2} \cdot (-\phi_A + \phi_B + \phi_C - \phi_D)$ $\phi_z = \frac{1}{2} \cdot (\phi_A - \phi_B + \phi_C - \phi_D)$

$\phi_s = \frac{1}{2} \cdot (\phi_A + \phi_B + \phi_C + \phi_D)$

FIGURE 6.7 The four linear combinations of hydrogen $1s$ atomic orbitals represented by expressions (6.4) through (6.7) that match the tetrahedral symmetry of the carbon $2s$ and $2p$ atomic orbitals in CH_4. Shading denotes phase—that is, the sign of the coefficient c_{ij}.

The electronic structure of methane is representative of most bioorganic compounds; all of the electrons in the highest occupied molecular orbital (*HOMO*) are paired. The pairing of all electrons in the *HOMO* makes them *closed-shell* compounds.

Atomic oxygen has 6 electrons in its valence atomic shell; when two oxygen atoms combine to form a molecule of dioxygen O_2, a total of 12 electrons

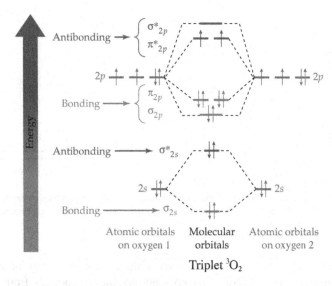

Triplet $^3O_2$

FIGURE 6.8 The molecular orbital diagram for ground state (triplet) dioxygen $^3O_2$ uses arrows to denote the electron spin state. The 6 electrons in each oxygen atom, a total of 12 electrons, do not completely fill the seven occupied molecular orbitals in dioxygen O_2.

are available to fill the molecular orbitals. The molecular orbital diagram for dioxygen O_2 appears in Figure 6.8. The two *HOMOs* have the same energy and are half filled with one electron in each; this is the lowest energy or *ground state* electron configuration. The two unpaired electrons in *triplet*[2] dioxygen $^3O_2$ make it an *open-shell* compound. Its electron configuration is fundamentally different from that of methane and other closed-shell compounds.

Molecular orbital theory forbids closed-shell compounds from reacting with open-shell compounds. The reaction is allowed if the closed-shell compound (e.g., methane) enters an excited state with an open-shell electron configuration. A molecule enters an excited state by absorbing sufficient energy to promote a single electron from the HOMO to the LUMO, changing its electron configuration. Groundstate triplet dioxygen can enter an excited singlet state by absorbing sufficient energy to promote one of its unpaired electrons to pair with the other electron (Figure 6.9), changing it to a closed-shell electron configuration.

The energy required to excite either the closed-shell organic compound (to an excited-state open-shell configuration) or open-shell dioxygen (to an

---

2. A triplet electron configuration has three possible orientations of the unpaired electrons in a magnetic field *H*: both unpaired electron spins aligned in the same direction as the field *H*, one spin aligned with and one aligned against the field *H* and both unpaired spins aligned in the opposite direction of the magnetic field *H*.

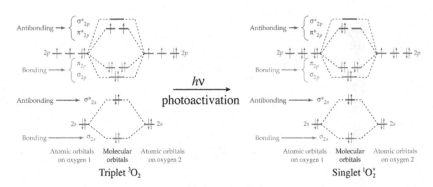

FIGURE 6.9 The molecular orbital diagrams showing excitation of ground state triplet $^3O_2$ to the lowest energy singlet $^1O_2^*$ excited state. The unpaired electrons in triplet $^3O_2$ become paired (i.e., a closed-shell configuration) in the excited state singlet $^1O_2^*$.

excited-state closed-shell configuration) is the activation energy needed to overcome the energy barrier that prevents organic compounds from being oxidized by dioxygen (Figure 6.10).

The rationale behind Figures 6.8 to 6.10 may seem difficult to grasp, but consider what it takes to start a fire where dioxygen is the oxidizer and wood tinder is the propellant. A magnifying glass can focus sufficient sunlight to heat wood tinder to its *auto-ignition* temperature. The auto-ignition temperature (or kindling point) is the temperature required to heat the propellant to

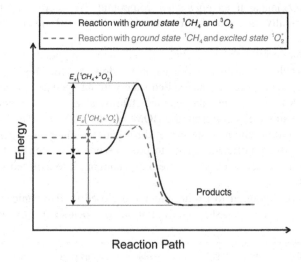

FIGURE 6.10 The ground state energy barrier $E_a(^1CH_4 \cdots ^3O_2)$ prevents oxidation of closed-shell (singlet) $^1CH_4$ by open-shell (triplet) $^3O_2$. The excited state energy barrier $E_a(^1CH_4 \cdots ^1O_2^*)$ is lower, allowing a reaction between closed-shell (singlet) $^1O_2^*$ and closed-shell (singlet) $^1CH_4$.

the open-shell excited state, removing the energy barrier blocking combustion by ground state (triplet $^3O_2$) dioxygen.

Bacteria, fungi, and other decomposers use enzyme catalysis to bypass the energy barrier blocking oxidation. Most of the chemical energy stored in biosphere organic compounds is released by enzyme-catalyzed biological oxidation reactions. Organic compounds represent kinetically stable chemical energy that organisms release at a rate matched to their metabolic needs.

6.3. SOIL CARBON

Belowground carbon decreases with increasing depth within the soil profile; relatively little carbon is found below the plant root zone in the intermediate vadose and saturated zones. This distribution reflects the source: photosynthetic carbon fixation by the plant community. Litter fall from standing plant biomass accumulates at the surface and may, especially in forest ecosystems, form an organic soil horizon (>18% organic carbon by weight). Belowground net primary production comes from the root system that supports a diverse decomposer community that, along with root biomass, represents total belowground biomass.

One technical criterion of soil includes all mineral grains, organic residue and biomass that pass through a 2 *mm* sieve. This definition excludes many arthropods, earthworms, and larger plant roots—to name a few—but does include much of the decomposer community. Soil biomass, regardless of whether it is from the plant community or the soil decomposer supported by net primary production, is the starting point of the carbon trajectory that ultimately leads to the humic colloids mentioned in the introduction.

6.3.1. Carbon Turnover Models

Net primary production—the total carbon fixed by photosynthesis during the growing season—ends up in one of three forms: carbon dioxide, biomass, and biological residue. Respiration of and biosynthesis derived from photosynthetic carbon represent investments in building and maintaining plant biomass. Biological residue, the third component, is the second stage of the soil carbon trajectory.

Ecologists have developed models to describe the turnover of biological residue—its mineralization to carbon dioxide, assimilation into decomposer biomass, and sequestration as belowground soil organic carbon. Hans Jenny (1941) proposed one of the simplest models (Eq. (6.8)) of organic carbon C_{soil} turnover in soils. Jenny originally applied his turnover model to soil organic nitrogen N_{soil} dynamics (Figure 6.11).

The Jenny soil organic carbon (or nitrogen) turnover model is a rate law (or rate equation) that combines a zero-order term—the annual addition of

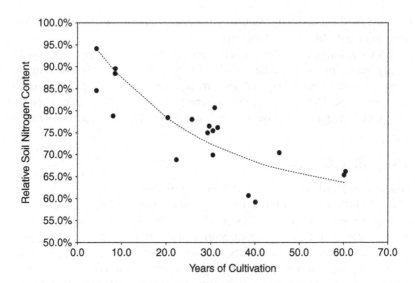

FIGURE 6.11 The accelerated mineralization of soil organic nitrogen during 60 years of agricultural cultivation resulted in a net 35% loss of soil nitrogen in midwestern U.S. soils. *Source: Reproduced with permission from Jenny, H., 1941. Factors of Soil Formation: A System of Quantitative Pedology. New York, McGraw Hill.*

plant residue A from net primary production—and a first-order term that describes the mineralization of soil organic carbon $k_M \cdot C_{soil}$[3]:

$$\frac{\Delta C_{soil}}{\Delta t} = A - k_M \cdot C_{soil} \qquad (6.8)$$

The units for each of the terms in Eq. (6.8) are $A[kg_{OC} \cdot m^{-2} \cdot y^{-1}]$, $C_{soil}[kg_{OC} \cdot m^{-2}]$ and $k_M [y^{-1}]$.

Henin and Dupuis (Henin, 1945) proposed a slightly different model (Eq. (6.9)) that adds the *isohumic coefficient* k_{ic}, a first-order rate constant for the conversion of net primary production C_{NPP} into soil organic carbon C_{soil}. The Jenny model (Eq. (6.8)) assumes that all of the residue input A will transform into soil organic carbon C_{soil}:

$$\frac{\Delta C_{soil}}{\Delta t} = k_{ic} \cdot C_{NPP} - k_M \cdot C_{soil} \qquad (6.9)$$

Turnover rate laws (Eqs. (6.8) and (6.9)) appear in *differential* form. The integral forms (Eqs. (6.10) and (6.11)) are better suited to fitting measured changes in soil organic carbon content $C_{soil}(t)$ as a function of time. The soil organic carbon content at the beginning of the turnover experiment is $C_{soil}(0)$:

---

3. The constant k_M is the first-order rate constant for mineralization.

$$Jenny\ (1941): \quad C_{soil}(t) = \left(\frac{A}{k_M}\right) + \left(\frac{k_M \cdot C_{soil}(0) - A}{k_M}\right) \cdot e^{-k_M \cdot t} \quad (6.10)$$

$$Henin\ \&\ Dupuis\ (1945):$$
$$C_{soil}(t) = \left(\frac{k_{ic} \cdot C_{NPP}}{k_M}\right) + \left(\frac{k_M \cdot C_{soil}(0) - k_{ic} \cdot C_{NPP}}{k_M}\right) \cdot e^{-k_M \cdot t} \quad (6.11)$$

The soil organic carbon content will converge on a steady-state content $C_{soil}(\infty)$ after a time interval that is long relative to the mineralization rate constant k_M:

$$Steady\text{-}State: \quad \frac{\Delta C_{soil}}{\Delta t} = 0$$

$$C_{soil}(\infty) = \frac{A}{k_M} = \frac{k_{ic} \cdot C_{NPP}}{k_M} \quad (6.12)$$

Substituting Eq. (6.12) into Eq. (6.10) results in a linearized expression for changes in soil organic carbon as it converges on its steady-state content:

$$\frac{C_{soil}(t) - C_{soil}(\infty)}{C_{soil}(0) - C_{soil}(\infty)} = e^{-k_M \cdot t}$$

$$\ln\left(\frac{C_{soil}(0) - C_{soil}(\infty)}{C_{soil}(t) - C_{soil}(\infty)}\right) = k_M \cdot t$$

$$\log_{10}\left(\frac{C_{soil}(0) - C_{soil}(\infty)}{C_{soil}(t) - C_{soil}(\infty)}\right) = \left(\frac{k_M}{\ln(10)}\right) \cdot t = \left(\frac{k_M}{2.303}\right) \cdot t \quad (6.13)$$

Table 6.1 lists data from four soil organic carbon turnover studies (Salter, 1933; Jenny, 1941; Myers, 1943; Haynes, 1955).

The soil organic carbon mineralization rate constant k_M and half-life $t_{1/2}$ for soil organic carbon for these midwestern U.S. agricultural soils (see Table 6.1)

TABLE 6.1 Rate Constant and Half-Life for Soil Organic Carbon in Agricultural Soil

Mineralization Rate Constant k_M [y^{-1}]	Half-Life $t_{1/2}$ [years]	Location	Reference
0.0608	11	Ohio (USA)	(Jenny, 1941)
0.052	13	Ohio (USA)	(Salter, 1933)
0.072	10	Ohio (USA)	(Haynes, 1955)
0.07	10	Kansas (USA)	(Myers, 1943)

are very similar. Both turnover models (Eqs. (6.10) and (6.11)) recognize two carbon pools: plant residue and soil organic carbon.

Woodruff (1949) proposed a model that subdivided soil organic carbon into several fractions j, each with a characteristic mineralization rate constant k_{Mj}. This representation of soil organic carbon is embodied in the mathematically complex *Rothamsted* carbon turnover model (Jenkinson, 1977, 1990). Plant material is subdivided into two forms—decomposable plant material *DPM* and resistant plant material *RPM*—and each type of plant residue has a characteristic rate constant for its conversion into soil organic carbon:

$$\text{Jenkinson and Rayner (1977):} \quad C_{soil}(t) = C_{DPM} \cdot e^{-k_{DPM} \cdot t} + C_{RPM} \cdot e^{-k_{RPM} \cdot t}$$

(6.14)

Data from a relatively short-term experiment measuring the mineralization of a one-time addition of ^{14}C labeled ryegrass appear in Figure 6.12, clearly showing the precipitous loss of decomposable plant material *DPM* during the first year followed by a more gradual loss of resistant plant material *RPM* over the remainder of the study.

The Rothamsted model divides soil organic carbon into two sub-pools: microbial biomass *BIO* and humus *HUM*. All biologically fixed carbon (net primary production *NPP*) flows through three pathways (Figure 6.13):

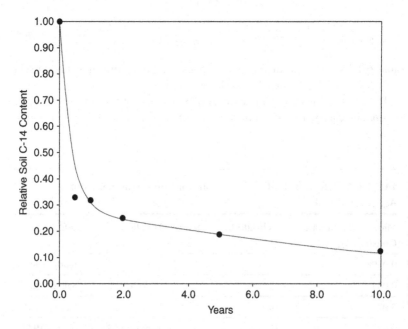

FIGURE 6.12 Decomposition of ^{14}C-labeled ryegrass under field conditions (Jenkinson, 1990) is fitted to Eq. (6.14).

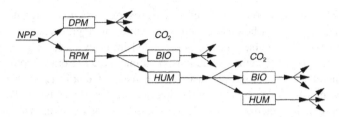

FIGURE 6.13 The carbon transformation pathways and carbon pools of the Rothamsted soil organic carbon turnover model: net primary production *NPP*, decomposable *DPM*, and resistant *RPM*, mineralization CO_2, assimilation *BIO*, and transformation into soil humus *HUM*. Each pathway has its characteristic transformation rate constant. *Source: Reproduced with permission from Jenkinson, D.S., Andrew, S.P.S., et al., 1990. The turnover of organic carbon and nitrogen in soil. Philosophical Transactions of the Royal Society of London Series B Biological Sciences 329 (1255), 361–368.*

mineralization into CO_2, assimilation into chemoorganotrophic biomass *BIO*, or transformation into soil humus *HUM*. Plant material initially partitions (Eq. (6.14)) between rapidly decomposing and slowly decomposing fractions; thereafter the carbon either is mineralized or partitions between the biomass of the soil decomposer community and soil humus.

6.3.2. Soil Carbon Pools

Scientists recognizing multiple soil carbon pools have developed methods designed to quantify the carbon in each pool. The scheme illustrated in Figure 6.14, which Zimmermann and colleagues (2007) used, is representative. Dissolved organic carbon *DOC* is defined operationally: any dissolved carbon passing a 0.45 μm filter membrane. Zimmermann and colleagues separate the particulate material into a coarse and fine fraction using a 63 μm

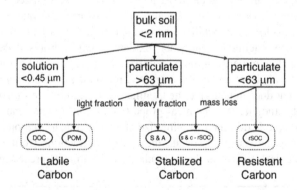

FIGURE 6.14 A representative fractionation procedure separates soil organic carbon into three pools: labile, stabilized, and resistant. *Source: Reproduced with permission from Zimmermann, M., Leifeld, J., et al., 2007. Quantifying soil organic carbon fractions by infrared-spectroscopy. Soil Biol. Biochem. 39, 224–231.*

sieve. The coarse (>63 μm) fraction consists of particulate organic matter *POM*, and organic carbon associated with stable soil aggregates (*A*) and sand (*S*) particles. The coarse fraction is separated into low (<1.8 $Mg \cdot m^{-3}$) and high (>1.8 $Mg \cdot m^{-3}$) density fractions. You may recall (from Chapter 2) that the mineral fraction is defined using the density of quartz (2.65 $Mg \cdot m^{-3}$). The low-density fraction is designated particulate organic matter *POM*, while the high-density fraction is stabilized through its association with mineral particles.

The fine particulate (<63 μm) fraction consists of any carbon associated with dispersed silt- and clay-sized mineral particles. The carbon associated with the fine particulate fraction is separated using chemical treatment with an oxidant or strong acid—typically 1 *M* sodium hypochlorite *NaOCl* (Kaiser, 2003; Zimmermann, 2007) or hot 6 *M HCl* (Trumbore, 1996; Six, 2002). Soil carbon in the fine particulate fraction oxidized by *NaOCl* is designated stabilized carbon (*s* and *c*, respectively), while that remaining after treatment is designated resistant carbon (*rSOC*).

The *labile carbon*[4] pool cannot be dated using ^{14}C radiocarbon methods. It is the fraction with the shortest residence time. Identifiable biochemical compounds in soils—amino acids, peptides, active enzymes, organic acids, carbohydrates, lipids, surfactants, siderophores, vitamins, growth factors, quorum sensing agents, and so on—belong to the labile carbon pool. Components of this soil carbon pool retain considerable activity, both chemical and biological, despite their relatively short residence time.

Stabilized carbon[5] is most susceptible to changes in land use and climate change (see Table 6.1). Its residence time is on the order of 10 years, sufficiently long to be dated using ^{14}C radiocarbon methods (Trumbore, 1996).

Resistant carbon[6] is the most stable soil carbon pool, with residence times on the order of hundreds to thousands of years (Baisden, 2002). Turnover of the resistant carbon pool is the most uncertain of the recognized soil carbon pools (Jenkinson, 1990), largely because the transformation rate is so slow.

Distinctions between stabilized carbon and resistant carbon are operationally defined by chemical treatments (oxidation and hydrolysis) designed to simulate the slow mineralization of stabilized carbon. Any soil carbon surviving harsh chemical treatment is designated resistant to mineralization, while the carbon that does not survive the treatment is susceptible to mineralization.

Chemists studying the chemical properties of soil organic matter rely on extraction methods designed to minimize chemical transformation. As a consequence, it is unwise to associate the chemical fractions (following) with carbon pools defined by resistance to chemical treatment (preceding).

---

4. This fraction is also called active carbon or annual carbon by other scientists (Parton, 1987).
5. The stabilized carbon pool is also called intermediate carbon or decadal carbon (Townsend, 1997).
6. This resistant carbon pool is also called passive carbon or millennial carbon.

6.4. DISSOLVED ORGANIC CARBON

The labile carbon pool consists of two fractions according to the fractionation scheme in Figure 6.14: dissolved organic carbon (*DOC*) and particulate organic carbon (*POM*). These soil carbon fractions represent organic material isolated without the use of chemical treatment.[7] This section identifies some of the bioorganic compounds dissolved in soil pore water. The bioorganic compounds dissolved in soil pore water are extremely labile, characterized by soil half-lives on the order of hours to days. Their short half-life does not diminish their importance because they are extremely reactive and continually replenished by soil organisms.

6.4.1. Organic Acids

Soil pore water, particularly in the rhizosphere,[8] contains a large variety of low molecular weight organic acids. Certain plant species actively secrete weak acids from their roots to increase nutrient availability (e.g., citric acid (Gardner, 1982), malic acid, fumaric acid, and oxalic acid). Other weak acids (e.g., the lignin precursors, Figure 6.15) leak from plant roots during cell wall biosynthesis. Bacteria and fungi release a variety of organic acids as fermentation by-products: succinic acid, butanoic acid, propionic acid, and lactic acid.

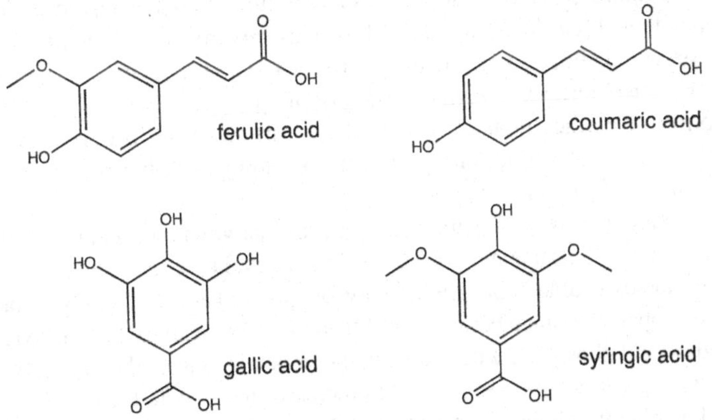

FIGURE 6.15   Lignin precursor molecules: ferulic acid, courmaric acid, gallic acid, and syringic acid.

7. Some authors use the term "water extractable organic carbon" (*WEOM*) interchangeably with *DOM* (Chantigny, 2003). Dissolved organic matter (*DOM*) encompasses all major elements (carbon, oxygen, nitrogen, etc.), while dissolved organic carbon (*DOC*) or dissolved organic nitrogen (*DON*) is based solely on carbon (or nitrogen) analysis.

8. "The rhizosphere is the narrow region of soil directly around roots. It is teeming with bacteria that feed on sloughed-off plant cells and the proteins and sugars released by roots" (Ingham, 1999).

Organic acids dissolved in soil pore water are on the order of 10–100 $mmol_c \cdot L^{-1}$ (Quideau, 1997). Proton dissociation constants and formation constants for metal-weak acid complexes for many of the common organic acids are included in the popular computer-based water chemistry models (see Chapter 5, Water Chemistry) or can be found in comprehensive chemistry databases (e.g., *Critically Selected Stability Constants of Metal Complexes*; U.S. National Institute of Standards & Technology).

6.4.2. Amino Acids

Free amino acids in soil pore water are on the order of 10^{-5} $mol \cdot L^{-1}$ (Jones, 2002; Jones, 2004; Hannam, 2003) and represent about 1 to 10% of the dissolved organic nitrogen (*DON*). The remaining *DON* is unidentifiable and presumably derived from soluble humic compounds (see following). The solution chemistry of free amino acids is best understood in the context of computer-based water chemistry models (see Chapter 5, Water Chemistry).

6.4.3. Extracellular Enzymes

Some enzymes remain active after the death of plant and microbial cells, associated with ruptured (lysed) cell debris or adsorbed to soil particles. Soil biochemists refer to these as *extracellular* enzymes. Many *endocellular* enzymes require cofactors, but active extracellular enzymes cannot be cofactor dependent (Pietramellara, 2002). Enzymes, regardless of whether they are endo- or extracellular, tend to lose activity outside a relatively narrow pH range because tertiary enzyme structure is pH sensitive. Apparently the adsorption of extracellular enzymes to soil particles stabilizes tertiary structure, at least active-site tertiary structure, extending both the active pH range and the lifetime in soil.

A recent review assayed extracellular enzyme activity in soil samples collected at 40 U.S. sites (Sinsabaugh, 2008). The assays evaluated the activity of seven enzyme classes (Table 6.2). The activity of glucosidase and cellobiohydrolase shows no apparent change between pH 4.0 and 8.5. The activity of N-acetylglucosaminidase and phosphatase decreases in the same pH range, while the activity of the remaining extracellular soil enzymes increases throughout the range.

Soil biochemists believe that extracellular enzymes explain soil organic matter turnover rates exceeding predictions based strictly on soil microbial biomass and respiration. There is no question that extracellular phosphatase enzymes are essential for phosphate release in soils (see "Chemical Composition" following).

6.4.4. Siderophores

Plants, fungi, and bacteria face identical challenges to meet their iron nutritional requirements. Kosman (2003) provides an excellent summary of these chemical challenges in the typical pH range of aerated soils: (1) Fe^{2+}

TABLE 6.2 Soil Extracellular Enzymes

Extracellular Enzyme	Reaction	Classification
β-1,4-glucosidase	Hydrolysis of terminal β-1,4-glucosyl residues, releasing β-D-glucose.	EC 3.2.1.21
cellobiohydrolase	Hydrolysis of 1,4-β-D-glucosidic linkages in cellulose, releasing cellobiose.	EC 3.2.1.91
β-N-acetylglucosaminidase	Random hydrolysis of N-acetyl-β-D-glucosaminide (1→4)-β-linkages in chitin.	EC 3.2.1.14
aminopeptidase	Hydrolysis of N-terminal amide linkages, releasing amino acid.	EC 3.4.11.1
acid (or alkaline) phosphatase	Hydrolysis of phosphate monoester, releasing phosphate.	EC 3.1.3.1
phenol oxidase	Hydroxylation of diphenols to semiquinones, releasing water.	EC 1.10.3.2
peroxidase	Oxidation (wide specificity) using H_2O_2.	EC 1.11.1.7

Source: Sinsabaugh, 2008.

spontaneously oxidizes to Fe^{3+}; (2) Fe^{3+} undergoes hydrolysis to form insoluble ferric oxyhydroxide minerals (Fe^{2+} has little tendency to hydrolyze, thereby explaining its higher inherent solubility); and (3) Fe^{3+} forms extremely stable complexes that are kinetically inert (Fe^{2+} forms weak complexes that are very labile to ligand exchange). Kosman places particular emphasis on the final chemical property; even if Fe^{3+} were not susceptible to hydrolysis (i.e., insoluble), the kinetic stability of its complexes would render them biologically inert. Plants, bacteria, and fungi rely on elaborate biological systems to solubilize, transport, and absorb ferric iron from soil pore water; central to this system is a family of compounds known as *siderophores*.

The list of siderophores in Table 6.3 is not comprehensive but is representative of the siderophores secreted by soil fungi and bacteria. Microbiologists report numerous siderophores in addition to those listed in Table 6.3; many remain uncharacterized, their structures unknown, and their detailed iron chemistry uncertain. Structures **1** and **2** illustrate the two major classes, respectively: *hydroxamate* and *catecholamide* (see Appendix 6A). Rhizoferrin (Structure **3**) is representative of the siderophore class known as *carboxylates*. Species of the *Poaceae* (or *Gramineae*) family are the only plants known to secrete phytosiderophores: mugineic acid (Structure **4**) and related compounds (nicotianamine, avenic acid, 2'-dehyxroxymugenic acid, 3-hydroxymugenic acid, etc.).

TABLE 6.3 Representative Fungal and Bacterial Siderophores

Siderophore	Class	Organism
Coprogen	Hydroxamate	Fungi
Ferrichrome	Hydroxamate	Fungi
Ferrioxamine	Hydroxamate	Bacteria
Ferrichrysin	Hydroxamate	Fungi
Ferrirubin	Hydroxamate	Fungi
Ferrirhodin	Hydroxamate	Fungi
Rhodotorulic acid	Hydroxamate	Fungi
Fusigen	Hydroxamate	Fungi
Putrebactin	Hydroxamate	Bacteria
Rhizoferrin	Carboxylate	Fungi
Aminochelin	Catecholamide	Bacteria
Protochelin	Catecholamide	Bacteria
Azotochelin	Catecholamide	Bacteria
Pyoverdine	Hydroxamate-catecholamide	Bacteria
Enterobactin	Hydroxamate-catecholamide	Bacteria
Salmochelin	Hydroxamate-catecholamide	Bacteria
Azotobactin	Hydroxamate-catecholamide	Bacteria
Aerobactin	Carboxylate-hydroxamate	Bacteria

Structure 1

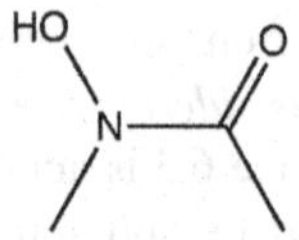

Structure 2

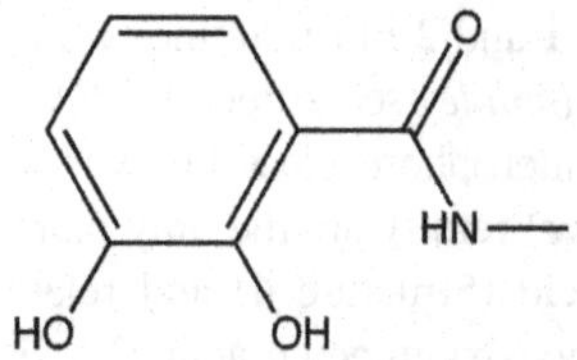

Structure 3

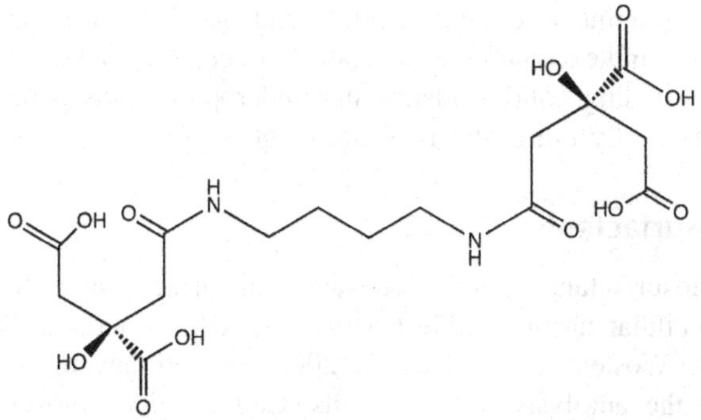

Structure 4

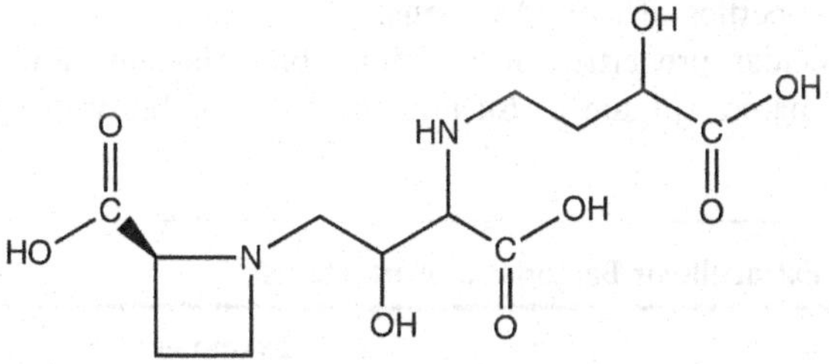

Fungi (Kosman, 2003; Philpott, 2006) rely on three uptake pathways to balance their iron nutritional requirements against iron cytotoxicity: a nonreductive siderophore uptake pathway, a reductive iron uptake pathway, and a direct ferrous uptake pathway. Plants, regardless of whether they secrete siderophores, employ two of these pathways: a reductive iron uptake pathway and a direct ferrous uptake pathway. Bacteria, regardless of whether they secrete siderophores, employ two of these pathways: a nonreductive siderophore uptake and a direct ferrous uptake pathway. The direct ferrous uptake pathways (plants, fungi, bacteria)—which are not operative under aerobic conditions when iron solubility is at its lowest—employ low-affinity transmembrane ion channels (or porins).

Poaceae use phytosiderophores to transport Fe^{3+} ions to root surfaces, where ferric reductase enzymes reduce Fe^{3+} to Fe^{2+} (Chaney, Brown, et al., 1972; Römheld, 1981). The labile ferrous complex releases Fe^{2+} to ion channels that transport it into the cytoplasm. The fungal reductive iron uptake pathway is functionally equivalent to the plant reductive iron uptake pathway.

The nonreductive siderophore uptake pathways in both bacteria and fungi rely on outer membrane receptors that bind the ferri-siderophore complex (Ferguson, 2002; Kosman, 2003; Philpott, 2006). Once the ferri-siderophore reaches the cytoplasm, it is reduced, and the labile ferrous complex releases Fe^{2+} to a cytoplasmic chaperone. As noted by Kosman (2003), Fe^{2+} is cytotoxic, making iron uptake and homeostasis a tightly regulated process in all living organisms.

The dissolved iron concentration in the pore water of neutral aerated soils consistently falls in the 1–10 μm range (Fuller, 1988; Grieve, 1990; Ammari,

2006), well above the saturation concentration of even the most soluble secondary ferric oxide minerals but within the range required to meet the nutritional requirements of plants, fungi, and bacteria. Weak organic acids do not form complexes sufficiently stable to account for the observed iron solubility, providing solid evidence that siderophore soil pore water concentrations typically fall in the 1–10 μm range.

6.4.5. Biosurfactants

Microbial biosurfactants in soil pore water and surface water are secreted as either extracellular agents (Table 6.4) with specific functions (Desai, 1997; Lang, 2002; Wosten, 1999; Wosten, 2001) or cell envelope constituents released by the autolysis of dying cells. Cell envelope biosurfactants are believed to influence cell attachment to environmental surfaces by altering the surface properties of individual cells.

The molecular properties of microbial biosurfactants arise from their amphiphilic[9] nature (Figure 6.16): limited water solubility (a result of their

TABLE 6.4 Extracellular Bacterial Biosurfactants

Biosurfactant	Examples
Glycolipids	Rhamnolipids, mycolic acids
Lipoproteins	Surfactin, visconsin, putisolvin
Proteins	Hydrophobins

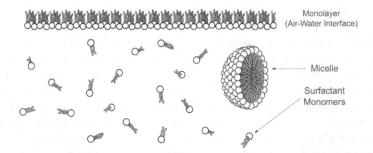

FIGURE 6.16 Surfactant molecules self-organize, forming films at the air-water interface and colloidal molecular aggregates (micelles) dispersed in water. Illustration adapted from Wikipedia.

9. Amphiphilic molecules have a primary structure that contains at least one significant segment that is strongly hydrophilic or polar and at least one significant segment that is strongly hydrophobic or nonpolar.

hydrophobic molecular segment) and a tendency to form micelles above a characteristic critical micelle concentration (*CMC*).

An intriguing consequence of extracellular biosurfactants is the enhanced solubility and bioavailability of hydrophobic or sparingly soluble organic contaminants in soil pore water (Cameotra, 2003; Mulligan, 2005). This behavior is intimately dependent on self-organization (see Figure 6.16) and—at least in the surface microlayer (see Appendix 6B)—does not require biosurfactant concentrations exceeding a *CMC*.

6.5. HUMIC SUBSTANCES

The elemental and chemical composition of soil organic matter and its chemical properties are virtually impossible to determine *in situ* because humic substances[10] are intimately associated with soil mineral colloids (Mikutta, 2006). This section describes a typical method for extracting soil organic matter with minimal chemical alteration and a fractionation scheme based on pH-dependent solubility and concludes with a summary of the elemental and chemical composition of soil organic matter extracts and their most salient chemical properties.

6.5.1. Extraction and Fractionation

The *International Humic Substance Society* (*IHSS*) maintains a collection of soil, aquatic, and peat humic substances extracted by a standard protocol. The protocol begins with an initial 1-hour extraction with 0.1 *M HCl* to dissolve an acid-soluble fulvic acid fraction. The second step employs 4-hour extraction 0.1 *M NaOH* to dissolve base-soluble humic substances, taking care to exclude dioxygen to minimize oxidation at high pH. The humic extract typically contains a considerable quantity of suspended minerals. The humic extract is acidified by adding 6 *M HCl* and allowed to stand for 12–16 hours.[11] The supernatant is the second acid-soluble fulvic acid fraction—which is combined with the first acid-soluble extract—while the precipitate is the humic acid fraction.

The humic acid fraction is dissolved a second time using 0.1 *M KOH* and sufficient *KCl* to make the final $K^+(aq)$ concentration 0.2 *M* (this step is designed to coagulate much of the entrained mineral colloids). The humic acid fraction is precipitated a second time using a combination of 0.1 *M HCl* and 0.3 *M HF* for 12 hours (this step is designed to dissolve any remaining silicate clays associated with the humic acids). The final cleanup involves

---

10. Soil and environmental chemists commonly designate natural organic matter suspended in fresh or marine water, associated with mineral particles in soils and sediments or preserved in peat as humic substances.

11. The standard method for protein hydrolysis refluxing 24 hours at 110°C in 6 *M HCl*.

adjusting to pH 7 and dialyzing to remove excess salts. The organic carbon that cannot be dissolved by this treatment protocol is designated humin.

Efforts to maximize the yield of organic carbon in the extract or to minimize chemical alteration by oxidation and hydrolysis have led to alternative extraction methods, some using organic solvents rather than aqueous solution, but the IHSS protocol is the most widely adopted and represents an acceptable compromise between yield and alteration. The fractions—humic acids, fulvic acids, and humin—are operationally defined: humic acids are soluble in basic solutions but precipitate at pH 1, fulvic acids are soluble at and above pH 1, and humin is the insoluble organic carbon residue (Rice, 1990).

6.5.2. Elemental Composition

The most comprehensive study of the elemental composition of humic substances is the report by Rice and MacCarthy (1991). They report composition probability distributions for five elements (C, H, O, N, and S) for a population of 215 soil humic acid samples, 127 soil fulvic acid samples, and 26 humin samples (soil, peat, aquatic, and marine sources pooled). Rice and MacCarthy also list statistics for humic and fulvic acids from peat, aquatic, and marine sources.

Most of the composition probability distributions appear to be symmetric or normal distributions, and the exceptions are indicated in Table 6.5. As a consequence, the arithmetic mean and standard deviation are suitable distribution statistics for most sample populations.

TABLE 6.5 Mean and Standard Deviation of Soil Humic Acids ($N = 215$), Soil Fulvic Acids ($N = 127$), and Humin ($N = 26$)

Soil Humic Acids	C, $g_C \cdot g_{OM}^{-1}$	H, $g_H \cdot g_{OM}^{-1}$	O, $g_O \cdot g_{OM}^{-1}$	N, $g_N \cdot g_{OM}^{-1}$	S, $g_S \cdot g_{OM}^{-1}$
Mean	0.554	0.048	0.360	0.036	0.008[†]
Standard Deviation	0.038	0.010	0.037	0.013	0.006[†]
Soil Fulvic Acids	C, $g_C \cdot g_{OM}^{-1}$	H, $g_H \cdot g_{OM}^{-1}$	O, $g_O \cdot g_{OM}^{-1}$	N, $g_N \cdot g_{OM}^{-1}$	S, $g_S \cdot g_{OM}^{-1}$
Mean	0.453	0.050	0.462	0.026	0.013
Standard Deviation	0.054	0.010	0.052	0.013	0.011
Humin	C, $g_C \cdot g_{OM}^{-1}$	H, $g_H \cdot g_{OM}^{-1}$	O, $g_O \cdot g_{OM}^{-1}$	N, $g_N \cdot g_{OM}^{-1}$	S, $g_S \cdot g_{OM}^{-1}$
Mean	0.561	0.055	0.347	0.037	0.004[†]
Standard Deviation	0.026	0.010	0.034	0.013	0.003[†]

Note: The statistics for sulfur (†) are geometric; all others are arithmetic. The sample populations for the sulfur analyses are consistently smaller because the sulfur content is rather low: humic acids ($N = 67$), soil fulvic acids ($N = 45$), and humin ($N = 16$).

Source: Rice, 1991.

TABLE 6.6 The Mean Mole Fraction $\overline{X}$ for Each of the Most Abundant Elements in Humic Acid, Fulvic Acid, and Humin

	$\overline{X}_C$	$\overline{X}_H$	$\overline{X}_O$	$\overline{X}_N$	$\overline{X}_S$	$\overline{N}_C$
Soil Humic Acids	0.387	0.400	0.189	0.022	0.002	+0.111
Soil Fulvic Acids	0.318	0.419	0.244	0.016	0.003	+0.364
Humin	0.371	0.434	0.172	0.021	0.001	−0.070

Note: The last column is the mean formal oxidation number of carbon $\overline{N}_C$.
Source: Rice, 1991.

Rice and MacCarthy found remarkably low variation in elemental composition within each population and no significant difference between humic acid and humin fractions. The only statistically significant composition difference between populations was the composition of the fulvic acid fraction relative to humic acid and humin fractions.

Most scientists reporting elemental composition quote mass fractions. The mole fractions in Table 6.6, computed from the data in Table 6.5, are more compelling because elements combine on a molar basis. The oxygen-to-nitrogen mole ratio in fulvic acids (≈ 16) is twice the ratio in humic acids and humin (≈ 8). The oxygen-to-carbon mole ratio in fulvic acids (≈ 0.8) is nearly twice the ratio in humic acids and humin (≈ 0.5). These results clearly indicate that the fulvic acid fraction has significantly higher oxygen content than the humic acid and humin fractions.

Using the results from Table 6.6, we can estimate the mean formal oxidation number (Example 6.1) of carbon $\overline{N}_C$ in each humic fraction. Chemical bonding details are irrelevant provided that nitrogen is assigned a formal oxidation number of (−3). It makes no difference whether hydrogen is assigned to H-O bonds or H-C bonds because, ultimately, the formal oxidation number of oxygen is (−2). The median formal oxidation number for carbon $\overline{N}_C$ in each of the three humic fractions is listed in the last column of Table 6.3.

6.5.3. Chemical Composition

Biochemists employ a hierarchy to describe the structure of complex biological molecules. The primary structure of a peptide is the amino acid sequence, while the primary structure of a glycan (or polysaccharide) is the saccharide sequence. Secondary structure refers to three-dimensional geometry of a particular segment of a biomolecule, while tertiary structure refers to the three-dimensional geometry of the entire biomolecule.

Example 6.1

Estimating the mean formal oxidation number of carbon $\bar{N}_C$ is one way to apply the elemental composition data of humic substances. In this example we use the mean composition of soil humic acids (Table 6.5), applying rules #3 and #4 from the protocol used in this book (*Clay Mineralogy & Clay Chemistry*, Appendix 3A): $N_H = (+1)$, $N_N = (-3)$ and $N_O = (-2)$. The estimate requires five steps to assign valence electrons to chemical bonds before estimating the formal oxidation number using the electronegativity principle.

Step 1: Divide the mass fraction of each element in soil humic acid fraction (Table 6.5) by the atomic mass of the element.

$$\frac{0.554 \ g_C \cdot g_{OM}^{-1}}{12.0107 \ g_C \cdot mol^{-1}} = 0.0461 \ mol_C \cdot g_{OM}^{-1}$$

$$\frac{0.048 \ g_H \cdot g_{OM}^{-1}}{1.0079 \ g_H \cdot mol^{-1}} = 0.0476 \ mol_H \cdot g_{OM}^{-1}$$

$$\frac{0.360 \ g_O \cdot g_{OM}^{-1}}{15.9994 \ g_O \cdot mol^{-1}} = 0.0225 \ mol_O \cdot g_{OM}^{-1}$$

$$\frac{0.036 \ g_N \cdot g_{OM}^{-1}}{14.0067 \ g_N \cdot mol^{-1}} = 0.0026 \ mol_N \cdot g_{OM}^{-1}$$

Step 2: Normalize the moles per gram of each element by the moles of carbon per gram from Step 1.

$$X_C = \frac{0.0461 \ mol_C \cdot g_{OM}^{-1}}{0.0461 \ mol_C \cdot g_{OM}^{-1}} = 1.000 \ mol_C \cdot mol_C^{-1}$$

$$X_H = \frac{0.0476 \ mol_H \cdot g_{OM}^{-1}}{0.0461 \ mol_C \cdot g_{OM}^{-1}} = 1.032 \ mol_H \cdot mol_C^{-1}$$

$$X_O = \frac{0.0225 \ mol_O \cdot g_{OM}^{-1}}{0.0461 \ mol_C \cdot g_{OM}^{-1}} = 0.488 \ mol_O \cdot mol_C^{-1}$$

$$X_N = \frac{0.0026 \ mol_O \cdot g_{OM}^{-1}}{0.0461 \ mol_C \cdot g_{OM}^{-1}} = 0.056 \ mol_O \cdot mol_C^{-1}$$

Step 3: Assign valence hydrogen electrons $(1 \cdot X_H)$ to *H-O* bonds, followed by *H-N* bonds, ending with *H-C* bonds. Assume all nitrogen is amide-*N*, assigning only one *H-N* bond per nitrogen. Oxygen has two valence electrons but remains attached to the humic molecules only if half of the oxygen electrons are assigned to *H-O* bonds.

$$n(H - O) = X_O = 0.488 \ mol_{H-O} \cdot mol_C^{-1}$$

$$n(H - N) = X_N = 0.056 \ mol_{H-N} \cdot mol_C^{-1}$$

$$n(H - C) = (1 \cdot X_H) - n(H - O) - n(H - N) = 0.488 mol_{H-C} \cdot mol_C^{-1}$$

Step 4: Assign valence carbon electrons $(4 \cdot X_C)$ to *C-O* bonds, followed by *C-N* bonds, ending with *C-C* bonds. Because most organic nitrogen is amide-*N* (see Figure 6.18), assign two *C-N* bonds per nitrogen.

$$n(C - O) = X_O = 0.488 \; mol_{C-O} \cdot mol_C^{-1}$$

$$n(C - N) = 2 \cdot X_N = 0.111 \; mol_{C-N} \cdot mol_C^{-1}$$

$$n(C - C) = (4 \cdot X_C) - n(H - C) - n(C - O) - n(C - N) = 2.913 mol_{C-C} \cdot mol_C^{-1}$$

Step 5: The formal oxidation number of carbon $\bar{N}_C$ in soil humic acid (Table 6.5) is found by accounting for electron loss in C-O and C-N bonds and the electron gain from H-C bonds. There is no need to count C–C bonds because they have no effect on $\bar{N}_C$.

Bond	Net Electron Transfer Relative to Carbon
$n(C - O)$	−0.488
$n(C - N)$	−0.111
$n(H - C)$	+0.488
N_C	+0.111

Atomic carbon has *four* valence electrons $(4 \cdot X_C)$ balanced by *four* protons in the nucleus. The net electron transfer relative to carbon means each carbon *loses* 0.111 electrons resulting in a net effective charge of +0.111 because nuclear charge remains unaffected by bonding.

$$\bar{N}_C = (n(C - O) + n(C - N))_{electron \; loss} - (n(H - C))_{electron \; gain}$$

$$\bar{N}_C = ((0.488 \; mol_{C-O} \cdot mol_C^{-1}) + (0.111 \; mol_{C-N} \cdot mol_C^{-1})) - (0.488 \; mol_{H-C} \cdot mol_C^{-1})$$

$$\bar{N}_C = +0.111$$

The key implication is this: assigning hydrogen to H-O bonds is equivalent to assigning carbon to C-O bonds. Assigning hydrogen to H-O bonds reduces the number of C-O bond required to satisfy the requirement: $N_O = (-2)$. Alternatively, assigning hydrogen to H-C bonds increases the number of C-O bond required to satisfy the requirement: $N_O = (-2)$.

Despite considerable research, we still have limited understanding of how the elements listed in Tables 6.5 and 6.6 combine to form chemical moieties[12] in humic substances. There is no evidence that humic molecules are polymers in the sense that peptides, glycans, and polynucleotides are polymers. Furthermore, we have no understanding of primary and secondary molecular structure in humic substances. Our understanding of the tertiary humic molecular structure is limited but probably more reliable than any other structural characteristic.

Our knowledge of chemical moieties—that is, *local* molecular structure—derives from a variety of chemical and instrumental methods: vibrational or infrared spectroscopy (IR), nuclear magnetic resonance spectroscopy (NMR), photoelectron absorption spectroscopy covering the spectrum from visible to x-ray, and mass spectrometry. The physical details of how these methods identify specific chemical moieties and quantify their abundance are beyond the scope of this book, but we will summarize the most reliable results.

---

12. "Chemical moiety is generally used to signify part of a molecule" (Muller, 1994).

Acidometric Titration

Acidometric titration is the most common method for quantifying the weak acid content of humic samples. Acidometric titration curves for two humic samples appear in Figure 6.17: Suwannee River (Georgia, USA) aquatic fulvic acid and Summit Hill (New Zealand) soil humic acid. The amount of base added during the titration $Q_{TOT} \left[mmol_c \cdot g_C^{-1}\right]$ is normalized by the mass of dissolved humic carbon.

The displacement of the fulvic acid titration curve in Figure 6.17 reveals the higher weak acid content of the fulvic acid sample relative to the humic acid sample, consistent with the significantly higher oxygen content of fulvic acid fractions relative to humic acid fractions.

The hydrolysis constants (K_{a1}, K_{a2}) determine the pH of the equivalence points, as indicated in Figure 6.17. The amount of each weak acid— $Q_1 \left[mmol_c \cdot g_C^{-1}\right]$ and $Q_2 \left[mmol_c \cdot g_C^{-1}\right]$, respectively—determines the position of the respective equivalence points along the Q_{TOT}-axis.

$$Q_{TOT} = \left(\frac{Q_1}{1 + (K_{a1} \cdot a_{H^+})^{1/n_1}} \right) + \left(\frac{Q_2}{1 + (K_{a2} \cdot a_{H^+})^{1/n_2}} \right) \tag{6.15}$$

Titration curves of dissolved humic molecules are significantly broadened compared to the titration curve typical of a weak acid with a single functional

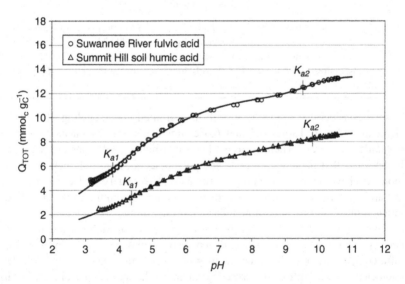

FIGURE 6.17 The acidometric titration curve of a fulvic acid and a humic acid sample from the International Humic Substance Society collection identifies two weak acid populations: a weak acid population with mean hydrolysis constant K_{a1} and a very weak acid population with mean hydrolysis constant K_{a2}. *Source: Reproduced with permission from Ritchie, J.D., Perdue, E.M., 2003. Proton-binding study of standard and reference fulvic acids, humic acids, and natural organic matter. Geo. Cosmo. Acta 67 (1), 85–96.*

group.[13] The two remaining parameters (n_1, n_2) express this broadening effect (see Figure 5 in (Katchalsky, 1947)). Katchalsky and Spitnik (Katchalsky, 1947) attribute a diffuse equivalence point to electrostatic interactions between closely spaced weak acid groups. Some humic chemists interpret equivalence point broadening to chemical heterogeneity of the weak and very weak acids.

The electrostatic effect, in its simplest terms, reduces the tendency of protons to dissociate from weak acid moieties when surrounded by chemically identical moieties that have already dissociated. In essence, the proximity of negatively charged (i.e., dissociated weak acid) sites exerts a collective attractive electrostatic force on protons, lowering the tendency of remaining weak acid moieties to release their protons.

Structure 5

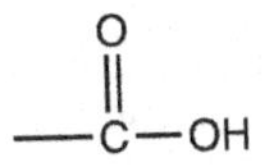

$$\overset{\displaystyle O}{\underset{\displaystyle }{\overset{\displaystyle \|}{-C}}}-OH$$

Structure 6

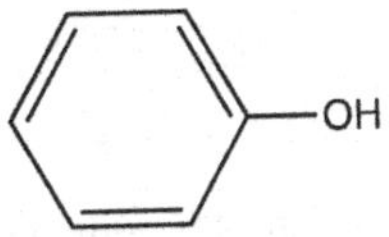

The smooth line through each titration curve in Figure 6.17 is the best fit by a six-parameter model (6.15) proposed by Katchalsky and Spitnik (1947). Model (6.15) represents the total acidity of a humic sample as the sum of a weak acid—nominally a carboxyl moiety (Structure 5) with hydrolysis constant K_{a1}—and a very weak acid—nominally a phenol moiety (Structure 6) with hydrolysis constant K_{a2}.

Appendix 6C illustrates how the elemental composition of a humic sample and the quantification of weak acid content by acidometric titration can be used to assess the chemical forms of oxygen in a humic sample. Regardless of whether the sample is a fulvic acid or a humic acid, a considerable fraction of the oxygen in humic samples cannot be assigned to weak acid moieties.

Nitrogen-15 Nuclear Magnetic Resonance Spectroscopy

What little we know about nitrogen in humic substances comes from N-15 ($^{15}_{7}N$) NMR. Nuclear magnetic resonance spectroscopy is an isotope-specific method. Although both stable nitrogen isotopes have a nuclear magnetic moment, and both absorb microwave radiation in an NMR experiment, each isotope presents distinct challenges when studying humic substances.

Nitrogen isotope $^{14}_{7}N$ is very abundant (99.63%), but the large electric quadrupole moment of its nucleus causes profound signal broadening, rendering $^{14}_{7}N$ of little practical use (Lambert, 1964). Nitrogen isotope $^{15}_{7}N$, though not susceptible to electric-quadrupole signal broadening, produces a relatively

---

13. Katchalsky and Spitnik (Katchalsky, 1947) report this behavior for polymeric acids.

TABLE 6.7 Nuclear magnetic properties of the most abundant elements in humic and fulvic acids

Isotope	Natural Abundance	Nuclear Spin	Nuclear Gyromagnetic Ratio $\gamma\ [MHz \cdot T^{-1}]$	Relative Sensitivity[†]
$^{1}_{1}H$	99.985%	+1/2	267.513	1.0000
$^{2}_{1}H$	0.015%	+1	41.065	$5.43 \cdot 10^{-7}$
$^{13}_{7}C$	1.10%	+1/2	67.262	$1.69 \cdot 10^{-4}$
$^{14}_{7}N$	99.63%	+1	19.331	$2.05 \cdot 10^{-5}$
$^{15}_{7}N$	0.37%	+1/2	−27.116	$2.11 \cdot 10^{-7}$
$^{17}_{8}O$	0.038%	−5/2	−36.264	$4.48 \cdot 10^{-7}$
$^{31}_{15}P$	100%	+1/2	108.291	$1.39 \cdot 10^{-4}$
$^{33}_{16}S$	0.75%	+3/2	20.534	$1.43 \cdot 10^{-8}$

[†]*Relative sensitivity is the product of the mean element abundance of soil humic acid (Rice, 1991), isotope abundance, and the cube of the gyromagnetic ratio $(\gamma_i/\gamma_{1_H})^3$, where i indicates isotope.*

weak signal—$2.11 \cdot 10^{-7}$ relative to $^{1}_{1}H$—in natural humic samples because both its low natural abundance (0.37%) (Lambert, 1964) and small gyromagnetic ratio γ (Table 6.7).

Thorn and Cox (Thorn, 2009) recorded the natural abundance N-15 NMR spectra of fulvic and humic acids, shown in Figure 6.18. The recording of these spectra required 56–560 hours, reflecting the low relative sensitivity of natural abundance N-15 NMR. Thorn and Cox (Thorn, 2009) assign the 119–120 ppm resonance peak to amide nitrogen typical of peptides (Structure **7**) and N-acetylated amino-polysaccharides (Structures **8** and **9**) and the 31–36 ppm to amine nitrogen found in amino sugars and terminal amino acids. Amide nitrogen typically accounts for 75 to 85% of the total organic nitrogen, terminal amine 6 to 8%, and unidentified nitrogen accounts for the remainder.

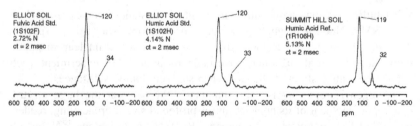

FIGURE 6.18 Nitrogen-15 nuclear magnetic resonance spectra of soil fulvic and humic acid samples from the IHSS collection. Chemical shifts are relative to the single N-15 resonance in the amino acid glycine. *Source: Reproduced with permission from Thorn, K.A., Cox, L.G., 2009. N-15 NMR spectra of naturally abundant nitrogen in soil and aquatic natural organic matter samples of the International Humic Substances Society. Org. Geochem. 40, 484–499.*

The survival of amide nitrogen in humic substances extracted by the IHSS method is significant. The acid phase of the IHSS extraction protocol represents conditions suitable for the acid-hydrolysis of labile amide linkages. Aluwihare and colleagues (2005) found that mild acid[14] hydrolysis (1 *M HCl*, 10-hour reflux at 90°C) converted about 24% of amide-*N* into amine-*N* via de-acetylation hydrolysis. They suggest that most of the amide in marine organic nitrogen is acetylated amino sugars (Structures **8** and **9**).

Structure 7. *Source: Reproduced with permission from Aluwihare, L.I., Repeta, D.J., et al., 2005. Two chemically distinct pools of organic nitrogen accumulate in the ocean. Science 308, 1007–1010.*

peptide amide moiety

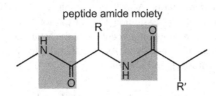

Structure 8. *Source: Reproduced with permission from Aluwihare, L.I., Repeta, D.J., et al., 2005. Two chemically distinct pools of organic nitrogen accumulate in the ocean. Science 308, 1007–1010.*

N-acetyl polysaccharide amide moiety

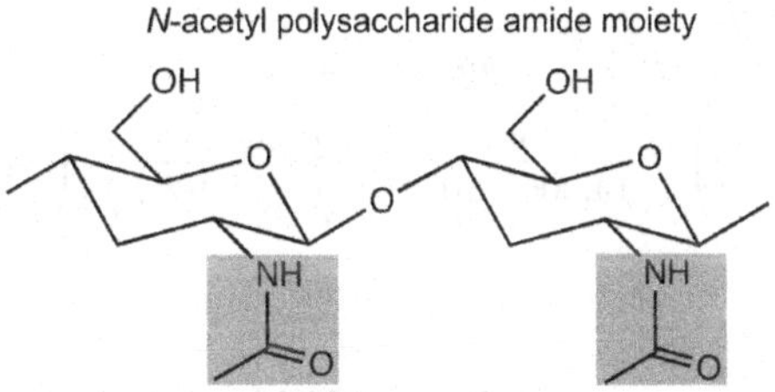

Peptidoglycan (Structure **9**)—the primary cell wall polymer in bacteria—is cross-linked by tetrapeptide segments besides containing acetylated amino sugars. Peptide hydrolysis releases free amino acids, while de-acetylation releases acetic acid from chitin and lactic acid from peptidoglycan; both hydrolysis reactions convert amide-*N* to amine-*N*.

Structure 9

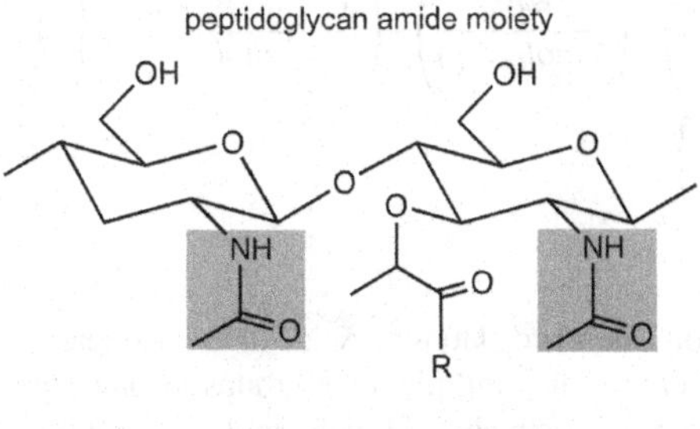

peptidoglycan amide moiety

14. Strong acid hydrolysis (6 *M HCl*, 12-hour reflux at 110°C) released hydrolyzes 21% of amide–*N* into free amino acids (Henrichs, 1985).

Although the significance of amide-N as the major form of humic nitrogen remains a topic of scientific discussion, it strongly suggests that microbial cell wall residue is a prominent secondary source of humic substances. The assimilation of plant carbon (and nitrogen) as a soil microbial biomass is consistent with the *Rothamsted* turnover model (see Figure 6.13) and represents an evolving understanding of the respective roles of plant and microbial biomass in the formation of humic substances.

Example 6.2

The Elliott Soil Standard Humic Acid (IHSS sample 1S102H) has the following carbon and nitrogen content. Assuming 15% of the total nitrogen is amine and the remaining 85% is amide, what percentage of the total Cu^{2+} binding capacity can be assigned to amine complexes? Assume that the total binding capacity is the sum of amine, carboxyl, and phenol.

Carbon, $kg_{carbon} \cdot kg^{-1}_{1S102H}$	Nitrogen, $kg_{nitrogen} \cdot kg^{-1}_{1S102H}$	Carboxyl, $mol_c \cdot kg^{-1}_{carbon}$	Phenol, $mol_c \cdot kg^{-1}_{carbon}$
0.5813	0.0414	8.28	1.87

amine groups

$$\left(\frac{6.21 \cdot 10^{-3} \, kg_{amine-N}}{kg_{1S102H}}\right) \cdot \left(\frac{mol_N}{14.0067 \cdot 10^{-3} \, kg_N}\right) \cdot \left(\frac{1 \, mol_{Cu}}{1 \, mol_N}\right) \cdot \left(\frac{63.546 \cdot 10^{-3} \, kg_{Cu}}{mol_{Cu}}\right)$$

$$= \left(\frac{0.028 \, kg_{Cu}}{kg_{1S102H}}\right)_{amine}$$

carboxyl groups

$$\left(\frac{8.28 \, mol_{carboxylic}}{kg_C}\right) \cdot \left(\frac{1 \, mol_{Cu}}{1 \, mol_{carboxyl}}\right) \cdot \left(\frac{63.546 \cdot 10^{-3} \, kg_{Cu}}{mol_{Cu}}\right) \cdot \left(\frac{0.5813 \, kg_C}{kg_{1S102H}}\right)$$

$$= \left(\frac{0.306 \, kg_{Cu}}{kg_{1S102H}}\right)_{carboxylic}$$

phenol groups

$$\left(\frac{1.87 \, mol_{phenolic}}{kg_C}\right) \cdot \left(\frac{1 \, mol_{Cu}}{1 \, mol_{carboxyl}}\right) \cdot \left(\frac{63.546 \cdot 10^{-3} \, kg_{Cu}}{mol_{Cu}}\right) \cdot \left(\frac{0.5813 \, kg_C}{kg_{1S102H}}\right)$$

$$= \left(\frac{0.069 \, kg_{Cu}}{kg_{1S102H}}\right)_{phenolic}$$

The Elliot Soil Standard Humic Acid (IHSS sample 1S102H) contains sufficient amine, carboxylic, and phenolic groups to bind 28 *mg*, 306 *mg*, and 69 *mg Cu* per gram, respectively, of humic acid, for a total binding capacity of 403 *mg Cu* per gram of humic acid.

Nitrogen-Containing Chemical Moieties in Humic Substances

Natural abundance nitrogen-15 NMR clearly identifies two nitrogen-containing chemical moieties in humic substances: amine and amide. The chemical significance is twofold. First, amide moieties have low affinity for bonding to trace metals, making amine moieties the only significant nitrogen-containing trace-metal bonding site in humic substances (Example 6.2). Second, amide moieties occur in peptides (Structure 7) and amino sugars (Structures 8 and 9). Peptide amides are labile to both extracellular soil protease enzymes and the IHSS extraction protocol, suggesting acetylated amino sugars may explain the slow rate of organic nitrogen turnover: soil organic nitrogen half-life on the order of 100 years (see Figure 6.11).

Phosphorus-31 Nuclear Magnetic Resonance Spectroscopy

Phosphorus-31 NMR results indicate that humic-associated phosphorus takes the form of phosphate mono- and diesters (Figure 6.19 and Structure 10). The cluster of resonance peaks in the range +3.5 ppm to +5.2 ppm is assigned to monoesters, while the broad peak in the range −0.5 ppm to −0.9 ppm is assigned to diesters. The resonance at +5.6 ppm is orthophosphate PO_4^{3-} that would be absent if the sample were thoroughly dialyzed to remove all small ions. The resonance peak at −5.0 ppm is pyrophosphate that often appears in P-31 NMR spectra of humic extracts.

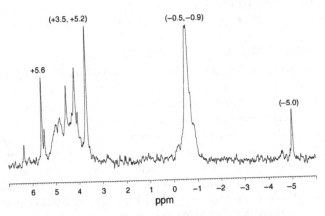

FIGURE 6.19 Phosphorus-31 nuclear magnetic resonance spectra of soil humic acid samples collected from the Bankhead National Forest (Alabama), courtesy of R. W. Taylor and T. Ranatunga. Chemical shifts are relative to the single P-31 resonance in phosphoric acid, an external reference.

Structure 10

$$\underset{HO}{\overset{O}{\underset{\|}{R\diagdown_{O}\diagdown_{\cdots\cdots}\overset{\|}{P}\diagup_{O}\diagup^{R}}}}$$

The final step in the alkaline IHSS humic extraction protocol removes by dialysis the excess salt that accumulates during the various extraction treatments. Dialysis traps humic molecules while allowing small ions to diffuse through the pores of the dialysis membrane. An indeterminate amount of the phosphorus in the initial extract is lost during dialysis as orthophosphate ion $PO_4^{3-}(aq)$. Some of orthophosphate in the dialysate derives from mineral colloids dispersed and ultimately dissolved during the extraction procedure; the remaining orthophosphate derives from the hydrolysis of phosphate mono- and diesters. Some humic chemists studying humic-associated phosphorus modify their extraction protocol to specifically minimize phosphate ester hydrolysis.

It is unclear whether the phosphate esters detected by *P*-31 *NMR* (Figure 6.19) derive from the labile soil carbon pool (i.e., microbial biomass and decomposable plant residue) or from esters that form during humification.

Phosphorus-Containing Chemical Moieties in Humic Substances

Natural abundance phosphorus-31 *NMR* identifies phosphate mono- and diesters as the dominant phosphorus chemical moieties in humic substances. Phosphate esters (Structure **10**) represent a significant source of phosphorus fertility in natural ecosystems. Extracellular phosphatase enzymes secreted by plant roots, soil bacteria, and soil fungi (Dodd, 1987; Kim, 1998) hydrolyze phosphate ester bonds, releasing orthophosphate ions into soil solution for plant and microbial uptake.

Sulfur K-edge X-Ray Absorption Spectroscopy

Our understanding of organosulfur in humic substances draws heavily on *x-ray absorption spectroscopy*. In many respects the level of our understanding of organosulfur is comparable to our understanding of organonitrogen, despite differences in chemistry and spectroscopic method.

Sulfur absorbs x-ray photons in a relatively narrow energy range above 2472.0 eV—the sulfur *K* x-ray absorption edge (or *K*-edge). X-ray absorption by sulfur at the *K*-edge excites an electron in the lowest energy atomic orbital—an electron in a 1*s* atomic orbital—to one of the lowest unoccupied molecular orbitals (LUMOs). The LUMO energy distribution mirrors the bonding interactions that determine the energy of each occupied molecular orbital (see Figures 6.6, 6.8, and 6.9).

Figure 6.20 displays the normalized x-ray absorption spectrum for *thiol* (–*SH*) sulfur in the amino acid L-cysteine. Peaks and oscillations near the

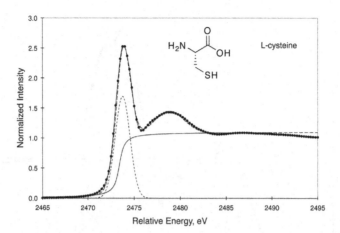

FIGURE 6.20   Normalized sulfur K-edge x-ray absorption spectrum of the thiol (-SH) sulfur in L–cysteine shows negligible x-ray absorption below the absorption edge (2472.5 eV) but considerable absorption above the edge.

edge—the x-ray absorption near-edge structure (XANES) or near-edge x-ray absorption fine structure (NEXAFS)—obscure the precise position of the x-ray absorption edge. Two key components of the sulfur K-edge XANES spectrum in Figure 6.20 are the absorption edge (an arctangent step function plotted as a solid line) and the *white-line* peak (a Gaussian function plotted as a dashed line) centered on the absorption edge.

The position of the edge step and both the position and intensity of the white-line peak are sensitive to the effective sulfur oxidation number. Formal oxidation numbers are useful when balancing electron transfer reactions because the reaction involves the transfer of an integral number of electrons from one molecule to another. Formal oxidation numbers are less reliable when interpreting XANES spectra because chemical bonding generally does not transfer a whole number of electrons between two atoms in the same compound.

By comparing the sulfur K-edge XANES spectra of natural samples with the XANES spectra of known organosulfur compounds, we can infer the presence of different sulfur oxidation states and their relative abundance in the humic sample. The peat fulvic acid fraction shown in Figure 6.21 contains what appear to be 6 distinct sulfur oxidation states. The estimated relative abundance of each sulfur oxidation state and its chemical name appear in Table 6.8.

Structure 11

$$-S-S-$$

Structure 12

$$-\overset{\overset{\textstyle O}{\|}}{C}-S-$$

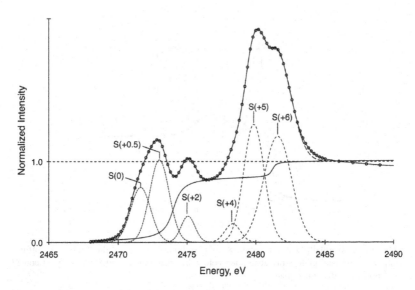

FIGURE 6.21 Normalized sulfur *K*-edge x-ray absorption spectra of sulfur in peat fulvic acid fraction (Loxely peat, Marcell Experimental Forest, Grand Rapids, MN; sample courtesy of P. R. Bloom).

TABLE 6.8 Formal Oxidation Numbers of Organosulfur Species and Their Relative Abundance in the Peat Fulvic Acid Fraction Shown in Figure 6.21

Sulfur Formal Oxidation Number	Relative Position of Absorption Edge	Relative Abundance
0.0	−0.4 *eV*	24%
+0.5	+1.0 *eV*	27%
+2	+3.5 *eV*	5%
+4	+5.5 *eV*	3%
+5	+8.5 *eV*	20%
+6	+10.0 *eV*	22%

The effective sulfur oxidation numbers listed in Figure 6.19 do not all match the formal oxidation numbers of representative organosulfur species listed in Table 6.9. The sulfur in disulfide (Structure **11**), thioester (Structure **12**), and thioether moieties all have a formal oxidation number of $S(0)$, assuming that the difference in Pauling electronegativity for carbon $\chi_C = 3.55$ and sulfur $\chi_S = 3.58$ is negligible.

TABLE 6.9 Formal Oxidation Numbers of Representative Organosulfur Species

Sulfur Formal Oxidation Number	Chemical Name	Relative Position of Absorption Edge
-0.5	Disulfide	-0.4 eV
$+0.5$	Thiols and thiol esters	$+1.0$ eV
$+2$	Thiosulfate	$+3.5$ eV
$+4$	Sulfone	$+5.5$ eV
$+5$	Sulfonate	$+8.5$ eV
$+6$	Sulfate esters	$+10.0$ eV

Thiol (-SH) would have a formal oxidation number of $S(-1)$, a result of the difference in Pauling electronegativity for hydrogen $\chi_H = 2.20$ and sulfur $\chi_S = 3.58$. The formal oxidation number assigned to thiosulfate (Structure **13**) is the average $S(+2)$ of the two sulfur atoms: $S(-1)$ and $S(+5)$. The sulfone sulfur (Structure **14**) is bonded to two oxygen atoms and two carbon atoms, resulting in a formal oxidation number of $S(+4)$. Sulfonate sulfur (Structure **15**) is identical to the $S(+5)$ sulfur in thiosulfate, while sulfate esters (Structure **16**) register a formal oxidation number of $S(+6)$.

Structure 13, 14

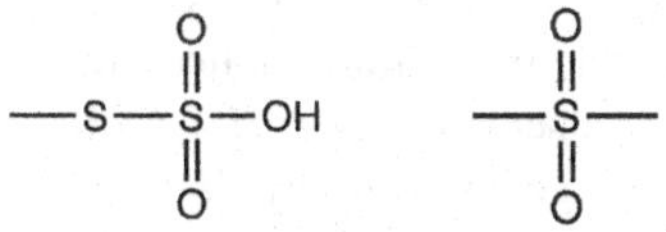

Structure 15, 16

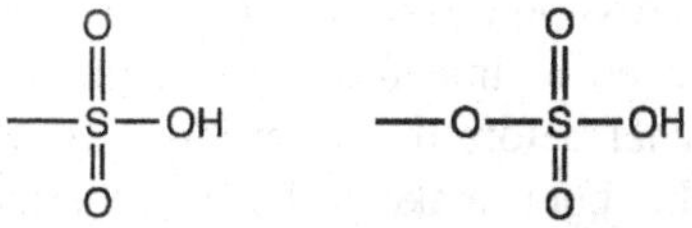

Chemical heterogeneity prevents scientists from distinguishing subtle variations in the chemical bonding of each sulfur oxidation state in humic substances. Sulfur K-edge XANES distinguishes little difference between thiol (-SH), disulfide (-SS-), and thioether (-S-) when all are present in a complex natural sample. Sulfur K-edge XANES allows humic chemists to quantify the effective oxidation states of sulfur in humic substances and to infer the identity of organosulfur moieties. What little we know about the chemical bonding of humic organosulfur moieties requires supplemental chemical information.

Sulfur-Containing Chemical Moieties in Humic Substances

Natural abundance sulfur K-edge XANES identifies multiple sulfur chemical moieties in humic substances representing a considerable range of formal oxidation numbers (see Figure 6.21). Extracellular soil sulfatase enzymes hydrolyze sulfate esters in a process resembling phosphate release by phosphatase enzymes in soil. Humic chemists have no explanation for the occurrence of sulfur moieties registering oxidation numbers greater than $S(+2)$ and less than $S(+6)$ because these intermediate forms do not correspond to the accepted mechanism for dissimilatory sulfate reduction (see Chapter 8).

Reduced organosulfur moieties have twofold importance: reducing agents and binding moieties for chalcophilic (sulfur-loving) trace metals. Szulczewski and colleagues (Szulczewski, 2001) found that "reduced sulfur" species in humic substances could reduce chromate CrO_4^{2-} to Cr^{3+}. Several scientists studying the binding of methylmercury CH_3Hg^+ by humic substances observed the formation of Hg-S bonds that originate from humic thiol ligands (Amirbahman, 2002; Qian, 2002; Yoon, 2005). Roughly one-third of the "reduced sulfur" in the humic samples appears to be capable of binding methylmercury CH_3Hg^+ cations. The unreactive "reduced sulfur" could be thio ether, disulfide, and thioester moieties.

Hydrogen-1 and Carbon-13 Nuclear Magnetic Resonance Spectroscopy

The assignment of proton or H-1 ($_1^1H$) NMR spectra derives largely from the carbon to which the protons are bonded, providing essentially the same chemical information found in C-13 ($_6^{13}C$) NMR spectra. The most practical distinction between H-1 NMR spectra and C-13 NMR spectra of humic substances is found in Table 6.4; natural abundance C-13 NMR spectroscopy is roughly 10,000-fold less sensitive than H-1 NMR.

The sensitivity advantage enjoyed by H-1 NMR enhances a particular C-13 NMR experiment: cross-polarization $^1H \rightarrow {}^{13}C$ solid-state NMR (Pines, 1973; Schaefer, 1976). Hydrogen-1 nuclei—rather than carbon-13 nuclei—absorb radiation in a cross-polarization $^1H \rightarrow {}^{13}C$ solid-state NMR experiment and are then coaxed to transfer their magnetic alignment to a dilute population of C-13 nuclei before the spectrum is recorded. Technical issues beyond the scope of this book make H-1 NMR the method of choice if the humic sample is dissolved, while cross-polarization $^1H \rightarrow {}^{13}C$ solid-state NMR is the method of choice for solid humic samples.

Regardless of the method, H-1 and C-13 NMR spectra of humic substances contain resonance peaks assigned to the following carbon moieties: aromatic, aliphatic, enols (alcohols or polysaccharides), and ketones (including carboxylic groups). Keeping in mind that these assignments pertain to the primary structure of humic molecules, NMR spectroscopy has yielded little reliable information on humic secondary molecular structure. In simple terms, we know the type of carbon moieties found in humic substances but are uncertain of their precise relative abundance or how these simple structures combine to form an entire molecule. Since humic substances are a mixture of humic molecules, each with a different secondary structure, it is unlikely we will ever be able to draw the structure of a humic molecule whose tertiary structure and molecular properties match the collective properties of a natural humic substance sample.

Quantifying Carboxyl Moieties in Humic Substances

Acidometric titration typically identifies two weak organic acid classes in humic extracts based on two equivalence points in the titration curves (see Figure 6.17). Carboxyl moieties estimated using NMR (Ritchie, 2008) overestimate titratable carboxyl acidity by roughly 40 to 50%. Nuclear magnetic resonance spectroscopy fails to distinguish carboxyl moieties from carboxyl ester and amide moieties, accounting for this overestimate. Phenol moieties cannot be reliably resolved using either H-1 or C-13 NMR. As in previous cases (N-15 NMR estimates of amine moieties and sulfur K-edge XANES estimates of thiol moieties), the NMR spectroscopic measurement overestimates chemically reactive moieties capable of binding metal ions or dissociating as a weak acid.

6.6. HUMIC COLLOIDS

Humic molecules dissolved in water exhibit properties similar to simple amphiphiles such as phospholipids. The formation of molecular-aggregate colloids above the critical micelle concentration (CMC) lowers the average diffusion rate of charge carriers in the electrical conductivity results plotted in Figure 6.22.

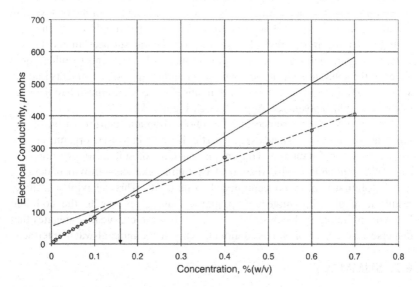

FIGURE 6.22 The critical micelle concentration of a humic acid solution ($\approx 0.16\%$(w/v)) appears as an inflection point (marked by an arrow) in a plot of the electrical conductivity of the molecular solution as a function of concentration.

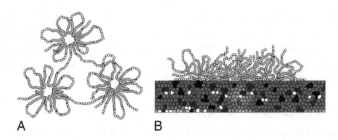

FIGURE 6.23 The molecular aggregates formed by segmented amphiphiles in aqueous solution (a) and at the mineral-water interface (b). Polar segments are white circles, and nonpolar segments are gray lines.

The adsorption of amphiphilic molecules at the air-water interface (see Figure 6.16) lowers the surface tension of water until a monolayer is formed, forcing molecules to aggregate as micelles in solution. Surface tension measurements of humic extract solutions consistently detect a CMC (Chen, 1978; Tschapek, 1978; Tschapek, Wasocusk, et al., 1981; Hayase, 1983; Yonebayashia, 1987; Kile, 1989; Shinozuka, 1991; Guetzloff, 1994; Wershaw, 1999).

It is unlikely, though not impossible, that humic molecules have a single polar end and a single nonpolar end (see Figure 6.16 and Appendix 6D). A more probable model for humic molecules would be a *segmented amphiphile*, a flexible molecule with alternating polar and nonpolar segments. Such molecules would fold and aggregate in water by presenting their polar segments at the aqueous contact and burying their nonpolar segments within their interior (Figure 6.23a).

Humic molecules have little tendency to form aggregates of the type seen in Figure 6.23a in the natural state because they adsorb strongly to soil mineral particles. Humic molecules self-organize on mineral surfaces in the natural state, folding in such a manner as to present their polar segments to either liquid water or the solid mineral surface, withdrawing their nonpolar segments into the aggregate interior to create nonpolar domains (Figure 6.23b) much like the interior of the lipid bilayer (see Appendix 6D, Figure 6D.1).

Evidence for the existence of hydrophobic domains capable of adsorbing nonpolar organic compounds (e.g., pesticides and industrial organic chemicals) appears in Chapter 9. The unique capacity of soil to adsorb nonpolar organic compounds is also displayed by humic extracts—provided that the humic solution is more concentrated than its CMC. This behavior is a direct result of amphiphilic molecular properties and gives rise to the colloidal behavior of humic molecules: their tendency to form colloidal aggregates that disperse in water or organized colloidal films at the mineral-water interface.

6.7. SUMMARY

Natural organic matter is a component of the terrestrial carbon cycle, representing both an important environmental substance and a vital component of the biological energy cycle. Photolithotrophs convert light energy into chemical energy during photosynthesis, storing this chemical energy in a host of bioorganic compounds comprising living biomass. The reduced carbon in bioorganic compounds is kinetically stable in the presence of ground state (triplet) dioxygen, a consequence of the peculiar electron configuration of the dioxygen molecule. The release of the chemical energy stored in natural organic matter requires enzymatic catalysis, firmly placing carbon turnover in the hands of a complex community of soil organisms.

The mathematical models quantifying organic carbon (and organic nitrogen) turnover have changed little in the past 70 years. The decomposition of soil organic carbon appears to follow a first-order rate law—a rate law familiar from general chemistry. Ecologists studying carbon turnover distinguish two or three carbon pools that turn over at different rates.

The labile carbon pool has a residence time of less than a year but contains many chemically and biologically active compounds. Though much of the dissolved organic carbon in pore water cannot be identified, environmental chemists have identified organic acids, free amino acids, active extracellular enzymes, siderophores, and biosurfactants (to say nothing of vitamins, growth factors, DNA fragments, etc.).

Chemically unidentifiable natural organic matter is collectively designated: humic substances. With the exception of the dissolved organic matter mentioned in the previous paragraph, humic substances occur largely as a colloidal film coating mineral particles. Environmental chemists routinely extract humic substances for chemical analysis and characterization. The elemental composition of humic extracts is remarkably invariant. Unfortunately, we know little about the primary or secondary structure of humic molecules beyond the identity of chemical moieties formed by the principle elements: carbon, oxygen, nitrogen, phosphorus, and sulfur. Our most comprehensive knowledge relates to humic molecular and colloidal properties rather than molecular structure.

Humic substances have a substantial capacity to adsorb ionic species from solution by ion exchange and complexation of metal ions to specific chemical moieties, a chemical property similar to mineral colloids in its effect on solution chemistry but arising from profoundly different chemical mechanisms. The affinity of humic substances for nonpolar organic compounds is a colloidal property that has no parallel in mineral colloids. Humic substances *in situ* are, in fact, hydrophilic colloids (see Appendix 6D) whose self-organized molecular arrangement results from the inherent amphiphilic properties of individual humic molecules.

APPENDIX 6A. HYDROXAMATE AND CATECHOLAMIDE SIDEROPHORE MOIETIES

Iron chemistry in the vadose zone and the water column of most freshwater and marine environments dramatically lowers iron biological availability. Bacteria and fungi have evolved complex systems to meet their iron nutritional requirements, including the secretion of siderophores—compounds that form extremely stable, water-soluble ferric iron complexes.

Biological activity is most intense under aerated conditions—throughout the water column in most freshwater and marine environments and above the saturated zone in terrestrial environments, precisely the conditions that favor ferric iron as the most stable oxidation state. Ferric iron solubility limits bacteria and fungal growth above pH 5, where total iron concentrations drop below 10^{-7}–10^{-8} M. The extremely low ferric iron solubility is a consequence of the Lewis acidity of hexaquoiron(3+), leading to its hydrolysis at low pH.

The Lewis acidity of $Fe(H_2O)_6^{3+}(aq)$ arises from the small ionic radius and large charge of the Fe^{3+} ion, the same physical characteristics that lower the ligand exchange rate of aqueous Fe^{3+} (Kosman, 2003). The slow rate of ligand exchange accounts for the high kinetic stability of Fe^{3+} complexes.

Ligand moieties that form two (or more) simultaneous bonds to Fe^{3+}—chelate complexes—further lower the ligand exchange rate because ligand exchange requires the simultaneous detachment of two (or more) ligand moieties. All natural siderophores contain bidentate ligand moieties; the entire ferrisiderophore complex is often hexa-dentate.

Structure A1

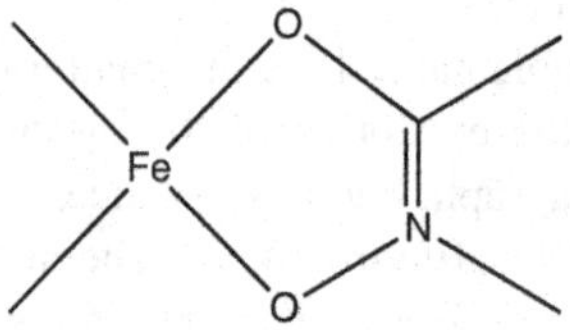

Structure A2

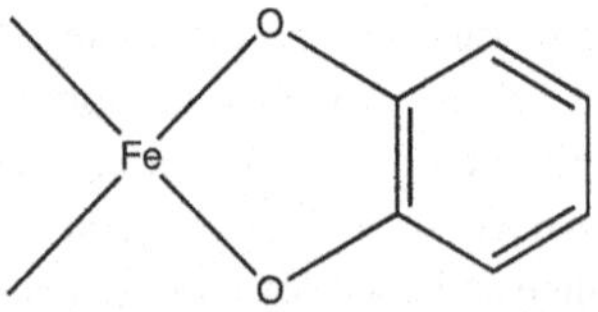

The two most common ligand moieties in natural siderophores are the hydro-xamate moiety (Structure **A1**) and the catecholate moiety (Structure **A2**). These two ligand moieties have several features in common: their bidentate geometry closely matches the preferred nonbonded $(O \cdot \cdot \cdot O)$ distance of the Fe^{3+} center by the relative length and bond angles in the ligand moiety, and both ligand moieties have bonding properties that permit fine-tuning of an already favorable geometry.

Structure A3

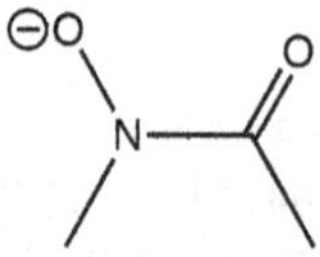

Structure A4

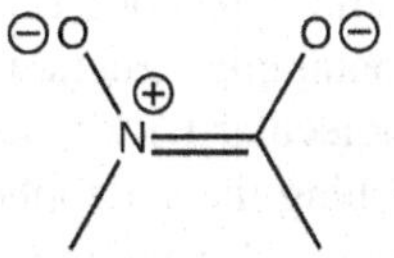

Bond resonance provides a mechanism for fine-tuning the geometry of the bidentate moiety to fit the preferred geometry of the Fe^{3+} center. The two most important hydroxamate resonance structures (Shanzer, 2009) are Structures **A3** and **A4**. Structure **A4** evenly distributes oxygen charge over the asymmetric hydroxamate ligand moiety, lengthens the $(C-O)$ bond, shortens the $(N-C)$ bond, and restricts free rotation by a partial $(N-C)$ double bond.

Siderophores based on bidentate catecholate moieties, with few exceptions, should be classified as catecholamide (Structure **2**). Intermolecular hydrogen bonding, common in catecholamide siderophores (Structures **A5–A7**), may permit fine-tuning of the bidentate geometry (Structure **A7**) (Hay, 2001).

Structure A5

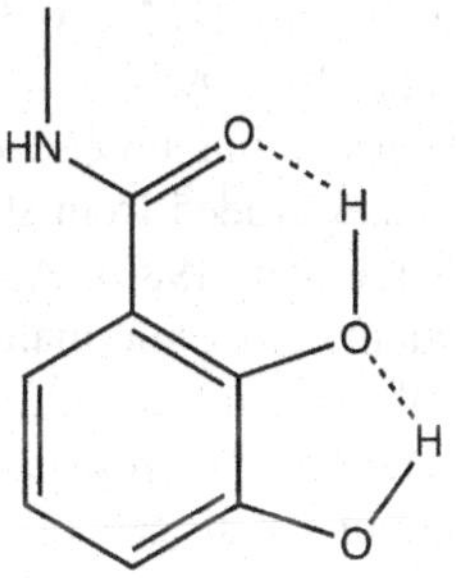

Structure A6

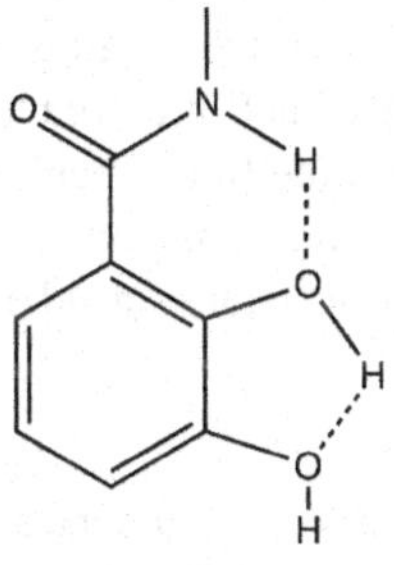

Structure A7

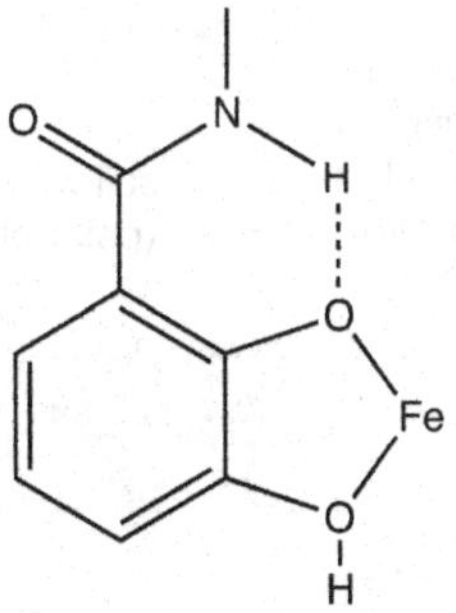

The bidentate ligand moieties are coupled together through the sidero-phore molecular backbone, containing structural moieties that increase water solubility. The siderophore molecular backbone also includes several hydro-gen-bonding moieties for attaching the siderophore to outer membrane recep-tors that transport the intact ferri-siderophore complex into the cytoplasm.

APPENDIX 6B. SURFACE MICROLAYERS

Chemical studies of freshwater and marine air-water interface date from the early 1970s (Duce, 1972; Andren, 1975; Liss, 1975; Mackin, 1980; Maguire, 1983; Meyers, 1983; Knulst, 1997; Knulst, 1998; Calace, 2007; Gasparovic, 2007; Wurl, 2008). These studies describe a surface microlayer—a chemically and biologically distinct layer ranging in thickness from a few tens to a few hundreds of nanometers—containing surface-active compounds (as evidenced by lower surface tension relative to pure water and compounds identified as fatty acids and fatty acid esters) and enriched in hydrophobic compounds relative to subsurface water (Duce, 1972; Andren, 1975; Maguire, 1983; Meyers, 1983; Knulst, 1997; Knulst, 1998).

Microbes living in the water column are one source of surface microlayer surfactants; soil humic substances eroded from surrounding land surfaces are another in freshwater bodies (Hunter, 1986). Andren and colleagues (1975) found that dissolved and particulate organic matter in the surface microlayer of Lake Mendota (Madison, Wisconsin) had ultraviolet-visible spectra indis-tinguishable from humic substances of terrestrial origin.

The organisms and conditions that generate the surface microlayer in freshwater and marine systems also exist in soils, as well as the humic colloi-dal film at the mineral-water interface (Figure 6B.1). The autolysis of dying bacteria and fungal cells releases cell membrane biosurfactants into soil pore water. Numerous bacteria and some fungi secrete exocellular biosurfactants as antibiotic or pathogenic agents and in response to the presence of hydrocarbon substrates. The microbial biosurfactants need not remain dissolved in pore

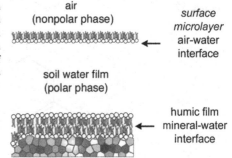

FIGURE 6B.1 The air-water interface in soils at water contents less than field capacity will be nearly equal to the total surface area of the soil. The water film thickness will be at most a couple of micrometers thick in coarse sand and less than a micrometer in finer-textured soils.

TABLE 6B.1 Select Properties of Lima Soil Series (Cayuga County, New York) and Savannah Soil Series (Clarke County, Mississippi)

Soil Series	Texture	Organic Carbon (wt%)	Surface Area, $m^2 \cdot g^{-1}$	Bulk Density ρ_B, $Mg \cdot m^{-3}$	Water Content[†] $\theta_{V(FC)}$, $m^3 \cdot m^{-3}$
Lima	Silt loam	2.99	6.04	1.10	0.14
Savannah	Sand	1.15	0.89	1.51	0.133

[†]Field capacity: $p_{FC} = -10\ kPa$ or $h_{FC} = -1.0\ m$.
Source: Nam, 1998.

water but can adsorb—along with dissolved humic substances (Herbert, 1995)—at the soil air-water interface. The following two examples provide estimates of the magnitude of the soil air-water interface and the water film thickness in gravity-drained soils.

Nam and colleagues (Nam, 1998) conducted a study of the relationship between soil organic matter content and the adsorption of phenanthrene. The organic carbon content and surface area of the soil are as reported by Nam and colleagues, but the remaining soil properties listed in Table 6B.1 are from the official soil series description.

Using soil bulk density ρ_B, we can estimate the surface area of each soil based on soil volume. The volume-based surface area of the Savannah soil series is given by expression (6B.1), which is an estimate of the total air-water interface in the Savannah soil series at field capacity.

$$Savannah\ Series: \left(\frac{0.89\ m^2}{10^{-6}\ Mg}\right) \cdot \left(\frac{1.51\ Mg}{m^3}\right) = 1.34 \cdot 10^6\ m^2 \cdot m^{-3} \quad (6B.1)$$

The volumetric water content at field capacity $\theta_{V(FC)}$ is a thin water film spread over the total soil surface area found by dividing the volumetric water content by the volume-based surface area (6B.2). The water film thickness at field capacity is remarkably thin.

$$Savannah\ Series: \left(\frac{0.13\ m^3 \cdot m^{-3}}{1.34 \cdot 10^6\ m^2 \cdot m^{-3}}\right) = 9.90 \cdot 10^{-8}\ m \approx 0.10\ \mu m \quad (6B.2)$$

Lenhart and Saiers (2004) studied the adsorption of natural organic matter at the air-water interface using a column packed with fine sand (300–355 nm). The volumetric water content θ_V was 0.12 $m^3 \cdot m^{-3}$ at a constant tension-head of $h_{tension} = -0.298\ m$. Given the uniform particle diameter of the sand (Accusand 40/60) and the precise water content measurements, we can easily estimate the mean volume-based area of the sand $5 \cdot 10^4 \, m^2 \cdot m^{-3}$ and water film thickness 2 μm.

APPENDIX 6C. HUMIC OXYGEN CONTENT AND TITRATABLE WEAK ACIDS

The Elliott Soil Standard Humic Acid (IHSS sample 1S102H) has the following composition: carbon and oxygen content, and carboxylic and phenolic acid content. What fraction of the total oxygen content cannot be assigned as weak acids?

Carbon, $kg_{carbon} \cdot kg^{-1}$	Oxygen, $kg_{oxygen} \cdot kg^{-1}$	Carboxyl, $mol_c \cdot kg^{-1}_{carbon}$	Phenol, $mol_c \cdot kg^{-1}_{carbon}$
0.5813	0.3408	8.28	1.87

carboxyl groups

$$\left(\frac{8.28 \, mol_{carboxylic}}{kg_C}\right) \cdot \left(\frac{2 \, mol_O}{1 \, mol_c}\right) \cdot \left(\frac{15.9994 \cdot 10^{-3} \, kg_O}{mol_O}\right) \cdot \left(\frac{0.5813 \, kg_C}{kg_{1S102H}}\right) = \frac{0.154 \, kg_{carboxylic-O}}{kg_{1S102H}}$$

phenol groups

$$\left(\frac{1.87 \, mol_{phenolic}}{kg_C}\right) \cdot \left(\frac{1 \, mol_O}{1 \, mol_c}\right) \cdot \left(\frac{15.9994 \cdot 10^{-3} \, kg_O}{mol_O}\right) \cdot \left(\frac{0.5813 \, kg_C}{kg_{1S102H}}\right) = \frac{0.017 \, kg_{phenolic-O}}{kg_{1S102H}}$$

$$\left(\left(\frac{0.154 \, kg_{carboxylic-O}}{kg_{1S102H}}\right) + \left(\frac{0.017 \, kg_{phenolic-O}}{kg_{1S102H}}\right)\right) \cdot \left(\frac{kg_{1S102H}}{0.3408 \, kg_O}\right) =$$

$$\left(\frac{0.171 \, kg_{weak \, acid-O}}{kg_{1S102H}}\right) \cdot \left(\frac{kg_{1S102H}}{0.3408 \, kg_O}\right) = \frac{0.503 \, kg_{weak \, acid-O}}{kg_O}$$

The titratable acidity of the Elliot Soil Standard Humic Acid (IHSS sample 1S102H) represents 50.3% of the total oxygen content. The remaining content must be assigned to some other chemical form.

APPENDIX 6D. HYDROPHOBIC AND HYDROPHILIC COLLOIDS

A liquid dispersion of colloidal particles is stable—meaning that the particles will not settle out—if the constant thermal motion of molecules in the liquid phase is sufficient to overcome the effects of gravity on the particles (see Chapter 9, Appendix 9A). Colloid chemists distinguish two different types of colloidal particles that disperse in liquids: hydrophobic and hydrophilic colloids. Hydrophobic colloids are solids—ultrafine mineral particles or clays—and hydrophilic colloids are molecular aggregates—composed of either humic substances or biosurfactants secreted by microbes.

 Micelle

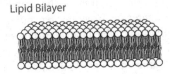 Lipid Bilayer

FIGURE 6D.1   Aqueous solution structures of amphiphilic molecular aggregates: micelle and lipid bilayer. *Modified from (Villarreal, 2007).*

Hydrophobic Mineral Colloids. Chapter 3 provides an adequate description of mineral (hydrophobic) colloids. It may seem trivial, but hydrophobic colloids have a well-defined surface that forms an interface between solid particles and the surrounding liquid (see Chapter 9, Figure 9.1); the surface is a barrier separating the interior of the hydrophobic colloid from the liquid. Adsorbate molecules attach to the surface but cannot penetrate the surface because the interior of a hydrophobic colloid does not behave like a fluid.[15]

Hydrophilic Organic Colloids. Organic (hydrophilic) colloids are molecular aggregates, not solid particles. Unlike hydrophobic colloids, molecular-aggregate colloids have a penetrable surface separating the interior of the colloid particle from the liquid. Adsorbates either attach to the particle surface or penetrate the liquid-colloid interface, entering the interior of molecular-aggregate colloid. The interior of molecular-aggregate colloids behaves much like a fluid.

The cell membrane of all living organisms is an aggregation of amphiphilic molecules called a lipid bilayer (Figure 6D.1). An amphiphile—also known as a surfactant or detergent—is a molecule that has both polar[16] and nonpolar[17] structures. The most abundant amphiphiles in cell membranes are phospholipids, the phosphate head is the polar structure, and the lipid tail is the nonpolar structure. The amphiphiles in a lipid bilayer aggregate to form two sheets, hence the term *bilayer*; the nonpolar tails oriented toward the center or interior of the bilayer; and the polar heads form the surface in contact with the liquid water. The bilayer is actually a liposome with a very large radius. Micelles are the smallest amphiphile aggregates, consisting of a fluid-like core formed by the nonpolar portion of the amphiphiles encased in a shell

---

15. The interlayer of swelling phyllosilicates is considered an extension of the surrounding solution. Solution cations exchange with interlayer cations and water molecules enter the interlayer to hydrate interlayer cations but the crystalline phyllosilicate layers do not imbibe interlayer cations and water molecules.

16. *Polar* in this context refers to the following molecular structures: weak inorganic acids (e.g., organosulfate esters), weak organic acids (e.g., carboxylic acid or phenolic acid groups), weak organic bases (e.g., primary amines), hydrogen-bond donors (e.g., alcohol groups, secondary amines), or hydrogen-bond acceptors (e.g., carbonyl groups). Polar structures usually involve extremely electropositive or electronegative elements bonded to carbon.

17. *Nonpolar* in this context refers to molecular structures in which carbon is bonded to carbon or hydrogen. Aliphatic structures are extremely nonpolar.

formed by the polar portion of the amphiphiles in contact with fluid water. The type of aggregate structure depends on the solution amphiphile concentration and the molecular characteristics of the amphiphile itself.

Detergents aggregate to form micelles that solubilize oils and other nonpolar organic liquids in water. Molecules of the nonpolar liquid dissolve or partition into the nonpolar interior of the micelle, causing it to swell. The aggregate structure accommodates both the nonpolar compound and water in a stable colloidal system. The nonpolar interior of all of these aggregates—micelles and bilayers—behaves like a fluid. The surface does not prevent the movement of molecules from aqueous solution into the nonpolar interior of the aggregate or from the aggregate interior into aqueous solution.

Problems

1. The International Humic Substances Society supplies reference samples to researchers (IHSS *Products* link at the website www.ihss.gatech.edu/). The *Chemical Properties* link leads you to data under the *Elemental Composition* and *Acidic Functional Groups* links.

   Using the oxygen content from the Elemental Composition page and the combined carboxyl and phenol content from the Acidic Functional Groups page, determine the oxygen percentage in the Summit Hill Soil Reference Humic Acid (IHSS sample 1R106H) attributable to carboxyl and phenol groups. What might be the chemical nature of the remaining oxygen?

   Solution

   Convert the carboxylic and phenol contents $[mol_c \cdot kg_C^{-1}]$ to the equivalent oxygen content $[kg_O \cdot kg_{HA}^{-1}]$.

   ### Summit Hill Soil Reference Humic Acid (1R106H)

Carbon, $kg_C \cdot kg_{HA}^{-1}$	Oxygen, $kg_O \cdot kg_{HA}^{-1}$	Carboxyl, $mol_c \cdot kg_C^{-1}$	Phenol, $mol_c \cdot kg_C^{-1}$
0.5400	0.3790	7.14	2.42

   carboxyl groups

   $$\left(\frac{7.14\ mol_{carboxylic}}{kg_C}\right) \cdot \left(\frac{2\ mol_O}{1\ mol_c}\right) \cdot \left(\frac{15.9994 \cdot 10^{-3}\ kg_O}{mol_O}\right) \cdot \left(\frac{0.5400\ kg_C}{kg_{HA}}\right)$$

   $$= \frac{0.123\ kg_{carboxylic-O}}{kg_{HA}}$$

   phenol groups

   $$\left(\frac{2.42\ mol_{phenolic}}{kg_C}\right) \cdot \left(\frac{1\ mol_O}{1\ mol_c}\right) \cdot \left(\frac{15.9994 \cdot 10^{-3}\ kg_O}{mol_O}\right) \cdot \left(\frac{0.5400\ kg_C}{kg_{HA}}\right)$$

   $$= \frac{0.021\ kg_{phenolic-O}}{kg_{HA}}$$

Add the oxygen content of the two weak acid moieties and divide by the total oxygen content of the sample.

$$
\left(\left(\frac{0.123\ kg_{carboxylic-O}}{kg_{1R106H}}\right) + \left(\frac{0.021\ kg_{phenolic-O}}{kg_{1R106H}}\right)\right) \cdot \left(\frac{kg_{1R106H}}{0.3790\ kg_O}\right)
$$

$$
= \left(\frac{0.144\ kg_{weak\ acid-O}}{kg_{1R106H}}\right) \cdot \left(\frac{kg_{1R106H}}{0.3790\ kg_O}\right) = \frac{0.381\ kg_{weak\ acid-O}}{kg_O}
$$

The titratable acidity of Summit Hill Soil Reference Humic Acid (IHSS sample 1R106H) represents 38% of the total oxygen content. The remaining content is probably alcohol oxygen from polysaccharides and acetal ether from the β(1-4) condensation of pentose and hexose sugar monomers into polysaccharides. Schnitzer also recognizes small quantities of carbonyl oxygen $C=O$.

2. The International Humic Substances Society supplies reference samples to researchers (IHSS Products link at the website www.ihss.gatech.edu/). The Chemical Properties link leads you to data on the Elemental Composition and Acidic Functional Groups.

   Using the oxygen content from the Elemental Composition page and the combined carboxyl and phenol content from the Acidic Functional Groups page, determine the percentage of oxygen content in the Pahokee Reference Humic Acid (IHSS sample 1R103H) attributable to carboxyl and phenol groups.

3. Table 1.2 in *The Chemical Composition of Soils* (Helmke, 2000) lists the geometric mean Cu content of world soils: 20 $\mu g_{Cu} \cdot g_{soil}^{-1}$. The soil organic carbon content of the Summit Hill (Christchurch, New Zealand) A-horizon is 4.3% (0.043 $\mu g_{OC} \cdot g_{soil}^{-1}$). Assume the metal binding capacity is the weak acid content of the Summit Hill Soil Reference Humic Acid. Compare the metal binding capacity of the A-horizon with the Cu content of the soil, assuming one Cu^{2+} per metal binding group. Is the metal binding capacity greater or less than the mean soil Cu content?

Solution
Convert the carboxylic and phenol contents $[mol_c \cdot kg_C^{-1}]$ to the equivalent copper content $[kg_{Cu} \cdot kg_{HA}^{-1}]$, using the weak acid content, the carbon content, and the atomic mass of copper.

$$
\left(\frac{7.14\ mol_{carboxylic}}{kg_C}\right) \cdot \left(\frac{1\ mol_{Cu}}{1\ mol_{carboxylic}}\right) \cdot \left(\frac{0.5400\ kg_C}{kg_{1R106H}}\right) = \left(\frac{3.86\ mol_{Cu}}{kg_{1R106H}}\right)_{carboxylic}
$$

$$
\left(\frac{2.42\ mol_{phenolic}}{kg_C}\right) \cdot \left(\frac{1\ mol_{Cu}}{1\ mol_{phenolic}}\right) \cdot \left(\frac{0.5400\ kg_C}{kg_{1R106H}}\right) = \left(\frac{1.31\ mol_{Cu}}{kg_{1R106H}}\right)_{phenolic}
$$

The Summit Hill Soil Reference Humic Acid (IHSS sample 1R106H) contains sufficient carboxylic and phenolic groups to bind 329 *mg Cu* per gram of humic acid.

$$\left(\left(\frac{3.86 \; mol_{carboxylic-Cu}}{kg_{1R106H}} \right) + \left(\frac{1.31 \; mol_{phenolic-Cu}}{kg_{1R106H}} \right) \right) \cdot \left(\frac{0.063546 \; kg_{Cu}}{mol_{Cu}} \right)$$

$$= \left(\frac{0.329 \; kg_{Cu}}{kg_{1R106H}} \right)$$

This represents a binding capacity of 26.2 *mg Cu* per gram of soil (for a soil with 4.3% organic matter). The potential *Cu* binding capacity of the organic matter in this soil is nearly 1000-fold higher than the geometric mean *Cu* content of world soils.

$$\left(\frac{0.329 \; kg_{Cu}}{kg_{1R106H}} \right) \cdot \left(\frac{kg_{1R106H}}{0.5400 \; kg_{OC}} \right) \cdot \left(\frac{0.043 \; kg_{OC}}{kg_{soil}} \right) = \frac{2.62 \cdot 10^{-2} \; kg_{Cu}}{kg_{soil}}$$

$$\frac{2.62 \cdot 10^{-2} \; kg_{Cu}}{kg_{soil}} = \frac{2.62 \cdot 10^{-2} \; g_{Cu}}{g_{soil}}$$

$$\left(\frac{2.89 \cdot 10^{-2} \; g_{Cu}}{g_{soil}} \right) \cdot \left(\frac{10^3 \; mg_{Cu}}{g_{Cu}} \right) = \frac{28.9 \; mg_{Cu}}{g_{soil}}$$

4. Table 1.4 in *The Chemical Composition of Soils* (Helmke, 2000) lists the geometric mean *Cu* content of U.S. peat soils (Histosols): 193 $\mu g_{Cu} \cdot g_{soil}^{-1}$. The organic matter content of peat soils is essentially 100%. Assume the metal binding capacity is identical with the Pahokee Peat Reference Humic Acid (IHSS sample 1R103H). Compare the metal binding capacity of this peat with the *Cu* content of U.S. peat soils, assuming one Cu^{2+} per metal binding group. Is the metal binding capacity greater or less than the mean soil *Cu* content?

5. The International Humic Substances Society supplies reference samples to researchers (IHSS Products link at the website www.ihss.gatech.edu/). The Chemical Properties link leads you to data on the Elemental Composition and Acidic Functional Groups.

   Using the oxygen content from the Elemental Composition page for the Suwannee River Standard Fulvic Acid (IHSS sample 1S101F), and assuming 10% of the total nitrogen is free amine (90% being amide), what percentage of the total Cu^{2+} binding capacity may be attributed to amine complexes.

6. The International Humic Substances Society supplies reference samples to researchers (IHSS Products link at the website www.ihss.gatech.edu/). The Chemical Properties link leads you to data on the Elemental Composition. Using the oxygen content from the Elemental Composition page for the Waskish Peat Reference Humic Acid (IHSS sample 1R107H), estimate the mean formal oxidation number of carbon $\overline{N}_C$. Assume the following formal oxidation states: $N_H = (+1)$, $N_O = (-2)$, and $N_N = (-3)$.

Chapter 7

Acid-Base Chemistry

Chapter Outline

7.1. INTRODUCTION

Environmental chemistry has a much different perspective on acid-base chemistry than the one you encountered in general chemistry. The most important difference is the *reference point* used to define acidity. General chemistry defines *acidity* and *basicity* relative to $a_{H^+} = 10^{-7}$ or pH = 7, the pH of pure water.

Environmental chemists define *acidity* and *alkalinity* relative to $a_{H^+} = 10^{-5.5}$ or pH = 5.5, the pH of pure water in equilibrium with an atmosphere containing carbon dioxide gas $CO_2(g)$ at a partial pressure of $p_{CO_2} = 10^{-3.41}$ *atm*. The environmental reference point acknowledges the ubiquity and importance of aqueous $CO_2(aq)$ in atmospheric, surface and soil pore water.

The environmental perspective on acid-base chemistry is also strongly influenced by two seemingly conflicting features of environmental acidity and alkalinity: the capacity of natural waters to resist changes in pH yet remain susceptible to shifts resulting from human activity ranging from land use and waste management to atmospheric emissions. Compounds dissolved in water and chemical reactions on the surfaces of mineral colloids buffer the pH of natural water, yet this buffer capacity (Appendix 7A) is finite and can be overwhelmed by acidity released by human activity. The magnitude and rate at which the pH of natural water decreases in response to the release of anthropogenic acidity determines the viability of natural ecosystems and alters human exposure to inorganic toxicants.

7.2. PRINCIPLES OF ACID-BASE CHEMISTRY

7.2.1. Dissociation: The Arrhenius Model of Acid-Base Reactions

Acid-base chemistry is fundamentally based on our understanding of acid-base reactions that, it turns out, can be understood in different ways. The most familiar is the *Arrhenius* concept that defines an *acid* as any chemical substance that dissociates to release hydrogen ions $H^+(aq)$—or more accurately hydronium ions $H_3O^+(aq)$—into aqueous solution:

$$HA(aq) + H_2O(l) \leftrightarrow A^-(aq) + H_3O^+(aq) \qquad (7.1)$$

A *base* is any chemical substance that dissociates to release hydroxide ions $OH^-(aq)$ into aqueous solution:

$$BOH(aq) \leftrightarrow B^+(aq) + HO^-(aq) \qquad (7.2)$$

The formation of water molecules when hydrogen ions $H^+(aq)$ and hydroxide ions $OH^-(aq)$ combine in solution is an acid-base reaction:

$$H^+(aq) + HO^-(aq) \leftrightarrow H_2O(l)$$
$$or \qquad\qquad (7.3)$$
$$H_3O^+(aq) + HO^-(aq) \leftrightarrow 2 \cdot H_2O(l)$$

The Arrhenius concept has an acid and a base reacting to form a *salt* and water:

$$\underset{acid}{HNO_3} + \underset{base}{NaOH} \rightarrow \underset{salt}{NaNO_3} + H_2O$$
$$or \qquad\qquad (7.4)$$
$$\underset{acid}{HNO_3} + \underset{base}{NaOH} \rightarrow Na^+(aq) + NO_3^-(aq) + H_2O(l)$$

7.2.2. Hydrogen Ion Transfer: the Brønsted-Lowery Model of Acid-Base Reactions

The *Brønsted-Lowery* concept defines an *acid* as any chemical substance that can donate hydrogen ions H^+ and defines a base as any chemical substance that can accept hydrogen ions. The Brønsted-Lowery concept has an acid and a base reacting to form conjugate bases (Eq. (7.5A)) and conjugate acids (Eq. (7.5B)) through hydrogen ion transfer reactions. The concept of acid-conjugate base and base-conjugate acid pairs appears in the following section.

$$\underset{\substack{\text{weak} \\ \text{acid}}}{HA(aq)} + \underset{\substack{\text{weak} \\ \text{base}}}{H_2O(l)} \leftrightarrow \underset{\substack{\text{conjugate} \\ \text{base}}}{A^-(aq)} + \underset{\substack{\text{conjugate} \\ \text{acid}}}{H_3O^+(aq)} \qquad K_a = \frac{a_{A^-} \cdot a_{H_3O^+}}{a_{HA}} \qquad (7.5A)$$

$$\underset{\substack{\text{weak} \\ \text{acid}}}{H_2O(l)} + \underset{\substack{\text{weak} \\ \text{base}}}{B(aq)} \leftrightarrow \underset{\substack{\text{conjugate} \\ \text{base}}}{OH^-(aq)} + \underset{\substack{\text{conjugate} \\ \text{acid}}}{BH^+(aq)} \qquad K'_a = \frac{a_{BH^+} \cdot a_{OH^-}}{a_B} \qquad (7.5B)$$

Acids $HA(aq)$ and conjugate acids $BH^+(aq)$ are both H^+ donors, while bases $B(aq)$ and conjugate bases $A^-(aq)$ are H^+ acceptors. The Brønsted-Lowery acid-base concept is more general than the Arrhenius concept because salt and water formation are not required.

The most important benefit of the Brønsted-Lowery acid-base concept is that it explains the dissociation of water in dilute solutions in the absence of either acids or bases:

$$\underset{\substack{\text{weak} \\ \text{acid}}}{H_2O(l)} + \underset{\substack{\text{weak} \\ \text{base}}}{H_2O(l)} \leftrightarrow \underset{\substack{\text{conjugate} \\ \text{base}}}{OH^-(aq)} + \underset{\substack{\text{conjugate} \\ \text{acid}}}{H_3O^+(aq)} \qquad K_w = 10^{-14.00} \qquad (7.6)$$

7.2.3. Conjugate Acids and Bases

The hydrolysis reaction of a base is rarely written as a hydronium-ion dissociation reaction (Eq. (7.7)) with acid dissociation constant K_a (Eq. (7.8)) but as the dissociation of the base to release a hydroxyl ion into solution (Eq. (7.9)) with base association constant K_b (Eq. (7.10)).

$$\underset{\substack{\text{conjugate} \\ \text{acid}}}{BH^+(aq)} + H_2O(l) \leftrightarrow \underset{\substack{\text{weak} \\ \text{base}}}{B(aq)} + H_3O^+(aq) \qquad (7.7)$$

$$K_a = \frac{a_B \cdot a_{H_3O^+}}{a_{BH^+}} \qquad (7.8)$$

$$\underset{\substack{\text{weak} \\ \text{base}}}{B(aq)} + H_2O(l) \leftrightarrow \underset{\substack{\text{conjugate} \\ \text{acid}}}{BH^+(aq)} + HO^-(aq) \qquad (7.9)$$

$$K_b = \frac{a_{BH^+} \cdot a_{HO^-}}{a_B} \qquad (7.10)$$

In fact, every acid-base reaction can be written as either a hydronium-ion dissociation reaction of an acid (or conjugate acid) (Eq. (7.7) or Eq. (7.11))—whose equilibrium constant is an acid dissociation constant K_a—or a hydroxyl-ion dissociation reaction of a conjugate base (Eq. (7.9) or Eq. (7.12))—whose equilibrium constant is a base association constant K_b.

$$\underset{\text{weak acid}}{HA(aq)} +H_2O(l) \leftrightarrow \underset{\text{conjugate base}}{A^-(aq)} +H_3O^+(aq) \tag{7.11}$$

$$\underset{\text{conjugate base}}{A^-(aq)} +H_2O(l) \leftrightarrow \underset{\text{weak acid}}{HA(aq)} +HO^-(aq) \tag{7.12}$$

Hydronium-ion dissociation expressions are related to a hydroxyl-ion dissociation reaction through the hydrolysis expression for water (Eq. (7.6)). The following two expressions use Eq. (7.6) to convert the hydronium ion dissociation constant of the conjugate acid $BH^+(aq)$ (Eq. (7.8)) into a hydroxyl ion dissociation constant of the base $B(aq)$ (Eq. (7.9)).

$$\frac{K_a}{K_w} = \left(\frac{a_B \cdot a_{H_3O^+}}{a_{BH^+}}\right) \cdot \left(\frac{1}{a_{OH^-} \cdot a_{H_3O^+}}\right)$$

$$\frac{K_a}{K_w} = \left(\frac{a_B}{a_{BH^+} \cdot a_{OH^-}}\right) = K_b^{-1}$$

Expression (7.13) relates the acid dissociation constant K_a to its corresponding base association constant K_b using the hydrolysis constant of water K_w:

$$K_a \cdot K_b = K_w \tag{7.13}$$

7.2.4. Defining Acid and Base Strength

The criteria that chemists use to distinguish strong acids from weak acids or strong bases from weak bases should be considered qualitative. Acid strength is generally quantified by the thermodynamic acid dissociation constant K_a for an acid-base reaction (Eq. (7.11)):

$$K_a = \frac{a_{A^-} \cdot a_{H_3O^+}}{a_{HA}} \tag{7.14}$$

Strong acids are identified by $pK_a \leq 1.0$, while weak acids are identified by $1 < pK_a < 7$. Base strength is quantified on the same scale using the thermodynamic acid dissociation constant K_a for acid-base reaction (7.7). Strong bases are characterized by $13 \leq pK_a$, and weak bases by $7 < pK_a < 13$.

Acidity and Basicity

Aqueous acid-base reactions—according to the Brønsted-Lowery acid-base concept—involve ions formed by the transfer of protons to or from water (Eq. (7.6)): $H_3O^+(aq)$ and $OH^-(aq)$. The participation of $H_3O^+(aq)$ and $OH^-(aq)$ in all aqueous acid-base reactions introduces an equilibrium constraint known as the *proton balance*. In Chapter 5 we introduced several equilibrium constraints to validate water chemistry simulations:[1] charge balance (electroneutrality condition), mass balances of all solutes other than water (mass conservation), and equilibrium activity products involving all solutes.

7.2.5. Water Reference Level

General chemistry defines *acidity* and *basicity* relative to a reference point determined by the proton balance of pure water (Eq. (7.15)). Subject to the equilibrium constant K_w for the spontaneous dissociation of water (Eq. (7.6)), the proton balance of pure water occurs at pH 7 (Eq. (7.16)), which we designate the water reference level:

$$Proton\ Balance\ of\ Pure\ Water: \quad C_{H_3O^+} = C_{OH^-} \qquad (7.15)$$

$$C_{H_3O^+} = C_{OH^-} = (K_w)^{1/2} = \left(10^{-14.00}\right)^{1/2} = 10^{-7.00} \qquad (7.16)$$

The water reference level (Eq. (7.16)) is the appropriate proton reference level for solutions where the only weak acid or base is water itself (Eq. (7.6)), and it remains the appropriate proton reference level even if the solution contains strong acids, strong bases, and neutral salts. Table 7.1 lists the conjugate acid-base pairs in solutions whose solutes are completely dissociated: strong acids, strong bases, or neutral salts formed by the acid-base neutralization of strong acids with strong bases. For example, the charge balance in solutions containing nitric acid HNO_3 and sodium hydroxide $NaOH$ is given by Eq. (7.17), regardless of the relative concentration of strong acid and base.

TABLE 7.1 Conjugate Acid-Base Pairs in Aqueous Solutions of Strong Acids, Strong Bases, and Neutral Salts

Weak Acid	Conjugate Base
H_3O^+	H_2O
H_2O	OH^-

---

1. See Chapter 5, Appendixes 5B and 5C for details.

Rearranging charge balance expression (7.17) to yield (7.18) allows us to associate pH > 7 with $C_{Na^+} > C_{NO_3^-}$ and pH < 7 with $C_{Na^+} < C_{NO_3^-}$.

$$C_{H_3O^+} + C_{Na^+} = C_{OH^-} + C_{NO_3^-} \qquad (7.17)$$

$$\left(C_{Na^+} - C_{NO_3^-}\right) = \left(C_{OH^-} - C_{H_3O^+}\right) \qquad (7.18)$$

7.2.6. The Aqueous Carbon Dioxide Reference Level

Environmental chemistry—recognizing the presence of aqueous carbon dioxide $CO_2(aq)$ in all natural waters in contact with Earth's atmosphere and its effect on the pH—defines *acidity* and *alkalinity* relative to a reference point determined by the proton balance of water in equilibrium with carbon dioxide $CO_2(g)$. This reference point is the aqueous carbon dioxide $CO_2(aq)$ reference level.

Table 7.2 lists the conjugate acid-base pairs in solutions containing carbonate solutions. Four species in Table 7.2 are boldface. Unlike solutions whose only conjugate acid-base pairs derive from water (see Table 7.1), carbonate solutions (see Table 7.2) encompass three proton reference levels: aqueous carbon dioxide $CO_2(aq)$, bicarbonate $HCO_3^-(aq)$, and carbonate $CO_3^{2-}(aq)$. The boldface solute species are those defining the aqueous carbon dioxide $CO_2(aq)$ reference level. Appendix 7D discusses the remaining two proton reference levels defined in carbonate solutions but not used by environmental chemists: the bicarbonate $HCO_3^-(aq)$ and carbonate $CO_3^{2-}(aq)$ reference levels.

All of the remaining solutes in solutions containing the weak acids depicted in Table 7.2 are completely dissociated: strong acids, strong bases, or neutral salts. For example, the charge balance in a solution containing nitric acid HNO_3, sodium hydroxide $NaOH$, and aqueous carbon dioxide $CO_2(aq)$ is given by the following expression, regardless of the relative concentration of strong acid and base:

$$C_{H_3O^+} + C_{Na^+} = C_{OH^-} + C_{NO_3^-} + C_{HCO_3^-} + 2 \cdot C_{CO_3^{2-}} \qquad (7.19)$$

TABLE 7.2 Conjugate Acid-Base Pairs in Aqueous Carbonate Solutions

Weak Acid	Conjugate Base
H_3O^+	H_2O
$CO_2(aq)$	HCO_3^-
HCO_3^-	CO_3^{2-}
H_2O	OH^-

Consider a carbonate solution in equilibrium with $CO_2(g)$ at a partial pressure of $3.87 \cdot 10^{-4}$ atm. The proton balance (expression (7.20)) for the aqueous carbon dioxide $CO_2(aq)$ reference level[2] at this specific $CO_2(g)$ partial pressure (using boldface species in Table 7.2) is satisfied at pH 5.5:

$$\left(C_{OH^-} + C_{HCO_3^-} + 2 \cdot C_{CO_3^{2-}} - C_{H_3O^+} \right) = 0 \tag{7.20}$$

Rearranging the charge balance expression (7.19), moving the proton balance (7.20) for the aqueous carbon dioxide $CO_2(aq)$ reference level to the right-hand side of the expression, yields the following:

$$\left(C_{Na^+} - C_{NO_3^-} \right) = \left(C_{OH^-} + C_{HCO_3^-} + 2 \cdot C_{CO_3^{2-}} - C_{H_3O^+} \right) \tag{7.21}$$

Expression (7.21) associates pH > 5.5 with $C_{Na^+} > C_{NO_3^-}$ and pH < 5.5 with $C_{Na^+} < C_{NO_3^-}$. Remember: The explicit reference pH for expression (7.20) depends on the $CO_2(g)$ partial pressure.

Figure 7.1 plots the conjugate acid-base species listed in Table 7.2 for a carbonate solution in equilibrium with $CO_2(g)$ at a partial pressure of $3.87 \cdot 10^{-4}$ atm. The concentration of carbonic acid $H_2CO_3(aq)$ is also plotted to emphasize its status as a minor solution species in carbonate solutions relative to aqueous carbon dioxide $CO_2(aq)$. Four proton reference levels are designated in Figure 7.1: water, aqueous carbon dioxide $CO_2(aq)$, bicarbonate $HCO_3^-(aq)$, and carbonate $CO_3^{2-}(aq)$.

Increasing the $CO_2(g)$ partial pressure has no effect on the water proton reference level and the bicarbonate proton reference level but shifts the carbon dioxide proton reference level to lower pH values and the carbonate proton reference level to higher pH values. Figure 7.2 illustrates the effect of increasing $CO_2(aq)$ partial pressure 100-fold $\left(p_{CO_2} = 3.87 \cdot 10^{-2} \ atm \right)$ relative to the partial pressure used in Figure 7.1.

7.3. SOURCES OF ENVIRONMENTAL ACIDITY AND BASICITY

Surface and pore water chemistry is dynamic, a result of atmospheric deposition, biological activity, and the chemical weathering of minerals. In this section we examine examples of each source and the associated chemical processes.

---

2. The proton balance defined for the aqueous carbon dioxide $CO_2(aq)$ reference level excludes—by definition—$CO_2(aq)$.

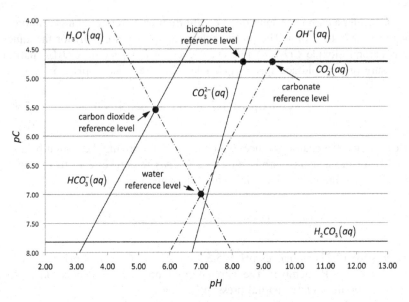

FIGURE 7.1   All concentrations are plotted as the negative logarithm of their concentrations pC. The $CO_2(aq)$ partial pressure: $p_{CO_2} = 3.87 \cdot 10^{-4}\ atm$.

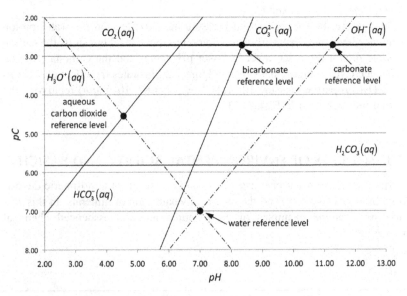

FIGURE 7.2   Increasing the $CO_2(g)$ partial pressure ($p_{CO_2} = 3.87 \cdot 10^{-2}\ atm$) alters the pH of the aqueous carbon dioxide and carbonate proton reference levels (see Figure 7.1).

7.3.1. Chemical Weathering of Rocks and Minerals

Geologists classify rocks according to the geological process that led to their formation—igneous, sedimentary, and metamorphic. The major rock classes are further subdivided by texture and mineralogy. Sedimentary rocks are subdivided into two major subclasses: *clastic* and *chemical*. Clastic sedimentary rocks form by the cementation of rock fragments: conglomerates, sandstones, and shales. Chemical sedimentary rocks result from either evaporation or biogenic precipitation from solution. The mineralogy of *evaporite* sedimentary rocks is predominantly chloride and sulfate minerals. *Precipitate* sedimentary rocks consist of biogenic carbonate and silicate minerals—calcite, aragonite, dolomite, and diatomite—derived from the exoskeleton of aquatic and marine organisms.

In the context of acid-base chemistry, geochemists group all igneous, metamorphic, and clastic sedimentary rocks into a single broad group: silicate rocks. Of the chemical sedimentary rocks, only the carbonates influence acid-base chemistry, and they represent the second broad group. As we will see following, environmental acid-base chemistry is indifferent to evaporite and diatomite rock mineralogy.

7.3.2. Silicate Rocks

The chemical weathering of silicate rocks is most easily understood if we denote rock composition as a *sum of oxides*. This representation neglects crystal structure distinctions, since crystal structure has a relatively minor effect when considering the global impact of chemical weathering on acid-base chemistry. Table 7.3 lists the oxide mole-per-mass fraction composition of Earth's crust. A detailed discussion of how the composition of rocks or minerals is converted from a mass fraction representation to a sum of oxides representation appears in Appendix 7B.

TABLE 7.3 Oxide Mole-per-Mass Composition of Earth's Crust

Oxide	Mole-per-Mass Fraction, $mol \cdot kg^{-1}$
$SiO_2(s)$	9.68
$Al_2O_3(s)$	1.54
$FeO(s)$	1.11
$CaO(s)$	1.16
$Na_2O(s)$	0.49
$MgO(s)$	1.14
$K_2O(s)$	0.24

Source: Greenwood and Earnshaw, 1997.

The formal oxidation state of iron in the crust is $Fe(II)$, but chemical weathering occurs under oxidizing conditions, favoring the oxidation of $Fe(II)$ to $Fe(III)$. Environmental acid-base chemistry is indifferent to this oxidation reaction:

$$2 \cdot FeO(s) + \frac{1}{2} \cdot O_2(g) \rightarrow Fe_2O_3(s) \qquad (7.22)$$

The *Jackson clay mineral weathering sequence* (see Table 3.2) lists aluminum, iron, and titanium oxides in the final stages of the mineral weathering sequence. The oxides $Al_2O_3(s)$ and $Fe_2O_3(s)$ behave as inert substances in the reaction depicting the chemical weathering of Earth's silicate crust. The only reactive substances in the reaction depicting the chemical weathering of Earth's silicate crust (expression (7.23)) are the alkali and alkaline earth oxides and $SiO_2(s)$:

$$9.68 \cdot SiO_2(s) + 1.54 \cdot Al_2O_3(s) + 0.56 \cdot Fe_2O_3(s)$$
$$+1.16 \cdot CaO(s) + 1.14 \cdot MgO(s) + 0.49 \cdot Na_2O(s) + 0.24 \cdot K_2O(s) + 6.06 \cdot H_2O(l)$$
$$\xrightarrow{\text{dissolution}} 9.68 \cdot SiO_2(aq) + 1.54 \cdot Al_2O_3(s) + 0.56 \cdot Fe_2O_3(s) + 1.16 \cdot Ca^{2+}(aq)$$
$$+1.14 \cdot Mg^{2+}(aq) + 0.99 \cdot Na^+(aq) + 0.47 \cdot K^+(aq) + 6.06 \cdot HO^-(aq)$$

$$(7.23)$$

Geochemists typically represent aqueous silicic acid $H_4SiO_4(aq)$ as $SiO_2(aq)$. Adding two moles of water to the left-hand side of Eq. (7.23) for each mole of $SiO_2(s)$ and replacing $SiO_2(aq)$ with $H_4SiO_4(aq)$ on the right-hand side, while more accurate chemically, has essentially no impact on environmental acid-base chemistry.

The acid-base significance of the chemical weathering of Earth's silicate crust (7.23) is the dissolution of the alkali and alkaline earth oxides. Charge neutrality demands one mole of $HO^-(aq)$ on the right-hand side of Eq. (7.23)—and one-half mole of $H_2O(l)$ on the left-hand side—for every mole of solute charge:

$$CaO(s) + H_2O(l) \xrightarrow{\text{dissolution}} Ca^{2+}(aq) + 2 \cdot HO^-(aq) \qquad (7.24)$$

Based on the criteria chemists use to distinguish strong and weak bases, both $CaO(s)$ and $MgO(s)$ are considered weak bases, but recognizing the qualitative nature of this distinction, we consider all alkali and alkaline earth oxides in the chemical weathering reaction (7.23) and Table 7.3 as strong bases. Clearly, the chemical weathering of silicate rocks (7.23) is a source of strong bases in natural waters.

Natural waters at the Earth surface absorb carbon dioxide from the atmosphere. Two processes cause $CO_2(aq)$ to dissolve in natural waters: the inherent water solubility of $CO_2(aq)$ (Eq. (7.25)) and the acid-base neutralization reaction (Eq. (7.26)) involving aqueous carbon dioxide and hydroxyl ions, the latter resulting from the chemical weathering of silicate minerals (Eq. (7.23)).

$$CO_2(g) \xrightarrow{\text{dissolution}} CO_2(aq) \tag{7.25}$$

$$\overset{\text{weak acid}}{CO_2(aq)} + \overset{\text{strong base}}{HO^-(aq)} \xrightarrow{\text{neutralization}} \overset{\text{conjugate base}}{HCO_3^-(aq)} \tag{7.26}$$

Taking into account acid-base reaction (7.26), the chemical weathering of silicate minerals (7.23) can be reformulated as the following—neglecting the inert oxides $Al_2O_3(s)$ and $Fe_2O_3(s)$:

$$9.68 \cdot SiO_2(s) + 3.03 \cdot H_2O(l) + 6.06 \cdot CO_2(g)$$
$$+ 1.16 \cdot CaO(s) + 1.14 \cdot MgO(s) + 0.49 \cdot Na_2O(s) + 0.24 \cdot K_2O(s)$$
$$\xrightarrow{\text{dissolution}} 9.68 \cdot SiO_2(aq) + 6.06 \cdot HCO_3^-(aq)$$
$$+ 1.16 \cdot Ca^{2+}(aq) + 1.14 \cdot Mg^{2+}(aq) + 0.99 \cdot Na^+(aq) + 0.47 \cdot K^+(aq)$$
$$\tag{7.27}$$

Garrels and Mackenzie (1971) observe that an estimate of solution bicarbonate derived from the chemical weathering of silicate rocks "can be achieved if it is assumed that, on the average, silicate minerals produce one bicarbonate ion and two silica ions for each molecule of carbon dioxide":

$$\text{silicate rocks}: \quad C_{HCO_3^-} \approx \frac{C_{SiO_2(aq)}}{2} \tag{7.28}$$

The *Garrels-Mackenzie* estimate of bicarbonate resulting from the chemical weathering of silicate minerals derives from the composition of Earth's crust (see Table 7.3) and the chemical weathering of silicate minerals (Eq. (7.27)).

7.3.3. Carbonate Rocks

The chemical weathering of carbonate rocks is another major factor in environmental acid-base chemistry. The chemical formula for calcite $CaCO_3(s)$—the principle mineral in limestone—can be reformulated as component oxides: $(CaO)(CO_2)(s)$. Calcite dissolution (Eq. (7.29)) releases both strong base and aqueous carbon dioxide. The aqueous carbon dioxide $CO_2(aq)$ combines with water (Eq. (7.30)) to form carbonic acid $H_2CO_3(aq)$:

$$\underset{\text{calcite}}{(CaO)(CO_2)(s)} + H_2O(l) \rightarrow Ca^{2+}(aq) + 2 \cdot \overset{\text{strong base}}{HO^-(aq)} + CO_2(aq) \tag{7.29}$$

$$CO_2(aq) + H_2O(l) \rightarrow \overset{\text{weak acid}}{H_2CO_3(aq)} \tag{7.30}$$

Because carbonic acid $H_2CO_3(aq)$ is a weak acid, carbonate $CO_3^{2-}(aq)$ and bicarbonate $HCO_3^-(aq)$ are soluble conjugate bases:

$$\overset{\text{conjugate base}}{CO_3^{2-}(aq)} + 2 \cdot H^+(aq) \rightarrow \overset{\text{weak acid}}{H_2CO_3(aq)} \tag{7.31}$$

The dissolution of calcite written as an acid-base reaction identifies calcite as a mineral conjugate base:

$$\underset{calcite}{\underset{\text{conjugate base}}{CaCO_3(s)}} + 2 \cdot H^+(aq) \rightarrow Ca^{2+}(aq) + \underset{\text{weak acid}}{H_2CO_3(aq)} \qquad (7.32)$$

Garrels and Mackenzie (1971) proposed an estimate of solution bicarbonate derived from carbonate rock dissolution (Eq. (7.33)). The total solution bicarbonate estimate (Eq. (7.34)) is the sum of estimates (7.28) and (7.33):

$$\textbf{carbonate rocks}: \quad C_{HCO_3^-} \approx 2 \cdot \left(C_{Ca^{2+}} + C_{Mg^{2+}} - C_{SO_4^{2-}} \right) \qquad (7.33)$$

$$\textbf{total bicarbonate}: \quad C_{HCO_3^-} \approx 2 \cdot \left(C_{Ca^{2+}} + C_{Mg^{2+}} - C_{SO_4^{2-}} \right)_{carbonates}$$

$$+ \left(\frac{C_{SiO_2(aq)}}{2} \right)_{silicates} \qquad (7.34)$$

7.3.4. Sulfide Minerals

The sulfate dissolved in natural waters (Eq. (7.33)) comes from two sources: the dissolution of gypsum $CaSO_4 \cdot 2H_2O(s)$ in evaporite sedimentary rocks and the oxidation of sulfide minerals. Sulfide oxidation is a natural source of sulfuric acid. The atmospheric chemistry of sulfur oxides and the resulting atmospheric deposition of sulfuric acid are deferred to a later section; here, we focus on the reaction between molecular dioxygen and sulfide ions.

The formal oxidation state[3] of sulfur in the disulfide ion S_2^{2-} found in the ferrous disulfide $FeS_2(s)$ minerals pyrite and marcasite is (–1). The sulfide ion S^{2-} found in troilite $FeS(s)$ has a sulfur formal oxidation state of (–2). Other elements form sulfide and disulfide minerals, but their relative abundance is much lower than iron, making them relatively insignificant contributors to the acid-base chemistry of natural waters.

The complete oxidation of both ferrous disulfide (Eq. (7.35)) and ferrous sulfide (Eq. (7.36)) minerals yields insoluble ferric oxide $Fe_2O_3(s)$ and sulfuric acid H_2SO_4:

$$4 \cdot \underset{\substack{ferrous\\disulfide}}{FeS_2\ (s)} + 15 \cdot O_2(g) + 8 \cdot H_2O(l) \rightarrow 2 \cdot \underset{\substack{ferric\\oxide}}{Fe_2O_3(s)} + 8 \cdot \underset{\substack{sulfuric\\acid}}{H_2SO_4} \quad (7.35)$$

$$4 \cdot \underset{\substack{ferrous\\sulfide}}{FeS\ (s)} + 9 \cdot O_2(g) + 4 \cdot H_2O(l) \rightarrow 2 \cdot \underset{\substack{ferric\\oxide}}{Fe_2O_3(s)} + 4 \cdot \underset{\substack{sulfuric\\acid}}{H_2SO_4} \quad (7.36)$$

---

3. See Chapter 3, Appendix 3A for details.

7.3.5. Evaporite Rocks

The congruent dissolution of gypsum $CaSO_4 \cdot 2H_2O(s)$ in sedimentary rocks generates a neutral salt solution:

$$CaSO_4 \cdot 2H_2O(s) \rightarrow Ca^{2+}(aq) + SO_4^{2-}(aq) + 2 \cdot H_2O(l) \qquad (7.37)$$
$$\text{gypsum}$$

Congruent dissolution reactions produce solutions where the mole ratio of the ions in solution is identical to their mole ratio in the solid (Eq. (7.37)). The highly soluble minerals typical of evaporite sedimentary rocks—halite $NaCl(s)$, sylvite $KCl(s)$, fluorite $CaF_2(s)$, and a host of borate minerals—precipitate from concentrated salt solutions or brines and dissolve congruently. The reaction depicting the chemical weathering of Earth's silicate crust (7.23) is an incongruent dissolution reaction because two of the constituent oxides—$Al_2O_3(s)$ and $Fe_2O_3(s)$—behave as inert substances and do not sustain a solute mole ratio identical to the composition of the solid.

We examined in this section the chemical weathering reactions and the acid-base implications of four types of rocks or minerals: silicates, sulfides, carbonates, and evaporite minerals. Chemical weathering of silicate rocks (Eq. (7.23)) and carbonate rocks (Eq. (7.29)) is the primary source of environmental basicity. Chemical weathering involving disulfide (or sulfide) oxidation and ferric iron oxide precipitation produces soluble sulfate ions and acidity (Eq. (7.35) and Eq. (7.36)). Congruent dissolution of evaporite rock minerals (Eq. (7.37)) is a source of soluble sulfate ions but has little impact on acid-base chemistry in natural waters. The acidity associated with the complete oxidation of ferrous sulfide minerals (Eq. (7.35) and Eq. (7.36)) results 16–27% from the hydrolysis of the weak acid hexaquairon(3+) during the precipitation of ferric iron oxides and 73–84% from the oxidation of sulfide itself.

7.4. ATMOSPHERIC GASES

7.4.1. Carbon Dioxide: Above Ground

The carbon dioxide content of Earth's atmosphere (Table 7.4) is $3.87 \cdot 10^{-4}$ or $10^{-3.41}$ atmospheres (*atm*). Of the ten most abundant gases in Earth's atmosphere, carbon dioxide $CO_2(g)$ is the most soluble in water (Table 7.4). Regardless of its water solubility, gaseous carbon dioxide $CO_2(g)$ is not necessarily in equilibrium with natural waters (Garrels and Mackenzie, 1971). The carbon dioxide dissolution reaction in water is given as follows with a Henry's Law constant $K_H = 3.31 \cdot 10^{-2} \ mol \cdot L^{-1} \cdot atm^{-1}$ (Zheng, Guo et al., 1997):

$$CO_2(g) \leftrightarrow CO_2(aq) \qquad (7.38)$$

The hydration of aqueous carbon dioxide $CO_2(aq)$ (Pietramellara, Ascher et al., 2002) yields carbonic acid $H_2CO_3(aq)$ in reaction (7.39), but the equilibrium constant $K_{hydration} = 1.18 \cdot 10^{-3}$ favors aqueous carbon dioxide $CO_2(aq)$ over carbonic acid $H_2CO_3(aq)$ by a ratio of 555-to-1:

TABLE 7.4 Composition and Henry's Law Constants of the Most Abundant Gases in Earth's Atmosphere

Gas		Volume %	$K_H[mol \cdot L^{-1} \cdot atm^{-1}]$
Nitrogen	N_2	78.084	$6.1 \cdot 10^{-4}$
Oxygen	O_2	20.946	$1.3 \cdot 10^{-3}$
Argon	Ar	0.934	$1.4 \cdot 10^{-3}$
Carbon dioxide	CO_2	0.0387	$3.3 \cdot 10^{-2}$
Neon	Ne	0.001818	$4.5 \cdot 10^{-4}$
Helium	He	0.000524	$3.8 \cdot 10^{-4}$
Methane	CH_4	0.000179	$1.4 \cdot 10^{-3}$
Krypton	Kr	0.000114	$2.5 \cdot 10^{-3}$
Hydrogen	H_2	0.000055	$7.8 \cdot 10^{-4}$
Nitrous oxide	N_2O	0.00003	$2.5 \cdot 10^{-2}$

Source: Sander, 2009.

$$CO_2(aq) + H_2O(l) \leftrightarrow H_2CO_3(aq) \tag{7.39}$$

Though many books identify carbonic acid $H_2CO_3(aq)$ as the weak acid in acid dissociation reactions (7.40), and others symbolize aqueous carbon dioxide $CO_2(aq)$ as the effective carbonic acid $H_2CO_3^*(aq)$, the most chemically accurate reaction is (7.41). The second acid dissociation step is given by reaction (7.42).

$$H_2CO_3(aq) \leftrightarrow H^+(aq) + HCO_3^-(aq) \quad K_{a1,H_2CO_3(aq)} = 3.64 \cdot 10^{-4} \tag{7.40}$$

$$CO_2(aq) + H_2O(l) \leftrightarrow H^+(aq) + HCO_3^-(aq) \quad K_{a1,CO_2(aq)} = 4.30 \cdot 10^{-7} \tag{7.41}$$

$$HCO_3^-(aq) \leftrightarrow H^+(aq) + CO_3^{2-}(aq) \quad K_{a2,CO_2(aq)} = 4.68 \cdot 10^{-11} \tag{7.42}$$

If we combine the hydration reaction (7.39) with the carbonic acid $H_2CO_3(aq)$ hydrolysis reaction (7.40), we obtain the aqueous carbon dioxide $CO_2(aq)$ acid dissociation constant $K_{a1,CO_2(aq)}$ for reaction (7.41). Carbonic acid $H_2CO_3(aq)$ is a significantly stronger acid—$K_{a1,H_2CO_3(aq)} = 3.64 \cdot 10^{-4}$—than aqueous carbon dioxide $CO_2(aq)$—$K_{a1,CO_2(aq)} = 4.30 \cdot 10^{-7}$.

$$K_{hydration} \cdot K_{a1,H_2CO_3(aq)} = \left(\frac{a_{H_2CO_3(aq)}}{a_{CO_2(aq)}} \right) \cdot \left(\frac{a_{H^+} \cdot a_{HCO_3^-}}{a_{H_2CO_3(aq)}} \right)$$

$$K_{a1,CO_2(aq)} = \left(1.18 \cdot 10^{-3} \right) \cdot \left(3.64 \cdot 10^{-4} \right) = \left(4.30 \cdot 10^{-7} \right)$$

7.4.2. Carbon Dioxide: Below Ground

Respiration by soil organisms—plant roots and soil microbes—combined with sluggish gas exchange between soil pores and the aboveground atmosphere result in $CO_2(g)$ partial pressures 10 to 100 times higher in soil pores relative to the aboveground atmosphere. The significance of variations in $CO_2(g)$ partial pressure is illustrated in Box 7.1. Aqueous carbon dioxide $CO_2(aq)$ is classified as a weak acid, but it remains an important geochemical weathering agent.

7.4.3. Sulfur Oxides

The two most important sulfur oxides in Earth's atmosphere—sulfur dioxide $SO_2(g)$ and sulfur trioxide $SO_3(g)$—enter the atmosphere through both natural and anthropogenic processes, the former being emissions originating in volcanic eruptions, and the latter being emissions resulting largely from the combustion of coal and petroleum.

Sulfide minerals in magma react with atmospheric oxygen at high temperatures characteristic of eruptions to produce $SO_2(g)$ and $SO_3(g)$ in volcanic emissions. Sulfide compounds in coal and petroleum undergo combustion reactions to produce the $SO_2(g)$ and $SO_3(g)$ in power plant and auto emissions. Thiophene (Structure 1), thioether (Structure 2), and thiol ($-SH$) represent organosulfur species in petroleum and coal (George and Gorbaty, 1989; Gorbaty, George et al., 1990; Waldo, Carlson et al., 1991; Waldo, Mullins et al., 1992). Coal and crude petroleum also contain sulfide and disulfide minerals.

Box 7.1

You can verify the impact of increasing p_{CO_2} by performing the following ChemEQL simulations using the two components H^+ and $CO_2(g)$. In the first simulation, designate the H^+ activity to be a simulation variable (select *total* when generating the input matrix)[4] and enter any concentration; designate $CO_2(g)$ pressure to be a simulation constant (select *free* when generating the input matrix) and enter p_{CO_2} = $3.87 \cdot 10^{-4}$ *atm* (the aboveground $CO_2(g)$ partial pressure). In the second simulation, set up the input matrix similar to the first simulation, except this time enter p_{CO_2} = $3.87 \cdot 10^{-2}$ *atm* (soil-pore $CO_2(g)$ partial pressure). The pH drops one full unit, from 5.5 to 4.5, and the total dissolved carbonate (aqueous carbon dioxide, bicarbonate plus carbonate) increases from $2.18 \cdot 10^{-5}$ *mol* $\cdot L^{-1}$ to $1.92 \cdot 10^{-3}$ *mol* $\cdot L^{-1}$.

---

4. See the *ChemEQL Manual* under the section "Set Up a Matrix."

Structure 1. Thiophene

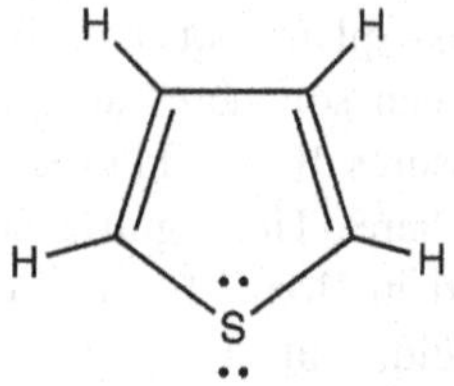

Structure 2. Thioether

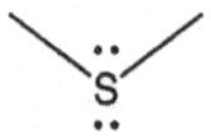

Sulfur dioxide $SO_2(g)$ is extremely water-soluble. The Henry's Law constant for Eq. (7.43) is $K_H \approx 1.2 \ mol \cdot L^{-1} \cdot atm^{-1}$ (Sander, 2009), but there is no evidence that sulfurous acid $H_2SO_3(aq)$ molecules exist in aqueous solution. Aqueous sulfur dioxide $SO_2(aq)$ has the acid dissociation constant (Eq. (7.44)) of a weak acid.

$$SO_2(g) \leftrightarrow SO_2(aq) \tag{7.43}$$

$$SO_2(aq) + H_2O(l) \leftrightarrow H^+(aq) + HSO_3^-(aq) \quad K_{a1, SO_2(aq)} = 1.5 \cdot 10^{-2} \tag{7.44}$$

Sulfur trioxide is also extremely water-soluble, combining with water (Eq. (7.45)) to form sulfuric acid $H_2SO_4(aq)$. Sulfuric acid has the acid dissociation constant (Eq. (7.46)) of a strong acid, but hydrogen sulfide $HSO_4^-(aq)$ (Eq. (7.47)) must be considered a weak acid comparable in strength to aqueous sulfur dioxide $SO_2(aq)$ (Eq. (7.44)).

$$SO_3(g) + H_2O(l) \leftrightarrow H_2SO_4(aq) \tag{7.45}$$

$$H_2SO_4(aq) + H_2O(l) \leftrightarrow H^+(aq) + HSO_4^-(aq) \quad K_{a1, SO_3(aq)} \approx 10^3 \tag{7.46}$$

$$HSO_4^-(aq) \leftrightarrow H^+(aq) + SO_4^{2-}(aq) \quad K_{a1, SO_3(aq)} = 1.02 \cdot 10^{-2} \tag{7.47}$$

The oxidation of sulfur dioxide $SO_2(g)$ to sulfur trioxide $SO_3(g)$ appears to be catalyzed by hydroxyl radicals $HO \cdot (g)$ in the atmosphere (Margitan, 1988):

$$\underset{\substack{hydroxyl \\ radical}}{SO_2(g) + HO \cdot (g)} \rightarrow \underset{\substack{hydroxysulfonyl \\ radical}}{HSO_3 \cdot (g)} \tag{7.48}$$

$$\underset{\substack{hydroxysulfonyl \\ radical}}{HSO_3 \cdot (g)} + O_2(g) \rightarrow SO_3(g) + \underset{\substack{hydrogen \ superoxide \\ radical}}{HOO \cdot (g)} \tag{7.49}$$

A major source of hydroxyl radicals $HO \cdot (g)$ is the photolysis of ozone (Eq. (7.50)) to produce photo-excited atomic oxygen $O^*(g)$. Photo-excited atomic oxygen $O^*(g)$ reacts with water molecules (Eq. (7.51)) to yield hydroxyl radicals $HO \cdot (g)$.

$$O_3(g) \xrightarrow{\text{sunlight}} O^*(g) + O_2(g) \tag{7.50}$$

$$O^*(g) + H_2O(g) \rightarrow 2 \cdot HO \cdot (g) \tag{7.51}$$
$$\underset{\substack{hydroxyl \\ radical}}{}$$

Another source of photo-excited atomic oxygen $O^*(g)$ is the photolysis of nitrogen dioxide $NO_2(g)$ (Li, Matthews et al., 2008):

$$\underset{\substack{nitrogen \\ dioxide}}{NO_2(g)} \xrightarrow{\text{sunlight}} \underset{\substack{nitrogen \\ monoxide}}{NO(g)} + O^*(g) \tag{7.52}$$

Alkaline compounds in atmospheric water droplets catalyzed the conversion of sulfur dioxide to sulfuric acid (Altshuller, 1973), resulting in a plateau of sulfuric acid levels in acid rain when sulfur dioxide levels exceed a certain threshold (Appendix 7C).

Some of the strong base released by chemical weathering (Eq. (7.24)) is neutralized by sulfuric acid deposition originating with sulfur oxides in the atmosphere and sulfide mineral weathering (Eqs. (7.35) and (7.36)). The product of this acid-base reaction is the following neutral salt solution that requires the sulfate correction in Eq. (7.33) used by Garrels and Mackenzie (1971):

$$\underset{\text{strong base}}{CaO(s)} + \underset{\text{strong acid}}{H_2SO_4} \xrightarrow{\text{neutralization}} Ca^{2+}(aq) + SO_4^{2-}(aq) + H_2O(l) \tag{7.53}$$

7.4.4. Nitrogen Oxides

Atmospheric chemists use the symbol NO_x to denote nitrogen monoxide $NO(g)$ and nitrogen dioxide $NO_2(g)$ in Earth's atmosphere. Anthropogenic nitrogen oxide emissions from the combustion of petroleum and coal are predominantly NO_x compounds.

Biogenic nitrogen oxide emissions—dinitrogen oxide[5] $N_2O(g)$ and nitrogen monoxide $NO(g)$—are the product of *denitrification* by soil bacteria. Denitrifying bacteria are anaerobic bacteria that rely on nitrate respiration (see Chapter 8) when soils are waterlogged.

Nitrogen compounds in coal and petroleum undergo combustion reactions to produce the nitrogen oxides NO_x in power plant and auto emissions. Pyridine (Structure **3**) and pyrrole (Structure **4**) represent organonitrogen species in petroleum and coal (Snyder, 1970; George and Gorbaty, 1989;

---

5. Dinitrogen oxide $N_2O(g)$ is a relatively inert gas in the troposphere but reacts with ultraviolet radiation in the stratosphere to form nitrogen monoxide and, ultimately, nitric acid.

Gorbaty, George et al., 1990; Waldo, Carlson et al., 1991; Waldo, Mullins et al., 1992).

Structure 3. Pyridine

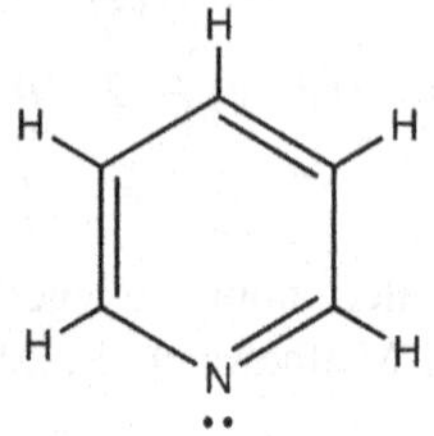

Structure 4. Pyrrole

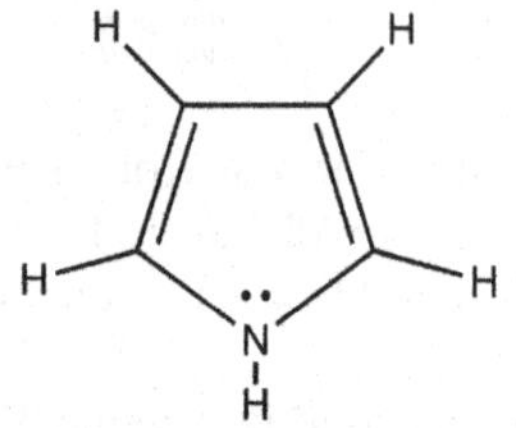

Some of the nitrogen dioxide $NO_2(g)$ in the atmosphere is produced by the oxidation of nitric oxide $NO(g)$:

$$2 \cdot \underset{N(2+)}{NO(g)} + O_2(g) \overset{oxidation}{\rightarrow} 2 \cdot \underset{N(4+)}{NO_2(g)} \tag{7.54}$$

The hydrolysis of nitrogen dioxide $NO_2(g)$ produces equal quantities of nitric $HNO_3(aq)$ and nitrous acid $HNO_2(aq)$. Reaction (7.55) is a *disproportionation* reaction because the net formal oxidation state of nitrogen (4+) remains unchanged:

$$2 \cdot \underset{N(4+)}{NO_2(g)} + H_2O(l) \xrightarrow{hydrolysis} \underset{N(5+)}{HNO_3(aq)} + \underset{N(3+)}{HNO_2(aq)} \tag{7.55}$$

Benkovitz and colleagues (1996) found that nitrogen emissions from anthropogenic and biogenic sources were roughly equal on the global scale, while current sulfur emissions are predominantly anthropogenic. The molar nitrogen-to-sulfur ratio of atmospheric emissions ranges (Figure 7.3) from $6.5 \cdot 10^4$ to $1.0 \cdot 10^{-4}$. Industrial and population centers with high sulfur oxide emissions are characterized by N/S ratios <0.01, while locations that are distant from industry and population centers with very low sulfur oxide emissions are characterized by N/S ratios >400. This variation in the nitrogen-to-sulfur ratio of atmospheric emissions reflects the importance of biogenic emissions.

Selecting data from Figures 7.4 and 7.5 for sulfate and nitrate wet deposition and converting units from $kg \cdot ha^{-1}$ to $mol \cdot ha^{-1}$, we can confirm the trend in the molar N/S depositional ratios reported by Benkovitz and colleagues (1996). The molar N/S depositional ratio for a site in western Kansas is 19 compared to a deposition ratio of 0.9 for a site in southeastern Ohio.

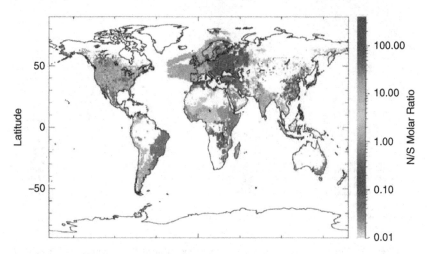

FIGURE 7.3   Mole ratio of nitrogen to sulfur in atmospheric emissions reflects the relative importance of industrial and biogenic emissions. *Source: Reproduced with permission from Benkovitz, C.M., Scholtz, M.T., et al., 1996. Global gridded inventories of anthropogenic emissions of sulfur and nitrogen. J. Geophys. Res. 101(D22), 29239–29253.*

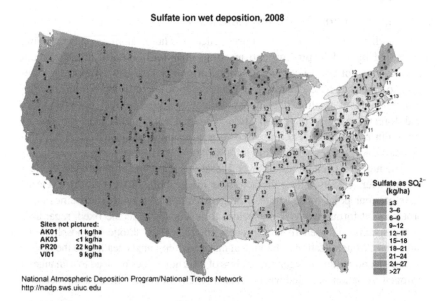

FIGURE 7.4   Sulfate wet deposition in the contiguous United States during 2008 (National Atmospheric Deposition Program, National Trends Network).

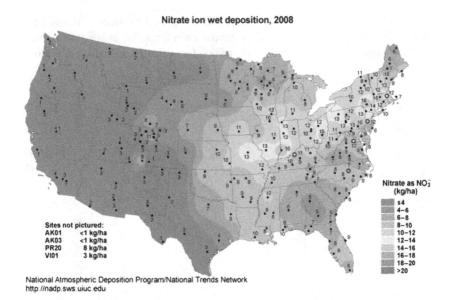

Nitrate ion wet deposition, 2008

National Atmospheric Deposition Program/National Trends Network
http://nadp.sws.uiuc.edu

FIGURE 7.5 Nitrate wet deposition in the contiguous United States during 2008 (National Atmospheric Deposition Program, National Trends Network).

7.5. AMMONIA-BASED FERTILIZERS AND BIOMASS HARVESTING

Nitrogen Cycling in Natural Landscapes

Soil acidification is a natural consequence of chemical mineral weathering promoted by soil respiration and leaching in humid climates. Plant nutrient uptake does not contribute to soil acidification because any components relevant to soil acid-base chemistry in standing biomass ultimately return to the soil when plant residue decomposes. Plant residue removal by agriculture and forestry exports plant ash from managed ecosystems, and the use of ammonia-based fertilizers, imported to replace nutrients lost by harvest, dramatically accelerates soil acidification.

The nitrogen cycle involves several reactions that transform nitrogen by electron transfer. Understanding the acid-base implications of the nitrogen cycle is, therefore, our point of departure. Nitrogen fixation (Eq. (7.56)) consumes eight protons as it forms two NH_4^+ molecules for every N_2 molecule fixed, regardless of the specific nitrogen fixation process (biological or abiotic—atmospheric or industrial). Mineralization of biological nitrogen compounds, leading to the assimilation of ammonium nitrogen (regardless of its origin), does not lead to a change in nitrogen oxidation state and has no impact on environmental acid-base processes:

$$\underset{N(0)}{N_2(g)} +8\cdot H^+ + 6\cdot e^- \xrightarrow{\text{nitrogen fixation}} 2\cdot \underset{N(3+)}{NH_4^+(aq)} \qquad (7.56)$$

Nitrification (Eq. (7.57)) is a biological oxidation process (see Chapter 8) that yields 20 protons as it forms $2\,NO_3^-$ molecules for every $2\,NH_4^+$ molecules oxidized. Nitrification is an essential nitrogen-cycle transformation. Plants, fungi, bacteria, and archaea must reduce NO_3^- to NH_4^+ (Eq. (7.58)) if they are to assimilate NO_3^- taken up from solution, reversing the nitrification reaction:

$$2\cdot NH_4^+(aq) + 6\cdot H_2O(l) \xrightarrow{\text{nitrification}} 2\cdot NO_3^-(aq) + 20\cdot H^+ + 16\cdot e^- \quad (7.57)$$
$$\underset{N(3+)}{} \qquad\qquad\qquad\qquad \underset{N(5+)}{}$$

$$2\cdot NO_3^-(aq) + 20\cdot H^+ + 16\cdot e^- \xrightarrow[\text{nitrate reduction}]{\text{assimilatory}} 2\cdot NH_4^+(aq) + 6\cdot H_2O(l)$$
$$\underset{N(5+)}{} \qquad\qquad\qquad\qquad\qquad \underset{N(3+)}{}$$
$$(7.58)$$

Denitrification (Eq. (7.59)) is a biological reduction process (see Chapter 8) that consumes 12 protons as it forms $1\,N_2$ molecule for every $2\,NO_3^-$ molecules reduced, completing the nitrogen cycle. The 20 protons consumed by nitrogen fixation (Eq. (7.56)) and denitrification (Eq. (7.59)) balance the 20 protons produced during nitrification (Eq. (7.57)).

$$2\cdot NO_3^-(aq) + 12\cdot H^+ + 10\cdot e^- \xrightarrow{\text{denitrification}} N_2(g) + 6\cdot H_2O(l) \quad (7.59)$$
$$\underset{N(5+)}{} \qquad\qquad\qquad\qquad\qquad\qquad \underset{N(0)}{}$$

At steady state, nitrogen fixation (Eq. (7.56)) replaces NO_3^- lost by leaching and denitrification. Nitrate leaching is equivalent to hydroxyl export from the soil where nitrate reduction occurs at the remote location (Eq. (7.60)). Keeping in mind most soils have little anion exchange capacity (humic colloids and clay minerals are cation exchangers), once nitrate has leached below the soil profile, there is little organic matter to sustain bacterial nitrate respiration (see Chapter 8). Remote nitrate reduction is most likely to occur in organic-rich sediments as groundwater discharges into a stream *hyporheic* (or lake *hypolentic*) zone.

$$2\cdot NO_3^-(aq) + 16\cdot e^- + 14\cdot H_2O(l) \xrightarrow[\text{nitrate reduction}]{\text{assimilatory}} 2\cdot NH_4^+(aq) + 20\cdot OH^-$$
$$\underset{N(5+)}{} \qquad\qquad\qquad\qquad\qquad\qquad \underset{N(3+)}{}$$
$$(7.60)$$

The only significant contributions to soil acidification from nitrogen cycling in natural landscapes arise from nitrate leaching losses (i.e., remote nitrate reduction). Agricultural landscapes export biomass and rely on fertilization to replace nutrients lost by export. Nitrate fertilizer import constitutes hydroxyl import—released during assimilation (Eq. (7.60))—while the import of ammonia-based fertilizer (urea, ammonia, and ammonium salts) constitutes acid import—released during nitrification (Eq. (7.56)).

7.6. CHARGE BALANCING IN PLANT TISSUE AND THE RHIZOSPHERE

Charge neutrality constrains ion uptake by plant roots, ion composition in plant tissue, and ion composition of the rhizosphere. The plant tissue charge balance, however, must account for postuptake assimilatory nitrate reduction (Eq. (7.60)) and sulfate reduction (Eq. (7.61)).

$$SO_4^{2-}(aq) +8 \cdot e^- + 5 \cdot H_2O(l) \xrightarrow{\substack{assimilatory \\ sulfate \ reduction}} HS^-(aq) +9 \cdot OH^- \quad (7.61)$$
$$\underset{S(6+)}{} \qquad \qquad \underset{S(2-)}{}$$

An excellent accounting of plant tissue and rhizosphere charge balancing in response to nitrate and ammonium uptake appears in van Beusichem and colleagues (1988). Unfortunately, the results appear in tabular form, making it difficult to visualize the contributions of each component to respective charge balance equations.

Figure 7.6 illustrates the cation and anion accounting required to balance charge in plant tissue and in the soil rhizosphere. The data are from an uptake experiment where nitrogen is supplied as the ammonium cation (van Beusichem, Kirby et al., 1988). Persistent cations (122.0 $cmol_c$ L^{-1})—Na^+, K^+, Ca^{2+}, and Mg^{2+}—and persistent anions (30.4 $cmol_c$ L^{-1})—Cl^- and $H_2PO_4^-$—appear on the left side of Figure 7.6. The ammonium-N in the experiment has been completely assimilated, which means that all NH_4^+ accumulated in plant tissue has been converted into organic nitrogen (330.8 $cmol_c$ L^{-1}). Sulfate is not persistent because a fraction is reduced to sulfide (Eq. (7.61)) and converted to bioorganic sulfur compounds. Total sulfur accumulated in plant tissue exists either as sulfate-S (32.1 $cmol_c$ L^{-1}) or assimilated organic sulfur (20.7 $cmol_c$ L^{-1}). Ion uptake in this experiment results in a net excretion of hydrogen ions by the roots into the rhizosphere (362.9 $cmol_c$ L^{-1}).

Turning our attention to the right-hand side of Figure 7.6, total accumulated cations in plant tissue is equal to the persistent cations (122.0 $cmol_c$ L^{-1}), while total accumulated anions (122.0 $cmol_c$ L^{-1}) is the sum of persistent anions (30.4 $cmol_c$ L^{-1}), conjugate bases (59.5 $cmol_c$ L^{-1}) derived from weak organic acids in plant tissue, and sulfate that has not been assimilated (32.1 $cmol_c$ L^{-1}). Reaction (7.12) relates the formation of organic conjugate bases in plant tissue to the consumption of hydroxyl ions, a reaction that is equivalent to the formation of bicarbonate when a strong base reacts with the weak acid aqueous carbon dioxide (Eq. (7.26)).

Total cation uptake is the sum of the persistent cations (122.0 $cmol_c$ L^{-1}) and the organic nitrogen (452.8 $cmol_c$ L^{-1})—which was taken up as NH_4^+ before being assimilated as organic nitrogen. Excess cation uptake—which must equal the hydrogen ions excreted by roots into the rhizosphere (362.9 $cmol_c$ L^{-1})—is equal to the total cation uptake minus the total accumulated anions.

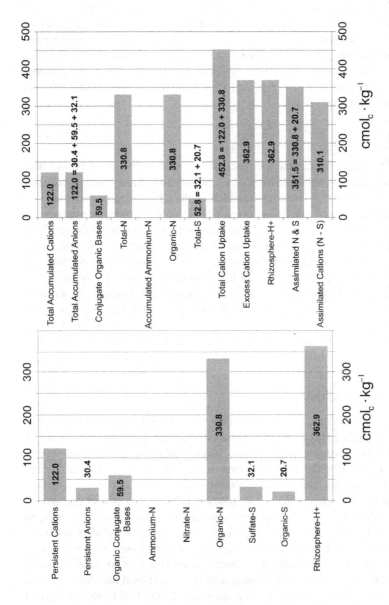

FIGURE 7.6 Cation and anion balances in plant tissue must account for the assimilation of both nitrate and sulfate. The ion components appear on the left, and summed components appear on the right, based on data from (*van Beusichem, Kirby et al., 1988*).

Net assimilated cations equals organic nitrogen—representing assimilated NH_4^+—minus organic sulfur—representing SO_4^{2-} anions lost by assimilation.

Plant species vary considerably in their tendency to accumulate cations and anions and to assimilate ammonium, nitrate, and sulfate. Pierre and Banwart (1973) devised a plant tissue parameter that rates the impact of biomass harvest on soil acidification. The Pierre-Banwart EB/N ratio compares the total hydroxyl content of plant tissue EB (total persistent cations minus total persistent anions equivalent to assimilated sulfate plus accumulated organic conjugate base in Figure 7.6) with the hydroxyl content derived from nitrogen assimilation N_{org}:

$$\frac{Excess\ Base}{N_{org}} = \frac{EB}{N} \tag{7.62}$$

If the EB/N ratio of harvested biomass is less than 1, then the export of base associated with remote denitrification (7.59) and assimilatory nitrate reduction (7.60) is less than the base counted as EB. If the EB/N ratio of harvested biomass is greater than 1, the loss of organic nitrogen in the biomass will increase acidity because the export of base associated with remote denitrification (7.59) and assimilatory nitrate reduction (7.60) is greater than the base counted as EB. Plants that have high rates of nitrogen assimilation relative to their tendency to accumulate excess base in their biomass have low EB/N ratios.

Barak and colleagues (1997) trace the impact of agriculture on soil acidification to two factors: nitrification (Eq. (7.56)) of nitrogen fertilizer in excess of the nitrogen assimilated in biomass—obviously nitrate fertilizers do not contribute to acidity arising from nitrification—and biomass harvest preventing organic conjugate bases (see Figure 7.6) in plant tissue from returning to the soil.

7.6.1. Water Alkalinity

Earlier in this chapter, we learned that water containing dissolved carbon dioxide requires a special reference point to distinguish the presence of strong acids and bases. Absent weak acids or bases, acidity and basicity are defined relative to pH 7 or the water reference level determined by the equilibrium constant K_w for the spontaneous dissociation of water (Eq. (7.6)).

Earth's atmosphere (Table 7.4) contains $CO_2(g)$ at a partial pressure of $p_{CO_2} = 3.87 \cdot 10^{-4}\ atm$, and, as a consequence, all natural waters are steeped in $CO_2(aq)$. The presence of $CO_2(aq)$ demands a p_{CO_2}-dependent reference point to distinguish the presence of strong acids and bases in aqueous solution: the aqueous carbon dioxide reference level (Eq. (7.20)). Environmental chemists distinguish between mineral acidity A_M and total alkalinity A_T relative to the aqueous carbon dioxide reference level. The following section extends the chemical concept of the aqueous carbon dioxide reference level to the geochemical concept of alkalinity A_T.

7.6.2. Carbonate Alkalinity

Carbonate alkalinity A_C—as it is defined by water chemists and geochemists—is simply an expression of the aqueous carbon dioxide proton reference level as it appears on the right-hand side of expression (7.21):

$$A_C = C_{HCO_3^-} + 2 \cdot C_{CO_3^{2-}} + C_{OH^-} - C_{H_3O^+} \tag{7.63}$$

Deffeyes (1965) observed that regrouping the charge balance expression for natural waters into the total charge concentration of permanent ions and the total charge concentration of changeable ions results in an expression where carbonate alkalinity A_C (Eq. (7.63)) is nothing more than the sum of the permanent-ion charges (Eq. (7.64)). Deffeyes simply extended (Eq. (7.19)) by including all of the major strong base cations—$Ca^{2+}(aq)$, $Mg^{2+}(aq)$, $Na^+(aq)$, and $K^+(aq)$—and all of the major strong acid anions—$SO_4^{2-}(aq)$ and $Cl^-(aq)$—in natural waters:

$$C_{H^+} + C_{Na^+} + C_{K^+} + 2 \cdot C_{Ca^{2+}} + 2 \cdot C_{Mg^{2+}}$$
$$= C_{Cl^-} + 2 \cdot C_{SO_4^{2-}} + \left(C_{HCO_3^-} + 2 \cdot C_{CO_3^{2-}} + C_{OH^-} \right) \tag{7.64}$$

$$A_C \equiv \left(C_{Na^+} + C_{K^+} + 2 \cdot C_{Ca^{2+}} + 2 \cdot C_{Mg^{2+}} \right) - \left(C_{Cl^-} + 2 \cdot C_{SO_4^{2-}} \right) \tag{7.65}$$

Expression (7.65) (which combines Eq. (7.63) and Eq. (7.64)) resembles an earlier empirical expression (7.33) proposed by Garrels and Mackenzie (1971). Regrouping (7.65) by taking (7.33) into account results in expression (7.66). The first term subtracts the chloride concentration from the two most abundant alkaline ions, correcting for the dissolution of halite $NaCl(s)$ and sylvite $KCl(s)$ in evaporite rocks. The second term is simply the right-hand side of the Garrels and Mackenzie expression, which subtracts the sulfate concentration from the two most abundant alkaline earth cations, corrected for the dissolution of the evaporite mineral gypsum.

$$A_C \equiv \left(C_{Na^+} + C_{K^+} - \underbrace{C_{Cl^-}}_{evaporites} \right) + 2 \cdot \left(C_{Ca^{2+}} + C_{Mg^{2+}} - \underbrace{C_{SO_4^{2-}}}_{evaporites} \right) \tag{7.66}$$

7.6.3. Silicate Alkalinity

Silicate alkalinity A_{Si} accounts for the contribution of aqueous silica (i.e., silicic acid $H_4SiO_4(aq)$) to the acid-base chemistry of natural waters. The conjugate base $H_3SiO_4^-$ is the only significant contributor to silicate alkalinity A_{Si} in natural waters:

$$A_{Si} \equiv C_{H_3SiO_4^-} + 2 \cdot C_{H_2SiO_4^{2-}} \approx C_{H_3SiO_4^-} \tag{7.67}$$

The dissolution reaction (7.68) of the silicon dioxide mineral opal—an amorphous polymorph of quartz—combined with the first aqueous silica acid dissociation reaction (7.69) yields an expression (7.70) for the activity of the conjugate base appearing in (7.67).

TABLE 7.5 Conjugate Acid-Base Pairs in Aqueous Silica Solutions

Weak Acid	Conjugate Base
H_3O^+	H_2O
$H_4SiO_4(aq)$	$H_3SiO_4^-(aq)$
$H_3SiO_4^-(aq)$	$H_2SiO_4^{2-}(aq)$
H_2O	OH^-

$$\underset{opal}{SiO_2(s)} + 2 \cdot H_2O(l) \leftrightarrow \underset{aqueous\ silica}{H_4SiO_4(aq)} \qquad K_{s0} = a_{H_4SiO_4} = 10^{-2.71} \qquad (7.68)$$

$$H_4SiO_4(aq) \leftrightarrow H^+(aq) + H_3SiO_4^-(aq) \qquad K_{a1} = \frac{a_{H^+} \cdot a_{H_3SiO_4^-}}{a_{H_4SiO_4}} = 10^{-9.84}$$

$$(7.69)$$

$$\underset{opal}{SiO_2(s)} + 2 \cdot H_2O(l) \leftrightarrow H^+(aq) + H_3SiO_4^-(aq)$$

$$(7.70)$$

$$K_{s0} \cdot K_{a1} = a_{H^+} \cdot a_{H_2SiO_4^{2-}} = 10^{-2.71} \cdot 10^{-9.84}$$

Table 7.5 lists the conjugate acid-base pairs in an aqueous silica solution. The boldface solute species are the same species that define the aqueous silica proton reference level:

$$\left(C_{OH^-} + C_{H_3SiO_4^-} + 2 \cdot C_{H_2SiO_4^{2-}} - C_{H_3O^+} \right) = 0 \qquad (7.71)$$

The boldface species designate the proton balance species of the aqueous silica $SiO_2(aq)$ reference level at pH 6.3, which is a bit less than pH unit higher than the aqueous carbon dioxide proton reference level used to define carbonate alkalinity A_C.

7.6.4. The Methyl Orange End-Point

Total alkalinity is the equivalent[6] concentration ($mol_c \cdot L^{-1}$) of all weak-acid conjugate bases in solution (carbonates and silicates).[7] The operational definition of *total alkalinity* is the *methyl orange* (CAS 547-58-0) *end-point* in a

---

6. An acid-base equivalent is the amount of a conjugate base that will combine with a mole of hydrogen ions H^+.

7. Phosphates and other weak acids are typically insignificant compared to the carbonates and silicates in freshwater systems. Borate conjugate acid-base pairs attain significant concentrations in marine systems.

strong acid titration. The color change associated with the methyl orange end-point begins at pH 4.4, near the second equivalence point for carbonate:

$$HIn\,(aq) \leftrightarrow H^+(aq) + In^-\,(aq) \qquad K_a = \frac{a_{H^+} \cdot a_{In^-}}{a_{HIn}} = 10^{-3.7} \qquad (7.72)$$
$$\underset{red}{} \qquad\qquad\qquad\quad \underset{yellow}{}$$

Alkalinity defined by the methyl orange end-point titration measures all of the conjugate bases in a water sample whose proton reference level is at a pH above 3.7. The most important conjugate base in groundwater besides the carbonate species is $H_3SiO_4^-\,(aq)$.

7.6.5. Mineral Acidity

Mineral acidity A_M (e.g., acid rain or acidic mine drainage water) applies to solutions whose pH is lower than solutions containing only aqueous carbon dioxide $CO_2(aq)$, indicating the presence of strong mineral acids, typically nitric or sulfuric acid. For solutions containing sodium hydroxide $NaOH$, nitric acid HNO_3, and aqueous carbon dioxide $CO_2(aq)$, mineral acidity A_M is simply the negative of *carbonate alkalinity A_C*:

$$Charge\ Balance: \quad C_{Na^+} + C_{H^+} = C_{HCO_3^-} + 2 \cdot C_{CO_3^{2-}} + C_{OH^-} + C_{NO_3^-}$$

$$CO_2(aq)\ Reference\ Level: C_{NO_3^-} - C_{Na^+}$$

$$= C_{H^+} - \left(C_{OH^-} + C_{HCO_3^-} + 2 \cdot C_{CO_3^{2-}} \right)$$

$$Mineral\ Acidity: \quad A_M \equiv -A_C \qquad (7.73)$$

7.7. MECHANICAL PROPERTIES OF CLAY COLLOIDS AND SOIL SODICITY

This section examines the impact of Na^+ ions on clay colloid chemistry. Given the relative abundance of the four most common alkali and alkaline earth cations in Earth's crust—sodium, potassium, magnesium, and calcium—soil pore water is typically dominated by Ca^{2+} ions unless saline groundwater discharge or mineral weathering in arid climates alters the pore water chemistry. In Chapter 3 we learned that smectite clay minerals were capable of osmotic swelling if saturated by either Na^+ or Li^+ ions (the latter being inconsequential in most soil or groundwater settings), but smectite clays were limited to crystalline swelling if saturated by larger alkali ions and divalent alkaline earth cations. This section extends the clay chemical impact of Na^+ ions beyond swelling behavior and considers the role of alkalinity in altering pore water chemistry.

We begin with a discussion of the impact of the clay fraction on soil mechanical and physical properties (soil strength, hydraulic conductivity, and erodibility) and continue with an examination of the effect that Na^+ ions have on these same properties. Finally, we discuss the pore water chemistry required to enrich the exchange complex in Na^+ ions while depleting it of Ca^{2+} and Mg^{2+} ions.

7.7.1. Clay Plasticity and Soil Mechanical Properties

The geotechnical properties of soil—the strength and hydraulic conductivity of the soil fabric and its resistance to particle detachment during erosion—depend on the relative amount of clay-size mineral particles. Clay-size mineral particles, regardless of mineralogy, exhibit behavior unlike granular-size (i.e., silt and sand) particles. The specific particle diameter separating the clay- and granular-size fractions varies somewhat among Earth science and engineering communities, but the primary distinction appears when the surface-to-volume ratio or surface area per unit mass has a measureable impact on the interactions between water and mineral surfaces.

Soil mechanics distinguishes between the granular-size fraction (silt plus sand: $2~\mu m < d < 2~mm$) and the clay-size fraction ($d < 2~\mu m$) because the granular-size fraction is never plastic, while the clay-size fraction exhibits plasticity over a range of water contents. The USDA defines *plasticity* as "the property of a soil that enables it to change shape continuously under the influence of applied stress and to retain that shape on removal of the stress" (Staff, 1987). In other words, a soil- or particle-size fraction is plastic if it can be molded (i.e., retain its shape) and deformed without cracking (i.e., change shape continuously). Soil scientists use the *ribbon test* (Box 7.2) as a field estimate of soil clay content. Here, a moist soil sample is molded into a ball and then pressed into a ribbon between the thumb and forefinger (Figure 7.7). Material composed solely of granular particles (silt and sand) is nonplastic and cannot be molded into a ribbon. Soil containing significant amounts of clay can be molded into a ribbon. As the clay-size fraction

---

Box 7.2 The Ribbon Test

Soil mineral particles are typically separated into three particle-size fractions: sand ($50~\mu m$ to $2.0~mm$), silt ($2~\mu m$ to $50~\mu m$), and clay ($d \leq 2~\mu m$)—soil being restricted to the solid fraction that passes a $2~mm$ sieve. Soil texture is usually a complex size distribution represented by the relative proportions of the three particle-size fractions (Staff, 1987): sand, silt, and clay.

A highly plastic soil can be molded into a ribbon greater than 5 cm in length without cracking. A medium plastic soil forms a ribbon 2.5 to 5 cm in length without cracking. A slightly plastic soil forms a ribbon less than 2.5 cm without cracking. Soil textural classes that are typically nonplastic include sand, loamy sand, and silt. These textural classes contain less than 7 to 15% clay, depending on silt content.

FIGURE 7.7 Molding moist clay between forefinger and thumb produces a clay ribbon.

increases—regardless of mineralogy—the force required to mold a soil sample and form a ribbon increases.

Plasticity arises from the interaction between water molecules and particle surfaces. These interactions are always present but have a discernible effect on soil mechanical properties when clay-size particles are present in sufficient quantities to influence the contact between granular particles. When the mean diameter of mineral particles, regardless of mineralogy, falls below 2 μm—corresponding to roughly 5 $m^2 \cdot g^{-1}$—the interaction of water with the mineral surface lead to the behavior we know as plasticity.

Nonswelling clays—iron oxides, aluminum oxides, and nonswelling clay minerals—exhibit more or less the same plasticity characteristics. Although it is possible to synthesize oxide particles in the laboratory with surface areas of 100–200 $m^2 \cdot g^{-1}$, oxide minerals rarely occur in the fine clay fraction ($d \leq 0.2$ μm) where the surface area exceeds 50 $m^2 \cdot g^{-1}$. Swelling clay minerals exhibit plasticity over a larger moisture range because the interlayer adds to the effective surface area. The typical surface area of a swelling clay mineral is 750 $m^2 \cdot g^{-1}$.

The geotechnical engineering community assesses the mechanical properties of soil and the clay-size fraction using Atterberg Limits: shrinkage limit, plastic limit, and liquid limit. Our focus will be on the plastic and liquid limits. Associated with the Atterberg Limits are several derived parameters: plasticity index, liquidity index, and activity index.

The geotechnical community expresses water content w as a fraction of the mineral-solids mass m_s (Figure 7.8). The plastic limit w_P is the lowest water content w at which clay or soil exhibits plastic behavior, the highest water content being the liquid limit w_L. Between w_P and w_L there is sufficient water for molded mineral particles to deform without cracking. Above w_L, plasticity is lost and the material behaves like a liquid, unable to retain a molded shape. Below w_P, soil containing sufficient clay for plasticity cannot be molded into a ribbon for lack of moisture.

The plasticity index I_P is a measure of the water content range over which the soil is plastic:

$$I_P = w_L - w_P \tag{7.74}$$

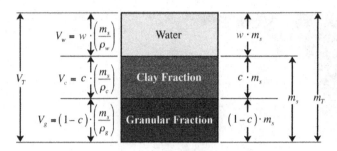

FIGURE 7.8 Mass-volume relations in a water-saturated soil mixture.

The activity index I_A normalizes the plasticity index I_P of a soil by its clay content. A plot of I_P as a function of clay content c (assuming the mineralogy of the clay remains unchanged) yields a straight line whose slope is the activity index I_A. Swelling clays have a significantly higher activity index I_A ($1 < I_A < 7$) than nonswelling clays ($I_A \leq 0.5$):

$$I_A = \frac{I_P}{100 \cdot c} \tag{7.75}$$

7.7.2. Clay Content and Granular Particle Contacts

We can estimate the amount of clay needed to prevent direct contact between granular mineral particles at water contents typical of a saturated soil. Referring to Figure 7.8, the volume available for water and clay in a soil consisting of clay and granular particles is the product of the void ratio of the granular fraction e_g and the total volume of all granular particles V_g:

$$V_{voids} = e_g \cdot V_g = e_g \cdot \left(\frac{(1-c) \cdot m_s}{\bar{\rho}_{mineral}} \right) \tag{7.76}$$

The void ratio e_g is related to bulk density ρ_{bulk} and mineral particle density $\bar{\rho}_{mineral}$ as follows:

$$e_g = \frac{V_{voids}}{V_s} = \frac{\bar{\rho}_{mineral} - \rho_{bulk}}{\rho_{bulk}} \tag{7.77}$$

The volume of water saturating all pores V_w and the volume occupied by clay V_c are both a function of the mass of the mineral solids m_s, the mass fraction (as defined in Figure 7.8) and the density of each component:

$$V_{voids} = V_w + V_c = \left(\frac{w[Mg \cdot Mg^{-1}] \cdot m_s[Mg]}{\rho_w[Mg \cdot m^{-3}]} \right) + \left(\frac{c[Mg \cdot Mg^{-1}] \cdot m_s[Mg]}{\bar{\rho}_{mineral}[Mg \cdot m^{-3}]} \right) \tag{7.78}$$

Setting these expressions equal to one another and simplifying yield the following expression for the amount of clay needed to prevent direct contact between granular particles:

$$\frac{(1-c)\cdot m_s \cdot e_g}{\bar{\rho}_{mineral}} = \left(\frac{w\cdot m_s}{\rho_w}\right) + \left(\frac{c\cdot m_s}{\bar{\rho}_{mineral}}\right)$$

$$c = \left(\frac{e_g}{1+e_g}\right) - \left(\frac{\bar{\rho}_{mineral}}{\rho_w + (e_g\cdot\rho_w)}\right)\cdot w = \left(\frac{e_g}{1+e_g}\right) - \left(\frac{\bar{\rho}_{mineral}}{1+e_g}\right)\cdot w$$

The void ratio e_g is typically 0.85, and the geotechnical community uses a mineral particle density of $\bar{\rho}_{mineral} = 2.75\,Mg\cdot m^{-3}$. The water content of granular particles at field capacity w_{FC} is on the order of 0.10 to 0.20:

$$c = (0.46) - w_{FC}\cdot(1.43) \tag{7.79}$$

The amount of clay c sufficient to coat granular particles and prevent direct contact (7.79) is on the order of 17% for a granular fraction consisting of silt and as high as 30% for a granular fraction consisting of sand.

The activity index I_A of the clay fraction—its tendency to confer plasticity on the whole soil—also depends on the exchangeable cations. In particular, clay activity index I_A tends to decrease as the clay fraction becomes saturated by Na^+ ions; activity index I_A increases as the exchange complex is dominated by Ca^{2+} (and to a lesser extent, Mg^{2+}) ions.

Soil fabric derives its load bearing and shear strength from the contact between mineral grains, and clay plasticity adds considerably to soil fabric strength. Furthermore, the pore structure in soil results from particle displacement during wetting-drying cycles, freezing-thawing cycles, and *bioturbation*.[8] Clay plasticity preserves an open soil fabric, promoting gas exchange between the soil and aboveground atmosphere and—of particular interest in this section—hydraulic conductivity.

The empirical hydraulic conductivity parameters in Chapter 2, Appendix 2F, represent the maximum hydraulic conductivity for each soil texture—that is, the clay fraction retains sufficient plasticity to maintain an open soil fabric with maximum pore size and, hence, hydraulic conductivity. Chemically altering soils by saturating the exchange complex with Na^+ ions lowers the activity index I_A of the clay fraction, resulting in significantly lower hydraulic conductivity values.

Figure 7.9 illustrates this behavior by showing the effect of two soil treatments on the activity index I_A of soil clay (Lebron, Suarez et al., 1994). The natural soils (filled circles) are saline-sodic, with Na^+ ions occupying 90% or more of the exchange complex and a pore water electrical conductivity EC in

---

8. Bioturbation is the rearrangement of solid particles in soil and sediment caused by the activity of organisms.

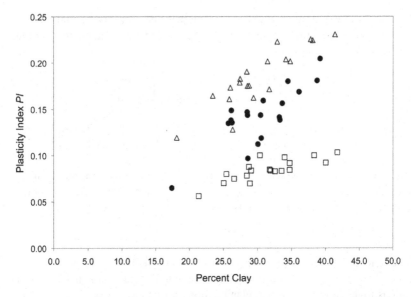

FIGURE 7.9   The activity index I_A of soil clay in 19 saline-sodic soils from Spain: (a) filled circles represent the untreated soil activity index I_A, (b) open triangles represent the activity index I_A for soils washed with calcium chloride to replace exchangeable Na^+ ions with Ca^{2+} ions, and (c) open squares represent the activity index I_A for soils washed with pure water to lower salinity. (*modified from Lebron, Suarez et al., 1994*).

the range of 6.2 to 40.5 $dS \cdot m^{-1}$. Replacing Na^+ ions with Ca^{2+} ions by washing the soils with calcium chloride solutions (open triangles) increases the activity index I_A of the clay fraction. Washing the soils with pure water to lower salinity reduces the fraction of Na^+ ions occupying the exchange complex to 20 to 30% and significantly lowers clay index I_A.

Geotechnical engineers and soil scientists (Resendiz, 1977; Lebron, Suarez et al., 1994) associate soil erodibility with both clay content and activity index I_A. Soil erosion begins with the detachment of soil particles by raindrop impact or the impact of fine windblown mineral particles bouncing across the soil surface. Soils with low clay contents or containing low-activity index I_A clay are more susceptible to detachment and thus are more erodible. The accumulation of sodium ions in the exchange complex of soil clays is known to increase soil erodibility.

7.7.3. Sodicity

Soil scientists apply several criteria to identify soils that either exhibit low hydraulic conductivity or are at risk for developing low hydraulic conductivity as a result of clay chemistry changes caused by Na^+ ions accumulating on the clay exchange complex. Geotechnical engineers have no universal term for

TABLE 7.6 USDA Natural Resource Conservation Service Sodicity Classification and Criteria

Soil Type	Exchangeable Sodium Percentage	Soil Water Electrical Conductivity*	Soil pH
Saline	$ESP < 15$	$EC_{sw} > 4dS \cdot m^{-1}$	$pH < 8.4$
Saline-Sodic	$ESP > 15$	$EC_{sw} > 4dS \cdot m^{-1}$	$pH < 8.4$
Sodic	$ESP > 15$	$EC_{sw} < 4dS \cdot m^{-1}$	$pH > 8.4$

*SI units for electrical conductivity are *siemens* per meter: $S \cdot m^{-1}$.

soils that lose strength when Na^+ ions accumulate on the clay exchange complex. Both *ESP* and clay activity index I_A are used to identify highly erodible or *dispersive* soils (Resendiz, 1977; Lebron, Suarez et al., 1994). The erodibility criterion based on clay activity index I_A does not explicitly assess the exchangeable Na^+ ion fraction. *Quick clays*[9] owe their susceptibility to liquefaction to Na^+ ion accumulation on the clay exchange complex, but the criteria used to assess quick clays do not explicitly assess the exchangeable Na^+ ion fraction.

The USDA Natural Resource Conservation Service employs three criteria—listed in Table 7.6—to identify sodicity risk: exchangeable sodium percentage, total dissolved salts as measured by electrical conductivity of soil pore water, and soil pH. We will examine the implications of each criterion to gain an understanding of the chemical processes at work in sodic soils.

7.7.4. Sodium-Ion Accumulation on the Clay Exchange Complex: ESP

The exchangeable sodium percentage *ESP* is a familiar ion exchange parameter: the equivalent fraction of exchangeable sodium $\tilde{E}_{Na^+}$ multiplied by 100 (Eq. (7.80)). A common alternative parameter is the exchangeable sodium ratio *ESR* (Eq. (7.81)).

$$ESP = \frac{m_{Na^+}}{\left(m_{Na^+} + 2 \cdot m_{Ca^{2+}} + 2 \cdot m_{Mg^{2+}}\right)} \cdot 100 = \tilde{E}_{Na^+} \cdot 100 \qquad (7.80)$$

---

9. Quick clays are typically clay-rich coastal soils that owe their relatively high exchangeable Na^+ equivalent fraction to their location and depositional history. Quick clays are usually deposited in coastal seawater followed by uplift above sea level. Quick clay formations remain relatively stable *in situ* but liquefy when remolded (i.e., disturbance of the soil fabric during excavation or construction of earthen structures). Liquefaction occurs because the *in situ* liquid limit w_L is metastable. Remolding disturbs the soil fabric, thereby eliminating metastability.

Soil and Environmental Chemistry

$$ESR = \frac{\tilde{E}_{Na^+}}{CEC - \tilde{E}_{Na^+}} = \frac{\tilde{E}_{Na^+}}{\left(\tilde{E}_{Ca^{2+}} + \tilde{E}_{Mg^{2+}}\right)} = \frac{(ESP/100)}{(1 - (ESP/100))} \quad (7.81)$$

The ESP is difficult to measure, leading the USDA Salinity Laboratory Staff (Allison, Bernstein et al., 1954) to search for an empirical relation between solution composition and the composition of the exchange complex. They discovered a linear relation between the solution *sodium adsorption ratio* SAR_P [10]—defined by expression (7.82)—and ESR. The solution SAR_P has units of $\sqrt{mmol_c \cdot L^{-1}}$ because the charge concentration of each ion is typically expressed in units of $mmol_c \cdot L^{-1}$.

$$SAR_P = \frac{n_{Na^+}[mmol_c \cdot L^{-1}]}{\sqrt{^1/_2 \cdot \left(n_{Ca^{2+}} + n_{Mg^{2+}}\right)}[mmol_c \cdot L^{-1}]} = \frac{c_{Na^+}[mmol \cdot L^{-1}]}{\sqrt{\left(c_{Ca^{2+}} + c_{Mg^{2+}}\right)}[mmol \cdot L^{-1}]}$$

$$(7.82)$$

Figure 7.10 plots the data from Allison and colleagues (1954) and lists the USDA Salinity Laboratory empirical (ESR, SAR_P) relation. Most current references simplify the original empirical expression by dropping the intercept and shortening

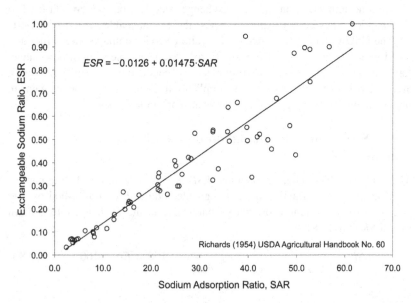

$ESR = -0.0126 + 0.01475 \cdot SAR$

Richards (1954) USDA Agricultural Handbook No. 60

Sodium Adsorption Ratio, SAR

Exchangeable Sodium Ratio, ESR

FIGURE 7.10 The exchangeable sodium ratio (see Eq. (7.81)) is proportional to the practical sodium adsorption ratio SAR_P for arid U.S. soils. *Source: Reproduced with permission from Allison, L.E., Bernstein, L., et al., 1954. Diagnosis and Improvement of Saline and Alkali Soils. U. S. S. Laboratory. Washington, D.C., United States Government Printing Office, 160.*

10. The SAR_P does not apply activity corrections to ion concentrations.

the proportionality factor to two significant digits. Relating ESR to SAR_P using expression (7.83) gives an alternative criterion for sodic soils: $SAR_P > 12$.

$$ESR \approx \left(1.5 \cdot 10^{-2}\right) \cdot SAR_P \qquad (7.83)$$

Appendix 7H summarizes $\left(Na^+, Ca^{2+}\right)$ cation exchange results for two closely related Wyoming smectite samples: American Petroleum Institute sample API-25 and Clay Minerals Society sample SWy-1. Appendix 7H also shows the effect of sodium exchange on the critical coagulation concentration of smectite (Oster, Shainberg et al., 1980). Figure 7.11 plots data points along the critical coagulation concentration curve (Oster, Shainberg et al., 1980) (see Figure 7H.1) using the same relation as that appearing in Figure 7.10. Figures 7.10 and 7.11 demonstrate that $\left(Na^+, Ca^{2+}\right)$ exchange by soil clays (Figure 7.10) is closely related to the $\left(Na^+, Ca^{2+}\right)$ exchange behavior of smectite clays (Figure 7.11).

The sodium adsorption ratio SAR_P (Eq. (7.82)) is related to the Gaines-Thomas exchange selectivity quotient (Eq. (7.84)) discussed previously in Chapter 4. Despite references to the Gapon exchange selectivity quotient by *USDA Handbook 60* (Allison, Bernstein et al., 1954), a quasi-linear relation between ESR (Eq. (7.81)) and SAR_P (Eq. (7.82)) appears in Figure 7.11. The subtle curve and negative intercept in the data plotted in Figure 7.11 are obscured by experimental and sample variation in Figure 7.10.

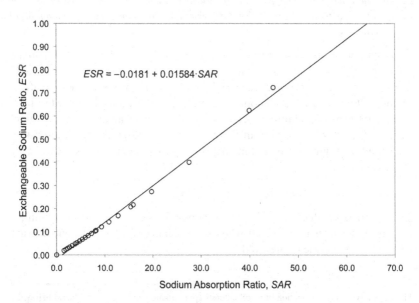

FIGURE 7.11 The exchangeable sodium ratio (Eq. (7.81)) is proportional to the practical sodium adsorption ratio SAR_P for $\left(Na^+, Ca^{2+}\right)$ cation exchange on Wyoming smectite API-25 (data points follow critical coagulation concentration curve from Oster and colleagues (1980) using $K_{GT} = 0.30$ plotted in Appendix 7H, Figure 7H.2).

$$\underset{exchanger}{2 \cdot Na^+} + \underset{solution}{Ca^{2+}(aq)} \leftrightarrow \underset{solution}{2 \cdot Na^+(aq)} + \underset{exchanger}{Ca^{2+}}$$

$$K_{GT}^c = \frac{\tilde{E}_{Ca^{2+}} \cdot C_{Na^+}^2}{\tilde{E}_{Na^+}^2 \cdot C_{Ca^{2+}}}$$

$$\left(\frac{K_{GT}^c \cdot \tilde{E}_{Na^+}^2}{\tilde{E}_{Ca^{2+}}} \right)^{1/2} = \frac{C_{Na^+}}{C_{Ca^{2+}}^{1/2}} = \frac{c_{Na^+}/1000}{\sqrt{c_{Ca^{2+}}}/\sqrt{1000}}$$

$$SAR_P = \sqrt{1000} \cdot \left(\frac{K_{GT}^c \cdot \tilde{E}_{Na^+}^2}{\tilde{E}_{Ca^{2+}}} \right)^{1/2} = \sqrt{1000 \cdot K_{GT}^c} \cdot \left(\frac{\tilde{E}_{Na^+}}{\sqrt{\tilde{E}_{Ca^{2+}}}} \right) \qquad (7.84)$$

7.7.5. Pore Water Electrical Conductivity EC_W

A contributing factor to sodification is the combination of low annual rainfall and high rates of evapotranspiration that concentrate soluble salts in soil pore water. The USDA Soil Taxonomy defines the *aridic* soil moisture regime as soils whose moisture-control section[11] is as follows:

Dry[12] in all parts for more than half of the cumulative days per year when the soil tem-perature at a depth of 50 cm from the soil surface is above 5°C and moist in some or all parts for less than 90 consecutive days when the soil temperature at a depth of 50 cm is above 8°C.

Empirical expression (7.85) is widely used by soil chemists to estimate ionic strength I from electrical conductivity EC_W (Griffin and Jurinak, 1973). Previously (see Figure 3.13 and Appendix 3D), we discussed the effect of ionic strength on the osmotic swelling of Na^+-saturated smectite clays. Figure 7H.1 offers additional evidence that the exchanger Na^+ equivalent fraction $\tilde{E}_{Na^+}$ correlates with an increased critical coagulation concentration (i.e., ionic strength at which coagulation begins to occur) for smectite clay suspensions.

$$I[mol \cdot L^{-1}] = 0.013 \cdot EC_W[dS \cdot m^{-1}] \qquad (7.85)$$

The electrical conductivity EC_W criterion (see Table 7.6) predicts clay miner-als will either disperse or swell when the exchanger Na^+ equivalent fraction $\tilde{E}_{Na^+}$ exceeds a critical threshold.

---

11. The moisture control section of a soil extends approximately (1) from 10 to 30 *cm* below the soil surface if the particle-size class of the soil is fine loamy, coarse-silty, fine-silty, or clayey; (2) from 20 to 60 *cm* if the particle-size class is coarse-loamy; and (3) from 30 to 90 *cm* if the particle-size class is sandy.

12. $p_{WP} \equiv -1500\ kPa$ or $h_{WP} = -153.0\ m$

7.7.6. Extreme Alkalinity: Soil pH>8.4

The pH criterion used by the USDA to assess sodicity risk (see Table 7.6), unlike the other two criteria, does not distinguish sodic from nonsodic behavior but serves as a predictor of sodicity risk. As we will see in this section, there are circumstances where soils exhibit low hydraulic conductivity characteristic of sodicity, meet both ESP (or SAR_P) and EC_W criteria, and yet still fail the pH > 8.4 criterion.

Pore water chemistry related to sodicity is best understood on a continuum between two end-points. The low-alkalinity end-point is the chemistry of calcite-saturated seawater or calcite-saturated saline groundwater. The alkalinity of the low-alkalinity end-point is derived solely from calcite dissolution because the other solutes are derived from neutral salts: chlorides and sulfates.

The major cations in seawater (Table 7.7) originate from chemical weathering of Earth's crust. The major anions result from acid deposition from volcanic activity—hydrochloric and sulfuric acid—and atmospheric

TABLE 7.7 Standard Mean Chemical Composition of Seawater: Ionic Strength $I = 0.72\ mol_c \cdot kg_{H_2O}^{-1}$, Total Alkalinity $A_C = 2.4 \cdot 10^{-3}\ mol_c \cdot kg_{H_2O}^{-1}$ and pH = 8.1

Species	Concentration, $mol \cdot kg_{H_2O}^{-1}$
Na^+	0.48616
Mg^{2+}	0.05475
Ca^{2+}	0.01065
K^+	0.01058
Sr^{2+}	0.00009
Cl^-	0.56576
SO_4^{2-}	0.02927
Br^-	0.00087
F^-	0.00007
HCO_3^-	0.00183
CO_3^{2-}	0.00027
$B(OH)_3^0$	0.00033
$B(OH)_4^-$	0.00010
OH^-	0.00001

Source: Dickson and Goyet, 1994.

CO_2. Acid deposition neutralizes most of the alkalinity resulting from chemical weathering of the silicate crust; most of the total alkalinity derives from calcite saturation.

Diluting seawater to $ECw = 4dS \cdot m^{-1}$ while maintaining calcite saturation results in pore water composition that satisfies the first two criteria in Table 7.6 but fails the pH criterion. The residual sodium carbonate RSC is negative, consistent with the low pH:

$$RSC = \left(C_{HCO_3^-} + 2 \cdot C_{CO_3^{2-}} \right) - \left(2 \cdot C_{Ca^{2+}} + 2 \cdot C_{Mg^{2+}} \right) \qquad (7.86)$$

The pore water chemistry in Table 7.8 is representative of the low-alkalinity end-point found in incompletely leached marine quick clay deposits and soils that develop sodicity from saline groundwater discharge or irrigation with saline groundwater.

The high-alkalinity end-point results solely from the chemical weathering of silicate rocks. Figure 7.12 illustrates the dissolution of Earth's silicate crust by reaction (7.23), where solution $Ca^{2+}(aq)$ activity $a_{Ca^{2+}}$ is constrained by calcite solubility. The process depicted in Figure 7.12 is equivalent to the irreversible dissolution of orthoclase $KAlSi_3O_8$ described by Helgeson and colleagues (1969). Rhoades and colleagues (1968) measured significant chemical weathering and increasing alkalinity when leaching arid-region soils with irrigation water. Much of the alkalinity recorded by Rhoades and colleagues, however, resulted from calcite dissolution.

TABLE 7.8 Seawater Diluted to Ionic Strength $I = 0.05 \; mol_c \cdot L^{-1}$ ($ECw = 4 \; dS \cdot m^{-1}$) While Maintaining Calcite Saturation and Soil Pore CO_2 Levels: $p_{CO_2} = 3.87 \cdot 10^{-2} \; atm$

Species	Concentration
Na^+	$3.49 \cdot 10^{-2} \; mol \cdot L^{-1}$
Mg^{2+}	$3.93 \cdot 10^{-3} \; mol \cdot L^{-1}$
Ca^{2+}	$3.92 \cdot 10^{-3} \; mol \cdot L^{-1}$
K^+	$7.06 \cdot 10^{-4} \; mol \cdot L^{-1}$
Cl^-	$4.07 \cdot 10^{-2} \; mol \cdot L^{-1}$
SO_4^{2-}	$2.10 \cdot 10^{-3} \; mol \cdot L^{-1}$
A_C	$8.11 \cdot 10^{-3} \; mol_c \cdot L^{-1}$
pH	6.96
SAR_p	$12.2 \; \sqrt{mmol \cdot L^{-1}}$
RSC	$-6.98 \cdot 10^{-3} \; mol_c \cdot L^{-1}$

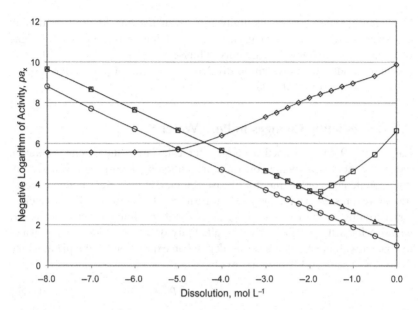

FIGURE 7.12   The irreversible dissolution of Earth's crust according to reaction (7.23) neglects the solubility of Al_2O_3, Fe_2O_3, and SiO_2. Calcium solubility is constrained by calcite solubility and atmospheric $CO_2(g)$ ($p_{CO_2} = 3.89 \cdot 10^{-4}\ atm$). Solution pH (diamond symbol) increases on the logarithmic scale once $10^{-5}\ mol \cdot L^{-1}$ of the crust has dissolved. Maximum calcium solubility (square symbol) is attained at pH 8.4 when the solution becomes saturated by calcite (dissolution $\approx 10^{-2}\ mol \cdot L^{-1}$) and then declines as pH and sodium solubility (triangle symbol) continue to increase.

Expression (7.65) (Deffeyes, 1965) defines alkalinity as the difference between the sum over the major alkali and alkaline earth cations—$Ca^{2+}(aq)$, $Mg^{2+}(aq)$, $Na^+(aq)$, and $K^+(aq)$—minus the sum over all major strong acid anions—$SO_4^{2-}(aq)$ and $Cl^-(aq)$—in natural waters. Calcite solubility imposes no global constrain on alkalinity.

Equation (7.65) implies another mechanism for increasing alkalinity without the need for chemical weathering of silicate rocks: conversion of $SO_4^{2-}(aq)$ to $OH^-(aq)$ by biological sulfate reduction. Whittig and Janitzky (1963) identified anaerobic dissimilatory sulfate reduction as an alternate source of hydroxyl ions in landscape settings where poor drainage, high levels of dissolved sulfate, and abundant soil organic matter are present:

$$\underset{S(6+)}{SO_4^{2-}(aq)} +8 \cdot e^- + 5 \cdot H_2O(l) \xrightarrow{\overset{dissimilatory}{sulfate\ reduction}} \underset{S(2-)}{HS^-(aq)} +9 \cdot OH^- \qquad (7.87)$$

Though conditions favoring dissimilatory sulfate reduction (7.87) are rare in arid environments and incompatible with the pH > 8.4 criterion (see Table 7.6), under most circumstances dissimilatory sulfate reduction could be a significant source of alkalinity in soils at *the low-alkalinity* end-point described earlier. Geotechnical engineers studying marine quick clays note that conditions favoring dissimilatory sulfate reduction can generate sufficient alkalinity to alter the univalent-to-divalent cation ratio of pore water in quick clay formations (Lessard, 1981).

7.7.7. Predicting Changes in Pore Water SAR

Langelier (1936) developed a method for calculating the calcite saturation index, the motivation being corrosion prevention by depositing and maintaining a calcite layer on the interior of cast iron pipes. Given the calcium concentration (expressed as the negative logarithm of the total soluble calcium concentration or $pCa = -\log(C_{Ca(Total)})$) and alkalinity (expressed as the negative logarithm of the carbonate alkalinity or $pA_C = -\log(A_C)$) of water in contact with the pipes, Langelier derived an expression for the pH at which the water is saturated by calcite: pH_c:

$$pH_c = \left(pK_{a2}^c - pK_{s0}^c\right) + pCa + pA_C \tag{7.88}$$

The first term on the right-hand side of expression (7.88) is the conditional hydrolysis coefficient (7.89) derived from Eq. (7.42); the second term is the conditional calcite solubility coefficient (7.90).

$$K_{a2,CO_2(aq)}^c = K_{a2,CO_2(aq)} \cdot \left(\frac{\gamma_{HCO_3^-}}{\gamma_{H^+} \cdot \gamma_{CO_3^{2-}}}\right) = \frac{C_{H^+} \cdot C_{CO_3^{2-}}}{C_{HCO_3^-}} \tag{7.89}$$

$$K_{s0}^c = K_{s0} \cdot \left(\frac{1}{\gamma_{Ca^{2+}} \cdot \gamma_{CO_3^{2-}}}\right) = C_{Ca^{2+}} \cdot C_{CO_3^{2-}} \tag{7.90}$$

The empirical formula for calculating an ion activity coefficient and the thermodynamic constants for carbonate hydrolysis[13] and calcite solubility[14] have changed since Langelier published his paper. Table 7.9 lists parameters calculated from data listed in Table 3 in Langelier's paper; the differences between the last two columns illustrate the effect of revised constants and current methods for calculating ion activity coefficients. Alkalinity A_C consistently increases and calcium solubility consistently decreases as pH increases above 8.4 (see Figure 7.12).

Bower (Bower, Wilcox et al., 1963; Bower, Ogata et al., 1968) used expression (7.91) to predict changes in the *SAR* resulting from calcite precipitation. Bower assumed that the soil pH buffer capacity immediately shifts irrigation water pH to the prevailing soil pH (assumed to be the pH of typical

---

13. $pK_{a2,CO_2(aq)} = 10.26$ in Langelier; $pK_{a2,CO_2(aq)} = 10.33$ in the ChemEQL database.
14. $pK_{s0} = 8.32$ in Langelier; $pK_{s0} = 8.48$ in the ChemEQL database.

TABLE 7.9 Negative Logarithm of the Ion Activity Product for Calcite Solubility $pIAP$, pH_c, and Calcite Saturation Index (pH–pH_c)

Water Source	$pIAP$	pH_c	(pH–pH_c)	(pH–pH_c)‡
Calaveras Reservoir	8.14	7.41	0.34	0.15
San Mateo	8.33	7.40	0.15	–0.12
Milpitas	8.31	7.65	0.17	0.13
Hayward	8.39	7.71	0.09	0.10
Crystal Springs Reservoir	8.72	8.44	–0.24	–0.07
Irvington	8.84	9.23	–0.38	–0.11
Berkeley	8.08	9.55	0.40	–0.12

Note: Calculated values are based on ChemEQL simulations, except column 4 (‡), which lists original published values.

Source: Langelier, 1936.

calcareous soils: 8.4) and used the Langelier Saturation Index to predict whether calcite precipitation would raise or lower SAR_{iw}. The pH at which calcite precipitates—pH_c—is calculated from the chemical composition of the irrigation water using expression (7.88).

$$adjSAR_{iw} = SAR_{iw} \cdot (1 + (8.4 - pH_c)) \tag{7.91}$$

Bower's $adjSAR_{iw}$ tends to overestimate changes in irrigation water SAR. Suarez (1981) proposed a more direct expression (7.92) for calculating changes in irrigation water SAR.[15] Expression (7.92) contains the predicted calcium concentration $c'_{Ca^{2+}} [mmol \cdot L^{-1}]$ at the soil pH, taking into account the soil $CO_2(g)$ partial pressure $p_{CO_2(g)}$ and the bicarbonate-to-calcium ratio $C_{HCO_3^-}/C_{Ca^{2+}}$ of the irrigation water.

$$SAR_{adj} = \frac{c_{Na^+} [mmol \cdot L^{-1}]}{\sqrt{c'_{Ca^{2+}} [mmol \cdot L^{-1}]}} \tag{7.92}$$

Water chemistry simulation applications such as ChemEQL and MINTEQ allow you to calculate $c'_{Ca^{2+}}$ without resorting to tables such as those published by Langelier and Suarez. An example appears in Appendix 7I.

---

15. Suarez (1981) includes the magnesium concentration $c_{Mg^{2+}} [mmol \cdot L^{-1}]$ in his original expression: $SAR_{ajd} = \dfrac{c_{Na^+}}{\sqrt{c_{Mg^{2+}} + c'_{Ca^{2+}}}}$

7.8. EXCHANGEABLE ACIDITY

The earliest studies of soil acidity (Vietch, 1902; Hopkins, Knox et al., 1903) found that the unidentified acids responsible for soil acidity could not be extracted by distilled water. Extracting soil acidity required adding some type of soluble salt to displace the insoluble acids. Vieth (1904) reported a comparison of his limewater (calcium hydroxide) method (Vietch, 1902) and the sodium-chloride extraction developed by Hopkins and colleagues (1903). Hopkins's sodium chloride extract released "considerable quantities of iron, alumina, and manganese" into solution that precipitated when the extract was titrated with sodium hydroxide to quantify acidity. Vietch (1904) found that extracting acid soils with a neutral sodium salt solution resulted in the "replacement of aluminum by sodium." He also found that potassium salts were more efficient at replacing aluminum than sodium.

Vietch coined the term *active acidity* to describe the acidity displaced by neutral salts such as sodium and potassium chlorides in Hopkins's method and *total apparent acidity* to describe the much greater acidity measured by his limewater (calcium hydroxide) method. The debate over soil acidity continued for another 60 years (Thomas, 1977), but no study during that span of years altered the conclusions one could easily draw—in hindsight—from the work of Vietch and Hopkins.

7.8.1. Exchangeable Calcium and Soil Alkalinity

Soils containing calcite are buffered at roughly pH 8.4 (see Appendix 7A); they resist becoming either more alkaline or more acidic as long as calcite remains in the soil. Although chemical weathering in humid climates will eventually remove all traces of the mineral calcite, the mineral and organic soil colloids (smectite clay minerals and humic substances) will remain largely saturated by Ca^{2+} ions, a consequence of the natural abundance of this element. Absent calcite, there is little to prevent a drop in soil pH. Though continued chemical weathering and leaching drive the pH ever lower, soil colloids remain Ca^{2+}-saturated until pH 6.5, the pH of minimum aluminum solubility.

As soil pH drifts below pH 6.5, aluminum solubility begins to gradually increase, displacing the exchangeable Ca^{2+} ions. As we have already seen, asymmetric ion exchange tends to favor the ion with higher valence, shifting the advantage decidedly in the favor of Al^{3+} (Figure 7.13).

7.8.2. Gibbsite Solubility

Example 5.16 in Chapter 5 examines the solubility of a common hydrous aluminum oxide mineral: gibbsite (Eq. (7.93)). Our model for exchangeable acidity is based on gibbsite solubility (Eq. (7.94)), but one could easily prepare similar

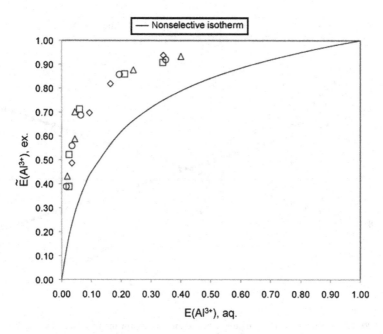

FIGURE 7.13   This graph plots the asymmetric (Ca^{2+}, Al^{3+}) exchange isotherm from Coulter and Talibudeen (1968): $K_{GT} \approx 40$.

models based on other naturally occurring aluminum oxide, hydroxide, or oxyhydroxide minerals.

$$Al(OH)_3(s) \leftrightarrow Al^{3+}(aq) + 3 \cdot OH^-(aq) \qquad (7.93)$$
$$\text{\small GIBBSITE}$$

$$K_{s0} = 1.29 \cdot 10^{-34} = a_{Al^{3+}} \cdot a_{OH^-}^3 \qquad (7.94)$$

Solubility expression (7.95) is a more convenient expression of pH-dependent gibbsite solubility. The net equilibrium constant (7.96) for reaction (7.95) results from dividing (7.94) by the water hydrolysis constant K_w (6) to the third power.

$$Al(OH)_3(s) + 3 \cdot H^+(aq) = Al^{3+}(aq) + 3 \cdot H_2O(l) \qquad (7.95)$$

$$K_{net} = K_{s0} \cdot K_w^{-3} = 10^{+8.11} = \frac{a_{Al^{3+}}}{a_{H^+}^3} \qquad (7.96)$$

Taking the logarithm of solubility expression (7.96) yields a linear expression (7.97) between the logarithm of $Al^{3+}(aq)$ activity and solution pH.

$$8.11 = \log(a_{Al^{3+}}) - 3 \cdot \log(a_{H^+})$$

$$\log(a_{Al^{3+}}) = 8.11 - 3 \cdot \text{pH} \qquad (7.97)$$

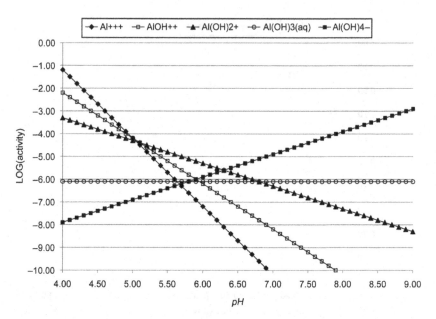

FIGURE 7.14 This graph plots the logarithm of ion activity for the five most abundant aluminum solution species as a function of pH from simulation using ChemEQL. The solubility of $Al^{3+}(aq)$ is constrained by gibbsite solubility, as discussed in Example 5.16, except here $H^{+}(aq)$ is treated as an independent variable covering the range depicted in the graph.

Figure 7.14 (see Figure 5.3) plots the activity of the most abundant solution aluminum species in equilibrium with gibbsite. The trivalent $Al^{3+}(aq)$ species is the most abundant aluminum species below pH 5, where aluminum solubility reaches millimolar concentrations. The univalent $Al(OH)_2^{+}(aq)$ species is the dominant species in a pH range—$5.10 < pH < 6.80$—where aluminum solubility remains below $3 \cdot 10^{-5}$ M. The divalent $Al(OH)^{2+}(aq)$ species is never the most abundant species.

7.8.3. The Role of Asymmetric (Al^{3+}, Ca^{2+}) Exchange

The relative abundance of univalent, divalent, and trivalent species as solubility increases is very important because these solution aluminum species become increasingly competitive with $Ca^{2+}(aq)$ for cation exchange sites as soil pH decreases. If we assume a representative soil pore water ionic strength of 10^{-3} M, the univalent $Al(OH)_2^{+}(aq)$ species is never competitive for asymmetric cation exchange with $Ca^{2+}(aq)$ in the pH range, where it is dominant because of its valence and low relative solubility. We can ignore symmetric $(Al(OH)^{2+}, Ca^{2+})$ cation exchange because $Al(OH)^{2+}(aq)$ is never the most abundant soluble aluminum species. By process of elimination, the only significant cation exchange reaction that is relevant to soil acidity is the

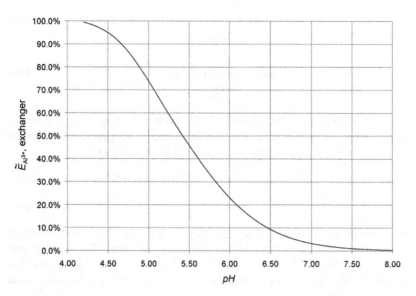

FIGURE 7.15 This graph plots the exchangeable Al^{3+} equivalent fraction $\tilde{E}_{Al^{3+}}$ for asymmetric (Al^{3+}, Ca^{2+}) exchange as a function of pH. Solution parameter C_0 (Appendix 7C) is restricted to $1.0 \cdot 10^{-3}\ mol_c \cdot L^{-1} < C_0 < 2.0 \cdot 10^{-3}\ mol_c \cdot L^{-1}$. The Gaines-Thomas selectivity coefficient is $K_{GT}^c = 160$ (Foscolos, 1968). Gibbsite solubility is simulated using ChemEQL as discussed in Example 5.16.

asymmetric (Al^{3+}, Ca^{2+}) cation exchange; the advantage lies with $Al^{3+}(aq)$ because of its valence and high relative solubility when soil pH drops below 5.

The complete solution of the cubic asymmetric (Al^{3+}, Ca^{2+}) exchange isotherm is found in Appendix 7E. The combined effect of pH-dependent aluminum solubility (Figure 7.14) and asymmetric (Al^{3+}, Ca^{2+}) exchange (Figure 7.13) are plotted in Figure 7.15.

7.8.4. Neutralizing Exchangeable Soil Acidity

Hydrolysis reactions (7.98), (7.99), and (7.100) identify exchangeable aluminum $Al^{3+}(ex)$ as a weak acid. Exchangeable calcium $Ca^{2+}(ex)$ is not a base according to acid-base chemistry, but is simply the cation associated with the true base $OH^-(aq)$. The accumulation of exchangeable aluminum $Al^{3+}(ex)$ portrayed in Figure 7.15 leads to the accumulation of an exchangeable weak acid that requires three equivalents of base to fully neutralize.

$$\underset{\text{weak acid}}{Al^{3+}}(aq) \leftrightarrow H^+(aq) + \underset{\text{conjugate base}}{Al(OH)^{2+}}(aq) \quad K_{a1} = 1.0 \cdot 10^{-5} \qquad (7.98)$$

$$Al(OH)^{2+} \underset{\text{weak acid}}{} (aq) \leftrightarrow H^+(aq) + Al(OH)_2^+ \underset{\substack{\text{conjugate} \\ \text{base}}}{} (aq) \quad K_{a2} = 7.9 \cdot 10^{-6} \qquad (7.99)$$

$$Al(OH)_2^+ \underset{\text{weak acid}}{} (aq) \leftrightarrow H^+(aq) + Al(OH)_3^0 \underset{\substack{\text{conjugate} \\ \text{base}}}{} (aq) \quad K_{a3} = 1.6 \cdot 10^{-7} \qquad (7.100)$$

Combining reactions (7.99) through (7.100) with the conjugate base hydrolysis reaction (7.32) yields the neutralization-exchange reaction (7.101) between exchangeable $Al^{3+}(ex)$ and the solid base calcite $CaCO_3(s)$.

$$2 \cdot Al^{3+}(ex) + 3 \cdot \underset{calcite}{CaCO_3(s)} \rightarrow 3 \cdot Ca^{2+}(ex) + 2 \cdot Al(OH)_3^0(aq) + 3 \cdot CO_2(aq)$$

$$(7.101)$$

The dissolution of calcite $CaCO_3(s)$ releases soluble $Ca^{2+}(aq)$ that undergoes asymmetric exchange (7.102) with exchangeable $Al^{3+}(ex)$. The displacement of exchangeable $Al^{3+}(ex)$ by soluble $Ca^{2+}(aq)$ is precisely the reaction induced by the limewater extraction used by Vietch (1902) to measure "total apparent acidity."

$$2 \cdot \underset{exchanger}{Al^{3+}} + 3 \cdot \underset{solution}{Ca^{2+}(aq)} \leftrightarrow 2 \cdot \underset{solution}{Al^{3+}(aq)} + 3 \cdot \underset{exchanger}{Ca^{2+}} \qquad (7.102)$$

Soluble $Al^{3+}(aq)$ reacts with bicarbonate $HCO_3^-(aq)$ and carbonate $CO_3^{2-}(aq)$—conjugate bases of aqueous carbon dioxide—in solution resulting in the precipitation of the aluminum hydroxide gibbsite $Al(OH)_3(s)$, the very precipitate described by Vietch.

7.9. SUMMARY

Environmental acidity and basicity arise from numerous sources: the dissolution of atmospheric gases in water and their subsequent wet deposition on the land surface, fertilization and harvest in agriculture, and chemical weathering of minerals in soils and aquifers. The criteria for measuring acidity or basicity in the environment must account for the dissolution of carbon dioxide in water and the production of bicarbonate ions—the conjugate base of the weak acid aqueous carbon dioxide. Bicarbonate ions are the basis of water alkalinity; the proton reference level used by environmental chemists shifts below pH 7 as hydroxyl ions are converted into bicarbonate ions.

Climatic conditions determine whether soils tend to be acidic or alkaline; the former are common in humic climates, while the latter are common in arid and semiarid climates. Land degradation can occur in arid and semiarid climates when the accumulation of sodium ions by clay minerals reduces the activity index I_A of soil clays, reducing soil mechanical strength and—as a consequence—soil-water hydraulic conductivity. Sodification is a complex process that involves cation exchange, pH-dependent calcite solubility, and extreme alkalinity that increases the sodium-to-calcium ratio in solution by lowering calcium solubility. At the other extreme, land degradation in humid climates can result in soil acidification. This process also involves interactions between pH-dependent changes in aluminum hydrous oxide solubility and cation exchange, leading to the accumulation of exchangeable aluminum ions. Exchangeable aluminum ions behave as weak acids, and their accumulation amounts to a chemically active but insoluble form of acidity in humid region soils.

APPENDIX 7A. BUFFER INDEX

A pH buffer is composed of one or more weak acids or bases (including conjugate acids and conjugate bases) that resist changes in pH when either strong acid or strong base is added. Buffer capacity is a measure of how much strong acid or strong base must be added to change pH by a fixed amount. *Buffering* is an important concept because a variety of chemical substances dissolved in surface and pore water of the saturated and vadose zones contribute to the soluble buffer capacity, poising the pH in these systems.

The buffer index β is defined in the following equations. Adding a strong base causes pH to increase, while adding a strong acid causes pH to decrease; the buffer index $\beta > 0$ regardless.

$$\beta = \frac{\Delta C_{base}}{\Delta pH} = -\frac{\Delta C_{acid}}{\Delta pH} \tag{7A.1}$$

$$\beta = \left(\frac{\Delta C_{base}}{\Delta a_{H^+}}\right) \cdot \left(\frac{\Delta a_{H^+}}{\Delta pH}\right) = (-2.303 \cdot a_{H^+}) \cdot \left(\frac{\Delta C_{base}}{\Delta a_{H^+}}\right) \tag{7A.2}$$

If the buffer is based on a generalized weak acid $HA(aq)$, then the complete dissociation of the strong base $NaOH$ and acid HNO_3 results in the following relations (neglecting activity corrections):

$$K_a^c = \frac{a_{H^+} \cdot C_{A^-}}{C_{HA}} \tag{7A.3}$$

$$K_w = 10^{-14} = a_{H^+} \cdot a_{OH^-} \tag{7A.4}$$

$$C_{base} = a_{OH^-} = C_{Na^+} \tag{7A.5}$$

$$C_{acid} = a_{H^+} = C_{NO_3^-} \tag{7A.6}$$

$$C_{buffer} = C_{HA} + C_{A^-} \tag{7A.7}$$

The preceding relations determine the concentration of the conjugate base $A^-(aq)$:

$$C_{buffer} = C_{A^-} + C_{HA} = C_{A^-} + \left(\frac{a_{H^+} \cdot C_{A^-}}{K_a^c}\right) \tag{7A.8}$$

$$C_{buffer} = C_{A^-} \cdot \left(1 + \left(\frac{a_{H^+}}{K_a^c}\right)\right) = C_{A^-} \cdot \left(\frac{K_a^c + a_{H^+}}{K_a^c}\right)$$

$$C_{A^-} = \frac{C_{buffer} \cdot K_a^c}{K_a^c + a_{H^+}} \tag{7A.9}$$

Total cation charge must equal total anion charge:

$$a_{H^+} + C_{Na^+} = a_{OH^-} + C_{NO_3^-} + C_{A^-} \tag{7A.10}$$

Solving the charge balance condition (7A.10) for C_{base} and replacing terms using Eqs. (7A.4), (7A.6), and (7A.9) result in Eq. (7A.11):

$$C_{base} = C_{Na^+} = C_{acid} + C_{A^-} + a_{OH^-} - a_{H^+}$$

$$C_{base} = C_{acid} + \left(\frac{C_{buffer} \cdot K_a^c}{K_a^c + a_{H^+}}\right) + \left(\frac{K_w}{a_{H^+}}\right) - a_{H^+} \tag{7A.11}$$

Applying the differential definition of buffer index β (7A.2) to expression (7A.11) yields the more tractable expression (7A.12):

$$\beta = (-2.303 \cdot a_{H^+}) \cdot \left(\frac{dC_{base}}{da_{H^+}}\right)$$

$$\beta = (-2.303 \cdot a_{H^+}) \cdot \frac{d}{da_{H^+}} \left(C_{acid} + \left(\frac{C_{buffer} \cdot K_a^c}{K_a^c + a_{H^+}}\right) + \left(\frac{K_w}{a_{H^+}}\right) - a_{H^+}\right)$$

$$\beta = (-2.303 \cdot a_{H^+}) \cdot \left(\frac{d}{da_{H^+}} \left(\frac{C_{buffer} \cdot K_a^c}{K_a^c + a_{H^+}}\right) + \frac{d}{da_{H^+}} \left(\frac{K_w}{a_{H^+}}\right) - \frac{da_{H^+}}{da_{H^+}}\right)$$

$$\beta = (-2.303 \cdot a_{H^+}) \cdot \left(-\left(\frac{C_{buffer} \cdot K_a^c}{(K_a^c + a_{H^+})^2}\right) - \left(\frac{K_w}{a_{H^+}^2}\right) - 1\right)$$

$$\beta = (2.303 \cdot a_{H^+}) \cdot \left(\left(\frac{C_{buffer} \cdot K_a^c}{(K_a^c + a_{H^+})^2}\right) + \left(\frac{K_w}{a_{H^+}^2}\right) + 1\right)$$

$$\beta = (2.303) \cdot \left(\left(\frac{C_{buffer} \cdot K_a^c \cdot a_{H^+}}{(K_a^c + a_{H^+})^2}\right) + \left(\frac{K_w}{a_{H^+}}\right) + a_{H^+}\right)$$

$$\beta = 2.303 \cdot \left(\frac{C_{buffer} \cdot K_a^c \cdot a_{H^+}}{(K_a^c + a_{H^+})^2} + \frac{K_w}{a_{H^+}} + a_{H^+}\right) \tag{7A.12}$$

The buffer index β (7A.12) depends on both buffer concentration C_{buffer} and the dissociation constant K_a^c of the weak acid used as buffer.

APPENDIX 7B. CONVERTING MASS FRACTION TO SUM-OF-OXIDES COMPOSITION

Table 7B.1 lists the mass fraction composition of Earth's outer crust in column 2. Dividing each entry in column 2 by the atomic mass (column 3) and multiplying by 1000 to convert from grams to kilograms yields the molar composition of each element (column 4). Multiplying each entry in column 4 by the valence of each element—iron is $Fe(II)$ in crustal rocks—we arrive at the moles of charge per gram.

Since the requirement of charge neutrality applies to Earth's crust, it should come as no surprise that the 10 most abundant elements in Earth's crust account for all but 0.6% of the charge.

Table 7.3 lists the oxide mole-per-mass $(mol \cdot kg^{-1})$ composition of Earth's crust (column 2) based on the listed oxides (column 1). The oxides are based on the valence of each element in crustal rocks. You will notice that the oxide mole-per-mass $(mol \cdot kg^{-1})$ composition (Table 7.1, column 2) is a multiple of mole-per-mass $(mol \cdot kg^{-1})$ composition in Table 7B.1 (column 4).

TABLE 7B.1 Composition of the Outer (Continental) Crust for Planet Earth

Atomic Symbol	Mass Fraction	Atomic Mass, $g \cdot mol^{-1}$	Moles, $mol \cdot kg^{-1}$	Moles of Charge, $mol_c \cdot kg^{-1}$
O	$4.55 \cdot 10^{-1}$	15.9994	28.4	−56.88
Si	$2.72 \cdot 10^{-1}$	28.0845	9.68	38.74
Al	$8.30 \cdot 10^{-2}$	26.9815	3.03	9.23
Fe	$6.20 \cdot 10^{-2}$	55.8452	1.11	2.22
Ca	$4.66 \cdot 10^{-2}$	40.0784	1.16	2.33
Na	$2.27 \cdot 10^{-2}$	22.9898	0.987	0.99
Mg	$2.76 \cdot 10^{-2}$	24.3051	1.14	2.27
K	$1.84 \cdot 10^{-2}$	39.0983	0.471	0.47

Source: Greenwood and Earnshaw, 1997.

APPENDIX 7C. SATURATION EFFECT: ATMOSPHERIC CONVERSION OF SULFUR TRIOXIDE TO SULFURIC ACID

Altshuller (1973) demonstrated the saturation effect that occurs when atmospheric sulfur dioxide concentrations exceed a certain threshold (Figure 7C.1). Below the threshold, the sulfur dioxide–to–sulfate ratio in the atmosphere of major cities increases linearly. Beyond the saturation threshold (80 μg m^{-3} in Figure 7C.1), atmospheric sulfuric acid levels cease to increase regardless of the sulfur dioxide concentration.

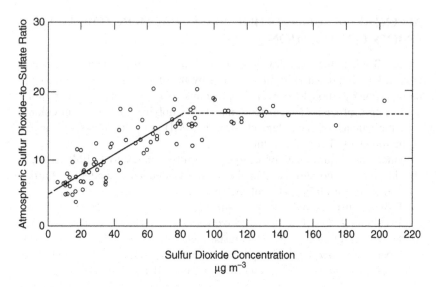

FIGURE 7C.1 The relationship between annual average sulfur dioxide concentration and the ratio of sulfur dioxide to sulfate for 18 U.S. cities. *Source: Reproduced with permission from Altshuller, A.P., 1973. Atmospheric sulfur dioxide and sulfate: Distribution of concentration at urban and nonurban sites in United States. Environ. Sci. Technol. 7, 709–712.*

Ammonia and other alkaline substances in the atmosphere catalyze the oxidation of sulfur dioxide to sulfuric acid (Scott and Hobbs, 1967). Once sulfur dioxide levels reach a certain threshold, the acidity from the sulfuric acid exceeds the alkalinity, resulting in a drop in the pH of water droplets in the atmosphere and effectively eliminating the further conversion of sulfur dioxide to sulfuric acid.

APPENDIX 7D. BICARBONATE AND CARBONATE REFERENCE LEVELS

The weak acids in carbonate solutions encompass three proton reference levels. Table 7D.1 (identical to Table 7.2) lists the species that appear in the proton balance defined for the aqueous carbon dioxide $CO_2(aq)$ reference level (boldface species).

Table 7D.2 lists the proton balance expression defined for the aqueous carbon dioxide $CO_2(aq)$ reference level (row 4) and is assembled from the species representing a proton surplus (row 2) and the species representing a proton deficit (row 3).

Table 7D.3 lists the species that appear in the proton balance defined for the bicarbonate $HCO_3^-(aq)$ reference level (boldface species).

Table 7D.4 lists the proton balance expression defined for the bicarbonate $HCO_3^-(aq)$ reference level (row 4) and is assembled from the species

---

TABLE 7D.1 Conjugate Acid-Base Pairs in Aqueous Carbonate Solutions

Weak Acid	Conjugate Base
H_3O^+	H_2O
$CO_2(aq)$	HCO_3^-
HCO_3^-	CO_3^{2-}
H_2O	OH^-

TABLE 7D.2 Proton Balance Relative to the Aqueous Carbon Dioxide $CO_2(aq)$ Reference Level

Proton Reference Level	$CO_2(aq)$
Proton Surplus	$C_{H_3O^+}$
Proton Deficit	$C_{OH^-} + C_{HCO_3^-} + 2 \cdot C_{CO_3^{2-}}$
Proton Balance	$C_{H_3O^+} = C_{OH^-} + C_{HCO_3^-} + 2 \cdot C_{CO_3^{2-}}$

TABLE 7D.3 Conjugate Acid-Base Pairs in Aqueous Carbonate Solutions

Weak Acid	Conjugate Base
H_3O^+	H_2O
$CO_2(aq)$	HCO_3^-
HCO_3^-	CO_3^{2-}
H_2O	OH^-

TABLE 7D.4 Proton Balance Relative to the Bicarbonate $HCO_3^-(aq)$ Reference Level

Proton Reference Level	$HCO_3^-(aq)$
Proton Surplus	$C_{H_3O^+} + C_{CO_2(aq)}$
Proton Deficit	$C_{OH^-} + C_{CO_3^{2-}}$
Proton Balance	$C_{H_3O^+} = C_{OH^-} + C_{CO_3^{2-}} - C_{CO_2(aq)}$

TABLE 7D.5 Conjugate Acid-Base Pairs in Aqueous Carbonate Solutions

Weak Acid	Conjugate Base
H_3O^+	H_2O
$CO_2(aq)$	HCO_3^-
HCO_3^-	CO_3^-
H_2O	OH^-

TABLE 7D.6 Proton Balance Relative to the Carbonate $CO_3^{2-}(aq)$ Reference Level

Proton Reference Level	Strong Base KOH
Proton Surplus	$C_{H_3O^+} + 2 \cdot C_{CO_2(aq)} + C_{HCO_3^-}$
Proton Deficit	C_{OH^-}
Proton Balance	$C_{H_3O^+} = C_{OH^-} - 2 \cdot C_{CO_2(aq)} - C_{HCO_3^-}$

representing a proton surplus (row 2) and the species representing a proton deficit (row 3). The proton surplus lists H_3O^+ (H_2O plus a H^+) and $CO_2(aq)$ (HCO_3^- plus a H^+). The proton deficit lists OH^- (H_2O minus a H^+) and CO_3^{2-} (HCO_3^- minus one H^+). The bicarbonate $HCO_3^-(aq)$ species does not appear in the proton balance in Table 7D.4.

Table 7D.5 lists the species that appear in the proton balance defined for the carbonate $CO_3^{2-}(aq)$ reference level (boldface species).

Finally, Table 7D.6 lists the proton balance expression defined for the carbonate $CO_3^{2-}(aq)$ reference level (row 4) and is assembled from the species representing a proton surplus (row 2) and the species representing a proton deficit (row 3). The proton surplus lists H_3O^+ (H_2O plus a H^+), $CO_2(aq)$ (CO_3^{2-} plus two H^+), and HCO_3^- (CO_3^{2-} plus one H^+). The proton deficit lists OH^- (H_2O minus a H^+). The carbonate $CO_3^{2-}(aq)$ species does not appear in the proton balance in Table 7D.6.

APPENDIX 7E. CALCULATING THE pH OF A SODIUM CARBONATE SOLUTION

The carbonate ion $CO_3^{2-}(aq)$ is an extremely important diprotic conjugate base in water chemistry. What is the pH of a 0.1 M solution of Na_2CO_3?

$$CO_3^{2-}(aq) + H_2O(l) \leftrightarrow HCO_3^-(aq) + OH^-(aq)$$

$$K_{b1} = 2.14 \cdot 10^{-4} \approx \frac{C_{OH^-} \cdot C_{HCO_3^-}}{C_{CO_3^{2-}}}$$

$$HCO_3^-(aq) \leftrightarrow CO_2(aq) + OH^-(aq)$$

$$K_{b2} = 2.33 \cdot 10^{-8} \approx \frac{C_{OH^-} \cdot C_{CO_2}}{C_{HCO_3^-}}$$

Because the second base association constant K_{b2} is much smaller than the first base association constant K_{b1}, the hydroxide ion $OH^-(aq)$ concentration results almost entirely from the first hydrolysis step. Assume, as a first approximation, that only the first step occurs. Use an equilibrium table to assign values to each component in equilibrium expression.

Stage	$CO_3^{2-}(aq)$	$\leftrightarrow$	$HCO_3^-(aq)$	+	$OH^-(aq)$
Initial	0.10		0		0
Reaction	$-x$		$+x$		$+x$
Equilibrium	$0.10 - x$		x		x

The equilibrium expression using concentrations from the preceding table follows:

$$K_{b1} = 2.14 \cdot 10^{-4} \approx \frac{C_{OH^-} \cdot C_{HCO_3^-}}{C_{CO_3^{2-}}}$$

$$K_{b1} \approx \frac{x \cdot x}{(0.10 - x)}$$

$$2.14 \cdot 10^{-4} \approx \frac{x \cdot x}{(0.10)}$$

$$x \approx \sqrt{2.14 \cdot 10^{-5}} = 4.63 \cdot 10^{-3}$$

Since $x \approx 4.63 \cdot 10^{-3}$, the approximation $(0.10 - x) \approx 0.10$ is acceptable.

$$C_{OH^-} \approx 4.66 \cdot 10^{-3} = 10^{-2.33}$$

$$pOH = 2.33$$

$$pH = 14 - pOH = 11.7$$

We estimated the pH of a 0.10 M sodium carbonate solution by assuming that the second hydrolysis step, the hydrolysis reaction that produces carbonic acid from bicarbonate, is negligible relative to the first hydrolysis step. The result confirms the validity of the simplifying assumption.

APPENDIX 7F. CALCULATING THE AQUEOUS CARBON DIOXIDE CONCENTRATION IN A WEAK BASE SOLUTION

Suppose we estimate the concentration of aqueous carbon dioxide $CO_2(aq)$, the product of the second hydrolysis step, from Appendix 7E.

$$HCO_3^-(aq) \leftrightarrow CO_2(aq) + OH^-(aq)$$

$$K_{b2} = 2.33 \cdot 10^{-8} \approx \frac{C_{OH^-} \cdot C_{CO_2}}{C_{HCO_3^-}}$$

To do this, we will construct an equilibrium table for the second hydrolysis step—the conversion of bicarbonate to aqueous carbon dioxide—to assign values to each component in equilibrium expression as before.

Stage	$HCO_3^-(aq)$	$\leftrightarrow$	$CO_2(aq)$	$+$	$OH^-(aq)$
Initial	$4.52 \cdot 10^{-3}$		0		$4.52 \cdot 10^{-3}$
Reaction	$-y$		$+y$		$+y$
Equilibrium	$4.52 \cdot 10^{-3} - y$		y		$4.52 \cdot 10^{-3} + y$

The equilibrium expression using concentrations from the preceding table follows:

$$K_{b2} \approx \frac{C_{OH^-} \cdot C_{H_2CO_3}}{C_{HCO_3^-}}$$

$$K_{b2} \approx \frac{\left(4.52 \cdot 10^{-3} + y\right) \cdot y}{\left(4.52 \cdot 10^{-3} - y\right)}$$

$$2.33 \cdot 10^{-8} \approx \frac{\left(4.52 \cdot 10^{-3}\right) \cdot y}{\left(4.52 \cdot 10^{-3}\right)}$$

$$y \approx 2.33 \cdot 10^{-8}$$

Because the second base association constant K_{b2} is relatively small, the second ionization step occurs to a *much* smaller extent than the first. This means the amount of $H_2CO_3(aq)$ and $OH^-(aq)$ produced in the second step (y) is much smaller than $4.52 \cdot 10^{-3}$ M. It is therefore reasonable that both $C_{HCO_3^-}$ and C_{OH^-} are very close to $4.52 \cdot 10^{-3}$ M.

It is important to realize the simplifying assumptions used in Appendixes 7E and 7F would be difficult to make when attempting to predict solute concentrations in a real water chemistry problem. An accurate calculation would dispense with simplifying assumptions and rely in ChemEQL to simulate the reactions.

APPENDIX 7G. ION EXCHANGE ISOTHERM FOR ASYMMETRIC (Ca^{2+}, Al^{3+}) EXCHANGE

The trivalent-divalent (Al^{3+}, Ca^{2+}) exchange isotherm is significantly more complex than the univalent-divalent (Na^+, Ca^{2+}) exchange isotherm because the former requires solving cubic equations throughout the exchange isotherm, while the latter involves relatively simple quadratic equations.

$$2 \cdot \underset{solution}{Al^{3+}(aq)} + 3 \cdot \underset{exchanger}{Ca^{2+}} \leftrightarrow 2 \cdot \underset{exchanger}{Al^{3+}} + 3 \cdot \underset{solution}{Ca^{2+}(aq)}$$

The conditional Gaines-Thomas selectivity expression for the asymmetric (Al^{3+}, Ca^{2+}) exchange reaction is given by expression (7G.1):

$$K_{GT}^c = \frac{[Ca^{2+}]^3 \cdot \tilde{E}_{Al^{3+}}^2}{\tilde{E}_{Ca^{2+}}^3 \cdot [Al^{3+}]^2} \tag{7G.1}$$

Solution parameter C_0 is the solution normality summed over all ions involved in the exchange (7G.2). The solution equivalent fractions are defined in expressions (7G.3) and (7G.4), which are substituted into Gaines-Thomas selectivity expression (7G.2) to yield new expression using only equivalent fractions (7G.5).

$$C_0 \equiv 3 \cdot [Al^{3+}] + 2 \cdot [Ca^{2+}] \tag{7G.2}$$

$$E_{Ca^{2+}} = \frac{2 \cdot [Ca^{2+}]}{3 \cdot [Al^{3+}] + 2 \cdot [Ca^{2+}]} = \frac{2 \cdot [Ca^{2+}]}{C_0} \tag{7G.3}$$

$$E_{Al^{3+}} = \frac{3 \cdot [Al^{3+}]}{3 \cdot [Al^{3+}] + 2 \cdot [Ca^{2+}]} = \frac{3 \cdot [Al^{3+}]}{C_0} \tag{7G.4}$$

$$K_{GT}^c = \left(\frac{\left(\frac{C_0 \cdot E_{Ca^{2+}}}{2} \right)^3 \cdot \tilde{E}_{Al^{3+}}^2}{\tilde{E}_{Ca^{2+}}^3 \cdot \left(\frac{C_0 \cdot E_{Al^{3+}}}{3} \right)^2} \right) \tag{7G.5}$$

$$K_{GT}^c = \left(\frac{9 \cdot C_0}{8} \right) \cdot \left(\frac{(1 - E_{Al^{3+}})^3 \cdot \tilde{E}_{Al^{3+}}^2}{E_{Al^{3+}}^2 \cdot (1 - \tilde{E}_{Al^{3+}})^3} \right) \tag{7G.5}$$

Expression (7G.5), after rearrangement, is revealed to be a cubic function (7G.6) whose independent variable is the Al^{3+} equivalent fraction on the exchanger $\tilde{E}_{Al^{3+}}$ and dependent variable is the Al^{3+} equivalent fraction in solution $E_{Al^{3+}}$.

$$\left(\frac{8 \cdot K_{GT}^c}{9 \cdot C_0} \right) = \frac{(1 - E_{Al^{3+}})^3 \cdot \tilde{E}_{Al^{3+}}^2}{E_{Al^{3+}}^2 \cdot (1 - \tilde{E}_{Al^{3+}})^3}$$

$$\left(\frac{8 \cdot K_{GT}^c \cdot E_{Al^{3+}}^2}{9 \cdot C_0} \right) \cdot \left(1 - 3 \cdot \tilde{E}_{Al^{3+}} + 3 \cdot \tilde{E}_{Al^{3+}}^2 - \tilde{E}_{Al^{3+}}^3 \right) = (1 - E_{Al^{3+}})^3 \cdot \tilde{E}_{Al^{3+}}^2$$

$$\left(1 - 3 \cdot \tilde{E}_{Al^{3+}} + 3 \cdot \tilde{E}_{Al^{3+}}^2 - \tilde{E}_{Al^{3+}}^3 \right) = \left(\frac{9 \cdot C_0 \cdot (1 - E_{Al^{3+}})^3}{8 \cdot K_{GT}^c \cdot E_{Al^{3+}}^2} \right) \cdot \tilde{E}_{Al^{3+}}^2 \tag{7G.6}$$

The term in parentheses on the right-hand side of (7G.6) contains the entire functional dependence of this cubic expression on the Al^{3+} equivalent fraction in solution $E_{Al^{3+}}$, leading to a new function (7G.7). Using function (7G.7), the cubic expression (7G.6) becomes (7G.8). The exchange isotherm is one of the roots of cubic function.

$$\beta_c \equiv \left(\frac{9 \cdot C_0 \cdot (1 - E_{Al^{3+}})^3}{8 \cdot K_{GT}^c \cdot E_{Al^{3+}}^2} \right) - 3 \tag{7G.7}$$

$$\tilde{E}_{Al^{3+}}^3 + \beta_c \cdot \tilde{E}_{Al^{3+}}^2 + 3 \cdot \tilde{E}_{Al^{3+}} - 1 = 0 \tag{7G.8}$$

Cubic function (7G.8) is a perfect cube when (7G.9)—or by extension (7G.10)—is true. Since (7G.10) is dependent on solution parameter C_0, exchange selectivity coefficient K_{GT}^c, and the solution equivalent fraction $E_{Al^{3+}}$, we must assume that (7G.10) is not satisfied in general.

$$\beta^2 = 9 \tag{7G.9}$$

$$\left(\left(\frac{9 \cdot C_0 \cdot (1 - E_{Al^{3+}})^3}{8 \cdot K_{GT}^c \cdot E_{Al^{3+}}^2} \right) - 3 \right)^2 = 9$$

$$\left(\frac{9 \cdot C_0 \cdot (1 - E_{Al^{3+}})^3}{8 \cdot K_{GT}^c \cdot E_{Al^{3+}}^2} \right)^2 - 6 \cdot \left(\frac{9 \cdot C_0 \cdot (1 - E_{Al^{3+}})^3}{8 \cdot K_{GT}^c \cdot E_{Al^{3+}}^2} \right) + 9 = 9$$

$$\left(\frac{9 \cdot C_0 \cdot (1 - E_{Al^{3+}})^3}{8 \cdot K_{GT}^c \cdot E_{Al^{3+}}^2} \right)^2 - 6 \cdot \left(\frac{9 \cdot C_0 \cdot (1 - E_{Al^{3+}})^3}{8 \cdot K_{GT}^c \cdot E_{Al^{3+}}^2} \right) = 0$$

$$\left(\frac{9 \cdot C_0 \cdot (1 - E_{Al^{3+}})^3}{8 \cdot K_{GT}^c \cdot E_{Al^{3+}}^2} \right) = 6 \tag{7G.10}$$

Define three new functions p (7G.11), q (7G.12) and r (7G.13).

$$p \equiv \frac{9 - \beta_c^2}{3} \tag{7G.11}$$

$$q \equiv \left(\frac{2 \cdot \beta_c^3}{27} \right) - (\beta_c + 1) \tag{7G.12}$$

$$r \equiv \left(\frac{p^3}{27} \right) + \left(\frac{q^2}{4} \right) \tag{7G.13}$$

The exchangeable equivalent fraction $\tilde{E}_{Al^{3+}}$—the only real root of (7G.8)—is given by (7G.14) where the value of r determines the appropriate expression for the term y in (7G.14).

$$\tilde{E}_{Al^{3+}} = y - \left(\frac{\beta_c}{3} \right) \tag{7G.14}$$

When $r \geq 0$ the term y in (7G.14) is given by (7G.15), using expressions u (7G.16) and v (7G.17).

$$y = (u + v) \tag{7G.15}$$

$$u = \left(\left(\frac{-q}{2}\right) + \sqrt{r}\right)^{1/3} \tag{7G.16}$$

$$v = \left(\left(\frac{-q}{2}\right) - \sqrt{r}\right)^{1/3} \tag{7G.17}$$

When $r < 0$ the term y in (7G.14) is given by (7G.18), using the expressions w (7G.19) and function α (7G.19), the latter in radians.

$$y = \left(2 \cdot \left(\frac{-p}{3}\right)^{1/2} \cdot \cos\left(\frac{\alpha}{3}\right)\right) \tag{7G.18}$$

$$w \equiv \frac{-q}{2 \cdot \sqrt{-(p/3)^3}} \tag{7G.19}$$

$$\alpha \equiv \cos^{-1}(w) \tag{7G.20}$$

A spreadsheet model of the asymmetric exchange isotherm (7G.8) requires a logical IF function—*IF(logical_test, [value_if_true], [value_if_false])*—to test the value of r (7G.13) and assign an appropriate value to y in (7G.14) using either (7G.15) or (7G.18) depending on the outcome of the logical test.

APPENDIX 7H. THE EFFECT OF (Na^+, Ca^{2+}) EXCHANGE ON THE CRITICAL COAGULATION CONCENTRATION OF MONTMORILLONITE

Oster and colleagues (1980) published a report correlating the concentration at which a colloidal suspension of montmorillonite clay (American Petroleum Institute *API-25*, montmorillonite, Upton, Wyoming) becomes unstable and coagulates.[16] Figure 7H.1 plots the critical coagulation concentration *CCC* as a function of the exchange equivalent fraction $\tilde{E}_{Na^+}$. Suspensions of this montmorillonite clay coagulate at solution concentrations exceeding the values represented by the curve plotted in Figure 7H.1 or remain dispersed at solution concentrations less than the values represented by the line. This phenomenon is related to the concentration-dependent osmotic swelling mentioned earlier (see Figure 3.13 and Appendix 3D). The *CCC* is very sensitive to the equivalent fraction $\tilde{E}_{Na^+}$ of exchangeable Na^+.

Changes in clay chemistry appear when Na^+ ions occupy a significant fraction of the exchange complex. Recall the equivalent fraction-dependent

---

16. The aggregation of colloidal particles is called coagulation or flocculation. Aggregates settle much more rapidly than unaggregated (dispersed) particles.

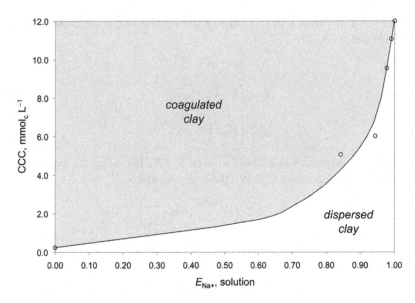

FIGURE 7H.1 The line represents the effect of exchangeable Na^+ on the critical coagulation concentration $(mmol_c \cdot L^{-1})$ of montmorillonite *API-25* (Upton, Wyoming). Experimental data plotted as open circles (*modified from Oster, Shainberg et al., 1980*).

selectivity coefficient for (Mg^{2+}, Ca^{2+}) exchange on the Libby Vermiculite (see Appendix 4C). Selectivity favors Ca^{2+} ions at the Ca^{2+}-rich limit of the exchange isotherm, while favoring Mg^{2+} ions at the Mg^{2+}-rich limit of the exchange isotherm. The shift in ion selectivity away from Ca^{2+} occurs when the equivalent fraction of exchangeable Mg^{2+} ions increases from trace levels to chemically significant levels, altering the character of the interlayer. Similarly, as the equivalent fraction of exchangeable Na^+ ions increases, the character of the interlayer gradually changes from an environment limited to crystalline swelling to an interlayer capable of osmotic swelling.

Oster and colleagues used a montmorillonite clay sample collected at the same site as that used by Sposito and colleagues (1983) in their study of asymmetric (Na^+, Ca^{2+}) exchange: Clay Minerals Society sample *SWy-1* (Upton, Wyoming). Figure 7H.2 plots the exchange isotherm reported by Sposito and colleagues; the exchange isotherm results from Oster and colleagues appear in Figure 7H.3.

Despite employing a significantly higher salt concentration; $(C_0 = 0.05 \ mol_c \cdot L^{-1})$ than Oster and colleagues, the Gaines-Thomas selectivity coefficient $K_{GT} = 0.30$ derived from fitting data from Sposito and colleagues provides an excellent fit of data reported by Oster and colleagues. The shift in position of the exchange isotherm going from Figure 7H.2 to 7H.3 results from the different salt concentrations used in the two studies.

The effect of (Na^+, Ca^{2+}) exchange on the *CCC* (see Figure 7H.1) and swelling behavior (see Figure 3.13) of smectite clays has profound implications

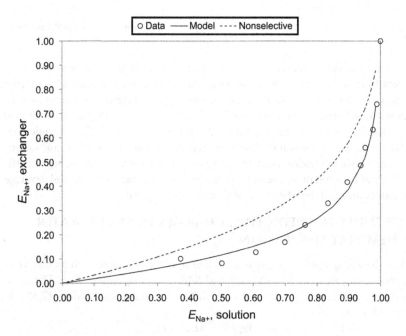

FIGURE 7H.2 The experimental exchange isotherms for asymmetric (Na^+, Ca^{2+}) exchange for Wyoming smectite SWy-1 (Sposito, Holtzclaw et al., 1983). The exchange isotherm representing Gaines-Thomas selectivity coefficient $K_{GT} = 0.30$ is plotted as a smooth line, and the nonselective exchange isotherm appears as a dashed line.

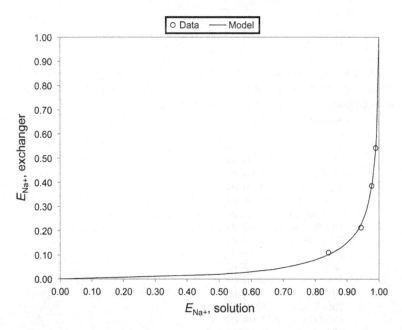

FIGURE 7H.3 The experimental exchange isotherms for asymmetric (Na^+, Ca^{2+}) exchange for Wyoming smectite API-25 (Oster, Shainberg et al., 1980). The exchange isotherm representing Gaines-Thomas selectivity coefficient $K_{GT} = 0.30$ is plotted as a smooth line.

for the hydraulic conductivity of arid region soils. Modest increases in the equivalent fraction $\tilde{E}_{Na^+}$ of exchangeable Na^+ reduces the plasticity index of smectite clays (and soil humic colloids), significantly lowering the activity index of soils containing smectite clays (Figure 7.9). The reduction of plasticity lowers the mechanical strength of the soil fabric and sharply decrease water hydraulic conductivity through the soil profile. Low hydraulic conductivity at the soil surface lowers water infiltration rates and increases runoff while also reducing the efficiency of both soil moisture recharge and internal drainage. Internal soil drainage controls the leaching of excess salts from the soil profile.

APPENDIX 7I. PREDICTING CHANGES IN SAR BY WATER CHEMISTRY SIMULATION

The following water analysis is from a groundwater sample with an electrical conductivity $EC_W = 0.77 \ dS \cdot m^{-1}$ and $SAR_P = 1.94\sqrt{mmol_c \cdot L^{-1}}$. The ionic strength estimated from $EC_W = 0.77 dS \cdot m^{-1}$ using expression (7.85) is $I = 1.04 \cdot 10^{-2} \ mol_c \cdot L^{-1}$.

Ion	ppm, mg L^{-1}	M, mol L^{-1}
Na+	76.00	3.31E-03
K+	3.00	7.67E-05
Mg++	40.00	1.65E-03
Ca++	50.00	1.25E-03
SO4–	77.00	8.02E-04
Cl–	139.00	3.92E-03
PO4– – –	0.01	1.05E-07
NO3– (N)	0.01	7.14E-07
HCO3–	228.80	3.75E-03
CO3–	0.00	0.00E+00

A ChemEQL water chemistry simulation that replaces component Ca++ with calcite (insert solid) and uses $pCO_2 = 3.5 \cdot 10^{-2} \ atm$ yields the following results (numerous minor species have been deleted).

Species	Activity
Ca++	4.33E-06
K+	6.80E-05
Mg++	9.71E-04
MgSO4	7.44E-05
MgHPO4	2.75E-05
Na+	2.94E-03
CO2(aq) (atm)	1.71E-03
HCO3– (atm)	1.33E-01
CO3– (atm)	1.13E-03
Cl–	3.49E-03
NO3–	1.43E-07
SO4–	4.51E-04
HPO4–	3.49E-05
CaCO3(aq)	5.01E-06
CaHCO3+	1.10E-05
OH–	1.82E-06
H+	5.50E-09

The adjusted sodium adsorption ratio calculated using (7.92) (Suarez, 1981) is $SAR_{adj} = 2.65$—assuming the pH is set by calcite solubility and $pCO_2 = 3.5 \cdot 10^{-2}$ atm—higher than the sodium adsorption ratio calculated from the water analysis: $SAR_w = 1.94$. This increase results from calcite precipitation. The Langelier calcite saturation $pH_C = 8.09$ is calculated from the preceding simulation results using the following expression:

$$pH_c = 1.85 - \log(a_{Ca^{2+}}) - \log(a_{HCO_3^-})$$
$$pH_c = 1.85 + 5.36 + 0.88$$

The adjusted sodium adsorption ratio calculated using Eq. (7.91) (Bower, Wilcox et al., 1963) is $adj\, SAR = 2.55$—assuming $pH = 8.4$—once again, higher than the sodium adsorption ratio calculated from the water analysis: $SAR_w = 1.94$.

Problems

1. Crandall and colleagues (1999) published a study of groundwater and river water mixing, listing dissolved salts in groundwater from a number of wells and sinkholes and surface water from the Suwannee and Little Springs rivers in Suwannee County, Florida. Under low flow (i.e., low rainfall) conditions, typical groundwater has the following composition: pH = 7.55, total $Ca = 40.0\ mg\ L^{-1}$, total $Mg = 5.2\ mg\ L^{-1}$, $Na = 3.3\ mg\ L^{-1}$, $K = 0.9\ mg\ L^{-1}$, total carbonate (as bicarbonate) = 190 $mg\ L^{-1}$, total chloride = 5.9 $mg\ L^{-1}$, total sulfate = 8.1 $mg\ L^{-1}$, and total silica = 6.3 $mg\ L^{-1}$.
Use ChemEQL to simulate the alkalinity of this groundwater sample.

Solution
Based on the results of this simulation, calculate the alkalinity of this solution using the definition of alkalinity A_T: Eqs. (7.63) and (7.67). NOTE: Subscript T indicates the *total* concentration of the species in the solution as measured, taking into account ion pair interactions.

$$A_T = (C_{bicarbonate} + 2 \cdot C_{carbonate}) + C_{silica} + (C_{hydroxyl} - C_{proton})$$

The ChemEQL simulation of groundwater with this composition yields the following solution species.

Species	Concentration, $mol\ L^{-1}$	Species	Concentration, $mol\ L^{-1}$
Ca++	9.57E-04	Mg(H3SiO4)2(aq)	2.26E-13
CaOH+	4.59E-09	Na+	1.44E-04
CaCO3(aq)	5.42E-06	NaOH	3.00E-11
CaHCO3+	2.71E-05	NaCO3−	1.20E-08
CaSO4	8.40E-06	NaHCO3	2.04E-07
CaH2SiO4(aq)	1.20E-12	NaSO4−	4.07E-08
CaH3SiO4+	9.13E-10	H2CO3	1.71E-04
Ca(H3SiO4)2(aq)	1.26E-13	HCO3−	2.90E-03
HSiO4−−−	1.22E-14	CO3−	5.90E-06
K+	2.30E-05	Cl−	1.66E-04

KOH	2.41E-12	SO4–	7.43E-05
KSO4–	9.20E-09	HSO4–	1.67E-10
Mg(OH)2(aq)	2.02E-17	H2SiO4–	1.73E-12
Mg++	2.07E-04	H3SiO4–	5.00E-07
MgOH+	2.17E-08	H4SiO4(aq) (SiO2)	1.05E-04
Mg4OH4++++	5.47E-25	H2(SiO4)2–	7.19E-13
MgCO3(aq)	6.72E-07	H4(SiO4)4– – – –	4.48E-22
MgHCO3+	5.34E-06	H6(SiO4)4–	3.13E-15
MgSO4	1.51E-06	OH–	3.80E-07
MgH2SiO4	3.05E-12	H+	2.82E-08
MgH3SiO4+	3.42E-10		

Subscript T indicates the total concentration of the species in the solution as measured. This is opposed to the *free* concentration, which does not account for ion pair species. The following formula includes the most important ion pairs contributing to alkalinity in the solution under consideration.

$$A_T = \left([HCO_3^-] + [Mg(HCO_3)^+] + [Ca(HCO_3)^+] + \left[Na(HCO_3)^0\right]\right)_{bicarbonate}$$
$$+ 2 \cdot \left([CO_3^{2-}] + \left[Ca(CO_3)^0\right] + \left[Mg(CO_3)^0\right] + [Na(CO_3)^-]\right)_{carbonate}$$
$$+ \left([H_3SiO_4^-] + [Ca(H_3SiO_4)^+] + [Mg(H_3SiO_4)^+]\right)_{silica}$$
$$+ 2 \cdot \left(\left[Ca(H_3SiO_4)_2^0\right] + \left[Mg(H_3SiO_4)_2^0\right]\right)_{silica} + [OH^-]$$
$$- \left([HSO_4^-] + [H^+]\right)$$

The ChemEQL simulation contains two significant figures. This means we can neglect species whose concentrations are three orders of magnitude lower than HCO_3^-, the most concentrated species.

$$A_T \approx \left(2.90 \cdot 10^{-3} + 2.71 \cdot 10^{-5}\right)_{bicarbonate} + 2 \cdot \left(5.90 \cdot 10^{-6} + 5.42 \cdot 10^{-6}\right)_{carbonate}$$
$$A_T \approx \left(2.90 \cdot 10^{-3} + 0.027 \cdot 10^{-3}\right) + \left(0.018 \cdot 10^{-3} + 0.012 \cdot 10^{-3}\right)$$
$$A_T \approx 2.94 \cdot 10^{-3}$$

2. Dai and colleagues (1998) studied the effect of acid rain on soils from the provinces of Hunan and Guangxi in southern China. The soil chemical properties relevant to acidity of the Hongmaochong Inceptisol (surface 0–10 *cm*) are pH (1 *M KCl*) = 3.91, CEC = 22.5 $cmol_c \cdot kg^{-1}$, exchangeable Al of 4.45 $cmol_c \cdot kg^{-1}$, and exchangeable base cations occupying 9.12% of the CEC. How much limestone would you need to eliminate the exchangeable aluminum in this soil? Is there any nonexchangeable aluminum in this soil? Does nonexchangeable aluminum contribute to soil acidity?

Solution
The exchangeable aluminum content of the soil amounts to $4.45 \cdot 10^{-2}$ moles of charge for every kilogram of soil. The following reactions demonstrate that each mole of charge (as exchangeable aluminum) is equivalent to a mole of exchangeable acidity.

$$Al^{3+}(ex) + 3 \cdot H_2O(l) \rightarrow Al(OH)_3^0 + 3 \cdot H^+(ex)$$

interlayer   interlayer   interlayer
                         polymeric Al

One mole of limestone (calcite) neutralizes two moles of acidity (i.e., three moles of calcite neutralizes the acidity from two moles of exchangeable aluminum).

$$CaCO_3(s) + 2 \cdot H^+(aq) \rightarrow Ca^{2+}(aq) + CO_2(g) + H_2O(l)$$

calcite
limestone

$$2 \cdot Al^{3+}(ex) + 3 \cdot CaCO_3 + 3 \cdot H_2O(l) \rightarrow 2 \cdot Al(OH)_3 + 3 \cdot Ca^{2+}(ex) + 3 \cdot CO_2(g)$$

interlayer   calcite   interlayer   interlayer
                                                  polymeric Al

Remember: The exchangeable aluminum content is expressed as moles of charge, eliminating the need to balance moles of exchangeable aluminum with moles of calcite as shown in the overall reaction.

Adding 2.23 grams of limestone (assuming the limestone is pure calcite) to each kilogram of soil would neutralize the acidity from the exchangeable aluminum.

$$4.45 \left(\frac{cmol_{c,\,Al}}{kg_{soil}} \right) \cdot 10^{-2} \left(\frac{mol_c}{cmol_c} \right) = 4.45 \cdot 10^{-2} \left(\frac{mol_{H^+}}{kg_{soil}} \right)$$

$$4.45 \cdot 10^{-2} \left(\frac{mol_{H^+}}{kg_{soil}} \right) \cdot \left(\frac{mol_{CaCO_3}}{2 \cdot mol_{H^+}} \right) = 2.23 \cdot 10^{-2} \left(\frac{mol_{CaCO_3}}{kg_{soil}} \right)$$

$$2.23 \cdot 10^{-2} \left(\frac{mol_{CaCO_3}}{kg_{soil}} \right) \cdot 100 \left(\frac{g_{CaCO_3}}{mol_{CaCO_3}} \right) = 2.23 \left(\frac{g_{CaCO_3}}{kg_{soil}} \right)$$

Nonexchangeable aluminum accounts for 71% of the CEC (16 $cmol_c$ kg^{-1}) but does not contribute to soil acidity and can be ignored when estimating the amount of limestone required.

3. The following table lists the chemical analysis of a groundwater sample from Waukesha, Wisconsin. The electrical conductivity of the water is 1.05 $dS \cdot m^{-1}$, and the pH is 8.0. Calculate the following: carbonate alkalinity, ionic strength, and SAR_P. Using ChemEQL, calculate SAR_{adj} using expression (7.92) for a soil pH of 8.4, and $p_{CO_2} = 3.5 \cdot 10^{-2}$ atm (approximate soil atmosphere).

Component	ppm, mg L^{-1}
Na+	79.0
K+	5.0
Mg++	38.0
Ca++	78.0
SO4–	85.0
Cl–	190.0
PO4– – –	0.01
NO3– (N)	0.01
HCO3–	250.7
CO3–	0.0

4. The subsurface soil chemical properties of a Rhodiudult from Guangxi Province (China) are (Dai, Liu et al., 1998) pH (1 M KCl) $= 4.0$ and CEC $= 29.8$ $cmol_c \cdot kg^{-1}$. Using Figures 7.13 and 7.14, estimate the exchangeable Al content $(cmol_c \cdot kg^{-1})$ of this soil.

Redox Chemistry

8.1. INTRODUCTION

Oxidation-reduction or *electron-transfer* reactions are extremely important in environmental chemistry, affecting the biological availability and mobility of many elements. *Redox* chemistry in the environmental context blends biochemistry and geochemistry in ways that can make the transition from what you learned in general chemistry challenging. The driving force behind redox reactions in the environment is a key component in respiration: the electron transport chain. Respiration is a complex series of electron-transfer reactions that ultimately couples to the environment, consuming electron acceptors ranging from molecular oxygen to carbon dioxide in order to release the chemical energy stored in reduced carbon compounds.

Environmental redox chemistry has at its source the oxidation of reduced carbon compounds— biomolecules, organic compounds, and organic matter serving as electron donors—by a variety of oxidizing agents—electron acceptors— through the respiration of living organisms. This means that the locus of most environmental redox reactions is the zone of biological activity, the zone of organic carbon accumulation.

The progress of environmental redox reactions and development of zones where certain redox processes dominate are both governed by the level of biological activity. The seeming absence of molecular oxygen—anoxia being conditions where the concentration of dissolved molecular oxygen is very low—is necessary but not sufficient for reducing conditions. The development of anoxia leads to changes in the active microbial population from communities that rely on aerobic respiration—molecular oxygen serving as the terminal electron acceptor—to anaerobic respiration, where other chemical substances replace molecular oxygen as the terminal electron acceptor supporting respiration.

This transition from one microbial community to another—from one type of respiration to another—sets the stage for the chemical reduction of the environment that will ultimately couple the redox reactions required for biological respiration to a host of redox reactions that occur simply because the redox potential is drawn down by anaerobic respiration.

This chapter is organized into three major sections. The first is a reprise of redox chemistry principles that are essential for a complete understanding of environmental chemistry, designed to bridge the gap between general chemistry and environmental chemistry. The second develops the methods used by geochemists to quantify and interpret redox conditions as they occur in the environment. The final section provides the mechanism that generates reducing conditions in soils and groundwater: anaerobiosis—microbial respiration in the absence of molecular oxygen. Respiration consists of a sequence of biological redox reactions organized as an electron transport chain coupled to select electron acceptors in the environment, and anaerobic respiration is the source of the compounds characteristic of reducing conditions.

8.2. REDOX PRINCIPLES

8.2.1. Formal Oxidation Numbers

Redox reactions are chemical reactions involving the transfer of one or more electrons from an electron donor—the reducing agent—to an electron acceptor—the oxidizing agent. General chemistry books typically discuss the assignment of formal oxidation numbers because it can be difficult in many cases to recognize whether electron transfer has occurred in a given chemical reaction and to identify electron donors and acceptors. Formal oxidation numbers are also used for electron accounting in redox reactions.

The formal oxidation number of an element in a compound is usually a poor estimate of the actual oxidation state of that element in this compound; subtle details in chemical bonding determine the true oxidation state. Regardless, an electron transfer reaction always results in the loss of one or more electrons from the electron donor compound and the acquisition of an equal number of electrons by the electron acceptor compound. The assignment of formal oxidation numbers is sufficiently accurate to demonstrate that electron transfer has occurred, to identify electron donors and acceptors, and to balance redox reactions.

Table 8.1 lists the chemical rules for assigning formal oxidation numbers to elements in ions and compounds. There are exceptions, but they are relatively inconsequential for environmental chemistry. Appendix 8A gives two examples to demonstrate the assignment of formal oxidation numbers.

TABLE 8.1 Rules for Assigning Formal Oxidation Numbers to Elements

Rule	Examples and Exceptions
1. The oxidation number of a free element is always (0).	The elements in $S(s)$ and $O_2(g)$ have a formal oxidation number of (0).
2. The oxidation number of a monatomic ion equals the charge of the ion.	The oxidation number of the ion $Na^+(aq)$ is (+1).
3. The oxidation number of hydrogen in a compound is usually (+1).	Exception: The oxidation number of hydrogen is (−1) in compounds containing elements that are less electronegative than hydrogen (Figure 8.1)—for example, alkali hydrides such as $NaH(s)$ or $CaH_2(s)$.
4. The oxidation number of oxygen in compounds is usually (−2).	Exception: Since F is more electronegative than O (Figure 8.1), its oxidation number takes precedence. The formal oxidation number of O in oxygen difluoride $OF_2(g)$ is (+2).
5. The oxidation number of a Group IA (alkali) element in a compound is (+1).	The oxidation number of K in the mineral orthoclase $KAlSi_3O_8(s)$ is (+1).
6. The oxidation number of a Group IIA (alkaline earth) element in a compound is (+2).	The oxidation number of Ca in the mineral anorthite $CaAl_2Si_2O_8(s)$ is (+2).
7. The oxidation number of a Group VIIA (halogen) element in a compound is (−1).	Exception: Since O is more electronegative than Cl (Figure 8.1), its oxidation number takes precedence. The formal oxidation number of Cl in $HOCl$ is (+1).
8. The sum of the oxidation numbers of all of the atoms in a neutral compound is (0).	The sum of the oxidation numbers for ascorbic acid $C_6H_8O_6$ is (0).
9. The sum of the oxidation numbers in a polyatomic ion is equal to the charge of the ion.	The sum of the oxidation numbers for $SO_4^{2-}(aq)$ is (−2).

Group (vertical)	1	2	3	4	5	6	7	8	9	10	11	12	13	14	15	16	17	18
Period (horizontal)																		
1	H 2.20																	He
2	Li 0.98	Be 1.57											B 2.04	C 2.55	N 3.04	O 3.44	F 3.98	Ne
3	Na 0.93	Mg 1.31											Al 1.61	Si 1.90	P 2.19	S 2.58	Cl 3.16	Ar
4	K 0.82	Ca 1.00	Sc 1.36	Ti 1.54	V 1.63	Cr 1.66	Mn 1.55	Fe 1.83	Co 1.88	Ni 1.91	Cu 1.90	Zn 1.65	Ga 1.81	Ge 2.01	As 2.18	Se 2.55	Br 2.96	Kr 3.00
5	Rb 0.82	Sr 0.95	Y 1.22	Zr 1.33	Nb 1.6	Mo 2.16	Tc 1.9	Ru 2.2	Rh 2.28	Pd 2.20	Ag 1.93	Cd 1.69	In 1.78	Sn 1.96	Sb 2.05	Te 2.1	I 2.66	Xe 2.60
6	Cs 0.79	Ba 0.89	*	Hf 1.3	Ta 1.5	W 2.36	Re 1.9	Os 2.2	Ir 2.20	Pt 2.28	Au 2.54	Hg 2.00	Tl 1.62	Pb 2.33	Bi 2.02	Po 2.0	At 2.2	Rn 2.2
7	Fr 0.7	Ra 0.9	**	Rf	Db	Sg	Bh	Hs	Mt	Ds	Rg	Uub	Uut	Uuq	Uup	Uuh	Uus	Uuo

Lanthanides *	La 1.1	Ce 1.12	Pr 1.13	Nd 1.14	Pm 1.13	Sm 1.17	Eu 1.2	Gd 1.2	Tb 1.1	Dy 1.22	Ho 1.23	Er 1.24	Tm 1.25	Yb 1.1	Lu 1.27
Actinides **	Ac 1.1	Th 1.3	Pa 1.5	U 1.38	Np 1.36	Pu 1.28	Am 1.13	Cm 1.28	Bk 1.3	Cf 1.3	Es 1.3	Fm 1.3	Md 1.3	No 1.3	Lr 1.291

FIGURE 8.1 Periodic table of Pauling electronegativity values.

8.2.2. Balancing Reduction Half Reactions

The primary utility of formal oxidation numbers comes when balancing redox reactions: verifying that electron transfer occurs, identifying electron donor and acceptor atoms, and balancing the electron count. General chemistry introduces the concept of the *reduction half reaction*:

$$\underset{acceptor}{A^s} + m \cdot H^+ + n \cdot e^- \rightarrow \underset{donor}{D^r} \tag{8.1}$$

Table 8.2 lists the steps you should follow when balancing a reduction half reaction (8.1). The balancing steps 2–5 (see Table 8.2) should be executed in the order given to avoid confusion and error (see Example 8.1).

TABLE 8.2 Steps for Balancing Reduction Half Reactions

Step	Example
1. Assign formal oxidation numbers to identify electron acceptor atom in the educt (left-hand) side and the electron donor in the product (right-hand) side.	$\underset{N(+1)}{N_2O} + \cdots \rightarrow \underset{N(-3)}{NH_4^+} + \cdots$
2. Balance atoms involved in electron transfer.	$\underset{N(+1)}{N_2O} + \cdots \rightarrow 2 \cdot \underset{N(-3)}{NH_4^+} + \cdots$

Continued

TABLE 8.2 Steps for Balancing Reduction Half Reactions—Cont'd

Step	Example
3. Determine the change in formal oxidation number and assign that integral number to the number of electrons n on the electron acceptor (or educt) side of the reaction equation (8.1).	$\underset{N(+1)}{N_2O} + 8 \cdot e^- + \cdots \rightarrow 2 \cdot \underset{N(-3)}{NH_4^+} + \cdots$
4. Balance the net charge $r = (s + m - n)$ by adding m protons H^+ to the left-hand side of the reaction equation (or an appropriate number of hydroxyl ions to the right-hand side).	$\underset{N(+1)}{N_2O} + 10 \cdot H^+ + 8 \cdot e^- \rightarrow 2 \cdot \underset{N(-3)}{NH_4^+} + \cdots$
5. Balance proton (or hydroxyl) mass by adding an appropriate number of H_2O.	$\underset{N(+1)}{N_2O} + 10 \cdot H^+ + 8 \cdot e^- \rightarrow 2 \cdot \underset{N(-3)}{NH_4^+} + H_2O$

Example 8.1

Balancing the reduction half reaction involving ascorbic acid (CAS 50-81-7, Structure 1) and dehydroascorbic acid (CAS 490-83-5, Structure 2) begins with step 1; the identity of electron donor and acceptor compounds may not be immediately evident from inspection.

Structure 1. Ascorbic acid

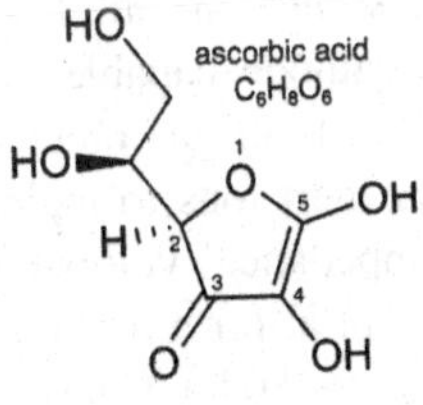

Structure 2. Dehydroascorbic acid

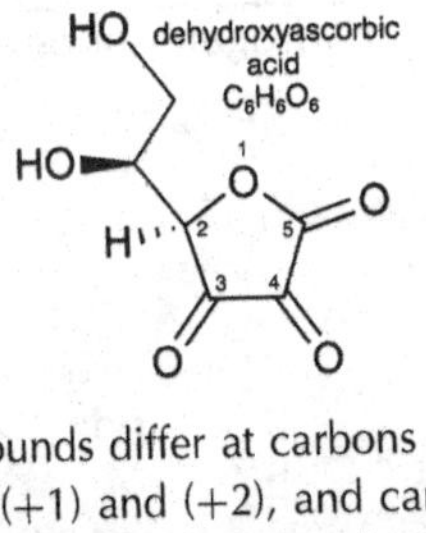

Step 1: These two compounds differ at carbons C_4 and C_5: the formal oxidation states of carbon C_4 are (+1) and (+2), and carbon C_5 are (+2) and (+3) in ascorbic and dehydroascorbic acid, respectively. This identifies dehydroascorbic acid as the electron acceptor and ascorbic acid as the electron donor. The reduction half reaction places the electron acceptor dehydroascorbic acid on

the left-hand (educt) side and the electron donor ascorbic acid on the right-hand (product) side.

Step 2: The stoichiometric coefficients on the electron donor and acceptor compounds are identical: 1.

Step 3: The net change in formal oxidation state summed over both carbons is $(+2)$. The required electron stoichiometric coefficient n is 2, as shown in Eq. (8.2):

$$C_6H_6O_6 + 2 \cdot e^- + \cdots = C_6H_8O_6 + \cdots \qquad (8.2)$$

electron acceptor   electron donor

Steps 4 and 5: Charge balance in expression (8.2) requires a proton stoichiometric coefficient m of 2, as shown in Eq. (8.3). The proton mass balance does not require any additional water molecules; reaction (8.3) is the fully balanced reduction half reaction.

Reaction

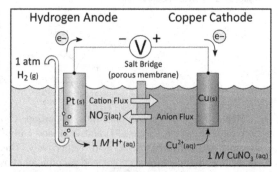

8.2.3. Reduction Half Reactions and Electrochemical Cells

Figure 8.2 illustrates the *standard galvanic cell*, a common illustration found in general chemistry books. The solutes are all $1\ M$, the electrode materials are pure elements (copper metal cathode capable undergoing oxidation to Cu^{2+} ions and an inert platinum anode), all gas pressures are 1 atmosphere. A salt bridge permits the diffusion of inert ions to maintain charge neutrality during electron transfer, and a high impedance[1] voltmeter measures the cell potential.

The standard reduction potential for a hydrogen half-cell E°_{SHE} is, by definition, equal to zero volts ($E^\circ_{SHE} \equiv 0.0\,V$). Consequently the cell potential E°_{cell}

FIGURE 8.2 A complete galvanic cell consists of a cathode and an anode; the cathode is a standard cupric $Cu^{2+}(1\ M)|Cu(s)$ half-cell, and the anode is a standard hydrogen $H_2(1\ atm)|H^+\ (1\ M)$ half-cell.

Hydrogen Anode	Copper Cathode
(e⁻) → $-$ (V) $+$ (e⁻)	
1 atm H₂ (g)	Salt Bridge (porous membrane)
Pt (s) Cation Flux ⇒	Cu(s)
NO₃⁻(aq) ⇐ Anion Flux	
→ 1 M H⁺ (aq)	Cu²⁺(aq)
	1 M CuNO₃ (aq)

---

1. A high impedance voltmeter requires a very small current to record the potential difference.

TABLE 8.3 Reduction Potentials for a Selection of Reduction Half Reactions

Reduction Half Reaction	Standard Biological Reduction Potential, $E_H^{\circ\dagger}$, mV	Standard Reduction Potential, E_H°, mV
$2 \cdot H^+(aq) + 2 \cdot e^- \rightarrow H_2(g)$	-414	0
$CO_2(g) + 8 \cdot H^+(aq) + 8 \cdot e^- \rightarrow CH_4(g) + 2 \cdot H_2O(l)$	-245	$+169$
$6 \cdot N_2(g) + 8 \cdot H^+(aq) + 6 \cdot e^- \rightarrow 2 \cdot NH_4^+(aq)$	-278	$+274$
$2 \cdot SO_4^{2-}(aq) + 10 \cdot H^+(aq) + 8 \cdot e^- \rightarrow S_2O_3^{2-}(aq) + 5 \cdot H_2O(l)$	-130	$+284$
$SO_4^{2-}(aq) + 8 \cdot H^+(aq) + 6 \cdot e^- \rightarrow S(s) + 4 \cdot H_2O(l)$	-61	$+353$
$Fe^{3+}(aq) + e^- \rightarrow Fe^{2+}(aq)$	$+771$	$+771$
$NO_3^-(aq) + 2 \cdot H^+(aq) + 2 \cdot e^- \rightarrow NO_2^-(aq) + H_2O(l)$	$+407$	$+821$
$NO_2^-(aq) + 2 \cdot H^+(aq) + e^- \rightarrow NO(aq) + H_2O(l)$	$+238$	$+1066$
$Fe(OH)_3(s) + 3 \cdot H^+(aq) + e^- \rightarrow Fe^{2+}(aq) + 3 \cdot H_2O(l)$	-262	$+981$
$O_2(g) + 4 \cdot H^+(aq) + 4 \cdot e^- \rightarrow 2 \cdot H_2O(l)$	$+815$	$+1229$

recorded in Figure 8.2 is equal to the standard reduction half-cell potential[2] E_H° of the other half-cell: $Cu^{2+}(1\,M)|Cu(s)$.

Most electron transfer reactions encountered in biology, geochemistry, and environmental chemistry cannot be replicated using galvanic cells of the type shown in Figure 8.2. Regardless, it is possible to use Gibbs energy relations to determine standard reduction half reaction potentials for all relevant electron transfer reactions. We will explain this in the following sections.

Standard conditions and standard reduction potentials E_H° (Table 8.3) are not found in nature. Biochemists typically tabulate standard biological reduction potentials $E_H^{\circ\dagger}$ that approximate physiological conditions. Most environmental chemists either tabulate or describe methods for computing standard environmental reduction potentials (which are identical to standard biological reduction potentials $E_H^{\circ\dagger}$) from standard reduction potentials.

8.2.4. The Nernst Equation

The *Nernst* equation, which you first encountered in general chemistry, allows you to calculate a half reaction potential under any nonstandard condition. This equation is derived from the Gibbs energy expression under nonstandard

---

2. The H subscript in the symbol for reduction half-cell potentials E_H° indicates that the potential is based on the hydrogen half-cell E_{SHE}°.

conditions (Eq. (8.4)) and the expression relating Gibbs energy to the half reaction potential (Eq. (8.5)).

$$\Delta G = \Delta G^{\circ} + R \cdot T \cdot \ln\left(\frac{a_{D'}}{a_{A^s} \cdot a_{H^+}^m}\right) = \Delta G^{\circ} + R \cdot T \cdot \ln(Q) \qquad (8.4)$$

$$\Delta G = -n \cdot F \cdot E \qquad (8.5)$$

Expression (8.4) represents the change in Gibbs energy ΔG in any chemical reaction. The symbols in Eq. (8.4) are reaction quotient Q, ideal gas constant R ($8.314472 \ 10^{-3} \, kJ \cdot mol^{-1} \cdot K^{-1}$), and the absolute temperature T (K). Expression (8.5) in the context of a redox reaction represents the change in Gibbs energy ΔG as electrons move from one electrostatic potential to another; the initial and final electrostatic potentials are those existing in the electron acceptor and donor compounds. The symbols in Eq. (8.5) are the Faraday constant F ($98.48534 \ kJ \cdot mol^{-1} \cdot V^{-1}$) and the stoichiometric electron coefficient n for the reduction half reaction.

Expression (8.5) applies to any balanced electron transfer reaction, including reduction half reactions. The electron transfer potential for a balanced electron transfer reaction is denoted E, while the electron transfer potential for the half reaction[3] is denoted E_H. Similarly, the change in Gibbs energy for a balanced electron transfer reaction is denoted ΔG, while the change in Gibbs energy for a reduction half reaction is denoted ΔG_H.

Setting Eq. (8.4) equal to (8.5) results in the seminal *Nernst* equation (8.6). Solving Eq. (8.6) for half reaction potentials yields the Nernst equation in its most familiar form: (8.7). Most expressions replace the natural logarithm with the logarithm to the base 10 (Eq. (8.8)): $\ln(x) = \ln(10) \cdot \log_{10}(x)$.

$$(-n \cdot F \cdot E_H) = (-n \cdot F \cdot E_H^{\circ}) + \left(R \cdot T \cdot \ln\left(\frac{a_{D'}}{a_{A^s} \cdot a_{H^+}^m}\right)\right) \qquad (8.6)$$

$$E_H = E_H^{\circ} - \left(\frac{R \cdot T}{n \cdot F}\right) \cdot \ln\left(\frac{a_{D'}}{a_{A^s} \cdot a_{H^+}^m}\right) \qquad (8.7)$$

$$E_H = E_H^{\circ} - \left(\frac{R \cdot T \cdot \ln(10)}{n \cdot F}\right) \cdot \log_{10}\left(\frac{a_{D'}}{a_{A^s} \cdot a_{H^+}^m}\right) \qquad (8.8)$$

At $298.15 \, K$ ($25°C$), Eq. (8.8) reduces to a simpler approximate expression (8.9):

$$\left(\frac{R \cdot T \cdot \ln(10)}{n \cdot F}\right) = \left(\frac{(8.314 \cdot 10^{-3} \, kJ \cdot mol^{-1} \cdot K^{-1}) \cdot T \cdot (2.303)}{n \cdot (96.49 \, kJ \cdot mol^{-1} \cdot V^{-1})}\right)$$

3. The convention we use is all half reactions are written as reduction half reactions (8.1).

$$E_{H,25°C} = E_H^\circ - \left(\frac{59.16\,mV}{n}\right)_{25°C} \cdot \log_{10}\left(\frac{a_{D'}}{a_{A^s} \cdot a_{H^+}^m}\right) \qquad (8.9)$$

The following two examples (Examples 8.2 and 8.3) demonstrate the calculation of standard biological reduction half potentials $E_H^{\circ\dagger}$ using the Nernst equation—either (8.8) or (8.9). A standard biological reduction potential $E_H^{\circ\dagger}$ requires that all components—other than $H^+(aq)$—have activities or partial pressures equal to 1 (i.e., $Q = 1$ or $\log_{10}Q = 0$).

Example 8.2

Calculate the standard biological reduction half potential $E_H^{\circ\dagger}$ for the reaction $H_2O|O_2(1\,atm), H^+(10^{-7}\,M)$ (see Table 8.3) using the Nernst equation.

The reaction $H_2O|O_2(1\,atm), H^+(1\,M)$ and its standard reduction half potential E_H° appear following. Under standard conditions the proton activity is $a_{H^+} = 1$.

$$O_2 + 4 \cdot H^+ + 4 \cdot e^- \rightarrow 2 \cdot H_2O \quad E_H^\circ = +1.229\ V$$

Under physiological conditions the reaction becomes $H_2O|O_2(1\,atm)$, $H^+(10^{-7}M)$, and the proton activity is $a_{H^+} = 10^{-7}$ because standard biological conditions are defined as pH = 7.

Substituting these conditions into the Nernst equation and simplifying yield the following expressions.

$$E_H^{\circ\dagger} = E_H^\circ - \left(\frac{R \cdot T}{4 \cdot F}\right) \cdot \ln\frac{1}{P_{O_2} \cdot \left(10^{-7}\right)^4}$$

$$E_H^{\circ\dagger} = E_H^\circ - \left(\frac{R \cdot T}{4 \cdot F}\right) \cdot \ln\left(\frac{1}{\left(10^{-7}\right)^4}\right) - \left(\frac{R \cdot T}{4 \cdot F}\right) \cdot \ln\left(\frac{1}{P_{O_2}}\right)$$

$$E_H^{\circ\dagger} = E_H^\circ - \left(\frac{4 \cdot R \cdot T \cdot \ln(10) \cdot (7)}{4 \cdot F}\right)$$

Applying the simplification in Eq. (8.9) yields the following, which appears in Table 8.3, column 3.

$$E_H^{\circ\dagger} = +1229\,mV - (59.16\,mV) \cdot (7)$$
$$E_H^{\circ\dagger} = +1229\,mV - 414.1\,mV = +814.9\,mV$$

Example 8.3 illustrates the calculation of standard biological reduction half potentials $E_H^{\circ\dagger}$ for the reaction $H^+(10^{-7}M)|H_2(1\,atm)$ using the Nernst equation. You will notice that it is very similar to the previous example.

Example 8.3

Calculate the standard biological reduction half potential $E_H^{\circ\dagger}$ for the reaction $H^+(10^{-7} M)|H_2(1\ atm)$ (see Table 8.3) using the Nernst equation.

The reaction $H^+(1M)|H_2(1\ atm)$ and its standard reduction half potential E_H° appear following. Under standard conditions proton activity is $a_{H^+} = 1$.

$$2 \cdot H^+ + 2 \cdot e^- \rightarrow H_2$$

Under physiological conditions the reaction becomes $H^+(10^{-7}M)|H_2(1\ atm)$, and the proton activity is $a_{H^+} = 10^{-7}$ because standard biological conditions are defined as pH $= 7$.

Substituting these conditions into the Nernst equation and simplifying yield the following expressions.

$$E_H^{\circ\dagger} = E_H^{\circ} - \left(\frac{R \cdot T}{2 \cdot F}\right) \cdot \ln \frac{p_{H_2}}{(10^{-7})^2}$$

$$E_H^{\circ\dagger} = E_H^{\circ} - \left(\frac{R \cdot T}{2 \cdot F}\right) \cdot \ln \frac{1}{(10^{-7})^2} - \left(\frac{R \cdot T}{2 \cdot F}\right) \cdot \ln p_{H_2}$$

$$E_H^{\circ\dagger} = E_H^{\circ} - \left(\frac{2 \cdot R \cdot T \cdot \ln(10) \cdot (7)}{2 \cdot F}\right)$$

Applying the simplification in Eq. (8.9) yields the following, which appears in Table 8.3, column 3.

$$E_H^{\circ\dagger} = 0.0\ mV - (59.16\ mV) \cdot (7)$$
$$E_H^{\circ\dagger} = -414.1\ mV$$

The general reduction half reaction (8.1) yields a general reaction quotient Q as shown in Eq. (8.10).

$$E_H = E_H^{\circ} - \left(\frac{R \cdot T}{n \cdot F}\right) \cdot \ln \left(\frac{a_{D^{(+m-n)}}}{a_A \cdot a_{H^+}^m}\right) \tag{8.10}$$

Rearranging expression (8.10) yields (8.11), which explicitly lists proton m and electron n coefficients.

$$E_H = \left(E_H^{\circ} - \left(\frac{m}{n}\right) \cdot \left(\frac{(2.303) \cdot R \cdot T \cdot pH}{F}\right)\right)$$
$$- \left(\frac{(2.303) \cdot R \cdot T}{n \cdot F}\right) \cdot \log_{10} \left(\frac{a_{D^{(+m-n)}}}{a_A}\right) \tag{8.11}$$

The first and second terms in Eq. (8.11) combined define the standard biological (or environmental) reduction half potential $E_H^{\circ\dagger}$ (Eq. (8.12)), regardless of the values assigned to the remaining terms in the reaction quotient Q. Expression (8.13) allows you to calculate a nonstandard reduction potential no matter what activity is assigned to other components in the reaction quotient Q, including gas partial pressures other than 1 atm.

$$E_H^{\circ\dagger} \equiv E_H^{\circ} - \left(\frac{m}{n}\right) \cdot \left(\frac{(2.303) \cdot R \cdot T \cdot (7)}{F}\right) \qquad (8.12)$$

$$E_H = E_H^{\circ\dagger} - \left(\frac{(2.303) \cdot R \cdot T}{n \cdot F}\right) \cdot \log_{10}Q_{\text{pH}=7} \qquad (8.13)$$

8.3. INTERPRETING REDOX STABILITY DIAGRAMS

One of the most important tasks facing an environmental chemist is assessing chemical conditions at a site and interpreting the implications of those conditions. Chapter 5 discussed the use of water chemistry simulations to identify minerals that potentially control water chemistry in environmental samples. Chapters 4 and 9 discuss the use of adsorption and exchange isotherms to quantify an insoluble but chemically active reserve that sustains solute concentrations without the involvement of precipitation reactions. The assessment and interpretation of redox conditions require chemical coordinates that characterize redox conditions and a means of interpreting the significance of those chemical coordinates.

Tables 8.3, 8.7, 8.9, 8.10, and 8.11 list reduction half reactions for numerous environmentally relevant redox reactions. All of these tables list standard biological reduction potential $E_H^{\circ\dagger}$, calculated from the standard reduction potential E_H° for each reduction half reaction. The only reduction half reaction in Table 8.3 where the standard biological reduction potential $E_H^{\circ\dagger}$ and the standard reduction potential E_H° are equal is the reduction of $Fe^{3+}(aq)$ to $Fe^{2+}(aq)$—the only reaction where $H^+(aq)$ does not appear as an educt. If reduction half reactions contain $H^+(aq)$ as an educt, then expression (8.11) predicts that the reduction potential will be pH-dependent.

The two chemical coordinates that chemists use to characterize redox conditions are the experimental potential E_H and pH. The electrochemist Marcel Pourbaix (1938) was the first to prepare *stability diagrams* to interpret redox conditions using (E_H, pH) coordinates. Pourbaix studied corrosion; corrosion scientists were the first to employ Pourbaix diagrams, but environmental chemists, soil chemists, and geochemists quickly adopted them to interpret geochemical and environmental redox conditions.

8.3.1. Environmental Redox Conditions

Baas-Becking and colleagues (1960) published a Pourbaix diagram plotting a host of experimental (E_H, pH) coordinates from scientific papers reporting environmental redox conditions (Figure 8.3). Few of the experimental (E_H, pH) coordinates fall outside of the pH range 4 to 8. The lower limit is

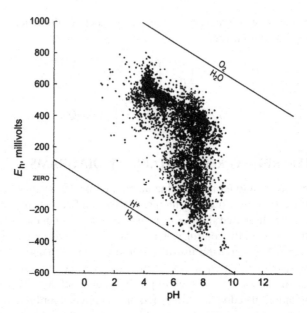

FIGURE 8.3 This Pourbaix diagram plots more than 6,200 experimental (E_H, pH) coordinates representative of the redox conditions found in the environment. *Source: Reproduced with permission from Baas-Becking, L.G. M., Kaplan, I.R., et al., 1960. Limits of the natural environment in terms of pH and oxidation-reduction potentials. J. Geol. 68, 243–284.*

the pH of maximum buffering by exchangeable *Al* ions (see Chapter 7), while the upper limit is the pH at which carbonate buffering resists further increases. Environmental reduction potentials E_H at the upper limit plot roughly parallel to the line defining the limit of water stability (Delahay, Pourbaix et al., 1950)—the line labeled $O_2|H_2O$ representing the O_2 reduction half reaction from Table 8.3—and never approach the lower limit of water stability—the line labeled $H_2|H^+$ representing the $H^+(aq)$ reduction half reaction from Table 8.3—below pH 6.

$$HCO_3^-(aq) + 7 \cdot H^+ + 8 \cdot e^- \rightarrow CH_4(g) + 2 \cdot H_2O \qquad E_H^{\circ\dagger} = -245 \; mV$$

(8.14)

Baas-Becking and colleagues (1960) attribute the significant increase in the range of environmental redox conditions centered on the alkaline pH limit to microbial sulfate respiration; however, there is no significant change in sulfate solubility in this pH range. An alternative interpretation supported by the bicarbonate reduction potential (Eq. (8.14)) would assign the dramatic expansion of reducing conditions at the alkaline pH limit to microbial carbonate respiration. Carbonate solubility increases dramatically above pH 6 (see Chapter 7), increasing both buffering capacity and the availability of the terminal electron donor required for carbonate respiration.

Regardless of why the range of reducing conditions recorded in soils expands with increasing pH, Figure 8.3 confirms the profound importance of anaerobic microbial respiration in defining environmental redox conditions.

8.3.2. Measuring Environmental Reduction Potentials Using Platinum Oxidation-Reduction Electrodes

Environmental reduction potentials are measured using some variation of a platinum combination oxidation-reduction potential ORP electrode, illustrated in Figure 8.4. The electrode is a compact version of the galvanic cell in Figure 8.2. The standard hydrogen half-cell (anode, Figure 8.2) is not practical for environmental measurements and is replaced by a reference half-cell such as the calomel reduction half reaction (8.60). The voltmeter employs a high-impedance design that requires a very small current to record an accurate potential.

$$Hg_2Cl_2(s) + 2 \cdot e^- \rightarrow 2 \cdot Hg(l) + 2 \cdot Cl^- \qquad E_H^\circ = +244.4 \, mV \qquad (8.15)$$

The platinum ORP combination electrode includes two porous ceramic junctions that function as salt bridges, allowing an ion current to flow between reference electrode and the external solution being measured. The electron current passes through a voltmeter in the circuit connecting the platinum sensing wire and the liquid mercury of the reference cell. Nordstrom and Wilde (1998) outline the limitations of measuring environmental reduction potentials using platinum ORP electrodes. These limitations are discussed in Appendix 8B.

The true redox potential of a soil that is rich in organic matter is highly reducing, but the reducing potential of natural organic matter is blocked by an activation barrier that prevents the transfer of electrons from organic matter to dioxygen (see Chapter 6). Aerobic respiration eliminates the aforementioned activation barrier through catabolism and the aerobic electron transport chain.

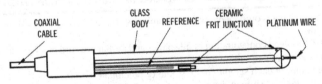

FIGURE 8.4 A typical platinum oxidation-reduction potential (ORP) electrode is a combination electrode. The reference half-cell potential in this case is based on the calomel reduction potential (8.15). The reduction potential is measured using a platinum wire as the second electrode.

8.3.3. Pourbaix Stability Diagrams: Preparation and Interpretation

This section describes the components in a Fe-O-C-H Pourbaix stability diagram to illustrate the essential features found in all Pourbaix diagrams. The preparation of accurate Pourbaix diagrams is beyond the scope of this book, since it requires an advanced understanding of chemical thermodynamics and redox chemistry. This section, however, will provide the foundation that will allow you to interpret virtually any Pourbaix diagram you will encounter, however complex.

Table 8.4 lists the features found in any complete Pourbaix diagram. Interpretation begins with identifying each of these features. Most of these features will be readily apparent, but some may be listed in the caption or text accompanying the diagram. The absence of these features makes interpretation impossible. Once you have correctly identified all of the essential features, the interpretation is surprisingly simple because Pourbaix diagrams provide a limited representation of the chemical system.

8.3.4. Water Stability Limits

The two boundaries in Figure 8.3 delineate the stability domain of water (see Delahay, Pourbaix et al., 1950). These boundaries are derived from the two reduction half reactions (8.16) and (8.17).

$$O_2(g) + 4 \cdot H^+(aq) + 4 \cdot e^- \rightarrow 2 \cdot H_2O(l) \qquad E_H^\circ = +1229 \, mV \qquad (8.16)$$

$$2 \cdot H^+(aq) + 2 \cdot e^- \rightarrow H_2(g) \qquad E_H^\circ = 0.0 \, mV \qquad (8.17)$$

Expression (8.18) is the Nernst expression for reduction half reaction (8.16) defining the boundary labeled $O_2|H_2O$ in Figure 8.3.

$$E_H = E_H^\circ - \left(\frac{R \cdot T}{n \cdot F}\right) \ln\left(\frac{1}{p_{O_2} \cdot a_{H^+}^4}\right) = E_H^\circ - \left(\frac{R \cdot T \cdot (\ln 10)}{4 \cdot F}\right) \log_{10}\left(\frac{1}{p_{O_2} \cdot a_{H^+}^4}\right)$$

$$(8.18)$$

TABLE 8.4 The Essential Features Needed to Interpret a Pourbaix Diagram

1. The elemental components whose chemical stability the diagram is designed to portray.
2. Total dissolved concentration of all solutes and partial pressure of all gases influencing the chemical stability of the defined system.
3. The nature of each stability field: solute or precipitate.
4. The nature of each boundary separating two fields: solute-solute, solute-precipitate, or precipitate-precipitate.

Factoring the second term on the right of expression (8.18) results in expression (8.19), containing three terms on the right: the standard reduction potential E_H°, the dependence on the O_2 partial pressure, and the dependence on proton activity. Notice that the stoichiometric coefficients for the electrons n and protons m are identical and cancel in the third term.

$$E_H = E_H^\circ + \left(\frac{59.16 \ mV}{4}\right) \cdot \log_{10}(p_{O_2}) + (59.16 \ mV) \cdot \log_{10}(a_{H^+})$$

$$E_H = E_H^\circ + \left(\frac{59.16 \ mV}{4}\right) \cdot \log_{10}(p_{O_2}) + (59.16 \ mV) \cdot (-\text{pH}) \tag{8.19}$$

$$E_H = 1229 \ mV - (59.16 \ mV) \cdot \text{pH} \tag{8.20}$$

The $O_2(g)|H_2O(l)$ stability boundary is usually drawn assuming an O_2 partial pressure of 1 atm (Eq. (8.20)). To demonstrate the effects of anoxia, expression (8.21) gives the pH-dependent reduction potential for the $O_2(g)|H_2O(l)$ couple at $p_{O_2} = 10^{-6} \ atm$. The intercept of Eq. (8.21) is shifted downward a mere 88.8 mV relative to the intercept of Eq. (8.20), a relatively minor shift considering the 700–800 mV potential recorded by a typical platinum ORP electrode E_{Pt} in aerated water (Appendix 8B).

$$E_H = (+1229 \ mV) + \left(\frac{(59.16 \ mV) \cdot (-6)}{4}\right) - (59.16 \ mV) \cdot \text{pH}$$

$$E_H = +1141 \ mV - (59.16 \ mV) \cdot \text{pH} \tag{8.21}$$

Expression (8.22) is the Nernst expression for reduction half reaction (8.17) defining the boundary labeled $H^+|H_2$ in Figure 8.3.

$$E_H = E_H^\circ - \left(\frac{R \cdot T \cdot (\ln 10)}{2 \cdot F}\right) \cdot \log_{10}\left(\frac{p_{H_2}}{a_{H^+}^2}\right) \tag{8.22}$$

Factoring the second term on the right of expression (8.22) results in expression (8.23), containing three terms on the right: the standard reduction potential E_H°, the dependence on the H_2 partial pressure (see Table 8.4, item 2), and the dependence on proton activity. The stoichiometric coefficients for the electrons n and protons m are identical and cancel in the third term. The $H^+(aq)|H_2(g)$ stability boundary is usually drawn assuming a H_2 partial pressure of 1 atm (8.24).

$$E_H = E_H^\circ - \left(\frac{59.16 \ mV}{2}\right) \cdot \log_{10}(p_{H_2}) + (59.16 \ mV) \cdot \log_{10}(a_{H^+})$$

$$E_H = E_H^\circ + \left(\frac{59.16 \ mV}{2}\right) \cdot \log_{10}(p_{H_2}) - (59.16 \ mV) \cdot \text{pH} \tag{8.23}$$

$$E_H = -(59.16 \ mV) \cdot \text{pH} \tag{8.24}$$

Boundary lines (8.20) and (8.24) are parallel and differ only in their intercept, which at pH 0 are the standard reduction potentials E_H° for the two reduction half reactions.

8.3.5. The Solute-Solute Reduction Boundary

The first boundary unique to the $Fe\text{-}O\text{-}C\text{-}H$ Pourbaix diagram we are preparing is the reduction equilibrium boundary for pH–independent reduction half reaction (8.25).

$$Fe^{3+}(aq) + e^- \rightarrow Fe^{2+}(aq) \qquad E_H^\circ = +771\ mV \qquad (8.25)$$

The Nernst expression for this half reaction is (8.26). Since this reduction half reaction is independent of pH, we can anticipate the boundary runs perpendicular to the E_H axis of our Pourbaix diagram.

$$E_H = E_H^\circ - \left(\frac{R \cdot T \cdot (\ln 10)}{n \cdot F}\right) \cdot \log_{10}\left(\frac{a_{Fe^{2+}}}{a_{Fe^{3+}}}\right)$$

$$E_H = (+771\ mV) - (59.16\ mV) \cdot \log_{10}\left(\frac{a_{Fe^{2+}}}{a_{Fe^{3+}}}\right) \qquad (8.26)$$

The convention used to define a solute-solute reduction equilibrium boundary is to draw the boundary where the activities of the two solutes are equal—in this case $a_{Fe^{2+}} = a_{Fe^{3+}}$. Expression (8.26) reduces to solute-solute reduction equilibrium boundary expression (8.27).

$$E_H = +771\ mV \qquad (8.27)$$

Figure 8.5 illustrates the Pourbaix diagram with the three boundaries (8.20), (8.24), and (8.27).

Solute $Fe^{3+}(aq)$ undergoes hydrolysis (8.28) and (8.29) as the pH increases.

$$Fe^{3+}(aq) + H_2O(l) \rightarrow Fe(OH)^{2+}(aq) + H^+(aq) \qquad K_{a1} = 10^{-2.19} \quad (8.28)$$

$$Fe(OH)^{2+}(aq) + H_2O(l) \rightarrow Fe(OH)_2^+(aq) + H^+(aq) \qquad K_{a2} = 10^{-3.50}$$

$$(8.29)$$

Solution specie $Fe(OH)^{2+}(aq)$ appearing in both (8.28) and (8.29) never reaches an activity where it accounts for 50% of the soluble Fe. Hydrolysis equilibrium involving the two dominant solute species is defined by reaction (8.30), which combines (8.28) and (8.29).

$$Fe^{3+}(aq) + 2 \cdot H_2O(l) \rightarrow Fe(OH)_2^+(aq) + 2 \cdot H^+(aq) \qquad K_{a1} \cdot K_{a2} = \beta_2 \quad (8.30)$$

The equilibrium expression for hydrolysis reaction (8.30) is given by expression (8.31).

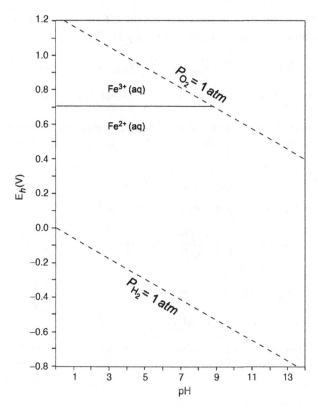

FIGURE 8.5   An incomplete Pourbaix diagram plotting the two water stability boundary lines, (8.20) and (8.24), and one solute-solute reduction equilibrium boundary (8.27). The positions of the water stability boundaries are determined by ($p_{O_2} = 1\ atm, p_{H_2} = 1\ atm$), as reported in the diagram.

$$\beta_2 = 10^{-5.69} = \frac{a_{Fe(OH)_2^+} \cdot a_{H^+}^2}{a_{Fe^{3+}}} \tag{8.31}$$

Since electron acceptor $Fe(OH)_2^+(aq)$ will exceed 50% of the total soluble Fe in the pH range > 2.84, we require a solute-solute reduction equilibrium boundary. This boundary is found by adding reduction half reaction (8.25) and hydrolysis reaction (8.30). The combined reaction is given by (8.32).

$$Fe(OH)_2^+(aq) + 2 \cdot H^+(aq) + e^- \rightarrow Fe^{2+}(aq) + 2 \cdot H_2O(l) \tag{8.32}$$

The reduction potential for this half reaction is calculated from the standard Gibbs energy ΔG_H° for reduction half reaction (8.32) using the standard Gibbs energy of formation ΔG_f° for each educt and product (8.33).[4] The source for

---

4. The standard Gibbs energy of formation ΔG_f° for $H^+(aq)$ is defined as $0.00\ kJ\ mol^{-1}$.

standard Gibbs energy of formation ΔG_f° values, formation constants, stability constants, and solubility constants is Lindsay (1979).

$$\Delta G_H^\circ = \left(\Delta G_{f,Fe^{2+}}^\circ + 2 \cdot \Delta G_{f,H_2O}^\circ\right) - \left(\Delta G_{f,Fe(OH)_2^+}^\circ + 2 \cdot \Delta G_{f,H^+}^\circ\right) \quad (8.33)$$

$$\Delta G_H^\circ = \left(\left(-91.21 \; kJ \; mol^{-1}\right) + 2 \cdot \left(-237.2 \; kJ \; mol^{-1}\right)\right) - \left(-458.7 \; kJ \; mol^{-1}\right)$$

The reduction potential—a function (8.5) of the standard Gibbs energy change ΔG_H° for the reduction half reaction (8.32)—is given in expression (8.34).

$$E_H^\circ = \frac{-\Delta G_H^\circ}{n \cdot F} = \frac{-(-106.9 \; kJ \; mol^{-1})}{1 \cdot (96.485 \; kJ \; mol^{-1}V^{-1})} = +1.108 \; V \quad (8.34)$$

The Nernst expression for this half reaction is (8.35).

$$E_H = E_H^\circ - \left(\frac{R \cdot T \cdot (\ln 10)}{n \cdot F}\right) \cdot \log_{10}\left(\frac{a_{Fe^{2+}}}{a_{Fe(OH)_2^+} \cdot a_{H^+}^2}\right)$$

$$E_H = (1108 \; mV) - (59.16 \; mV) \cdot \log_{10}\left(\frac{a_{Fe^{2+}}}{a_{Fe(OH)_2^+}}\right)$$
$$+ 2 \cdot (59.16 \; mV) \cdot \log_{10}(a_{H^+}) \quad (8.35)$$

The convention used to define solute-solute reduction equilibrium boundaries is to draw the boundary where the activities of the two solutes are equal—in this case $a_{Fe^{2+}} = a_{Fe(OH)_2^+}$. Expression (8.35) reduces to boundary expression (8.36).

$$E_H = (1108 \; mV) - (118.3 \; mV) \cdot pH \quad (8.36)$$

8.3.6. The Solute-Solute Hydrolysis Boundary

Solute hydrolysis is a common equilibrium included in Pourbaix diagrams. Solute $Fe^{3+}(aq)$ undergoes hydrolysis (8.28) through (8.30), as noted in the preceding section. Solving expression (8.31) for the proton activity results in expression (8.37). Following the convention used to derive (8.27), we will draw the solute-solute hydrolysis equilibrium boundary where the activities of the two solutes are equal—in this case $a_{Fe^{3+}} = a_{Fe(OH)_2^+}$.

Expression (8.37) reduces to boundary expression (8.38). Since this hydrolysis reaction is independent of E_H, boundary (8.38) runs perpendicular to the pH axis of our Pourbaix diagram.

$$a_{H^+}^2 = \beta_2 \cdot \left(\frac{a_{Fe^{3+}}}{a_{Fe(OH)_2^+}}\right) = (10^{-5.69}) \cdot \left(\frac{a_{Fe^{3+}}}{a_{Fe(OH)_2^+}}\right) \quad (8.37)$$

$$pH = 2.84 \quad (8.38)$$

8.3.7. The Solute-Precipitate Boundary

Solute precipitation is a common equilibrium included in Pourbaix diagrams. Solute $Fe^{3+}(aq)$ undergoes hydrolysis as the pH increases until precipitation occurs in a pH range where $Fe(OH)_2^+(aq)$ is the most abundant ferric solute (8.39). Solute-precipitate equilibrium involving the dominant specie $Fe(OH)_2^+(aq)$ is defined by reaction (8.40), which combines hydrolysis reactions (8.28) and (8.29) with solubility reaction (8.39).

$$Fe(OH)_3(s) + 3 \cdot H^+(aq) \rightarrow Fe^{3+}(aq) + 3 \cdot H_2O(l) \qquad K_{s0} \cdot K_w^{-3} = 10^{+3.54}$$
$$(8.39)$$

$$Fe(OH)_3(s) + H^+(aq) \rightarrow Fe(OH)_2^+(aq) + H_2O(l) \qquad K_{s2} \qquad (8.40)$$

$$K_{s2} = \left(\frac{a_{Fe^{3+}}}{a_{H^+}^3}\right) \cdot \left(\frac{a_{Fe(OH)^{2+}} \cdot a_{H^+}}{a_{Fe^{3+}}}\right) \cdot \left(\frac{a_{Fe(OH)_2^+} \cdot a_{H^+}}{a_{Fe(OH)^{2+}}}\right) = 10^{-2.15}$$

$$a_{H^+} = \frac{a_{Fe(OH)_2^+}}{10^{-2.15}} \qquad (8.41)$$

The position of the solute-precipitate equilibrium boundary (8.41) depends on the activity of solute $a_{Fe(OH)_2^+}$, the numerator. The selection of this value relates to the second item in Table 8.4 that must be identified before you can interpret the diagram. In this example, the total soluble iron concentration is chosen to be $C_{Fe,total} = 10^{-7.00}$ M, so $a_{Fe(OH)_2^+} = 10^{-7.00}$. The solute-precipitate equilibrium boundary for this choice is (8.42).

$$a_{H^+} = \frac{10^{-7.00}}{10^{-2.15}} = 10^{-4.85} \qquad (8.42)$$

The condition $C_{Fe,total} = 10^{-7.00}$ M (Table 8.4) is reported by labeling the Pourbaix diagram itself (Figure 8.6).

Carbon dioxide gas dissolved in water will cause solute $Fe^{2+}(aq)$ to precipitate as siderite $FeCO_3(s)$ under alkaline conditions. This solute-precipitate equilibrium boundary depends on the carbon dioxide partial pressure, as shown in reaction (8.43).

$$\underset{\text{siderite}}{FeCO_3(s)} + 2 \cdot H^+(aq) \rightarrow Fe^{2+}(aq) + CO_2(g) + H_2O(l) \qquad (8.43)$$

The equilibrium solubility expression (8.44) is rearranged to obtain expression (8.45).

$$K_{sp} = \left(\frac{a_{Fe^{2+}} \cdot p_{CO_2}}{a_{H^+}^2}\right) = 10^{+7.92} \qquad (8.44)$$

$$a_{H^+}^2 = a_{Fe^{2+}} \cdot p_{CO_2} \cdot K_{sp}^{-1}$$

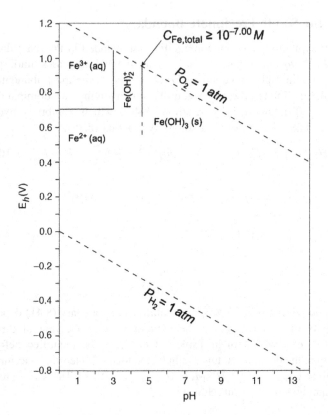

FIGURE 8.6 An incomplete Pourbaix diagram displaying two water stability boundary lines, one solute-solute reduction equilibrium boundary (8.27), a solute-solute hydrolysis equilibrium boundary (8.33), and a solute-precipitate equilibrium boundary (8.42). The total soluble iron concentration is $C_{Fe,\,total} = 10^{-7.00}\,M$.

$$\mathrm{pH} = \left(\frac{1}{2}\right) \cdot \left((-\log_{10}(a_{Fe^{2+}})) + (-\log_{10}(p_{CO_2})) + (7.92)\right) \qquad (8.45)$$

The solute-precipitate equilibrium boundary (8.46) is found by applying our earlier condition on total soluble $C_{Fe,\,total} = 10^{-7.00}\,M$ and choosing $p_{CO_2} = 10^{-3.45}\,atm$ (the atmospheric carbon dioxide level at mean sea level):

$$\mathrm{pH} = \left(\frac{1}{2}\right) \cdot ((7.00) + (3.45) + (7.92)) = 9.19 \qquad (8.46)$$

8.3.8. The Solute-Precipitate Reduction Boundary

Adding reduction half reaction (8.25) to solubility reaction (8.39) generates a reduction half reaction (8.47) where the educt is a solid—amorphous $Fe(OH)_3(s)$—and the product is the solute $Fe^{2+}(aq)$. Reduction half reactions of this type are common in environmental systems.

$$Fe(OH)_3(s) + 3 \cdot H^+(aq) + e^- \rightarrow Fe^{2+}(aq) + 3 \cdot H_2O(l) \qquad (8.47)$$

You will not find the reduction potential for (8.47) tabulated in any database but it is readily computed using the same approach applied to (8.28). The reduction potential for this half reaction is calculated from the standard Gibbs energy ΔG_H° for reduction half reaction (8.47) from the standard Gibbs energy of formation ΔG_f° for each educt and product (8.48). The standard reduction potential E_H° is given by (8.49).

$$\Delta G_H^{\circ} = \left(\Delta G_{f, Fe^{2+}}^{\circ} + 3 \cdot \Delta G_{f, H_2O}^{\circ} \right) - \left(\Delta G_{f, Fe(OH)_3(s)}^{\circ} + 3 \cdot \Delta G_{f, H^+}^{\circ} \right) \quad (8.48)$$

$$\Delta G_H^{\circ} = \left((-91.21 \; kJ \; mol^{-1}) + 3 \cdot (-237.2 \; kJ \; mol^{-1}) \right) - (-708.1 \; kJ \; mol^{-1})$$

$$E_H^{\circ} = \frac{-\Delta G_H^{\circ}}{n \cdot F} = \frac{-(-94.6 \; kJ \; mol^{-1})}{1 \cdot (96.485 \; kJ \; mol^{-1}V^{-1})} = +0.981 \, V \qquad (8.49)$$

The Nernst expression for this half reaction is (8.50). Substituting the standard reduction potential E_H° (8.49) into (8.50) and separating terms for the contributions from $Fe^{2+}(aq)$ and $H^+(aq)$ yield expression (8.51).

$$E_H = E_H^{\circ} - \left(\frac{RT \cdot (\ln 10)}{1 \cdot F} \right) \cdot \log_{10} \left(\frac{a_{Fe^{2+}}}{a_{H^+}^3} \right) \qquad (8.50)$$

$$E_H = (981 \; mV) - (59.16 \; mV) \cdot \log_{10}(a_{Fe^{2+}}) - 3 \cdot (59.16 \; mV) \cdot \text{pH} \quad (8.51)$$

Every Pourbaix diagram that includes reduction half reactions where an educt or product is a solid will result in an expression much like (8.51). The position of the solute-precipitate reduction equilibrium boundary (8.51) depends on the activity of solute $Fe^{2+}(aq)$—the second term on the right.

Earlier, when drawing the solute-precipitate equilibrium boundary (8.42), the following condition was adopted: $C_{Fe, total} = 10^{-7.00} \, M$. This condition also applies to (8.51): $a_{Fe^{2+}} = 10^{-7.00}$ or $\log_{10}(a_{Fe^{2+}}) = -7.00$. The solute-precipitate reduction equilibrium boundary for this choice is (8.52) and is plotted in Figure 8.7.

$$E_H = (981 \; mV) - (59.16 \; mV) \cdot (-7.00) - (177.5 \; mV) \cdot \text{pH}$$

$$E_H = (1395 \; mV) - (177.5 \; mV) \cdot \text{pH} \qquad (8.52)$$

8.3.9. The Precipitate-Precipitate Reduction Boundary

Reduction equilibrium can involve two solids. In this case the electron acceptor is amorphous ferric hydroxide $Fe(OH)_3(s)$, and the electron donor is the ferrous carbonate mineral siderite $FeCO_3(s)$. The reduction half reaction includes $CO_2(g)$ as an educt, making this reaction dependent on p_{CO_2}:

$$Fe(OH)_3(s) + CO_2(g) + H^+(aq) + e^- \rightarrow FeCO_3(s) + 2 \cdot H_2O(l) \quad (8.53)$$

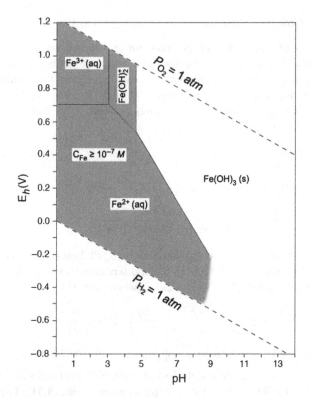

FIGURE 8.7 An incomplete Pourbaix diagram plotting the following equilibrium boundaries: solute-solute reduction equilibrium boundary (Eq. (8.27)), solute-solute hydrolysis equilibrium boundary (Eq. (8.38)), solute-precipitate equilibrium boundary (Eq. (8.42)), and solute-precipitate reduction equilibrium boundary (Eq. (8.48)). The domains where $C_{Fe, total} \geq 10^{-7.00} M$ are shaded gray.

The reduction potential for half reaction (8.53) is calculated from the standard Gibbs energy of formation ΔG_f° for each educt and product (8.54). The standard reduction potential E_H° is given by (8.55).

$$\Delta G_H^\circ = \left(\Delta G_{f, FeCO_3(s)}^\circ + 2 \cdot \Delta G_{f, H_2O}^\circ \right) \\ - \left(\Delta G_{f, Fe(OH)_3(s)}^\circ + \Delta G_{f, CO_2(g)}^\circ + \Delta G_{f, H^+}^\circ \right) \quad (8.54)$$

$$\Delta G_H^\circ = ((-677.6 \ kJ \ mol^{-1}) + 2 \cdot (-237.2 \ kJ \ mol^{-1})) \\ - ((-708.1 \ kJ \ mol^{-1}) + (-394.4 \ kJ \ mol^{-1}))$$

$$E_H^\circ = \frac{-\Delta G_H^\circ}{n \cdot F} = \frac{-(-49.43 \ kJ \ mol^{-1})}{1 \cdot (96.485 \ kJ \ mol^{-1} V^{-1})} = 0.5123 V \quad (8.55)$$

The Nernst expression for this half reaction is (8.56). Substituting the standard reduction potential E_H° (8.55) into (8.56) and separating terms for the contributions from $CO_2(g)$ and $H^+(aq)$ yield expression (8.57).

$$E_H = E_H^\circ - \left(\frac{R \cdot T \cdot (\ln 10)}{n \cdot F}\right) \cdot \log_{10}\left(\frac{1}{p_{CO_2} \cdot a_{H^+}}\right) \qquad (8.56)$$

$$E_H = (512.3\ mV) + (59.16\ mV) \cdot \log_{10}(p_{CO_2}) - (59.16\ mV) \cdot \mathrm{pH} \qquad (8.57)$$

Earlier, when drawing the solute-precipitate equilibrium boundary (8.52), the following condition was adopted: $p_{CO_2} = 10^{-3.45}\ atm$. This condition also applies to (8.57): $\log_{10}(p_{CO_2}) = -3.45$. The precipitate-precipitate reduction equilibrium boundary for this choice is (8.58).

$$E_H = (308.2\ mV) - (59.16\ mV) \cdot \mathrm{pH} \qquad (8.58)$$

8.3.10. Simple Rules for Interpreting Pourbaix Diagrams

The complete Pourbaix diagram in Figure 8.8 was earlier identified as a *Fe-O-C-H* Pourbaix diagram because these four elements are the only components displayed in the diagram (Table 8.4, item 1). Adding other components will introduce new solutes and solids, altering the appearance of the diagram. A *Fe-O-C-H* Pourbaix diagram has no value when interpreting Fe chemistry of a system containing components besides O-C-H.

The position of numerous boundary lines depends on solute concentrations and gas partial pressures: boundary (8.20) depends on p_{O_2}, boundary (8.24) depends on p_{H_2}, boundaries (8.46) and (8.58) depend on p_{CO_2}, and boundaries (8.42), (8.46), and (8.52) depend on the total soluble iron concentration $C_{Fe,total}$ (Table 8.4, item 2).

Experimental (E_H, pH) coordinates plotting to the left of boundaries (8.42), (8.46), and (8.52) encompass conditions where $C_{Fe,total} \geq 10^{-7}\ M$. Experimental (E_H, pH) coordinates plotting to the right of these same boundaries cover conditions where $C_{Fe,total} \leq 10^{-7}\ M$. Notice that domains where the solute concentration exceeds the defined limit (Figure 8.8) are labeled as solutes: $Fe^{3+}(aq)$, $Fe(OH)_2^+(aq)$, and $Fe^{2+}(aq)$. Domains where the solute concentration is less than the defined limit (Figure 8.8) are labeled as solids: $Fe(OH)_3(s)$ and $FeCO_3(s)$. This labeling convention (Table 8.4, item 3) is a reminder regarding the total solute concentration.

Each boundary represents a specific type of equilibrium (Table 8.4, item 4): solute-solute reduction, solute-solute hydrolysis, solute-precipitate reduction, and precipitate-precipitate reduction.

- Experimental (E_H, pH) coordinates plotting along solute-solute reduction boundaries indicate that the activities of the electron donor and the electron acceptor are equal.
- Experimental (E_H, pH) coordinates plotting along solute-solute hydrolysis boundaries indicate that the activities of the two solute species separated by the boundary are equal.

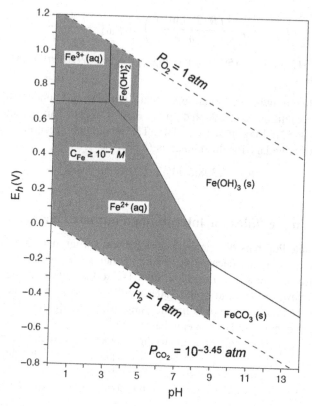

FIGURE 8.8 A complete Pourbaix diagram plotting the following equilibrium boundaries: solute-solute reduction equilibrium boundary (8.27), solute-solute hydrolysis equilibrium boundary (8.33), solute-precipitate equilibrium boundaries (8.42) and (8.52), solute-precipitate reduction equilibrium boundary (8.48), and precipitate-precipitate reduction equilibrium boundary (8.58). The domains where $C_{Fe, total} \geq 10^{-7.00} M$ are shaded gray.

- Experimental (E_H, pH) coordinates plotting along solute-precipitate boundaries indicate that the activity of the solute species is equal to the defined limit and the designated solid controls solubility. For example, (E_H, pH) coordinates along boundary (8.42) indicate that $Fe(OH)_3(s)$ is the solid controlling iron solubility, iron solubility is at the designated limit $(C_{Fe, total} = 10^{-7} M)$, and the most abundant solute is $Fe(OH)_2^+ (aq)$.
- Experimental (E_H, pH) coordinates plotting along precipitate-precipitate reduction boundaries indicate that solute activities are below the designated level and the two solids separated by the boundary coexist. When (E_H, pH) coordinates plot on either side of the boundary, one of the two solids must transform into the other by reduction or oxidation (depending on which side of the boundary the (E_H, pH) coordinate lies). For example,

(E_H, pH) coordinates along boundary (8.58) represent conditions where $Fe(OH)_3(s)$ and $FeCO_3(s)$ coexist and iron solubility is below the designated limit $(C_{Fe,total} = 10^{-7} M)$. If experimental (E_H, pH) coordinates plot on the reducing side of boundary (8.58) (i.e., below), then $Fe(OH)_3(s)$ is unstable and will transform into siderite under prevailing conditions $(p_{CO_2} = 10^{-3.45} atm)$. If experimental (E_H, pH) coordinates plot on the oxidizing side of boundary (8.58) (i.e., above), then $FeCO_3(s)$ is unstable and will transform into $Fe(OH)_3(s)$ under prevailing conditions.

8.4. MICROBIAL RESPIRATION AND ELECTRON TRANSPORT CHAINS

Section (8.3) provides the chemical basis for interpreting redox conditions in soils and the saturated zone, but it does not supply the mechanism and driving force behind the development of reducing conditions in these environments. Experimental (E_{Pt}, pH) coordinates reflect the profound impact of biological activity. Absent biological respiration, the reducing potential of natural organic matter remains blocked (see Chapter 6). Respiration, particularly microbial anaerobiosis, removes the activation barriers blocking the oxidation of organic matter and releases reduced solutes that a platinum *ORP* electrode can detect as a voltage displacement (see Figure 8.3).

Planet Earth—metallic core, mantle, and crust as a whole—is chemically reduced. Even though oxygen is the most abundant element in crustal rocks, the most abundant transition elements are not completely oxidized. The ten most abundant elements in Earth's (Table 8.5) crust account for 99.48% of the mass and, based on the listed formal oxidation numbers, 99.86% of the charge. The most probable formal oxidation number for iron is $(+2)$, indicating that the formation of the crust resulted in the partial oxidation of iron from the elemental state found in the core.

The present atmosphere (mean sea level) has the following volume percent composition: 78.084% N_2, 20.946% O_2, 0.9340% Ar, and 0.0387% CO_2. The primordial atmosphere contained an estimated 1% of present O_2 levels ($\approx 0.2\% O_2$), most probably the product of abiotic water photolysis (Berkner and Marshall, 1965). Primordial O_2 levels were believed to be sufficient for aerobic respiration as early as the beginning of the Paleozoic Era 542 million years ago. Estimated atmospheric CO_2 levels early in the Paleozoic Era were in the range 0.42–0.60% or 11- to 15-fold higher than today.

Present O_2 levels are the result of photosynthesis. The reduction half reactions for carbon dioxide (8.59) and oxygen (8.60) are combined to yield the net redox reaction for biological carbon fixation to form glucose (8.61).

$$6 \cdot CO_2 + 24 \cdot H^+ + 24 \cdot e^- \rightarrow \underset{glucose}{C_6H_{12}O_6} + 6 \cdot H_2O \quad E_H^{\circ\dagger} = -404 \, mV \quad (8.59)$$

TABLE 8.5 Abundance of Elements in Crustal Rocks of Earth and Charge Balance Based on Formal Oxidation States.

Element	Mass Fraction	$mol \cdot kg^{-1}$	Formal Oxidation Number	$mol_c \cdot kg^{-1}$	Cumulative % Mass
O	0.4550	28.439	−2	−56.877	45.50%
Si	0.2720	9.685	+4	38.739	72.70%
Al	0.0830	3.076	+3	9.229	81.00%
Fe	0.0620	1.110	+2	2.220	87.20%
Ca	0.0466	1.163	+2	2.325	91.86%
Na	0.0227	0.987	+1	0.987	94.13%
Mg	0.0276	1.137	+2	2.274	96.89%
K	0.0184	0.471	+1	0.471	98.73%
Ti	0.0063	0.132	+4	0.528	99.37%
P	0.0011	0.036	+5	0.181	99.48%

Source: Greenwood and Earnshaw, 1997.

$$6 \cdot O_2 + 24 \cdot H^+ + 24 \cdot e^- \rightarrow 6 \cdot H_2O \quad E_H^{\circ \dagger} = +814.9 \, mV \tag{8.60}$$

$$6 \cdot CO_2 + 6 \cdot H_2O \rightarrow \underset{glucose}{C_6H_{12}O_6} + 6 \cdot O_2 \quad E^{\circ \dagger} = -1219mV; \Delta G^{\circ}\dagger = +2823 \, kJ \cdot mol^{-1} \tag{8.61}$$

Carbon dioxide fixation occurs when the enzyme *rubisco* (ribulose 1,5-bisphosphate carboxylase) catalyzes the conversion (8.61) of *RuBP* (ribulose 1,5-bisphosphate), carbon dioxide, and water into *PGA* (3-phosphoglyceric acid). The energy required for this process (8.61) is provided by the capture of light energy in earlier steps of the photosynthetic process.

$$\underset{RuBP}{C_5H_8O_5(PO_4)_2} + CO_2 + H_2O \xrightarrow{rubisco} 2 \cdot \underset{PGA}{C_3H_4O_4(PO_4)} + O_2 + 2 \cdot H^+ + H_2PO_4^- \tag{8.62}$$

The chemical energy captured during photosynthesis is released during respiration—the reverse of reaction (8.61).

TABLE 8.6 Classification of Living Organisms: Carbon Source and Energy Source

| | | Primary Energy Source | |
		Light	Oxidation of Electron Donors
Carbon Source	Carbon dioxide	Photolithotrophs	Chemolithotrophs
	Organic carbon	Photoorganotrophs	Chemoorganotrophs

Table 8.6 lists a classification of living organisms. Plants and other photosynthetic organisms are classified as *photolithotrophs* that assimilate carbon dioxide. *Chemolithotrophic bacteria* assimilate carbon dioxide by oxidizing a variety of inorganic electron donors (ferrous iron Fe^{2+}, thiosulfate $S_2O_3^{2-}$, elemental sulfur, ammonium NH_4^+, nitrite NO_2^-, and dihydrogen H_2). Chemolithotrophic bacteria make an insignificant contribution to net carbon dioxide assimilation but play a prominent role in environmental redox cycling.

Organotrophic organisms cannot assimilate carbon dioxide but instead assimilate organic compounds produced by lithotrophic organisms. Chemoorganotrophs—all eukaryotic organisms that are not photolithotrophs and most prokaryotes—derive their energy from the oxidation of organic compounds.

When lithotrophs assimilate carbon dioxide, they generate small organic molecules used to synthesize complex biomolecules that comprise living biomass while simultaneously capturing the chemical energy necessary for anabolic (i.e., biosynthetic) processes. Both lithotrophs and organotrophs release stored chemical energy by catabolic pathways that disassemble large molecules into fragments and oxidize the fragments.

8.4.1. Catabolism and Respiration

The release of chemical energy stored in organic compounds occurs in two stages: catabolism and respiration. Reduction leading to carbon dioxide release occurs during catabolism: the electron donor is the organic compound being catabolized, while the electron acceptor is the oxidized form of *nicotinamide adenine dinucleotide NAD$^+$* (CAS 53-84-9, Structure 3) or a similar compound (e.g., flavin adenine dinucleotide FAD or nicotinamide adenine dinucleotide phosphate $NADP^+$).

Structure 3. NAD^+

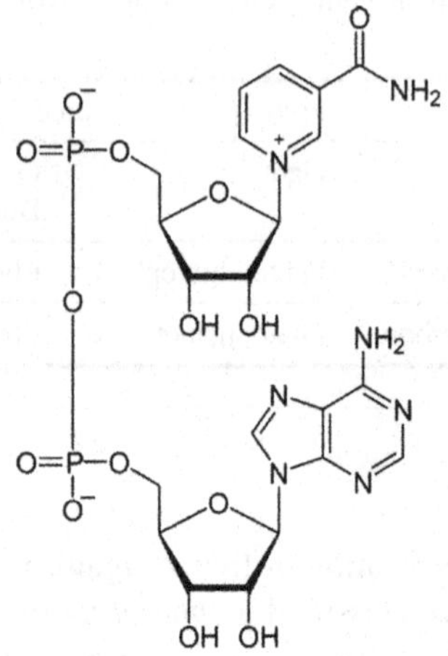

Structure 4. NADH

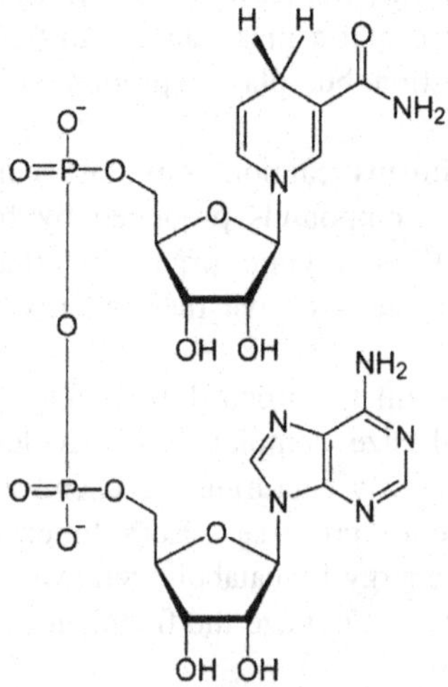

$$NAD^+ + H^+ + 2 \cdot e^- \rightarrow NADH \qquad (8.63)$$

Catabolism reduces NAD^+ to $NADH$ (CAS 606-68-8, Structure 4) in reduction half reaction (8.63). For example, glucose catabolism involves three processes: glycolysis, pyruvate oxidation, and the tricarboxylic acid cycle. Reduction half reaction (8.64) represents glucose catabolism by this pathway.

$$6 \cdot CO_2 + 24 \cdot H^+ + 24 \cdot e^- \rightarrow \underset{glucose}{C_6H_{12}O_6} + 6 \cdot H_2O \qquad (8.64)$$

Combining reduction half reactions (8.63) and (8.64) illustrates glucose dissimilation to carbon dioxide (8.65). Notice that the electron acceptor is NAD^+

(8.63), not O_2: the electron transport chain, and not the catabolic pathway, defines the terminal electron acceptor.

$$\underset{glucose}{C_6H_{12}O_6} + 12 \cdot \underset{\substack{nicotinamide \\ adenine \\ dinucleotide}}{NAD^+} + 6 \cdot H_2O \rightarrow 6 \cdot CO_2 + 12 \cdot NADH + 12 \cdot H^+ \quad (8.65)$$

Reduced nicotinamide adenine dinucleotide *NADH* functions as a redox cofactor. Reduced *NADH* is an electron carrier, shuttling electrons from various catabolic pathways to the electron transport chain, and functions as the electron donor feeding electrons into the electron transport chain.

While all organisms employ numerous catabolic pathways, bacteria have evolved a broader array of catabolic pathways than other organisms, making them important players in the environmental degradation of organic contaminants. Bacteria also draw upon a more flexible electron transport chain, providing them with respiratory alternatives absent in eukaryotes. Many bacteria continue to respire in the absence of O_2—a process called anaerobic respiration—by utilizing a variety of terminal electron acceptors as the need arises.

Anaerobiosis eliminates the activation barrier impeding the abiotic oxidation of nitrate and sulfate through the specialized electron transport chains of nitrate- and sulfate-reducing bacteria. While platinum ORP electrodes may have limited sensitivity to these redox couples, some of the reduction products (ferrous iron, thiosulfate, etc.) are sufficiently electroactive and soluble to yield measureable platinum ORP electrode voltage displacements.

8.4.2. Electron Transport Chains

Aerobic respiration uses an electron transport chain whose electron donor is *NADH* (8.62) and whose terminal electron acceptor is O_2 (8.65).

$$O_2 + 4 \cdot H^+ + 4 \cdot e^- \rightarrow H_2O \quad (8.66)$$

The net redox reaction in aerobic respiration (8.67) combines reduction half reactions for the electron donor (8.63) and the terminal electron acceptor (8.66).

$$6 \cdot O_2 + 12 \cdot NADH + 12 \cdot H^+ \rightarrow 12 \cdot H_2O + 12 \cdot NAD^+ \quad (8.67)$$

Aerobic glucose dissimilation (8.68) combines glucose dissimilation (8.65) with aerobic respiration (8.67). The $NAD^+|NADH$ couple (8.63) does not appear in (8.68) because it is recycled between the catabolic pathway and electron transport chain.

$$\underset{glucose}{C_6H_{12}O_6}\,(aq) + 6 \cdot O_2(g) \rightarrow 6 \cdot CO_2(aq) + 6 \cdot H_2O(l) \quad (8.68)$$

The glucose–to–carbon dioxide reduction half reaction (8.64) does not list a reduction potential. It would be impossible to construct a galvanic cell similar to the one shown in Figure 8.2 to measure this potential, but we can calculate the reduction potential using Gibbs energy expression (8.5).

We begin with (8.69)—an expression you will recognize from general chemistry—the change in Gibbs energy for reduction half reaction (8.64) under standard conditions ΔG_H° being the sum of the standard Gibbs energy of formation for each product minus the sum of the standard Gibbs energy of formation for each educt (Amend and Plyasunov, 2001; Amend and Shock, 2001).

$$\Delta G_H^\circ = \left(\Delta G_{f,glucose}^\circ + 6 \cdot \Delta G_{f,water}^\circ\right) - \left(6 \cdot \Delta G_{f,CO_2}^\circ + 24 \cdot \Delta G_{f,H^+}^\circ\right) \quad (8.69)$$

$$\Delta G_H^\circ = ((-915.90 \, kJ \cdot mol^{-1}) + 6 \cdot (-237.18 \, kJ \cdot mol^{-1}))$$
$$- (6 \cdot (-385.98 \, kJ \cdot mol^{-1}) + 24 \cdot (0.0 \, kJ \cdot mol^{-1}))$$

$$\Delta G_H^\circ = -23.100 \, kJ \cdot mol^{-1} \quad (8.70)$$

The change in Gibbs energy under standard conditions ΔG_H° for the glucose–to–carbon dioxide reduction half reaction (8.64) is (8.70). Applying (8.5) to (8.70) yields the glucose–to–carbon dioxide half reaction standard reduction potential (8.71).

$$E_H^\circ = \frac{-\Delta G_H^\circ}{n \cdot F}$$

$$E_H^\circ = \left(\frac{2.3100 \cdot 10^4 \, J \cdot mol^{-1}}{24 \cdot (9.6485 \cdot 10^4) \, J \cdot V^{-1} \cdot mol^{-1}}\right) = +9.976 \, mV \quad (8.71)$$

The glucose–to–carbon dioxide half-reaction *standard biological* reduction potential (8.72) requires expression (8.12).

$$E_H^{\circ\dagger} = (+9.976 \, mV) - \left(\frac{24}{24}\right) \cdot (+414.1 \, mV) = -404.1 \, mV \quad (8.72)$$

The calculation illustrated in expressions (8.69) through (8.72) is quite general. We used this approach earlier in this chapter to demonstrate the preparation and interpretation of redox stability diagrams.

Aerobic Electron Transport Chains

The reduction half reactions listed in Table 8.7 are components of the electron transport chain typical of aerobic bacteria and prokaryotes. The electron donor for the sequence is typically *NADH* or *NADPH*, and the terminal electron acceptor is O_2.

The reactive center in iron sulfur proteins and the cytochromes—*b*, *c*, *a*, and a_3—is ferrous iron in iron sulfide clusters and heme groups, respectively. Subtle changes in the iron bonding environment determine the reduction potential $E_H^{\circ\dagger}$ of each species.

TABLE 8.7 Standard Biological Reduction Half Reactions of the Aerobic Electron Transport Chain

Reduction Half Reaction	Reduction Potential, $E_H^{\circ\dagger}$, mV
$NAD^+ + H^+ + 2 \cdot e^- \rightarrow NADH$	-320
$NADP^+ + H^+ + 2 \cdot e^- \rightarrow NADPH$	-324
$FMN + 2 \cdot H^+ + 2 \cdot e^- \rightarrow FMNH_2$	-212
$Fe_2^{3+} S_2 + 2 \cdot e^- \rightarrow Fe_2^{2+} S_2$ _P.putidia_ $\qquad$ _P.putidia_	-30
cytochrome $b + e^- \rightarrow$ cytochrome b $\quad$ oxidized $\qquad\qquad$ reduced	$+75$
ubiquinone $+ 2 \cdot H^+ + 2 \cdot e^- \rightarrow$ ubiquinol	$+100$
cytochrome $c + e^- \rightarrow$ cytochrome c $\quad$ oxidized $\qquad\qquad$ reduced	$+235$
cytochrome $a + e^- \rightarrow$ cytochrome a $\quad$ oxidized $\qquad\qquad$ reduced	$+290$
cytochrome $a_3 + e^- \rightarrow$ cytochrome a_3 $\quad$ oxidized $\qquad\qquad$ reduced	$+305$
$O_2 + 4 \cdot H^+ + 4 \cdot e^- \rightarrow H_2O$	$+815$

The half reactions listed in Table 8.7 combine to yield the redox reaction sequence of the aerobic electron transport chain, where _NADH_ (or _NADPH_) serves as the electron donor (Table 8.8). These reactions release the chemical energy carried to the electron transport chain by the electron donor in relatively small increments.

Figure 8.9 illustrates the aerobic electron transport chain as a simplified redox cascade. Electrons enter the redox cascade from the electron donor _NADH_ and release energy as they cascade over each reduction step until they reach the terminal electron acceptor, which defines the terminus of the cascade. The floor of the redox cascade representing the aerobic electron transport chain is half reaction for reduction of O_2 to water (8.66).

Although the reduction potential of each step in the redox cascade is more positive than the preceding step, using the Gibbs energy expression (8.5) requires each step in the cascade to lie at a more negative Gibbs energy than the preceding step. The complete redox reaction represented by the aerobic electron transport chain in Figure 8.9 is reaction (8.67), and the total Gibbs energy $\Delta G^{\circ\dagger}$ change associated with the complete cascade of electrons from _NADH_ to O_2 in (8.67) is $-435.8\ kJ \cdot mol^{-1}$.

Chemolithotrophic eukaryotes utilize a variety of inorganic electron donors (Table 8.9) as a source of chemical energy—ferrous iron Fe^{2+}, thiosulfate $S_2O_3^{2-}$, elemental sulfur, ammonium NH_4^+, nitrite NO_2^-, and H_2. These

TABLE 8.8 Reduction Reactions of the Aerobic Electron Transport Chain

$NADH + H^+ + FMN \xrightarrow{1} NAD^+ + FMNH_2$

$FMNH_2 + 2 \cdot [Fe_4S_4]^{2+} \xrightarrow{2} FMN + 2 \cdot [Fe_4S_4]^+ + 2 \cdot H^+$

$2 \cdot [Fe_4S_4]^+ + 2 \cdot H^+ + Q \xrightarrow{3} 2 \cdot [Fe_4S_4]^{2+} + QH_2$

$QH_2 + 2 \cdot$ cytochrome $b \xrightarrow{4} Q + 2 \cdot H^+ + 2 \cdot$ cytochrome b
　　　　　　　　oxidized　　　　　　　　　　　　　　　reduced

$2 \cdot$ cytochrome $b + 2 \cdot$ cytochrome $c \xrightarrow{5} 2 \cdot$ cytochrome $b + 2 \cdot$ cytochrome c
　　reduced　　　　　　oxidized　　　　　　　　oxidized　　　　　　reduced

$2 \cdot$ cytochrome $c + 2 \cdot$ cytochrome $a \xrightarrow{6} 2 \cdot$ cytochrome $c + 2 \cdot$ cytochrome a
　　reduced　　　　　　oxidized　　　　　　　　oxidized　　　　　　reduced

$2 \cdot$ cytochrome $a + 2 \cdot$ cytochrome $a_3 \xrightarrow{7} 2 \cdot$ cytochrome $a + 2 \cdot$ cytochrome a_3
　　reduced　　　　　　oxidized　　　　　　　　oxidized　　　　　　reduced

$4 \cdot$ cytochrome $a_3 + 4 \cdot H^+ + O_2 \xrightarrow{8} 4 \cdot$ cytochrome $a_3 + 2 \cdot H_2O$
　　reduced　　　　　　　　　　　　　　　　　oxidized

Symbols: flavin mononucleotide *FMN*, iron-sulfur protein $[Fe_4S_4]$, ubiquinone Q.

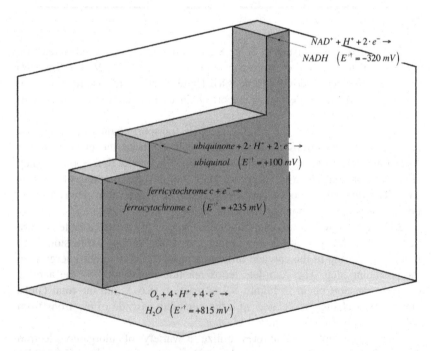

FIGURE 8.9 The simplified redox cascade for the aerobic electron transport chain begins with the electron donor *NADH* and ends with the terminal electron acceptor O_2.

TABLE 8.9 Standard Biological Reduction Half Reactions of Electron Donors Used by Chemolithotrophic Bacteria

Reduction Half Reaction	Reduction Potential, $E_H^{+\circ}$, mV
$2 \cdot H^+(aq) + 2 \cdot e^- \rightarrow H_2(g)$	-414
$2 \cdot SO_4^{2-}(aq) + 10 \cdot H^+(aq) + 8 \cdot e^- \rightarrow S_2O_3^{2-}(aq) + 5 \cdot H_2O(l)$	-233
$SO_4^{2-}(aq) + 8 \cdot H^+(aq) + 6 \cdot e^- \rightarrow S(s) + 4 \cdot H_2O(l)$	-199
$NO_3^-(aq) + 2 \cdot H^+(aq) + 2 \cdot e^- \rightarrow NO_2^-(aq) + H_2O(l)$	$+407$
$NO_3^- + 10 \cdot H^+ + 8 \cdot e^- \rightarrow NH_4^+ + 3 \cdot H_2O$	$+363$
$Fe^{3+}(aq) + e^- \rightarrow Fe^{2+}(aq)$	$+771$
$O_2(g) + 4 \cdot H^+(aq) + 4 \cdot e^- \rightarrow 2 \cdot H_2O(l)$	$+815$

inorganic electron donors feed electrons directly into the aerobic electron transport chain instead of relying on the catabolism of organic compounds for chemical energy. Each inorganic electron donor couples to the electron transport chain at one of three points, defined by the reduction half reaction potential of the electron donor and the electron chain half reaction (see Table 8.7). The highest entry point uses NAD^+ or $NADP^+$ as electron acceptors (reaction 1, Table 8.8). The next entry point uses ubiquinone as the electron acceptor (reaction 3, Table 8.8). The lowest level, the entry point with the least energy recovery, uses ferrous cytochrome c as the electron acceptor (reaction 5, Table 8.8). These levels correspond to successively more positive redox potentials and, therefore, decreased potential differences relative to the O_2, the terminal electron acceptor. The redox cascades are similar to the one in Figure 8.9, all with the same floor, but each begins at progressively lower entry points and releases less energy.

Facultative and Obligate Anaerobes

Reaction 4 in Table 8.8 (ubiquinol reduction to ubiquinone) in the aerobic electron transport chain is particularly problematic for aerobic organisms because electrons divert prematurely to O_2 at this stage. The normal electron transfer process begins with the oxidation of ubiquinol QH_2, delivering one electron to two different one-electron acceptors, one being cytochrome b (see Table 8.7) in a bifurcated pathway. Disruption occurs when the one-electron transfer to cytochrome b fails, leaving an unstable *ubisemiquinone radical* $Q^{\bullet-}$. The ubisemiquinone radical $Q^{\bullet-}$ reacts with O_2 to form *superoxide anion radical* $O_2^{\bullet-}$.

The occurrence of this electron transport disruption has resulted in the evolution of two key enzymes in aerobic organisms: *superoxide dismutase*—an enzyme that swiftly and safely converts the superoxide anion radical

$O_2^{\bullet-}$ into hydrogen peroxide (8.73)—and *catalase*—an enzyme that decomposes hydrogen peroxide into water and O_2 (8.74).

$$2 \cdot O_2^{\bullet-} + 2 \cdot H^+ \xrightarrow[\text{dismutase}]{\text{superoxide}} O_2 + H_2O_2 \qquad (8.73)$$

$$2 \cdot H_2O_2 \xrightarrow[\text{catalase}]{} O_2 + 2 \cdot H_2O \qquad (8.74)$$

Facultative anaerobic bacteria possess superoxide dismutase, enabling them to tolerate disruption of the aerobic electron transport chain in reaction (8.4) by decomposing the reactive oxygen species it produces. *Obligate anaerobic* bacteria lack superoxide dismutase and, as a consequence, cannot tolerate significant levels of O_2.

Fermentative Anaerobic Bacteria

One anaerobic respiratory process is *fermentation*. Fermentation, unlike other anaerobic respiratory processes, does not use the electron transport chain. Organisms capable of fermentation—certain fungi and many bacteria—are facultative anaerobes. During fermentation the electron transport chain is inactive; oxidation occurs during catabolism.

Structure 5. Pyruvic acid

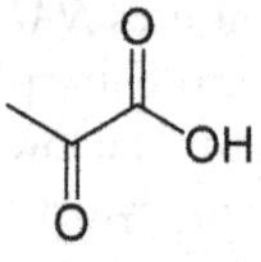

Structure 6. Lactic acid

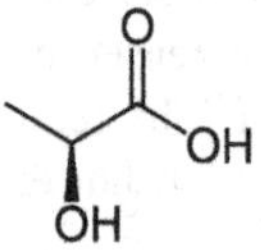

The first stage of glucose catabolism (glycolysis) produces two molecules of pyruvic acid ($CH_3(CO)COOH$, CAS 127-17-3, Structure 5). Lactic acid fermentation (8.75) uses *NADH* as an electron donor and pyruvic acid as an electron acceptor to produce lactic acid ($CH_3(COH)COOH$, CAS 598-82-3, Structure 6).

$$2 \cdot CH_3(CO)COOH + NADH + H^+ \rightarrow 2 \cdot CH_3(COH)COOH + NAD^+$$
$$(8.75)$$

Alcohol (ethanol) fermentation (8.76) uses *NADH* as an electron donor and pyruvic acid as an electron donor to produce ethanol (CH_3CH_2OH, CAS 64-17-5). Pyruvic acid, however, does not appear in (8.76), and the electron donor appears to be acetaldehyde ($CH_3(CO)H$, CAS 75-07-0).

$$CH_3(CO)H + NADH + H^+ \rightarrow CH_3CH_2OH + NAD^+ \qquad (8.76)$$

Pyruvic acid is the true electron donor in alcohol fermentation but its transformation to *acetealdehyde* by *pyruvate decarboxylase* (8.77) cannot be classified as a redox or electron transfer reaction.

$$CH_3(CO)COOH \rightarrow CH_3(CO)H + CO_2 \qquad (8.77)$$

Bacteria use other fermentation pathways to produce many of the low molecular weight organic acids found in the pore water of soils and sediments.

Nitrate Respiratory Chain of Nitrate-Reducing Bacteria

The electron transport chain of nitrate-reducing bacteria utilizes most of the aerobic electron transport chain, diverting electrons from cytochrome c to nitrate and eliminating reactions 6 through 8 in Table 8.8. The use of nitrate as the terminal electron acceptor, however, requires an enzyme system that transfers electrons from cytochrome c to nitrate. The reduction of nitrate to N_2 involves the transfer of five electrons per nitrogen. Since biological redox reactions typically are one- and two-electron reductions, the complete reduction of nitrate to N_2 cannot occur in a single redox reaction (Table 8.10).

Figure 8.10 illustrates the components of the nitrate reduction system in nitrate-reducing bacteria. The first step—the reduction of nitrate to nitrite—is catalyzed by a nitrate reductase *NAR* bound to plasma membrane. The nitrite reductase *NIR*—reducing nitrite to nitric oxide—is located in the *periplasm* because nitric oxide *NO* is cytotoxic. The remaining steps occur in the periplasm: nitric oxide reduction to nitrous oxide N_2O and nitrous oxide reduction to N_2.

Figure 8.11 illustrates the anaerobic electron transport chain of nitrate-reducing bacteria as a simplified redox cascade.[5] Electrons enter the redox

TABLE 8.10 Standard Biological Reduction Half Reactions of the Nitrate Respiratory Chain

Reduction Half Reaction	Reduction Potential, $E°†, mV$
$NO_3^- + 2 \cdot H^+ + 2 \cdot e^- \rightarrow NO_2^- + H_2O$	+407
$NO_2^-(aq) + 2 \cdot H^+(aq) + e^- \rightarrow NO(aq) + H_2O(l)$	+238
$2 \cdot NO(aq) + 2 \cdot H^+(aq) + 2 \cdot e^- \rightarrow N_2O(aq) + H_2O(l)$	+1285
$N_2O(aq) + 2 \cdot H^+(aq) + 2 \cdot e^- \rightarrow N_2(g) + H_2O(l)$	+1403
$2 \cdot NO_3^-(aq) + 12 \cdot H^+(aq) + 10 \cdot e^- \rightarrow N_2(g) + 6 \cdot H_2O(l)$	+748

5. The aerobic redox cascade (Figure 8.9) is used in Figures 8.11 and 8.12 to illustrate the submergence of the electron transport chain when alternative electron acceptors replace O_2.

Gram-negative Bacteria

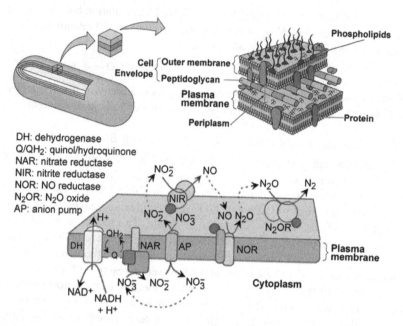

DH: dehydrogenase
Q/QH₂: quinol/hydroquinone
NAR: nitrate reductase
NIR: nitrite reductase
NOR: NO reductase
N₂OR: N₂O oxide
AP: anion pump

FIGURE 8.10 Components of the nitrate reduction system of nitrate-reducing bacteria are located in the inner or plasma membrane and the periplasm. *Source: Reproduced with permission from Zumft, W.G., 1997. Cell biology and molecular basis of denitrification. Microbiol. Mol. Biol. Rev. 61 (4), 533–616*

cascade from the electron donor *NADH* and release energy as they cascade to the level set by the electron acceptor, defined in this example as the half reaction for reduction of nitrate to nitrite. The complete redox reaction represented by the aerobic electron transport chain in Figure 8.11 is reaction (8.78), and the total Gibbs energy change $\Delta G^{\circ\dagger}$ associated with the complete cascade of electrons from *NADH* to O_2 is $-217.9 \ kJ \cdot mol^{-1}$ per *NADH*. The submerged portion of the electron transport chain makes no contribution to the Gibbs energy change $\Delta G^{\circ\dagger}$ because nitrate terminates the cascade at a higher level than aerobic respiration (see Figure 8.9). The Gibbs energy change $\Delta G^{\circ\dagger}$ for electron transfer from *NADH* to N_2 is $-205.0 \ kJ \cdot mol^{-1}$ per *NADH*.

$$2 \cdot NO_3^- + 5 \cdot NADH + 7 \cdot H^+ \rightarrow N_2 + 6 \cdot H_2O + 5 \cdot NAD^+ \qquad (8.78)$$

Ferric Iron Respiratory Chain of Metal-Reducing Bacteria

Most iron-reducing bacteria are capable of using several terminal acceptors besides ferric iron (Ruebush, Brantley et al., 2006), but their distinguishing characteristic is their ability to grow under anaerobic conditions where the

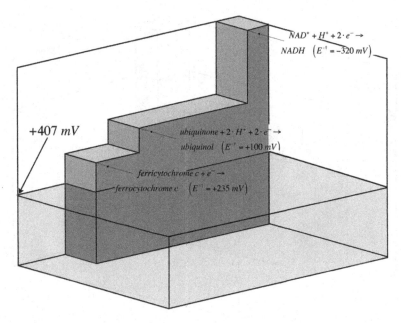

$$NAD^+ + \underline{H}^+ + 2 \cdot e^- \rightarrow$$
$$NADH \quad \left(E^{-1} = -320\,mV\right)$$

$$+407\,mV$$

$$ubiquinone + 2 \cdot H^+ + 2 \cdot e^- \rightarrow$$
$$ubiquinol \quad \left(E^{-1} = +100\,mV\right)$$

$$ferricytochrome\,c + e^- \rightarrow$$
$$ferrocytochrome\,c \quad \left(E^{-1} = +235\,mV\right)$$

FIGURE 8.11 The simplified redox cascade for the anaerobic electron transport chain in nitrate-reducing bacteria begins with electron donor *NADH* and ends with the electron acceptor nitrate.

sole electron acceptor is either *Fe(III)* or *Mn(IV)*. The relative natural abundance of iron and manganese makes *Fe(III)* a more abundant terminal electron acceptor in the environment, hence the use of *iron respiratory chain* and *iron-reducing* bacteria.

The iron respiratory chain faces a singular challenge: the terminal electron acceptor is insoluble and cannot be taken up into the cytoplasm like nitrate, sulfate, or carbonate. There is considerable evidence for membrane-bound terminal ferric reductase enzymes (Lovley, 1993) that overcome the need for a soluble terminal electron acceptor, relying on direct contact between cell surface and insoluble ferric minerals. Some studies (Lovley, Holmes et al., 2004) suggest that some iron-reducing species—*Shewanella* and *Geothrix*—may secrete soluble electron shuttles capable of reducing ferric oxide minerals without coming into direct physical contact. Microbiologists studying the iron respiratory chain have identified several components of an electron transport chain that couples to outer-membrane-bound terminal ferric reductase enzymes. Identification of soluble electron shuttles, however, remains elusive.

One of the more unusual features of the terminal ferric reductase is its location. The position of reductase enzymes in the nitrate respiratory chain (see Figure 8.10) in the inner or cytoplasmic membrane and the periplasm—the location of the cell wall of Gram negative bacteria separating the inner and outer membranes—also applies to the reductase enzymes of

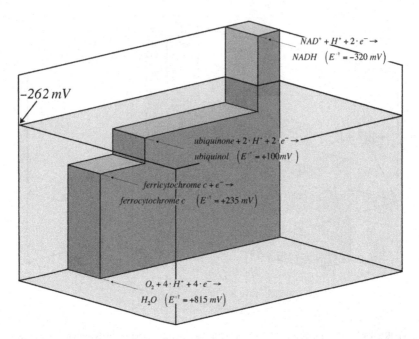

FIGURE 8.12 The simplified redox ladder for the anaerobic electron transport chain in iron-reducing bacteria begins with electron donor *NADH* and ends with the electron acceptor ferric iron in mineral form $Fe(OH)_3(s)$.

the sulfate and carbonate respiratory chains. The placement of the terminal ferric reductase, however, is on the outer surface of the outer membrane (DiChristina, Moore et al., 2002; Myers and Myers, 2002).

Figure 8.12 shows the simplified redox cascade for the iron respiratory chain. The level in the pool has risen to $-262\ mV$, submerging much of the aerobic redox cascade and limiting the recovery of chemical energy in electron donor *NADH*.

The complete redox reaction represented by the aerobic electron transport chain in Figure 8.12 is reaction (8.79), and the total Gibbs energy change $\Delta G^{\circ\dagger}$ associated with the complete cascade of electrons from *NADH* to Fe^{2+} is $-46.51\ kJ \cdot mol^{-1}$ per *NADH*. The submerged portion of the electron transport chain makes no contribution to the Gibbs energy $\Delta G^{\circ\dagger}$ change because $Fe(OH)_3(s)$ sets the floor of the cascade at a higher level.

$$2 \cdot Fe(OH)_3(s) + NADH + 5 \cdot H^+ \rightarrow 2 \cdot Fe^{2+} + NAD^+ + 6 \cdot H_2O \quad (8.79)$$

Sulfate Respiratory Chain of Sulfate-Reducing Bacteria

Table 8.11 lists reduction half reactions and potentials of what is considered the likely reduction pathway from sulfate SO_4^{2-}—the terminal electron acceptor of

---

TABLE 8.11 Standard Biological Reduction Half Reactions of the Sulfate Respiratory Chain

Reduction Half Reaction	Reduction Potential, $E°†$, mV
$APS + H^+ + 2 \cdot e^- \rightarrow HSO_3^- + AMP$	-60
$HSO_3^- + 6 \cdot H^+ + 6 \cdot e^- \rightarrow HS^- + 3 \cdot H_2O$ hydrogen sulfite hydrogen sulfide	-76
$APS + 7 \cdot H^+ + 8 \cdot e^- \rightarrow HS^- + 3 \cdot H_2O + AMP$	-72

---

the sulfate respiratory chain—to hydrogen sulfide HS^-—the final product of dissimilatory sulfate reduction. Sulfate-reducing bacteria are obligate anaerobes, unlike the facultative anaerobes: iron-reducing and nitrate-reducing bacteria.

The reduction half reaction of the electron donor for this respiratory chain $NADH$ is -320 mV (see Table 8.7), which represents a significant hurdle: the Gibbs energy change for sulfate reduction by NADH is *positive* and, therefore, the reaction would absorb rather than releases energy.

$$NADH + SO_4^{2-}(aq) + 2 \cdot H^+(aq) = NAD^+ + HSO_3^-(aq) + H_2O(aq)$$
$$\Delta G°† = +38.76 \ kJ \cdot mol^{-1} \tag{8.80}$$

All bacteria capable of growth when the sole electron acceptor is sulfate SO_4^{2-} activate sulfate for assimilatory and dissimilatory reduction by forming 5′-adenylyl sulfate ($C_{10}H_{14}N_5O_{10}PS$, CAS 485-84-7, Structure 7) from adenosine-5′-triphosphate ATP and sulfate in reaction (8.81), coupled to the hydrolysis of a high-energy phosphate ester bond (8.82) in guanosine-5′-triphosphate GTP (Liu, Ya Suo et al., 1994).

$$ATP^{3-} + SO_4^{2-} \xrightarrow[\text{ATP sulfurylase}]{} APS^{2-} + \underset{\text{pyrophosphate}}{HP_2O_7^{3-}} \tag{8.81}$$

$$GTP^{3-} + H_2O \rightarrow GDP^{2-} + H_2PO_4^- \tag{8.82}$$

Structure 7. 5′-adenylyl sulfate *APS*

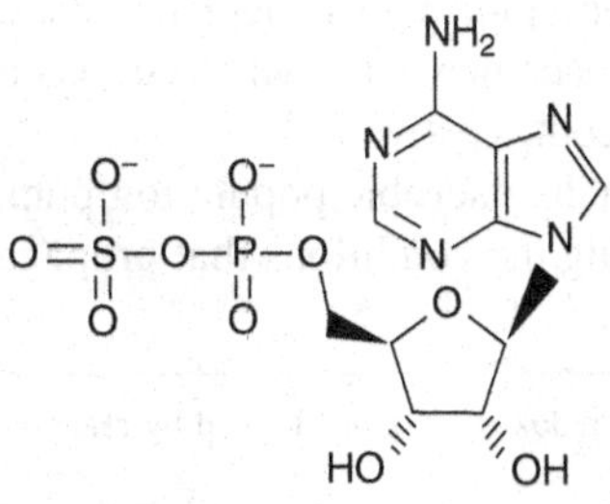

Activation of the sulfate lowers the potential for the reduction of sulfate to hydrogen sulfite to −60 mV (see Table 8.11) and allows the remaining reduction steps to follow. The Gibbs energy change $\Delta G^{\circ\dagger}$ for electron transfer from $NADH$ to $HS^-(aq)$ is −22.92 $kJ \cdot mol^{-1}$ per $NADH$ (accounting for the energy cost of activation).

Methanogenic Anaerobic Archaea

There are no carbonate-reducing bacteria; methanogenic organisms belong to the domain *archaea* (Thauer, 1998). Bacterial fermentation supports methanogenesis by converting organic matter into acetate CH_3COO^-, dihydrogen H_2 and either formate $HCOO^-$ or bicarbonate HCO_3^-. Methanoarchaea derive metabolic energy by either disproportionating[6] one-carbon compounds (e.g., reactions (8.83) through (8.85)) or using dihydrogen H_2 to reduce bicarbonate HCO_3^- (8.14).

$$CH_3COO^-(aq) + H^+(aq) \rightarrow CO_2(aq) + CH_4(g) \qquad (8.83)$$

$$4 \cdot HCOO^-(aq) + 4 \cdot H^+(aq) \rightarrow 3 \cdot CO_2(aq) + CH_4(g) + 2 \cdot H_2O(l) \quad (8.84)$$

$$4 \cdot CH_3COH(aq) \rightarrow CO_2(aq) + 3 \cdot CH_4(g) + 2 \cdot H_2O(l) \qquad (8.85)$$

Other carbon compounds used by methanoarchaea include carbon monoxide, methylamines $N(CH_3)_n[1 \leq n \leq 3]$, and methylsulfide $HSCH_3$.

8.4.3. Environmental Redox Conditions and Microbial Respiration

The energy yield from each of these respiratory chains is progressively less. As a result, metabolic activity by facultative anaerobic bacteria is lower than that of aerobic bacteria; the activity of obligate anaerobic bacteria is the lowest of all. Reducing conditions in the environment are usually the result of anaerobic microbial respiration. Biologically generated reducing conditions develop in water-saturated soil and sediments when there is abundance of carbon, a suitable terminal electron acceptor, and temperatures favoring biological activity.

Anoxia develops as aerobic organisms deplete pore water of dissolved O_2, setting the stage for the sequential rise and decline in the activity of nitrate-reducing bacteria, followed by iron-reducing, sulfate-reducing, and, ultimately, methanoarchaea. Each succeeding population drives the reducing potential lower as each electron acceptor pool is depleted and a new terminal electron acceptor is sought. With each new terminal electron acceptor, a new population of anaerobic bacteria becomes active.

The respiration of each anaerobic population pumps electron donors into the environment, reducing the conditions that propagate throughout the zone

---

6. Disproportionation is not a redox reaction as defined by reaction (8.1).

of biological activity as abiotic electron transfer reactions reduce compounds that bacteria do not utilize for respiration. Percolating water carries the electron donors generated by anaerobic bacteria—and those generated by abiotic electron transfer reactions—beyond the zone of biological activity into aquifers, where carbon levels are too low to sustain appreciable metabolism.

The fate of nitrate in groundwater illustrates the role of anaerobic respiration on nitrate levels. Nitrate readily leaches through soil because ion exchange capacity of soils is overwhelmingly cation exchange capacity. Denitrifying bacteria in water-saturated soil reduce nitrate, but the likelihood of that happening drops dramatically once the nitrate leaches into the subsoil and underlying surficial materials, where organic carbon levels and biological activity tend to be quite low. Nitrate in the groundwater persists until it enters a surface water recharge (*hyporheic*) zone in a wetland or sediment zone of a stream or lake. Biological activity in these zones is much higher because these zones are carbon rich. Dissolved nitrate rapidly disappears as nitrate-reducing bacteria in the recharge zone utilize the nitrate to metabolize available carbon.

8.5. SUMMARY

This chapter began with a review of redox principles from general chemistry: formal oxidation numbers, balancing redox and reduction half reactions, and the Nernst equation. These are the basic tools needed to calculate standard biological (or environmental) reduction potentials $E_H^{\circ\dagger}$ and to understand biological electron transfer reactions.

Section 8.3 focused on the measurement and interpretation of redox conditions by plotting experimental (E_{red}, pH) coordinates on Pourbaix diagrams. Table 8.4 summarizes the essential features common to Pourbaix diagrams and the foundation for interpreting reducing conditions using (E_{red}, pH) coordinates.

Reducing conditions in the environment result from a key biological process: respiration (Baas-Becking, Kaplan et al., 1960). Section 8.4 explained how catabolism and electron transport are fundamentally chemical reduction processes that release the chemical energy stored in bioorganic compounds. Numerous catabolic pathways feed electrons into electron transport chains that harvest the chemical energy released during catabolism.

Since biological activity requires an energy (i.e., carbon) source and suitable temperatures, reducing conditions develop when and where there is a significant biological activity. Anaerobic respiration effectively pumps reactive electron donors into the environment, depending on the type of anaerobic respiration. These electron donors—especially thiosulfate, hydrogen sulfide, and ferrous iron—transfer electrons to a host of electron acceptors through abiotic reduction reactions. The mobility and biological availability of many inorganic contaminants are sensitive to prevailing redox conditions. Solubility increases under reducing conditions for some contaminants and decreases for others. Regardless, prevailing redox conditions determine environmental risk for many contaminants.

APPENDIX 8A. ASSIGNING FORMAL OXIDATION NUMBERS

Example 8A.1

The formal oxidation numbers of carbon atoms in phenol (CAS 108-95-2, Structure 8) distinguish between carbon C_1 bonded to oxygen and the remaining five carbon atoms bonded to hydrogen.

PHENOL	
Bonding	*Charge Transfer*
$C_1 \leftrightarrow C_2$	0
$C_1 \leftrightarrow C_6$	0
$C_1 \rightarrow O$	+1
Formal Oxidation Number	$N_{C_1} = +1$
Bonding	*Charge Transfer*
$C_2 \leftrightarrow C_1$	0
$C_2 \leftrightarrow C_3$	0
$C_2 \leftarrow H$	−1
Formal Oxidation Number	$N_{C_2} = N_{C_3} = N_{C_4} = N_{C_5} = N_{C_6} = -1$

The formal oxidation sum for phenol verifies that Rule 8 is not violated.

$$N_{total} = 6 \cdot N_H + N_O + \sum_i N_{C_i}$$

$$N_{total} = 6 \cdot (+1) + (-2) + N_{C_1} + \sum_{i=2}^{6} N_{C_i}$$

$$N_{total} = 6 - 2 + (+1 - 5) = 0$$

Structure 8. Phenol

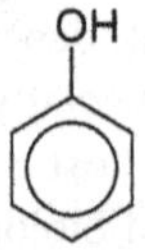

Example 8A.2

The formal oxidation numbers of carbon atoms in acetic acid (CAS 64-19-7, Structure 9) distinguish between carbon C_1 bonded to oxygen and the carbon atoms bonded to hydrogen.

The effect of relative electronegativity (see Figure 8.1) is the same as the example for phenol. Rules 3 and 4 (see Table 8.1) apply to O and H.

ACETIC ACID	
Bonding	*Charge Transfer*
$C_1 \leftrightarrow C_2$	0
$C_1 \rightarrow O_{carbonyl}$	+2
$C_1 \rightarrow O_{hydroxyl}$	+1
Formal Oxidation Number	$N_{C_1} = +3$
Bonding	*Charge Transfer*
$C_2 \leftarrow H$	−3
$C_2 \leftrightarrow C_1$	0
Formal Oxidation Number	$N_{C_2} = -3$

Structure 9. Acetic acid

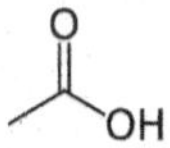

The formal oxidation sum for acetic acid verifies that Rule 8 is not violated.

$$N_{total} = 4 \cdot N_H + 2 \cdot N_O + \sum_i N_{C_i}$$
$$N_{total} = 4 \cdot (+1) + 2 \cdot (-2) + (N_{C_1} + N_{C_2})$$
$$N_{total} = 4 - 4 + (+3 - 3) = 0$$

APPENDIX 8B. CONVERTING (pe, pH) REDOX COORDINATES INTO (E_H, pH) COORDINATES

Substituting the half-cell reduction potential under standard conditions E_H° into the Nernst equation (8.7) yields a relation between the half-cell reduction potential E_H and the conceptual redox parameter pe. Redox parameter pe is not recognized by the International Union of Pure and Applied Chemists (IUPAC). Deriving the relation between E_H and pe serves a single purpose: it allows readers to covert pe values reported in the scientific literature into E_H values. The computer-based water chemistry model *ChemEQL*, for example, uses redox parameter pe to plot (pe, pH) Pourbaix diagrams.

The general reduction half reaction is given by (8B.1).

$$\underset{acceptor}{A^s} + m \cdot H_3O^+ + n \cdot e^- \rightarrow \underset{donor}{D^r} \tag{8B.1}$$

The development of conceptual redox parameter pe begins with equilibrium expression (8B.2) for the reduction half reaction (8B.1) that includes electron activity a_{e^-}. Expression (8B.2) is not consistent with convention for writing reduction half reactions (8B.1). Chemists list the electron e^- and its stoichiometric coefficient n in reduction half reactions as a convenient means of tracking electron transfer between the donor D^r and the acceptor A^s. One can imagine other conventions for writing reduction half reactions that dispense with listing electrons e^- as an educt.

$$K_{red} = \frac{a_{D^r}}{a_{A^s} \cdot a_{H^+}^m \cdot a_{e^-}^n} = \left(\frac{a_{D^r}}{a_{A^s} \cdot a_{H^+}^m} \right) \cdot a_{e^-}^{-n} \tag{8B.2}$$

The convention for writing reduction half reaction (8B.1) was adopted with the understanding that listing the electron e^- as an educt *did not imply the existence of free electrons*. Electron activity a_{e^-} (8B.2) in solution has no physical basis; electrons do not exist as free particles, unlike hydronium ions $H_3O^+(aq)$.

Taking the *base-10* logarithm of all of the terms in (8B.2) results in expression (8B.3).

$$\log_{10}(K_{red}) = \log_{10}\left(\frac{a_{D^r}}{a_{A^s} \cdot a_{H^+}^m} \right) - n \cdot \log_{10}(a_{e^-})$$

$$-n \cdot \log_{10}(a_{e^-}) = \log_{10}(K_{red}) - \log_{10}\left(\frac{a_{D^r}}{a_{A^s} \cdot a_{H^+}^m} \right) \tag{8B.3}$$

Expression (8B.4) defines the conceptual redox parameter pe by analogy with the definition of pH.

$$-\log_{10}(a_{e^-}) \equiv pe \tag{8B.4}$$

Notice that the Nernst equation (8B.5) for reduction half reaction (8B.1) does not list the electron as an educt.

$$E_H = E_H^\circ - \left(\frac{R \cdot T \cdot \ln(10)}{n \cdot F} \right) \cdot \log_{10}\left(\frac{a_{D^r}}{a_{A^s} \cdot a_{H^+}^m} \right) \tag{8B.5}$$

Expression (8B.6) relates the standard change in Gibbs energy ΔG° to the equilibrium constant K for any chemical reaction.

$$\Delta G^{\circ} = -R \cdot T \cdot \ln(K) \qquad \text{(8B.6)}$$

Natural logarithms are converted to *base-10* logarithms using the following relation: $\ln(x) = \ln(10) \cdot \log_{10}(x)$. Applying this conversion to (8B.6) yields (8B.7), upon rearrangement.

$$\log_{10}(K_{red}) = -\frac{\Delta G_H^{\circ}}{R \cdot T \cdot \ln(10)} \qquad \text{(8B.7)}$$

Using expressions (8.4) and (8.5) from earlier, we can write expression (8B.8).

$$\log_{10}(K_{red}) = \left(\frac{n \cdot F \cdot E_H^{\circ}}{R \cdot T \cdot \ln(10)} \right) \qquad \text{(8B.8)}$$

Rearranging expression (8B.5) and combining it with (8B.8) yields (8B.9).

$$\left(\frac{n \cdot F \cdot E_H}{R \cdot T \cdot \ln(10)} \right) = \log_{10}(K_{red}) - \log_{10}\left(\frac{a_{D^r}}{a_{A^s} \cdot a_{H^+}^m} \right) \qquad \text{(8B.9)}$$

The two terms on the right side of (8B.9) also appear in (8B.3). Combining (8B.3), (8B.4), and (8B.9) yields (8B.10), defining the relation between the conceptual redox parameter *pe* and E_H.

$$\left(\frac{n \cdot F \cdot E_H}{R \cdot T \cdot \ln(10)} \right) = n \cdot pe$$

$$E_H = \left(\frac{R \cdot T \cdot \ln(10)}{F} \right) \cdot pe \qquad \text{(8B.10)}$$

The equivalence between the experimentally measurable half-cell reduction potential under general conditions E_H and the conceptual redox parameter *pe* is often written using an numerical parameter (8B.11) appropriate for 25°C (298.15 K).

$$E_{H, 25°C} = (59.16 \, mV)_{25°C} \cdot pe \qquad \text{(8B.11)}$$

APPENDIX 8C. LIMITATIONS IN THE MEASUREMENT OF THE ENVIRONMENTAL REDUCTION POTENTIAL USING PLATINUM ORP ELECTRODES

Aqueous solutions saturated with O_2 under standard conditions ($p_{O_2} = 1 \, atm$, pH = 0) typically result in a 700–800 mV experimental potential E_{Pt} when using a platinum *ORP* electrode (Garrels and Christ, 1965; Stumm and

Morgan, 1970) rather than the expected $E_H^{\circ} = 1229 \; mV$ (see Table 8.3). The discrepancy results from the detailed electron transfer mechanism between the platinum electrode and O_2 molecules adsorbed on its surface and the slow kinetics of the electron transfer process. A thin layer of platinum oxide forms when O_2 is being reduced by electrons at the electrode surface (Liang and Juliard, 1965). This oxide layer *passivates* the platinum metal, inhibiting electron transfer by forming an insulating layer and altering the experimental potential. The net effect of the reduction mechanism and kinetics is a zero electrode current over a considerable potential range (Stumm and Morgan, 1970).

Numerous electrochemically active redox couples are dissolved in any environmental sample. The kinetics, mechanism, and reversibility of electron transfer and the extent to which each couple deviates from its equilibrium state ensure that any experimental electrode potential E_{Pt} represents a composite (i.e., mixed) electron transfer potential between all of these species and the platinum electrode surface.

TABLE 8C.1 Limitations on Experimental Platinum *ORP* Electrode Measurements of Reduction Potentials in Environmental Samples

1. Many redox species exhibit no tendency to transfer electrons at the platinum metal electrode surface (e.g., nitric oxide *NO*, dinitrogen gas N_2, sulfate SO_4^{2-}, and methane CH_4) and, therefore, fail to contribute to the measured potential E_{Pt}.

2. A platinum *ORP* electrode will record the potential of a couple only if the electron transfer reaction is reversible. Rosca and Koper (2005) and de Vooys (de Vooys, Koper et al., 2001a, b) discovered at least two pathways for nitric oxide *NO* reduction at a platinum electrode, one yielding nitrous oxide N_2O and the other yielding ammonia NH_3.

3. The electron transfer rate at the platinum metal electrode surface is a function of solution activity for redox couples where one or both of the electron donor and acceptor species are solutes. Solutes such as $Fe^{3+}(aq)$ and $Mn^{4+}(aq)$, and their hydrolysis products, are extremely insoluble over a considerable pH range, rendering the measured potential E_{Pt} insensitive to significant variations in the redox status of these couples.

4. Many redox couples do not attain equilibrium under environmental conditions—regardless of the reversibility or kinetics of the half reaction at the platinum metal electrode surface—because of slow electron transfer kinetics *in situ*. The experimental reduction potential E_{Pt} fails to represent the equilibrium redox status of these couples.

Problems

1. Determine the formal oxidation number of nitrogen in pyrrole (CAS 109-97-7) and calculate the formal oxidation state sum for the molecule.

Pyrrole

PYRROLE	
Bonding	*Charge Transfer*
$N_1 \leftarrow C_2$	
$N_1 \leftarrow C_5$	
$N_1 \leftarrow H$	
Formal Oxidation Number	$N_{N_1} =$
Bonding	*Charge Transfer*
$C_2 \rightarrow N_1$	
$C_2 \leftrightarrow C_3$	
$C_2 \leftarrow H$	
Formal Oxidation Number	$N_{C_2} = N_{C_5} =$
Bonding	*Charge Transfer*
$C_3 \leftrightarrow C_2$	
$C_3 \leftrightarrow C_2$	
$C_3 \leftarrow H$	
Formal Oxidation Number	$N_{C_3} = N_{C_4} =$

Solution

PYRROLE	
Bonding	*Charge Transfer*
$N(1) \leftarrow C(2)$	−1
$N(1) \leftarrow C(5)$	−1
$N(1) \leftarrow H$	−1

PYRROLE

Formal Oxidation Number	$N_{N(1)} = -3$
Bonding	*Charge Transfer*
$C(2) \rightarrow N(1)$	$+1$
$C(2) \leftrightarrow C(3)$	0
$C(2) \leftarrow H$	-1
Formal Oxidation Number	$N_{C(2)} = N_{C(5)} = 0$
Bonding	*Charge Transfer*
$C(3) \leftrightarrow C(2)$	0
$C(3) \leftrightarrow C(4)$	0
$C(3) \leftarrow H$	-1
Formal Oxidation Number	$N_{C(3)} = N_{C(4)} = -1$

$$N_{total} = 2 \cdot N_{C(2)} + 2 \cdot N_{C(3)} + 5 \cdot N_H + N_N$$
$$N_{total} = 2 \cdot (0) + 2 \cdot (-1) + 5 \cdot (+1) + (-3) = 0$$

2. The incomplete reduction half reaction for the dechlorination of DDT (dichloro-diphenyl-trichloroethane, *CAS* 50-29-3) to DDD (dichloro-diphenyl-dichloroethane, *CAS* 72-54-8) appears following.

dichloro-diphenyl-trichloroethane

dichloro-diphenyl-dichloroethane

Determine the formal oxidation number of the chlorinated methyl carbon in both DDT and DDD (show your work, please). Balance the reduction half reaction (electron balance, charge balance, and mass balance) by inserting the missing coefficients and product.

(DDT (1,1,1-trichloro-2,2-di(4-chlorophenyl)ethane, CAS 50-29-3)	
Bonding	*Charge Transfer*
$C_1 \rightarrow Cl$	
$C_1 \leftrightarrow C_2$	
Formal Oxidation Number	$N_{C_1} =$

(DDD (1-chloro-4-[2,2-dichloro-1-(4-chlorophenyl)ethyl]benzene, CAS 72-54-8)

Bonding	Charge Transfer
$C_1 \rightarrow Cl$	
$C_1 \leftrightarrow C_2$	
$C_1 \leftarrow H$	
Formal Oxidation Number	$N_{C_1} =$

Solution

DDT (1,1,1-trichloro-2,2-di(4-chlorophenyl)ethane, CAS 50-29-3)

Molecular Charge

Bonding	Charge Transfer
$C(1) \rightarrow Cl$	+3
$C(1) \leftrightarrow C(2)$	0
Formal Oxidation Number	$N_{C(1)} = +3$

DDD (1-chloro-4-[2,2-dichloro-1-(4-chlorophenyl)ethyl]benzene, CAS 72-54-8)

Molecular Charge

Bonding	Charge Transfer
$C(1) \rightarrow Cl$	+2
$C(1) \leftrightarrow C(2)$	0
$C(1) \leftarrow H$	−1
Formal Oxidation Number	$N_{C(1)} = +1$

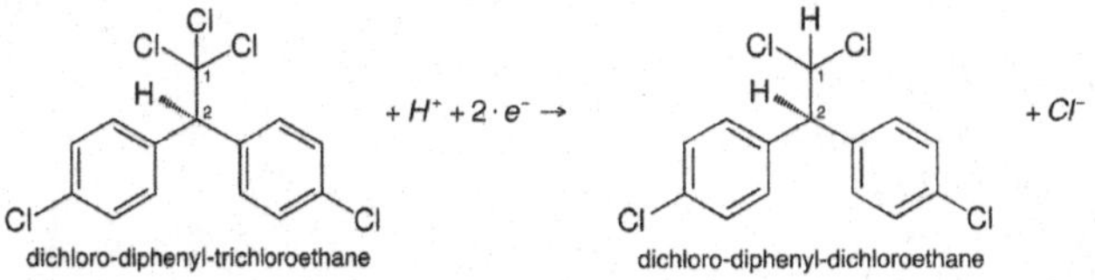

dichloro-diphenyl-trichloroethane $+ H^+ + 2 \cdot e^- \rightarrow$ dichloro-diphenyl-dichloroethane $+ Cl^-$

3. Given the relation between the Standard Gibbs Energy of Formation ΔG_f° of the educts and products (below) and the half–reaction reduction potential $(\Delta G_h^\circ = -nFE_h^\circ)$, calculate the reduction potential for DDT dechlorination half reaction in Problem 1.

Compound	ΔG_f°
DDT(aq)	$+274.9\ kJ \cdot mol^{-1}$
DDD(aq)	$+238.7\ kJ \cdot mol^{-1}$
$H^+(aq)$	$0.0\ kJ \cdot mol^{-1}$
$Cl^-(aq)$	$-131.3\ kJ \cdot mol^{-1}$

4. The following (E_H, pH) Pourbaix diagram shows the stability fields for several chromium species. This diagram was prepared using a total soluble chromium concentration of 50 ppb (9.62 10^{-7} $mol\ L^{-1}$), the USEPA Maximum Contaminant Limit for Cr.

   Interpret the chemical significance of each stability field and boundary.

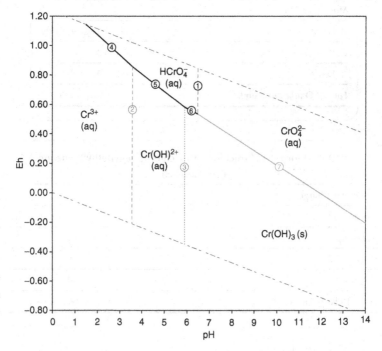

Field $Cr^{3+}(aq)$:
Field $Cr(OH)^{2+}(aq)$:
Field $HCrO_4^-(aq)$:
Field $CrO_4^{2-}(aq)$:
Field $Cr(OH)_3(s)$:
Boundary (1):
Boundary (1):
Boundary (3):
Boundary (4):
Boundary (5):
Boundary (6):

Chapter 9

Adsorption and Surface Chemistry

Chapter Outline

9.1. INTRODUCTION

Ion exchange represents a chemical process common in the natural environment but rarely mentioned in general chemistry text books. The ion exchange process involves the exchange of one or more ions from a solution with one or more ions bound to sites on the surface of a solid particle or a macromolecule—hereafter referred to as the *adsorbent*. Ion exchange is a special case of a more general surface chemistry process: *adsorption*.

Adsorption and adhesion share the same root, and both imply binding to a surface. An adsorption process involves the transfer of a solute from solution to adsorbent surface site surface. The solute does not have to be an ion, and adsorption does not require the exchange of a molecule or ion bound to an adsorbent surface site with another molecule or ion in solution.

Ion exchange retains ions at surface sites through the *electrostatic* force between the ion and one or more sites bearing an opposite electrostatic charge. Adsorption, being a more general chemical process, retains molecules or ions—*adsorbates*—through the same chemical bonds or intermolecular forces you would have learned about in general chemistry.

Protons, for instance, are bound to water molecules and hydroxyl ions in solution—and at mineral surfaces—through a covalent bond. Protons also form *hydrogen bonds* between adsorbate molecules and surface sites, the adsorbate and the adsorbent acting as a hydrogen-bond donor-receptor pair whose roles are interchangeable between adsorbate and surface site. Transition metal ions form *coordinate-covalent bonds* when lone-pair electrons on a *ligand* atom (e.g., oxygen or nitrogen in a molecule or ion) overlap with vacant *d*-orbitals of the metal ion. As with hydrogen bonds, the adsorbate and the adsorbent constitute an electron-pair donor (the *ligand* or *Lewis base*) and an electron-pair receptor (the *transition metal ion* or *Lewis acid*), whose roles are interchangeable between adsorbate and adsorbent site. Weak electrostatic forces also bind molecular dipoles in adsorbate molecules to dipoles of the adsorbent. These *van der Waals forces* act in concert, becoming the dominant binding-force when the adsorbate is a large neutral molecule.

9.2. MINERAL AND ORGANIC COLLOIDS AS ENVIRONMENTAL ADSORBENTS

Adsorption processes, though ubiquitous, become apparent only when the surface-to-volume ratio of the adsorbent is extremely high. This condition is commonly satisfied in environmental systems where clay-size mineral particles (Figure 9.1) or organic matter molecules are present in significant amounts. Table 9.1 lists the size range and specific surface area for the three

FIGURE 9.1 These electron micrographs depict fine grains of the minerals kaolinite $Al_4(Si_4O_{10})(OH)_8$ (L; (Roe, 2007)) and hematite Fe_2O_3 (R; Contributors, 2008). *The images are courtesy of the Clay Minerals Group of the Mineralogical Society.*

TABLE 9.1 Specific Surface Area of Spherical Particles with Diameters Defining Each USDA Particle-Size Class (Particle density is 2.65 $Mg\ m^{-3}$, the typical density of mineral grains in Earth's crust)

Particle Class	Diameter	Specific Area, m^2g^{-1}
Very coarse sand	2 mm	4.53E-03
Coarse sand	1.0 mm	9.06E-03
Medium sand	0.50 mm	1.81E-02
Fine sand	0.25 mm	3.62E-02
Very fine sand	0.10 mm	9.06E-02
Coarse silt	50 μm	1.81E-01
Clay	<2 μm	4.53E+00

TABLE 9.2 Specific Surface Area of USDA Texture Classes (Grain Density = 2.65 $Mg\ m^{-3}$) and Percentage of the Specific Surface Area from the Clay-Size Fraction

Texture Class	Sand	Silt	Clay	Specific Area, $m^2\ g^{-1}$	Percent Surface Area from Clay Fraction
Clay	0.20	0.20	0.60	5.62	97%
Silt loam	0.20	0.70	0.10	1.55	59%
Sandy loam	0.65	0.25	0.10	1.16	78%
Loam	0.40	0.40	0.20	2.19	83%

particle-size classes (sand, silt, and clay), and Table 9.2 lists the specific surface area of several soil textural classes. Soil textural classes represent the particle-size distributions commonly found in soils and sediments.

Tables 9.1 and 9.2 demonstrate the importance of clay-size mineral particles to the specific surface area of surficial material: soils, sediments, and aquifers. A similar case could be made for organic matter that, in the context of this chapter, is considered the *organic colloidal fraction* in soils and sediments.

One further example demonstrates how solute adsorption by the clay-size fraction can significantly alter pore-water composition. The typical adsorbing capacity of any mineral surface is on the order of 1 $\mu mol\ m^{-2}$. From Tables 9.1 and 9.2 we can see that the typical specific surface area of soils and sediments is on the order of $2\ m^{-3}g^{-1}$. Given a typical 20% volumetric water content of soils at field capacity and a typical bulk density of $1.3\ Mg\ m^{-3}$, a cubic meter of soil will hold about $200\ L$ and have dry mass of $1.3 \cdot 10^6\ g$ with a specific surface area of $2.6 \cdot 10^6\ m^2$, most from the clay fraction (see Table 9.2).

The adsorbing capacity of mineral particles in a cubic meter of soil is about 2 moles, which means that adsorption by the mineral fraction in soil can remove 0.01 moles per liter $(2\ mol\ m^{-3}/200\ L\ m^{-3})$ from solution. Clearly the adsorption of $0.01\ M$ represents a significant fraction of the total solute concentration in pore water. The capacity of adsorption processes to alter pore-water composition is a direct result of the extremely high specific surface area of soil colloids.

Clays and humic substances are profoundly different colloids, displaying markedly different adsorption behavior. To grasp these differences, we must first understand how an adsorption experiment is performed and the general appearance of the experimental results.

9.3. THE ADSORPTION ISOTHERM EXPERIMENT

In this section we describe the *adsorption isotherm* experiment, one of many adsorption experiments designed by surface chemists. The adsorption isotherm experiment requires a series of solutions containing the same mass of suspended adsorbent $m_{adsorbent}$ and the same volume of water V. Varying the initial concentration C_i of solute S—the potential adsorbate— imposes the experimental treatment.

Figure 9.2a depicts the initial state of a clay adsorbent suspension. During the equilibration stage the suspension is agitated to ensure ample opportunity for the solution containing adsorbate to contact the adsorbent (Figure 9.2b). The duration of the equilibration stage (Figure 9.2b) depends on how rapidly the adsorbate binds to the clay, usually a few minutes to several hours or days. The final stage (Figure 9.2c) requires separation of suspended clay from the solution for each chemical treatment in the series, usually by filtration or accelerated sedimentation in a centrifuge (see Appendix 9A). The supernatant solutions contain varying final concentrations C_f of the solute S (the adsorbate), representing the quantity of the solute S remaining in solution after

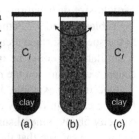

FIGURE 9.2 Measuring the adsorption isotherm by allowing a series of solutions containing varying concentrations (a) of a solute—$S(aq)$—to adsorb on the clay adsorbent, equilibration during agitation (b), and separation of adsorbent clay from solution (c).

adsorption has reached equilibrium. The amount of solute S adsorbed per mass of adsorbent $q(C)$ is the difference between the initial and final solute concentrations multiplied by the total volume V and divided by mass of adsorbent in the suspension $m_{adsorbent}$:

$$q(C) = \frac{(C_i - C_f) \cdot V}{m_{adsorbent}} \tag{9.1}$$

Figure 9.3 depicts the equilibrium state when adsorption is measured using a *dialysis cell*. If the adsorbent is a molecular adsorbent such as humic substances (see Chapter 6, section 6.5.1), it is impossible to separate the solution from the adsorbent by filtration or centrifugal sedimentation. Separation is achieved by *equilibrium dialysis*.

The appropriate dialysis membrane is permeable to solute S (the adsorbate) but impermeable to the molecular adsorbent. A series of dialysis cells (Figure 9.3) are prepared, each containing the same total volume of aqueous solution V, the same mass of molecular adsorbent $m_{adsorbent}$—confined to one-half of the cell by a semipermeable membrane—and varying initial concentrations C_i of solute S. Adsorption of the solute S to the confined adsorbent prevents a fraction of the solute from diffusing across the membrane. The free solute concentration is the same on both sides of the membrane because it can freely diffuse across the membrane, but the total concentration is higher in the half containing the adsorbent.

The amount of solute adsorbed per mass of adsorbent $q(C)$ is the difference between the initial and final solute concentrations divided by the mass of adsorbent in the suspension (9.1). The final solute concentration is either the solute concentration in the supernatant solution (see Figure 9.2) or the solute concentration inside of the dialysis chamber lacking adsorbent (Figure 9.3, left).

Table 9.3 lists the results from a typical adsorption isotherm. The adsorbent was a calcareous soil from Grand Bahama Island. The solute—orthophosphate $HPO_4^{2-}(aq)$—is a divalent molecular anion. The data are plotted in Figure 9.4. The adsorption isotherm displays a steep initial slope that decreases steadily toward what appears to be a zero slope representing an adsorption maximum. Adsorption isotherms showing this behavior are called *L-type* adsorption isotherms.

Solute   Solute + Adsorbent

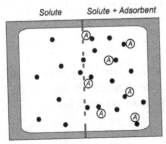

FIGURE 9.3 A semipermeable membrane (dashed line, center) confines the adsorbent (open circles labeled "A," right) from diffusing into the left half of the cell. The free solute (filled circles) concentration is the same on both sides of the membrane.

TABLE 9.3 Adsorption Isotherm for Orthophosphate $HPO_4^{2-}(aq)$ by 0.25 g of a Calcareous Soil

$C_{HPO_4^{2-},f} \; (mg \cdot L^{-1})$	$q_{HPO_4^{2-}} \; (mg \cdot kg^{-1})$
2.00	17.82
2.75	21.85
4.15	25.67
5.50	29.34
9.70	35.11
14.50	40.32

Source: Data from R. W. Taylor, Alabama A & M University.

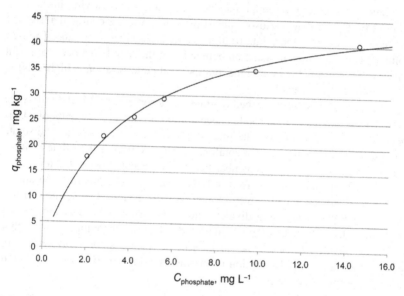

FIGURE 9.4 This graph plots the adsorption isotherm of the orthophosphate $HPO_4^{2-}(aq)$ by a calcareous soil (Grand Bahama Island, Commonwealth of The Bahamas). The estimated adsorption maximum is $q_{max} = 50.0 \; mg \; kg^{-1}$.

Environmental chemists commonly encounter linear or *C-type* adsorption isotherms (9.2) where the slope apparently remains constant throughout the solute concentration range used in the adsorption isotherm experiment.

$$q(C) = K_{soil} \cdot C_f \qquad (9.2)$$

Lambert (1968) reported linear isotherms for the adsorption of the insecticide *chlorfenvinphos* (CAS 470-90-6; Structure **1**) by two California soil series: Yolo silty clay loam and Hanford (Ripperdan) fine sandy loam. The adsorption isotherms for chlorfenvinphos (Structure **1**) by these two soils appear in Figure 9.5.

Structure 1. Chlorfenvinphos

The data are from Lambert (1968), who relied on chlorfenvinphos labeled with the radioisotope ^{14}C, so concentration is reported as radioactivity without specified units.

These linear adsorption isotherms are reminiscent of the results one would obtain from a process in which a solute partitions between two immiscible solvents: water and an organic solvent. This behavior led Lambert (1965) to hypothesize the adsorption of organic compounds by soil "corresponds to partition of the compound into the organic matter." The Yolo soil clearly adsorbs much more of the insecticide than the Hanford soil (Figure 9.5), but if the

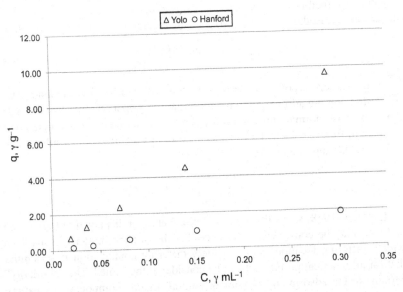

FIGURE 9.5   This graph plots the adsorption isotherm of the insecticide chlorfenvinphos (CAS 470-90-6) by two California soils: Yolo silty clay loam and Hanford (Ripperdan) fine sandy loam.

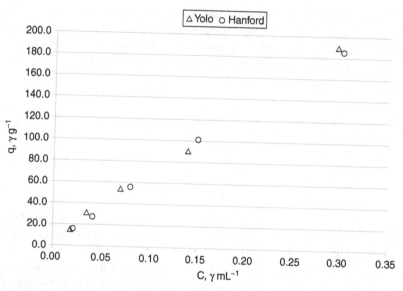

FIGURE 9.6 This graph plots the adsorption isotherm of the insecticide chlorfenvinphos (CAS 470-90-6) by two California soils: Yolo silty clay loam (5% organic matter) and Hanford (1% organic matter) fine sandy loam. The data, from Lambert (1968), are normalized by the organic matter content of the two soils.

quantity adsorbed is normalized by the organic matter content of the soil (9.3), the adsorption isotherms for the two soils are essentially identical (Figure 9.6).

$$q(C) = \frac{(C_i - C_f)}{(m_{organic\ matter}/V)} \tag{9.3}$$

The Lambert hypothesis has been widely adopted by environmental chemists studying the adsorption of neutral organic compounds by soils and sediments because normalizing the adsorption isotherms for the same compound on many different soils and sediments typically results in a plot similar to that in Figure 9.6 and summarized in Eq. (9.4).

$$q(C) = \frac{(C_i - C_f)}{(m_{OM}/V)} = K_{OM} \cdot C_f \tag{9.4}$$

Lambert (1968) states that the adsorption behavior displayed in Figures 9.5 and 9.6 "may be construed as evidence that the organic matter of soils from widely differing locations and of widely differing composition is so similar that for all practical purposes it can be considered the same." The "similarity" pertains to the adsorption behavior of neutral organic compounds by natural organic matter and should not be construed to apply to other properties of soil organic matter.

9.4. HYDROPHOBIC AND HYDROPHILIC COLLOIDS

A dispersion of colloidal particles in a liquid is stable, meaning that the particles will not settle out, because the constant thermal motion of molecules in the liquid is sufficient to overcome the effects of gravity on the particles (Appendix 9A). The two types of colloidal particles—*hydrophobic* and *hydrophilic* colloids—are found in soils and sediments. Hydrophobic colloids are fine-clay-size mineral particles, while hydrophilic colloids are humic substances and bioamphiphiles secreted by microbes (see Appendix 6A).

Humic molecules exhibit amphiphilic properties similar to simple amphiphiles such as phospholipids. It is unlikely, though not impossible, that humic molecules have a single polar end and a single nonpolar end. A more probable structure would resemble an amphiphilic block-copolymer (see Figure 6.23), a flexible linear molecule with alternating flexible polar and nonpolar segments. Such molecules would fold and aggregate in water by presenting their polar segments at the aqueous contact and burying their nonpolar segments within the interior.

Bacteria secrete a variety of bioamphiphiles—glycolipids, lipopeptides, and so on—that form micelles (see Appendix 6A). These bioamphiphiles play a variety of roles, including increasing the biological availability of nonpolar organic contaminants for degradation. Humic amphiphiles have little tendency to form aggregates of the type seen in Figure 9.7a, except when concentrated humic extract solutions are prepared in the laboratory. Otherwise, humic molecules aggregate at mineral surfaces (Figure 9.7b), folding in such a manner as to present their polar segments to either water or the mineral surface and withdrawing their nonpolar segments into the aggregate interior to create nonpolar domains much like the interior of the micelles and colloidal films pictured in Figures 9.7a and b, respectively.

9.5. INTERPRETING THE ADSORPTION ISOTHERM EXPERIMENT

Figures 9.4 and 9.5 illustrate the two types of adsorption isotherms most commonly encountered by environmental chemists. Figure 9.4 is a nonlinear *L-type* isotherm exhibiting an apparent adsorption maximum. Figure 9.5 is a linear *C-type* isotherm with no apparent adsorption maximum.

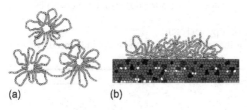

(a)   (b)

FIGURE 9.7 The probable structure of molecular aggregates formed by block copolymer amphiphiles in aqueous solution (a) and at the mineral-water interface (b); polar segments are open circles and nonpolar segments are flexible gray lines.

L-type adsorption isotherms result from site-limited adsorption processes. Adsorbent particles bind adsorbate molecules at a limited number of sites. The adsorption maximum, which is characteristic of L-type isotherms, represents the saturation of a limited number of adsorption sites. C-type adsorption isotherms result when the adsorbent has solvent-like properties, which is characteristic of molecular-aggregate adsorbents. Adsorption processes that produce C-type isotherms are comparable to the adsorbate dissolving (or partitioning) into solvent-like domains found within hydrophilic colloids. The adsorption isotherm, consequently, reveals the fundamental nature of the adsorption process, which is determined by the dominant interaction between the adsorbate and the adsorbent.

The following sections present models for both types of adsorption isotherms. Nonlinear L-type adsorption isotherms are represented by a particular two-parameter function. Linear C-type adsorption isotherms require a single parameter, represented by the slope of the isotherm; the physical requirements of the adsorption process demand that the isotherm pass through the origin (i. e., zero intercept).

9.5.1. The Langmuir Adsorption Model

Irving Langmuir (1916) derived the model commonly used to represent *site-limiting* adsorption processes. The assumptions upon which this model was originally based (Table 9.4) are idealized and generally do not apply to site-limiting adsorption processes involving environmental colloids. The use of the Langmuir model generated considerable debate within the environmental chemistry community, particularly the appropriateness of using a model whose foundational assumptions are not satisfied. We can apply this model to parameterize nonlinear adsorption isotherms, provided that we carefully avoid assigning certain physical interpretations of the L-type isotherm parameters.

Langmuir (1916; 1918) based his theory of site-limited adsorption on a kinetic model. The rate that molecules desorb from a solid surface, by his

---

TABLE 9.4 The Assumptions Used by Langmuir to Derive the Nonlinear *L-type* Gas Adsorption Isotherm Expression

- The solid surface has a fixed number of adsorption sites. Depending on the temperature and gas pressure, adsorbed molecules occupy a fraction θ of the sites, while the remainder $(1 - \theta)$ is vacant.
- Only one adsorbed molecule can occupy each site.
- The enthalpy of adsorption ΔH_{ads} is the same for all sites and is not influenced by the fraction θ of sites occupied.

reasoning, is proportional to the fraction of occupied sites θ (9.5). The proportionality constant k_D is the rate constant for desorption:

$$rate\ of\ desorption:\quad k_D \cdot \theta \qquad (9.5)$$

The rate that molecules adsorb is proportional to the fraction of unoccupied states $(1 - \theta)$ and the adsorbate concentration C when we apply the Langmuir model to adsorption from solution. The proportionality constant for adsorption k_A (9.6) is the rate constant for adsorption from solution. Langmuir (1916) developed the theory for gas molecule adsorption using the partial pressure of the adsorbent instead of the concentration in solution.

$$rate\ of\ adsorption:\quad k_A \cdot C \cdot (1 - \theta) \qquad (9.6)$$

Equilibrium (9.7) is the state where the rates of adsorption and desorption are equal.

$$equilibrium\ state:\quad k_D \cdot \theta = k_A \cdot C \cdot (1 - \theta) \qquad (9.7)$$

Langmuir (1916) solved this equation for θ, defining K_{ac} as the adsorption coefficient, which is actually the rate constant ratio k_A/k_D:

$$k_D \cdot \theta = (k_A \cdot C) - (k_A \cdot C \cdot \theta)$$

$$\theta = \frac{(K_{ac} \cdot C)}{1 + (K_{ac} \cdot C)} \qquad (9.8)$$

Plotting θ as a function of the adsorbate concentration C illustrates an important characteristic of the Langmuir adsorption isotherm: adsorption approaches a plateau as the occupied sites θ approach saturation (Figure 9.8).

Langmuir (1918) replaced the symbol θ with the ratio between the mass concentration of the adsorbate on the adsorbent as defined by Eq. (9.1) divided by the mass concentration at the adsorption maximum q_{max} where adsorption sites are completely saturated:

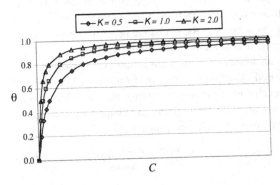

FIGURE 9.8 The L-type adsorption isotherm derived by Langmuir (1916) plots the fraction θ of adsorption sites occupied by adsorbate molecules as a function of the adsorbate concentration C using three different values for the adsorption coefficient parameter K_{ac}.

$$\theta = \frac{q}{q_{max}} = \frac{(K_{ac} \cdot C)}{1 + (K_{ac} \cdot C)}$$

$$q = \frac{(q_{max} \cdot K_{ac} \cdot C)}{1 + (K_{ac} \cdot C)} \tag{9.9}$$

Take another look at Figures 9.4 and 9.8. In the high adsorbate concentration range, specifically wherever $(K_{ac} \cdot C) \gg 1$, then $\theta \approx 1$ or $q \approx q_{max}$. On the other hand, in the low adsorbate concentration range, specifically wherever $(K_{ac} \cdot C) \ll 1$, both θ and q tend to be proportional to the adsorbate concentration C.

Two parameters determine the adsorption isotherm: the adsorption coefficient K_{ac} (i.e., the rate constant ratio) and the adsorption maximum q_{max}. The significance of the adsorption maximum q_{max} in the site-limiting L-type isotherms is self-evident, but the adsorption coefficient K_{ac} requires further explanation. Langmuir (1916) developed his kinetic adsorption model subject to stringent conditions (see Table 9.4), enabling him to associate the rate constant ratio with a kinetic definition of an equilibrium constant. This association provides a thermodynamic basis for the adsorption coefficient K_{ac} under the conditions imposed by Langmuir (1916; 1918):

$$\frac{k_A}{k_D} \equiv K_{ac} = e^{-\Delta G_{ads}/RT} \tag{9.10}$$

Langmuir's conditions (see Table 9.4) are not met for site-limited adsorption processes encountered by environmental chemists. At most we can use adsorption coefficients K_{ac} to compare the relative adsorption affinity of different adsorbates on the same environmental colloid or the same adsorbate on different environmental colloids.

9.5.2. Ion Exchange Adsorption Isotherms

Ion exchange is an adsorption process, but the physical characteristics are very different from those giving rise to the L-type isotherm described by Langmuir (1916). Symmetric exchange reactions yield isotherms that do not change with salt concentration—linear if the exchange process is nonselective (see Figure 4.9) and curved if it is selective (see Figure 4.7).

There are cases where an asymmetric exchange isotherm closely resembles the Langmuir site-limited adsorption isotherm. Consider a hypothetical nonselective asymmetric exchange isotherm involving the univalent cation Na^+ and the divalent cation Ca^{2+} (Eq. (9.11)) and the Gaines-Thomas selectivity expression for the asymmetric exchange process (Eq. (9.12)).

$$\underset{solution}{2 \cdot Na^+(aq)} + \underset{exchanger}{Ca^{2+}} \longleftrightarrow \underset{exchanger}{2 \cdot Na^+} + \underset{solution}{Ca^{2+}(aq)} \tag{9.11}$$

$$K_{GT}^c = \frac{[Ca^{2+}] \cdot \tilde{E}_{Na^+}^2}{\tilde{E}_{Ca^{2+}} \cdot [Na^+]^2} \quad (9.12)$$

The asymmetric exchange isotherm (Eq. (9.13)) for reaction (9.11) is a quadratic equation that depends on the normality C_0 of the electrolyte solution and the selectivity of the exchange process K_{GT}^c (Eq. (9.14)).

Asymmetric Exchange Isotherm

$$\tilde{E}_{Na^+} = \frac{-\beta \pm \sqrt{\beta^2 + 4 \cdot \beta}}{2} \quad (9.13)$$

$$\beta \equiv \frac{2 \cdot C_0 \cdot K_{GT}^c \cdot E_{Na^+}^2}{1 - E_{Na^+}} \quad (9.14)$$

Figure 9.9a is a plot of the exchange isotherm for a nonselective asymmetric (9.11) univalent-divalent ion exchange process, plotting the fraction of exchange sites occupied by the divalent cation as a function of the concentration of the divalent cation in solution. This plot clearly resembles the plot of a prototypical L-type isotherm in Figure 9.4 with a relatively large adsorption coefficient K. A plot of the exchange isotherm for the univalent cation (Figure 9.9b) does not remotely resemble the isotherm in Figure 9.4.

9.5.3. Linear Adsorption or Partitioning Model

As just mentioned, linear C-type adsorption isotherms result from adsorption processes where the interaction between adsorbate and adsorbent does not involve binding to a limited number of sites. Instead, the adsorbent behaves similar to a solvent, a characteristic of hydrophilic (or molecular-aggregate) adsorbents. The adsorption process yielding C-type linear adsorption isotherms involves molecular-aggregate adsorbents and nonpolar adsorbate molecules. The adsorbate is thought to partition from water into nonpolar solvent-like domains found within the molecular-aggregate colloid.

The slope of the C-type linear adsorption isotherm is often called the *distribution* or *partition coefficient*:

$$K_D[L \cdot kg^{-1}] = \frac{quantity\ adsorbed\ per\ mass\ adsorbent}{concentration\ dissolved\ in\ solution} = \frac{q[g \cdot kg^{-1}]}{C[g \cdot L^{-1}]} \quad (9.15)$$

Substances that have a very low affinity for the environmental adsorbent have distribution coefficients near zero. Substances with very high distribution coefficients K_D tend to be nonpolar organic compounds with low water solubility.

As pointed out earlier, L-type nonlinear adsorption processes exhibit behavior resembling (9.15) at low adsorbate concentration C range, provided

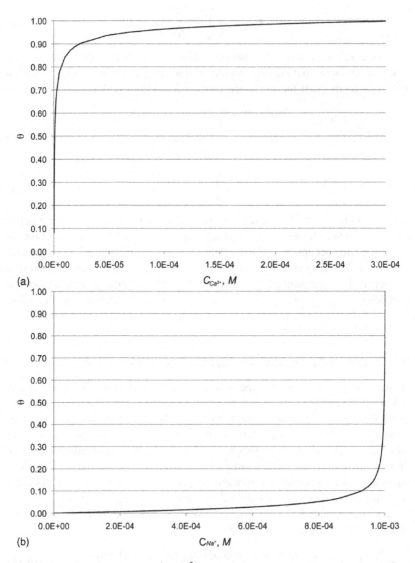

(a)

(b)

FIGURE 9.9 The asymmetric $Na^+ - Ca^{2+}$ ion exchange isotherm plots the fraction θ of exchange sites occupied by (a) Ca^{2+} ions as a function of the Ca^{2+} ion concentration $C_{Ca^{2+}}$ and (b) Na^+ ions as a function of the Na^+ ion concentration C_{Na^+} assuming nonselective exchange ($K_{GT}^c = 1$) and an ionic strength $I = 0.001$.

$(K_{ac} \cdot C) \ll 1$, as do asymmetric ion exchange processes in the same low adsorbate concentration range. True C-type linear adsorption processes in environmental chemistry involve a particular type of adsorbent—hydrophilic (molecular aggregate) colloids (see Figures 9.5 and 9.6)—and a particular type of adsorbate—nonpolar organic molecules. True C-type linear adsorption is fundamentally a partitioning process; a nonpolar organic compound distributes itself between two solvents: water and the solvent-like nonpolar domains formed when amphiphiles aggregate in solution (see Figure 9.7a) or on clay surfaces (see Figure 9.7b).

Small nonpolar compounds typically have a high vapor pressure and readily evaporate from water rather than adsorb to molecular-aggregate colloids. Vapor pressure decreases as molecular weight increases, increasing the tendency of nonpolar organics to adsorb to humic colloids. Water solubility correlates with the overall polarity of organic compounds and is a major factor determining the distribution tendency between aqueous solution and the nonpolar domains of molecular-aggregate colloids.

Lambert (1965) assumed that the adsorption of uncharged (i.e., nonpolar) organic molecules by soil "corresponds to partition of the compound into organic matter of the soil" and chose to base the adsorption of organic compounds by soil on the organic matter content (Eq. (9.16)). Lambert cited earlier studies of pesticide adsorption by soil with varying organic matter contents that confirm a C-type linear adsorption isotherm.

$$K_{OM}\left[L \cdot kg_{OM}^{-1}\right] = \frac{quantity\ adsorbed\ per\ mass\ organic\ matter}{concentration\ dissolved\ in\ solution} = \frac{q\left[g \cdot kg_{OM}^{-1}\right]}{C\left[g \cdot L^{-1}\right]}$$

$$(9.16)$$

While Lambert's simple partitioning model is useful and the generalization that K_{OM} for a given nonpolar compound is a good predictor of the whole-soil distribution coefficient K_D, it did not provide a basis for predicting the K_{OM} based on the molecular structure of the compound. Lambert (1967; 1968) applied the *linear free energy principle* to predict changes in K_{OM} from adding a substituent or otherwise making small changes in molecular structure. The basis for his linear free energy relationship for predicting K_{OM} is based on the notion (Langmuir, 1925) that a major determinant of solubility is the energy of creating a cavity in the solvent sufficient to accommodate the solute—the *volume-energy concept*.

Adding a bulky substituent, for instance, to a nonpolar molecule imposes little or no energy cost to creating a solute cavity in a nonpolar domain within a molecular-aggregate colloid (or a nonpolar solvent). The energy cost is substantial if the solute is nonpolar and the solvent is water because water molecules are bound together through hydrogen bonds in the liquid state. The hydrogen bond network in water must be broken to create a cavity for a nonpolar molecule.

Consequently, any change in molecular structure that increases the volume of a nonpolar molecule will decrease water solubility, increasing the tendency of that compound to partition from water into the nonpolar domains provided by the organic matter coating mineral grains in the soil or sediment. Similarly, any change in molecular structure that increases the nonpolar surface area of a complex molecule—molecules with a mixture of polar and nonpolar groups—will have a similar effect: decreased water solubility and increased distribution coefficient K_{OM} (Eq. (9.16)).

This model for predicting the distribution coefficient K_{OM} anticipates increasing water solubility and decreasing distribution coefficients K_{OM} (Eq. (9.16)) if changes in molecular structure add polar groups. Adding polar groups to a nonpolar molecule reduces the energy cost of creating a solute cavity in water. Lambert (1968) found a linear correlation between the *parachor value P* of similar nonpolar organic compounds and the distribution coefficient K_{OM}. The parachor value P is a measure of the molecular volume for a compound.

These principles underlying C-type linear adsorption of nonpolar organic compounds by humic colloids were subsequently confirmed and extended by Chiou and colleagues (1979) and others (see Doucette and Andren, 1987, 1988; Shiu, Doucette et al., 1988). Some of these later studies correlate the molecular volume and total molecular surface area of nonpolar organic compounds with the *octanol-water partition coefficient* K_{OW}. Octanol-water partition coefficients K_{OW} interest environmental chemists because K_{OW} correlates with fat solubility of a compound and the tendency for biomagnification (Box 9.1).

Box 9.1 Biomagnification and Bioaccumulation

The Office of Pollution Prevention and Toxics (U.S. Environmental Protection Agency) maintains an Internet site called the PBT Profiler (Center, 2006) that estimates a variety of important environmental characteristics of organic compounds, principally persistence, bioaccumulation, and toxicity. The Office of Pollution Prevention and Toxics defines *biomagnification* as "the concentration of a chemical to a level that exceeds that resulting from its diet." Biomagnification is quantified using the biomagnification factor *BMF*, where C_B indicates the concentration in an organism (e.g., zooplankton, fish, etc.) and C_A the total concentration in the organism's diet.

$$BMF \equiv \frac{C_B}{C_A}$$

Bioaccumulation encompasses both bioconcentration and biomagnification, where the former relates to uptake from all sources and the latter specifically relates to diet. *Bioconcentration* is quantified using the bioconcentration factor *BCF*, where C_B indicates the concentration in an organism (e.g., zooplankton, fish, etc.) and C_W the total concentration in water.

$$BCF \equiv \frac{C_B}{C_W}$$

Mackay and Fraser (2000) reviewed models of bioaccumulation and methods for estimating *BCF*. A popular approach draws on the extensive octanol-water partition coefficients K_{OW} database for organic chemicals. Mackay (1982)

> **Box 9.1 Biomagnification and Bioaccumulation—cont'd**
>
> proposed a simple, one-parameter estimator of *BCF* for rainbow trout (*Oncorhynchus mykiss*) based on a linear regression with K_{OW} for nonpolar organic compounds.
>
> $$BCF = 0.048 \cdot K_{OW}$$
>
> Mackay observed that the proportionality constant "can be regarded as an effective octanol content of the fish and is related, or equal, to the lipid content" of the trout.

9.6. VARIABLE-CHARGE MINERAL SURFACES

Environmental acidity is a complex phenomenon that we discuss in Chapter 7. Suffice it to say at this point, pore water pH in soils, sediments, and aquifers covers a relatively large range. Natural biological and geochemical processes can result in pore water acidities as low as pH 4 and as high as pH 9. Chemists observed that ion adsorption, both cations and anions, by environmental colloids varied significantly with pore water pH. Some of this could be explained by the hydrolysis of weak acid groups in natural organic matter, but weak acid hydrolysis of organic matter could not explain all of the variability that the chemists measured. Mineral colloids, the clay fraction, account for much of the observed variability in ion adsorption in soil and sediments.

Parks and de Bruyn (1962) were the first to suggest that the adsorption of H^+ by mineral colloid surfaces could be described by a hydrolysis reaction with a characteristic equilibrium constant. They believed that a charge-free mineral surface occurs at a characteristic pH depending on the nature of the mineral colloid. The hypothetical charge-free mineral surface develops a positive surface charge through the adsorption of H^+ ions—as illustrated following, with the neutral amine $-NH_2$ designated as the conjugate base:

$$-NH_3^+ + H_2O(l) \xleftrightarrow{K_a} -NH_2 + H_3O^+(aq) \tag{9.17}$$

$$\underset{\substack{\text{conjugate} \\ \text{acid}}}{} \qquad \underset{\substack{\text{weak} \\ \text{base}}}{}$$

The hypothetical charge-free mineral surface develops a negative surface charge through H^+ ion desorption—as illustrated in Eq. (9.18), with the neutral carboxyl $-COOH$ designated as the weak acid. The desorption of H^+ ions, however, is equivalent to the adsorption of OH^- ions—as shown in reaction (9.19), where $-COO^-$ is designated as the conjugate base and $K_a' \cdot K_b' = K_w$.

$$\underset{\text{weak acid}}{-COOH} + H_2O(l) \xleftrightarrow{K_a'} \underset{\substack{\text{conjugate} \\ \text{base}}}{-COO^-} + H_3O^+(aq) \tag{9.18}$$

$$\underset{\substack{\text{weak} \\ \text{acid}}}{-COOH} + OH^-(aq) \xleftrightarrow{K_b'} \underset{\substack{\text{conjugate} \\ \text{base}}}{-COO^-} + H_2O(l) \tag{9.19}$$

The model proposed by Parks and de Bruyn, and later Stumm and colleagues (Stumm, Huang et al., 1970), is shown in reactions (9.20) and (9.21). The hypothetical charge-free mineral surface develops a positive charge through the adsorption of H^+ ions (9.20), with the neutral site $-Me-OH$ designated as the weak base and the positively charged site $-Me-OH_2^+$ as the corresponding conjugate acid. The hypothetical charge-free mineral surface develops a negative charge through the desorption of H^+ ions (9.21), with the neutral site $-Me-OH$ now designated as the weak acid and the negatively charged site $-Me-O^-$ as the corresponding conjugate base.

$$\underset{conjugate\ acid}{-Me-OH_2^+} +H_2O(l) \overset{K_1}{\longleftrightarrow} \underset{weak\ base}{-Me-OH} +H_3O^+(aq) \qquad (9.20)$$

$$\underset{weak\ acid}{-Me-OH} +H_2O(l) \overset{K_2}{\longleftrightarrow} \underset{conjugate\ base}{-Me-O^-} +H_3O^+(l) \qquad (9.21)$$

The following two sections describe the experiments that measure pH-dependent ion adsorption by mineral colloids and pH-dependent surface charge, the latter being a major determinant of pH-dependent ion adsorption.

9.7. THE ADSORPTION ENVELOPE EXPERIMENT: MEASURING pH-DEPENDENT ION ADSORPTION

The design of a pH-dependent ion adsorption experiment is very similar to the previously described adsorption isotherm experiment. In both experiments the chemists must separate the adsorbent from the solution by either filtration or accelerated sedimentation in a centrifuge (see Figure 9.2) or equilibrium dialysis (see Figure 9.3) in order to analyze the solution.

The major experimental design difference relates to the treatments applied to the suspension. An adsorption isotherm experiment is designed to measure adsorption as a function of varying the adsorbate concentration; control of solution pH may or may not be imposed, and measurement of equilibrium solution pH may or may not be recorded. A pH-dependent adsorption experiment explicitly adjusts solution pH so that the equilibrium pH of the treatments adequately spans the desired pH range; the initial adsorbate concentration is identical for all pH treatments.

A decrease in adsorbate concentration caused by precipitation of a mineral rather than adsorption should be considered possible when designing a pH-dependent ion adsorption experiment. The simplest way to evaluate whether precipitation can explain the observed change in adsorbate ion concentration is to analyze the equilibrium solution and perform a water chemistry simulation to identify whether precipitation involving the adsorbate solute is likely.

For instance, suppose the experiment is designed to measure the adsorption of phosphate by the aluminum hydroxide mineral gibbsite $Al(OH)_3(s)$. A well-designed pH-dependent phosphate adsorption experiment would measure the following at equilibrium: total dissolved phosphate concentration and total dissolved aluminum concentration. The researcher would then perform a water chemistry simulation for each treatment based on the equilibrium solution concentration, calculating the saturation index of each aluminum phosphate mineral that could potentially form. The researcher can reasonably assign the decrease in phosphate concentration to an adsorption process for all treatments where the saturation indices for all potential aluminum phosphate minerals are less than 1. All other treatments where the saturation indices of potential aluminum phosphate minerals are 1 or greater represent conditions where precipitation of aluminum phosphate minerals is possible and might account for the decrease in phosphate solution concentrations.

9.7.1. Adsorption Edges

Typical cation adsorption envelopes appear in Figure 9.10, which shows a rapid increase in the percent adsorbed in a narrow pH range—typically 2 pH units from negligible to nearly complete adsorption from solution. Many environmental chemists reporting cation adsorption envelopes similar to those in Figure 9.10 may also report an *adsorption edge*, defined as the pH at which cation adsorption is 50% of the initial cation concentration. The adsorption edges for $Co(II)$ and $Cu(II)$ by imogolite are pH 6.0 and pH 7.7, respectively.

Typical anion adsorption envelopes appear in Figure 9.11, which shows a rapid increase in the percent adsorbed in a narrow pH range—typically 2 pH units from negligible to nearly complete adsorption from solution. While cation adsorption is negligible when the solution pH is acidic, anion adsorption is negligible when the solution pH is basic. The adsorption edges for selenate SeO_4^{2-} and selenite SeO_3^{2-} by goethite are pH 4.4 and pH 9.8, respectively.

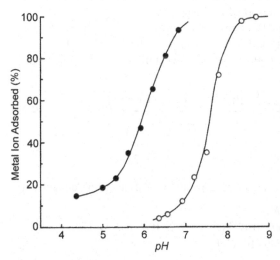

FIGURE 9.10 This graph plots the pH-dependent adsorption results from Clark and McBride (1984). The initial adsorbate ion concentrations were 250 μM: $Co(II)$ (open symbol) and $Cu(II)$ (filled symbol). The adsorbent suspension contains $5\,g\,L^{-1}$ of synthetic imogolite $Al_2SiO_3(OH)_4(s)$ and an ionic strength I of 0.15 M as $Ca(NO_3)_2$.

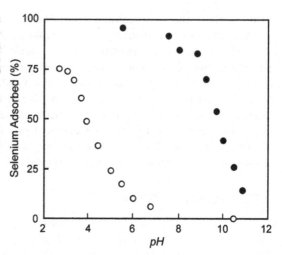

FIGURE 9.11 This graph plots the pH-dependent adsorption results from Su and Swarez (1997). The initial adsorbate ion concentrations were 100 μM: selenate SeO_4^{2-} (open symbol) and selenite SeO_3^{2-} (filled symbol). The adsorbent suspension contains 4 $g\,L^{-1}$ of synthetic goethite $FeO(OH)(s)$ and an ionic strength I of 0.1 M as $NaCl$.

9.7.2. Measuring pH-Dependent Surface Charge

The design of a pH-dependent surface charge experiment is essentially a titration experiment that is otherwise similar to the previously described pH-dependent ion adsorption experiment. The proton-surface charge experiment measures the relative adsorption of protons H^+ or hydroxyls OH^- by the colloid over the desired range. The design of a proton-surface charge experiment involves titrating the colloid suspension with a strong acid solution (to cover the acid pH range) and a strong base solution (to cover the basic pH range). The acid concentration C_A and base concentration C_B of the titrating solutions must be precisely measured because protons H^+ are being adsorbed and hydroxyls OH^- are being desorbed in the process.

Parks and de Bruyn (1962) proposed the following definition of proton surface charge σ_H (9.22) (units: Coulomb per square meter $C \cdot m^{-2}$), where S_{clay} is the clay (i.e., hydrophobic colloid) surface concentration (9.23) (units: square meter per liter $m^2 \cdot L^{-1}$).

$$\frac{\sigma_H}{F} = \frac{\left(C_{SA} - 10^{-pH}\right) - \left(C_{SB} - 10^{14-pH}\right)}{S_{clay}} \tag{9.22}$$

$$S_{clay}\left[\frac{m^2}{L}\right] = A_{clay}\left[\frac{m^2}{g}\right] \cdot M_{clay}\left[\frac{g}{L}\right] \tag{9.23}$$

The Faraday constant F in Eq. (9.22) converts mole of charge to Coulomb units.

$$F = e \cdot N_A$$
$$F = \left(1.602 \cdot 10^{-19}\,C\right) \cdot \left(6.022 \cdot 10^{23}\,mol^{-1}\right)$$
$$F = 9.649 \cdot 10^4\,C \cdot mol^{-1}$$

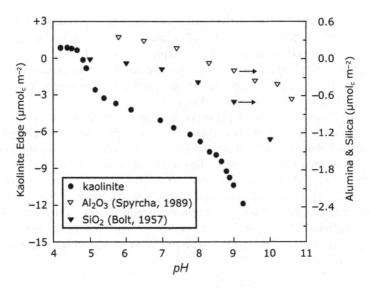

FIGURE 9.12 This graph plots the pH-dependent proton surface charge σ_H of kaolinite $Al_4Si_4O_{10}(OH)_8$ from Zhou and Gunter (1992), with additional data published previously by Bolt (1957) and Sprycha (1989).

Zhou and Gunter (1992) measured the pH-dependent proton surface charge of kaolinite $Al_4Si_4O_{10}(OH)_8$, comparing their results with earlier pH-dependent proton surface charge σ_H measurements of silica (Bolt, 1957) and alumina (Sprycha, 1989). Kaolinite is a layer aluminosilicate, with each layer composed of a gibbsite sheet and a silica sheet (see Figure 3.9), and develops surface charge at its edge surface. Figure 9.12 demonstrates that the kaolinite layer-edge develops a considerably greater surface charge than its component oxides. The kaolinite pH-dependent proton surface charge σ_H cannot be considered the additive sum of the two components.

9.7.3. Proton Surface Charge Sites

Soils and sediments are environments exposed to significant chemical weathering, a geochemical process that transforms primary minerals in rocks into the secondary minerals constituting the mineral colloid fraction. These secondary clays are typically oxide minerals, and the outermost atomic layer at the mineral-water interface consists of oxygen ions at various stages of protonation. Hence, reactions that confer surface charge (9.20) and (9.21) are a reasonable representation of the surface charging process.

The nature of the surface sites in the Langmuir model (Langmuir, 1916) and the pH-dependent surface charge model (Stumm and Morgan, 1970) are fundamentally quite similar. Both imply site-limited adsorption processes involving an adsorbate molecule binding to a site whose defining characteristic is the affinity for the site for the adsorbate. Implicit is the notion that each adsorbate molecule occupies a finite "footprint" on the adsorbent surface that determines the adsorbate surface density at the adsorption maximum. You can consider the pH-dependent proton surface charge σ_H (see Figure 9.12) as a plot of the pH-dependent proton adsorption.

If the Parks and de Bruyn pH-dependent surface charge model represented in reactions (9.20) and (9.21) is chemically reasonable, then the limiting proton surface charge σ_H should approximate the O^{2-} ion surface density typical of oxide mineral surfaces. The Pauling ionic radius of an O^{2-} ion is 140 pm, which is equivalent to 62,000 $pm^2 \cdot ion^{-1}$ and a surface density of 24 $\mu mol \cdot m^{-2}$ for a plane of close-packed O^{2-} spheres. Schindler and Kamber (1968) estimate the area density of proton adsorption sites on silica gel—colloidal hydrous $SiO_2(s)$—to be 4.0 $\mu mol \cdot m^{-2}$. The limiting proton surface charge σ_H of kaolinite (see Figure 9.12) is comparable, reaching 12 $\mu mol_c \cdot m^{-2}$ at pH 9.

9.8. VALENCE BOND MODEL OF PROTON SITES

The crystal structure of goethite contains two distinct O^{2-} ions (Structure 2): O_I is four-coordinate (three Fe^{3+} ions and one H^+ ion), and O_{II} is three-coordinate (three Fe^{3+} ions plus a hydrogen bond).

Structure 2. Oxygen coordination in goethite

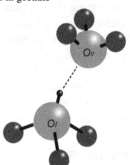

Hydrogen bonding plays an important role in goethite crystal structure, causing $Fe - O_I$ bonds to be longer than $Fe - O_{II}$ bonds: 210.5 pm and 193.9 pm (Szytula, Burewicz et al., 1968; Yang, Lu et al., 2006), respectively. The actual bond valence sum ζ for the two oxygen ions in goethite (Structure 2), based on the empirical bond valence function of Brown and Altermatt (1985) and the goethite crystal data of Yang and colleagues (2006) listed in Table 9C.1 are $\zeta_I = 3 \cdot (0.61\ v.u.) + (0.02\ v.u.) = 1.85\ v.u.$ and $\zeta_{II} = 3 \cdot (0.39\ v.u.) + (0.99\ v.u.) = 2.16\ v.u.$, respectively (see Table 9.5).

At the goethite surface the coordination of most O^{2-} ions decreases dramatically relative to Structure 2. Russell and colleagues (1974) described a

TABLE 9.5 Bond Lengths and Bond Valences for the Mineral Goethite

Bond	Bond Length	r_{ij}	Bond Valence
$Fe - O_I$	210.3 pm	175.9 pm	$s_{Fe-O_I} = 0.61$ v.u.
$Fe - O_{II}$	194.4 pm	175.9 pm	$s_{Fe-O_{II}} = 0.39$ v.u.
$H - O_{II}$	88.5 pm	88.2 pm	$s_{H-O_I} = 0.99$ v.u.
$H \cdots O_{II}$	231.1 pm	88.2 pm	$s_{H\cdots O_I} = 0.02$ v.u.

Note: Brown and Altermatt (1985): $s_{ij} = e^{(r_0 - r_{ij})/B}$; $B = 37$ pm
Source: Based on the refinement by Yang et al., 2006.

surface site model (Structure **3**) for goethite $FeO(OH)(s)$ that identified three different O^{2-} ion sites at the surface. Site A O^{2-} ions are coordinated by a single Fe^{3+} ion, Site B O^{2-} ions are coordinated by three Fe^{3+} ions, and Site C O^{2-} ions are coordinated by two Fe^{3+} ions.

Structure 3. Oxygen coordination at the surface of goethite. *Source: Reproduced with permission from Russell, J.D., Parfitt, R. L., et al., 1974. Surface structures of gibbsite goethite and phosphated goethite. Nature 248, 220–221.*

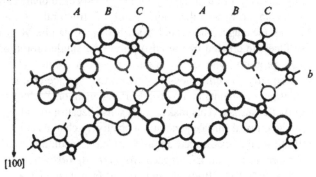

Site B O^{2-} ions have the same coordination as O_{II} ions in the interior of goethite, making these the least likely to undergo hydrolysis reactions (9.20) and (9.21), as suggested by Parks and de Bruyn. Site C O^{2-} ions would tend to bind a single proton as a neutral site and release a proton through a reaction similar to (9.21). Russell and colleagues (1974) identify Site A O^{2-} ions as likely to bind two protons to form a positively charged site through a reaction similar to (9.20).

The model proposed by Russell and colleagues would have one proton bound to all three sites when the goethite surface has a net zero surface charge; Site A OH would bear a negative –0.5 charge, Site B OH would bear a positive +0.5 charge, and Site C OH would charge neutral.

$$\zeta_A = \sum_i \frac{z_i}{v_i} = \left(\frac{3}{6}\right) + \left(\frac{1}{1}\right) = 1.5 \; v.u.$$

$$\zeta_B = \sum_i \frac{z_i}{v_i} = \left(\frac{3}{6}\right) + \left(\frac{3}{6}\right) + \left(\frac{3}{6}\right) + \left(\frac{1}{1}\right) = 2.5 \; v.u.$$

$$\zeta_C = \sum_i \frac{z_i}{v_i} = \left(\frac{3}{6}\right) + \left(\frac{3}{6}\right) + \left(\frac{1}{1}\right) = 2.0 \; v.u.$$

This assignment is based on the Pauling (1929) electrostatic valence principle:

In a stable coordination structure the electric charge of each anion tends to compensate the strength of the electrostatic valence bonds reaching to it from the cations at the centers of the polyhedra of which it forms a corner.

The electrostatic valence sum for Site A OH is 1.5 instead of 2.0, so this site registers a net –0.5 site charge. The electrostatic valence sum for Site B OH is 2.5 instead of 2.0, making this site a net +0.5 site charge.

An alternative model appears in Appendix 9C. This model creates a neutral surface by cleaving through the crystal along a specific plane and adsorbing water molecules to restore the coordination of structural cations exposed at the surface by cleavage. The model described in Appendix 9C shows that a charged surface site created by the adsorption of a single proton or desorption of a single proton is best seen as a pair of oxygen species—O^{2-} or OH^- or H_2O—that equally share the positive or negative surface charge.

Charge neutrality is maintained by coadsorbing an anion X^- for each proton H^+ adsorbed by the surface. This makes good sense because the adsorption of a proton H^+ by the mineral surface generates a positively-charged anion exchange site that will attract the coadsorbed anion X^-. Charge neutrality is also maintained by coadsorbing a cation M^+ for each proton H^+ desorbed from the surface. Desorption of a proton H^+ from the mineral surface generates a negatively-charged cation exchange site that will attract the coadsorbed cation M^+.

9.8.1. Interpreting pH-Dependent Ion Adsorption Experiments

Figure 9.10 plots the pH-dependent adsorption of $Co(II)$ and $Cu(II)$ by the aluminosilicate imogolite, a tubular mineral similar to kaolinite. The solution chemistry of these two components is quite different. Component $Co(II)$ exists predominantly as the divalent cation $Co^{2+}(aq)$ throughout the entire pH range. The divalent cation $Cu^{2+}(aq)$ predominates below pH 7.6, and the divalent cation $Cu_2(OH)_2^{2+}(aq)$ predominates over a narrow pH range before the neutral species $Cu(OH)_2^0(aq)$ becomes the major solution species at pH 8. Despite the fact that both adsorbates are divalent cations below pH 7.6, the imogolite surface has a markedly higher affinity for adsorbing

$Cu^{2+}(aq)$. This difference in adsorption behavior cannot be explained by cation valence.

Figure 9.11 plots the pH-dependent adsorption of selenate SeO_4^{2-} and selenite SeO_3^{2-} by goethite. The solution chemistry of these two components is also quite different. Selenate exists predominantly as the divalent anion SeO_4^{2-} throughout the entire pH range. Selenite exists predominantly as the neutral species $H_2SeO_4^0(aq)$ below pH 3.5, the univalent anion $HSeO_4^-(aq)$ predominates over the pH range 3.5 to 7.6, and the divalent anion $SeO_4^{2-}(aq)$ becomes the major solution species at pH 7.6.

While the decrease in selenate adsorption with increasing pH appears to correlate with the decrease in negative surface charge (Su and Suarez, 1997), nearly all of the selenite in solution adsorbs to a negatively charged goethite surface while the dominant solution species is neutral $H_2SeO_4^0(aq)$ and substantial selenite adsorption occurs in the pH range 9 to 11, where the goethite surface either has a net zero surface charge or is negatively charged and the dominant selenite solution species is a divalent anion SeO_3^{2-}. Selenium oxyanion adsorption behavior cannot be explained by anion valence.

Numerous studies report similar results where the adsorption of solution species by mineral surfaces does not correlate with surface charge. This type of adsorption behavior is clearly not an ion exchange process, though ion exchange and electrostatic forces may contribute to the adsorption process. Apparently the adsorption affinity of a particular mineral surface for a given component is determined by the chemistry of both reactants—surface site and adsorbate. Surface chemists have yet to develop a predictive model for the adsorption of inorganic components (neutral or charged species) by mineral surfaces based on the chemical properties of both reactants.

The chemical state and the properties of a mineral surface site are more difficult to characterize than solution species. Measurements of proton surface charge (see Figure 9.12) have proven insufficient to characterize adsorption affinity. In the past few decades, most adsorption research by environmental chemists has focused on characterizing the chemical structure of the adsorbate-surface site complex—the chemical product that forms when adsorbate binds to the surface of an adsorbent. Perhaps once chemists have collected sufficient chemical information about surface complexes, they will understand what to measure in order to predict adsorption affinity.

9.9. SURFACE COMPLEXES

Huang and Stumm (1973), Stumm and colleagues (Stumm, Hohl et al., 1976), and Schindler and colleagues (Schindler, Furst et al., 1976) extended earlier proton adsorption models to include ion adsorption by mineral surfaces. In much the same way that earlier proton adsorption models depict the binding of protons to the surface by forming a complex with a surface site (9.20), ion adsorption binds to sites at the mineral-water interface through electrostatic forces (see Appendix 9C, Structures **C4** and **C5**) and ligand-exchange bonds.

Structure 4. Monodentate Complex. *Source: Reproduced with permission from Barger, J.R., Brown, G.E., et al., 1997. Surface complexation of Pb(II) at oxide-water interfaces: I. XAFS and bond-valence determination of mononuclear and polynuclear Pb(II) sorption products on aluminum oxides. Geochim. Cosmochim. Acta 61, 2617–2637.*

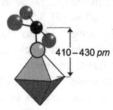

410–430 pm

nonbonded d(Pb···Al)

Structure 5. Bidentate Complex. *Source: Reproduced with permission from Barger, J.R., Brown, G.E., et al., 1997. Surface complexation of Pb(II) at oxide-water interfaces: I. XAFS and bond-valence determination of mononuclear and polynuclear Pb(II) sorption products on aluminum oxides. Geochim. Cosmochim. Acta 61, 2617–2637.*

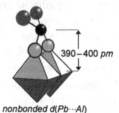

390–400 pm

nonbonded d(Pb···Al)

A ligand-exchange reaction occurs when an oxygen or hydroxyl ion at the mineral surface displaces water molecules surrounding a metal cation to form an inner-sphere surface complex. The mono-dentate surface complex in Structure **4** represents the probable structure for about half of the 2.0 $\mu mol \cdot m^{-2}$ $Pb(II)$ adsorbed to $\alpha - Al_2O_3(s)$ at pH 7 (Barger, Brown et al., 1997). The bidentate surface complex (Structure **5**) represents the probable structure of the remaining half of the $Pb(II)$ surface complexes under those conditions. Barger and colleagues estimate the bond length $d(Pb - O)$ and the nonbonded internuclear distance $d(Pb \cdots Al)$ using scattering data from X-ray absorption spectroscopy.

Barger and colleagues assumed 3-fold coordination for the observed $d(Pb - O)$ bond distance of 232 pm. $Pb(II)$ coordination in the mineral litharge $PbO(s)$ is 4-fold, and the mean $d(Pb - O)$ bond distance is 231 pm, so Structures **4** and **5** were modified to show the most probable $Pb(II)$ coordination environment; the nonbonded internuclear distance $d(Pb \cdots Al)$ is not affected as long as the $d(Pb - O)$ bond distance remains the same.

X-ray absorption methods offer two independent estimates of coordination number: the number of oxygen atoms directly bonded to $Pb(II)$, which is a direct estimate based on the intensity of the scattering peak (Figure 9.13) for the first

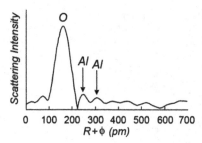

FIGURE 9.13   This graph plots the X-ray absorption scattering intensity at the Pb L_3 absorption edge from 2.0 $\mu mol \cdot m^{-2}$ $Pb(II)$ adsorbed to $\alpha - Al_2O_3(s)$ at pH 7 *Source: Reproduced with permission from Barger, J.R., Brown, G.E., et al., 1997. Surface complexation of Pb(II) at oxide-water interfaces: I. XAFS and bond-valence determination of mononuclear and polynuclear Pb(II) sorption products on aluminum oxides. Geochim. Cosmochim. Acta 61, 2617-2637.*

atomic shell, and an indirect estimate based on the bond valence (Brown and Altermatt, 1985), which is calculated from $d(Pb - O)$ bond distance. Bond distance measurements from X-ray absorption methods, based on the position of the scattering peaks (Figure 9.13), are consistently more accurate than coordination number measurements, making the latter approach for estimating coordination number the most reliable.

The X-ray absorption scattering curve (Figure 9.13) is equivalent to a radial distribution function, showing the intensity of X-ray scattering by atoms surrounding the atom that absorbs the X-ray radiation. Since the scattering curve in Figure 9.13 is from X-ray absorption at the Pb L_3 absorption edge (i.e., 13,055 eV), the origin is occupied by $Pb(II)$. The first scattering peak, labeled O for oxygen atoms in the first atomic shell, is the most intense (the distance R in the scattering curve is not the actual bond distance because X-ray scattering introduces a phase shift ϕ that depends on the scattering atom). Two other scattering peaks, labeled Al for aluminum atoms in the second atomic shell, also appear in Figure 9.13. The first corresponds to the shorter bi-dentate Structure **5** and the second to the longer mono-dentate Structure **4**.

Anions also form surface complexes by displacing a hydroxyl or water molecule at the mineral surface and water molecules surrounding the hydrated anion (Structure **6**).

Structure 6. Phosphate (shaded circles) inner-sphere complex with Site C at the goethite (100) surface. *Source: Reproduced with permission from Russell, J.D., Parfitt, R.L., et al., 1974. Surface structures of gibbsite goethite and phosphated goethite. Nature 248, 220–221.*

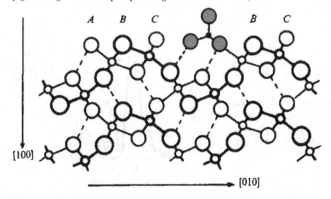

Structure 7. Outer-sphere complex of *Pb(II)* adsorbed to α–Al₂O₃(s). *Source: Reproduced with permission from Barger, J.R., Towle, S.N., et al., 1996. Outer-sphere Pb(II) adsorbed at specific sites on single crystal alpha-alumina. Geochim. Cosmochim. Acta 60, 3541–3547.*

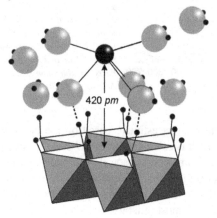

Structure **7** illustrates an outer-sphere complex where the ion—*Pb(II)* in this case—retains its entire first hydration shell and binds to the mineral surface through a combination of electrostatic forces and hydrogen bonds between the surface and water molecules hydrating the adsorbed ion.

Surface chemists using X-ray absorption spectroscopy and other methods have also collected evidence of surface precipitates forming at the mineral-water interface under conditions in which water chemistry simulations fail to predict precipitation. The X-ray scattering curve in Figure 9.14 represents a surface complex between *Dy(III)*, a lanthanide element whose chemistry resembles trivalent actinide elements, and surface sites on the hydrous aluminum oxide boehmite $\gamma-AlOOH(s)$.

Phosphate ions adsorb to all environmental clay surfaces under acid conditions. The presence of the phosphate dramatically increases the affinity of the boehmite surface for *Dy(III)*, producing a surface complex whose structure is very

FIGURE 9.14 This graph plots the X-ray absorption scattering intensity at the *Dy* L_3 absorption edge from 1.0 $\mu mol \cdot m^{-2}$ *Dy(III)* adsorbed to $\gamma - AlO(OH)(s)$ boehmite at pH 7. *Source: Reproduced with permission from Yoon, S.J., Helmke, P.A., et al., 2002. X-ray absorption and magnetic studies of trivalent lanthanide ions sorbed on pristine and phosphate-modified boehmite surfaces. Langmuir 18, 10128–10136.*

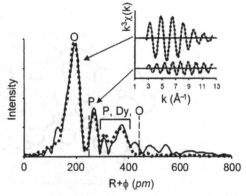

similar to the structure of known dysprosium phosphate crystalline solids (Yoon, Helmke et al., 2002). This is only one example of many where the surface complex formed when inorganic ions adsorbed at mineral surfaces resemble a precipitate rather than the surface complexes illustrated in Structures **4–7**.

9.10. SUMMARY

This chapter began with a description of mineral and organic colloidal adsorbents commonly found in soils, sediments, aquifers, and suspended surface waters. These colloidal adsorbents have sufficient capacity, a direct consequence of their high surface-to-volume ratio, to adsorb compounds dissolved in water to significantly alter pore water chemistry.

Environmental chemists typically encounter two types of adsorption processes: site-limited and partitioning. The two processes have distinctly different adsorption isotherms, L-type and C-type, respectively. L-type adsorption isotherms characterize the adsorption of most compounds by mineral colloids or cation adsorption by humic colloids, but C-type isotherms are restricted to the adsorption of neutral organic compounds by humic colloids and other molecular-aggregate colloids. Neutral organic compounds partition into nonpolar domains within molecular-aggregate colloids, giving rise to their distinctive adsorption isotherm.

Both the capacity and the affinity of mineral colloids, with the exception of layer silicate clays, to adsorb inorganic and organic ions depend on pH. The pH-dependent adsorbing behavior of mineral colloids results from the adsorption and desorption of protons, leading to the development of surface charge. The effect of pH-dependent surface charge on the adsorbing behavior of mineral colloids is well documented but poorly understood.

The current state of our understanding comes from studies that determine the structure and composition of the surface complex that forms when inorganic and organic ions bind to mineral surfaces. In some cases the chemical structure of the surface complex resembles the simple model similar to the model for proton binding to oxygen atoms at the mineral surface, but in other cases the chemical structure of the surface complex has a composition that is incompatible with the simple site-binding model.

The ultimate significance of surface chemistry and adsorption for environmental chemists can be traced back to the larger question of biological availability and toxicity. Plants and animals take up compounds dissolved in water much more readily than compounds that are precipitated as insoluble solids or adsorbed to the surfaces of environmental colloids. Adsorption processes immobilize potential toxicants, lowering their uptake by living organisms and their movement with groundwater. Adsorption also ensures that toxicants adhering to environmental colloids pose a continuing health risk because it represents a reservoir that continually releases toxicants into pore water for biological uptake.

APPENDIX 9A. PARTICLE SEDIMENTATION RATES IN WATER: STOKES'S LAW

Solid particles suspended in a liquid settle under the force of gravity. The rate of sedimentation depends on particle buoyancy (the difference between the density of the particles and the density of the liquid), the viscosity of the liquid and the size of the particles.

The net force (9A.1) causing a particle suspended in a liquid to settle is the difference between the gravitational F_g force and buoyancy F_b force. The buoyancy force F_b is a function of: particle volume V, particle density $\rho_{particle}$, and the density ρ_{fluid} of the suspending fluid.

$$F_g - F_b = V_{particle} \cdot \Delta\rho \cdot g \tag{9A.1}$$

$$\Delta\rho = \left(\rho_{particle} - \rho_{fluid}\right) \tag{9A.2}$$

A frictional force f—proportional to the viscosity η of the fluid—opposes the settling of the suspended particle (9A.3). The particle will initially accelerate to a terminal velocity v at which point the frictional force equals the net settling force and the particle begins to settle at a constant velocity v.

$$V_{particle} \cdot \Delta\rho \cdot g = f \cdot v \tag{9A.3}$$

G. G. Stokes (1851) derived an equation (9A.4) for the frictional force f acting on a spherical particle with radius R. Combining the settling velocity of a spherical particle (9A.3) with the frictional expression (9A.4) yields (9A.5).

$$f = 6\pi \cdot \eta \cdot R \tag{9A.4}$$

$$\left(\frac{4\pi}{3}R^3\right) \cdot \Delta\rho \cdot g = (6\pi \cdot \eta \cdot R) \cdot v \tag{9A.5}$$

Solving (9A.5) for the sedimentation velocity we have (9A.6), the Stokes sedimentation velocity expression used to determine particle size.

$$v = \left(\frac{2}{9}\right)\left(\frac{R^2 \cdot \Delta\rho \cdot g}{\eta}\right) \tag{9A.6}$$

Example

The viscosity η of water at 20°C is 1.002 10^{-3} $kg\ m^{-1}\ s^{-1}$. The acceleration due to gravity g is 9.802 $m\ s^{-2}$. The densities of water and silicate minerals at 20°C are 1.00×10^3 $kg\ m^{-3}$ and 2.65 10^3 $kg\ m^{-3}$, respectively. The following table lists the sedimentation rate of spherical silicate particles in water at 20°C.

Radius, m	Sedimentation Velocity, $m\ s^{-1}$
1.0×10^{-6}	3.59×10^{-6}
1.0×10^{-7}	3.59×10^{-8}
1.0×10^{-8}	3.59×10^{-10}
1.0×10^{-9}	3.59×10^{-12}

The sedimentation rate is on the order of one particle diameter per second when the radius is 1 μm at the upper limit of the colloidal range and has decreased to on the order of one-thousandth of a particle diameter when the radius is 1 nm at the lower limit of the colloidal range. In other words, a particle with a radius of 1 nm would settle 1 mm in 10^9 seconds or roughly 30 years. Such slow settling rates make colloidal suspensions "stable" on the timescale of most laboratory experiments.

APPENDIX 9B. LINEAR LANGMUIR EXPRESSION

The non-linear *L-type* adsorption isotherm expression (9B.1) given here is identical to expression (9.9) given earlier in section 9.5.1.

$$q = \frac{q_{max} \cdot K_{ac} \cdot C}{1 + (K_{ac} \cdot C)} \tag{9B.1}$$

Non-linear expression (9B.1) is rearranged to yield a linearized expression (9B.2) where the slope $(q_{max})^{-1}$ and intercept $(q_{max} \cdot K_{ac})^{-1}$ contain the same parameters appearing in the non-linear expression. This approach was popular at a time when non-linear optimization was not a practical option.

$$\left(\frac{1}{q}\right) = \frac{1 + (K_{ac} \cdot C)}{q_{max} \cdot K_{ac} \cdot C}$$

$$\left(\frac{1}{q}\right) = \left(\frac{1}{q_{max} \cdot K_{ac} \cdot C}\right) + \left(\frac{K_{ac} \cdot C}{q_{max} \cdot K_{ac} \cdot C}\right)$$

$$\left(\frac{1}{q}\right) = \left(\frac{1}{q_{max} \cdot K_{ac} \cdot C}\right) + \left(\frac{1}{q_{max}}\right)$$

$$\left(\frac{C}{q}\right) = \left(\frac{C}{q_{max} \cdot K_{ac} \cdot C}\right) + \left(\frac{C}{q_{max}}\right)$$

$$\left(\frac{C}{q}\right) = (q_{max} \cdot K_{ac})^{-1} + (q_{max}^{-1}) \cdot C \tag{9B.2}$$

Parameter fitting involves linear regression of a linearized dataset $(C/q, C)$ followed by extraction of the two *L-type* isotherm parameters (q_{max}, K_{ac}) from the slope (9B.3) and intercept (9B.4).

$$q_{max} = \frac{1}{slope} \tag{9B.3}$$

$$K_{ac} = \frac{slope}{intercept} \tag{9B.4}$$

APPENDIX 9C. HYDROLYSIS MODEL OF PROTON SITES

It is possible to generate a model of any mineral surface by cleaving the crystal structure parallel to a specific crystal face. The example here is the same (100) goethite surface described by Russell and colleagues (Russell, Parfitt et al., 1974). The goethite crystal structure, viewed along the b crystal axis, appears in Structure **C1**. Oxygen ions O^{2-} appear as open circles, hydroxyl ions OH^- as shaded circles, water molecules H_2O as large black-filled circles, and protons as small filled circles in Structures **C1–C4**.

Structure C1. Goethite crystal structure (a- and c-axes are in plane of the page). *Source: Reproduced with permission from Russell, J.D., Parfitt, R.L., et al., 1974. Surface structures of gibbsite goethite and phosphated goethite. Nature 248, 220–221.*

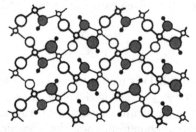

Structure **C2** illustrates the cleavage along the goethite (100) crystal plane to create two identical (100) surfaces. The arrows show the final position of oxygen ions O^{2-} following cleavage. One quarter of the Fe^{3+} coordination polyhedra on each surface is 5-coordinate after this imaginary cleavage.

Structure C2. Cleavage along the (100) crystal plane of goethite. *Source: Reproduced with permission from Russell, J.D., Parfitt, R.L., et al., 1974. Surface structures of gibbsite goethite and phosphated goethite. Nature 248, 220–221.*

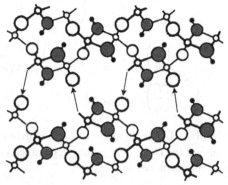

Parks and de Bruyn (1962) described the mechanism for the development of surface charge as follows:

The mechanism by which the surface charge is established may be viewed qualitatively as a two-step process: surface hydration followed by dissociation of the surface "hydroxide." The hydration step may be visualized as an attempt by the exposed surface cations to complete their coordination shell of nearest neighbors. . . . The net result is that the surface now is covered by a hydroxyl layer with the cations buried below the surface.

The adsorption of water molecules at the tail of each arrow in structure **C2** restores all Fe^{3+} coordination polyhedra to 6-coordinate without depositing charge on the newly formed surface. This is illustrated in Structure **C3**. Additional water molecules would adsorb to the oxide surface without directly coordinating Fe^{3+} ions through hydrogen bonding. Water molecules adsorbed by hydrogen bonding represent the first layer of water molecules hydrating the oxide surface.

Structure **C3** differs from Structure **4** (Russell, Parfitt et al., 1974) because a water molecule bridges between the structural water molecule at Site α in Structure **C3** and the structural oxygen at Site χ. Structure **C3** is explicitly charge neutral because the water molecule α transfers one proton through the bridging water molecule to the structural oxygen χ.

The bond valence sum for each of the three sites is calculated using the electrostatic valence principle (Pauling, 1929) in equations (9C.1) through (9C.3).

$$\zeta_\alpha = \left(\frac{3}{6}\right)_{Fe-O} + \left(\frac{1}{1}\right)_{H-O} = 1.5 \ v.u. \qquad (9C.1)$$

$$\zeta_\beta = \left(\frac{3}{6}\right)_{Fe-O} + \left(\frac{3}{6}\right)_{Fe-O} + \left(\frac{3}{6}\right)_{Fe-O} + \left(\frac{1}{1}\right)_{H-O} = 2.5 \ v.u. \qquad (9C.2)$$

$$\zeta_\chi = \left(\frac{3}{6}\right)_{Fe-O} + \left(\frac{3}{6}\right)_{Fe-O} + \left(\frac{1}{1}\right)_{H-O} = 2.0 \ v.u. \qquad (9C.3)$$

Structure C3. The neutral (100) surface of goethite with adsorbed water molecules. *Source: Reproduced with permission from Russell, J.D., Parfitt, R.L., et al., 1974. Surface structures of gibbsite goethite and phosphated geothite. Nature 248, 220–221.*

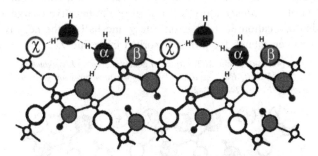

Structure **C3**—the model for a charge neutral goethite surface—is our point of reference. Titrating the surface by adding a strong acid *HX* will increase the tendency for the surface to adsorb protons by Site α (9C.4), as depicted in Structure **C4**.

$$O_\alpha \quad +H^+(aq) + X^-(aq) = O_\alpha H^+(aq) + X^-(aq) \qquad (9C.4)$$
$$\zeta_\alpha = 1.5 \ v.u. \qquad\qquad\qquad\qquad \zeta_\alpha = 2.5 \ v.u.$$

The state of Site χ remains the same as in Structure **C3**, with a bond valence sum (9C.3) of 2.5 v.u. Half of the positive charge acting on the co-adsorbed anion $X^-(aq)$ is on Site α and half of the negative charge is on Site χ even though the proton adsorbed at Site α.

Structure C4. The proton-rich (100) surface of goethite. *Source: Reproduced with permission from Russell, J.D., Parfitt, R.L., et al., 1974. Surface structures of gibbsite goethite and phosphated goethite. Nature 248, 220–221.*

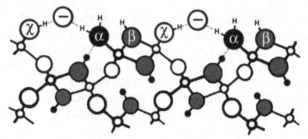

Proton adsorption, by the structural hydroxyl (Site α in Structure **C3**) converting it to a water molecule (Site α in Structure **C4**), generates a positively charged surface that will co-adsorb anions X^- to maintain charge balance.

Titrating the surface by adding a strong base *MOH* will increase the tendency for the surface to desorb protons from Site χ (9C.5), as depicted in Structure **C5**. Removing protons from Structure **C3** through proton desorption (Structure **C5**) creates a negatively charged surface that will coadsorb cations to maintain charge balance.

$$O_\chi + M^+(aq) + OH^-(aq) = O_\chi^-(aq) + M^+(aq) + H_2O(l) \qquad (9C.5)$$
$$\zeta_\chi = 2.5 \ v.u. \qquad\qquad\qquad \zeta_\chi = 1.5 \ v.u.$$

The state of Site α remains the same as in Structure **C3**, with a bond valence sum (9C.1) of 1.5 valence units (v.u.). Half of the negative charge acting on the co-adsorbed cation $M^+(aq)$ is on Site χ and half of the negative charge is on Site α even though the proton desorbed from Site χ.

Structure C5. The proton-deficient (100) surface of goethite. *Source: Reproduced with permission from Russell, J.D., Parfitt, R.L., et al., 1974. Surface structures of gibbsite goethite and phosphated goethite. Nature 248, 220–221.*

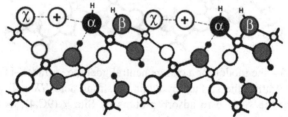

A proton-rich (positively charged) oxide surface (Structure **C4**) develops as protons bind to surface oxygen ions. Similarly, a proton-deficient (negatively charged) oxide surface (Structure **C5**) develops as protons desorb from water molecules on the surface. The proton surface site is best considered as a water-hydroxyl pair (Structure **C4**) or a hydroxyl-oxygen pair (Structure **C5**) on goethite. In a similar analysis of gibbsite surfaces, where the Al^{3+} coordination polyhedra form sheets rather than chains, the proton-rich surface consists of water-water pairs, the neutral surface as water-hydroxyl pairs, and the proton-deficient surface as hydroxyl-hydroxyl pairs.

Problems

1. A sorption experiment of the organophosphate insecticide *Supona* by the Yolo soil series yielded the results listed in the following table. (The insecticide was labeled with radioactive C-14, so the activity is in arbitrary units.)

C_{Supona}, activity mL^{-1}	q_{Supona}, activity g^{-1}
0.0182	12.04
0.0358	26.18
0.0709	47.12
0.1406	91.10
0.2896	194.5

Estimate the coefficient, K_{OM}, for *Supona* sorption on the Yolo soil, based on a linear fit of the data. Express K_{OM} in units of milliliters per gram of organic matter.

Solution
A linear fit of sorbed activity q_{Supona} as a function of C_{Supona} activity in solution yields a slope of 667: $K_{OM} = 667 \ g \ mL^{-1}$.

2. The Yolo soil in problem 1—from a University of California-Davis experiment station near Sacramento, California—contains 5.3% organic matter on a dry-weight basis. Using your results from problem 1, determine the K_D for *Supona* sorption on the Yolo soil. Express K_D in units of milliliters per gram of (whole) soil.

Solution

The relationship between K_{OM} and K_D is given following.

$$K_D = f_{OM} \cdot K_{OM}$$
$$K_D = (0.053) \cdot (667)$$
$$K_D = 35.4$$

3. The following data are from an experiment measuring phosphate adsorption by a soil. The experimental data consist of the quantity of phosphate remaining in solution at steady state, $C_P (\mu g_P \cdot mL^{-1})$, and the quantity of substance phosphate-phosphorus on the surface at steady state, $q_P (\mu g_P \cdot g_{soil}^{-1})$.

$C_P (\mu g_P \cdot mL^{-1})$	$q_P (\mu g_P \cdot g_{soil}^{-1})$
0.040	35.62
0.141	49.38
0.344	62.71
0.463	75.21
1.001	92.71
1.821	108.33
5.984	146.04
11.09	168.75
17.26	172.71
22.53	187.92

An effective way to estimate the two Langmuir adsorption isotherm parameters (q_{max} and the affinity parameter K_D) is by using a linearized Langmuir equation:

$$\left(\frac{C}{q}\right) = \alpha \cdot C + \beta$$
$$\alpha \equiv q_{max}^{-1}$$
$$\beta \equiv (K_{ac} \cdot q_{max})^{-1}$$

Plot the data from the experiment using the preceding linearized Langmuir equation, perform a linear fit of the linearized data, and determine the two Langmuir adsorption isotherm parameters: q_{max} and K_{ac}.

Solution

A linear fit $(C_P \cdot q_P^{-1})$ as a function of the soluble concentration C_P yields a slope α of 5.331E-3 and an intercept β of 4.628E-3. The two Langmuir isotherm parameters q_{max} and K_D are defined using the preceding relations.

$$\alpha \equiv q_{max}^{-1}$$

$$q_{max} = \alpha^{-1} = \left(5.331 \cdot 10^{-3}\right)^{-1} = 187.6 \ \mu g_P \cdot g_{soil}^{-1}$$

$$\beta \equiv \left(K_D \cdot q_{max}\right)^{-1}$$

$$K_{ac} = \frac{\alpha}{\beta} = \frac{4.628 \cdot 10^{-3}}{5.331 \cdot 10^{-3}} = 1.152$$

4. The following data (courtesy of T. Ranatunga and R. W. Taylor, Alabama A & M University) are from an experiment measuring Pb^{2+} adsorption at pH 4 by the nonswelling clay mineral kaolinite with a surface area of $9.6 \ m^2 \cdot g^{-1}$. The experimental data consist of the quantity of Pb^{2+} remaining in solution at steady state, $C_{Pb^{2+}} \left(mol_{Pb^{2+}} \cdot L^{-1}\right)$, and the quantity of Pb^{2+} adsorbed by kaolinite at steady state, $q_{Pb^{2+}} \left(\mu mol_{Pb^{2+}} \cdot m_{kaolinite}^{-2}\right)$.

$C_{Pb^{2+}} \left(mol_{Pb^{2+}} \cdot L^{-1}\right)$	$q_{Pb^{2+}} \left(\mu mol_{Pb^{2+}} \cdot m_{kaolinite}^{-2}\right)$
4.83E-05	2.52E-07
9.65E-05	3.92E-07
1.45E-04	4.48E-07
1.93E-04	5.17E-07
2.41E-04	5.60E-07
2.90E-04	5.98E-07
3.38E-04	6.29E-07
3.86E-04	6.81E-07
4.34E-04	6.75E-07
4.83E-04	7.17E-07

An effective way to estimate the two Langmuir adsorption isotherm parameters (q_{max} and the affinity parameter K_D) is by using a linearized Langmuir equation:

$$\left(\frac{C}{q}\right) = \alpha \cdot C + \beta$$

$$\alpha \equiv q_{max}^{-1}$$

$$\beta \equiv \left(K_{ac} \cdot q_{max}\right)^{-1}$$

Plot the data from the experiment using the preceding linearized Langmuir equation, perform a linear fit of the linearized data, and determine the two Langmuir adsorption isotherm parameters: q_{max} and K_{ac}.

Solution

A linear fit $\left(C_{Pb^{2+}} \cdot q_{Pb^{2+}}^{-1}\right)$ as a function of the soluble concentration $C_{Pb^{2+}}$ yields a slope α of 1.12E+6 and an intercept β of 1.51E+2. The two Langmuir isotherm parameters q_{max} and K_{ac} are defined using the preceding relations.

Chapter 10

Risk Assessment

Chapter Outline

10.1. INTRODUCTION

A major impetus for the growth of environmental chemistry dates from the jarring discovery in the 1960s and 1970s of human health hazards caused by environmental pollution. Prior to the modern environmental movement, key amendments to the Federal Food, Drugs and Cosmetic Act (U.S. Congress,

Soil and Environmental Chemistry.
© 2012 Elsevier Inc. All rights reserved.

1938)—the Miller Pesticide Amendments (U.S. Congress, 1954), the Food Additives Amendment (U.S. Congress, 1958), and the Color Additives Amendment (U.S. Congress, 1960)—were enacted to mitigate exposure to potential carcinogens in food. This formed a basis for developing the risk assessment paradigm employed by the U.S. federal government (Figure 10.1) that expanded risk assessment beyond food to include air, water, and soil.

Chemical pollution of the environment results from the release of compounds that are toxic to humans. Chronic human toxicity can take many forms (Table 10.1) that invariably originate with either the impairment of normal

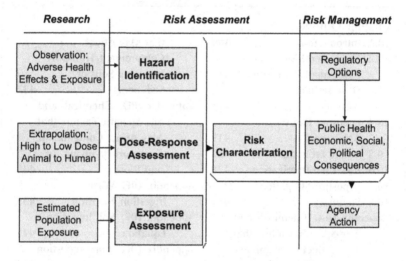

FIGURE 10.1 The Federal Risk Assessment Paradigm. *Source: Reproduced with permission from National Research Council, 1983. Risk Assessment in the Federal Government: Managing the Process. Washington, D.C., National Academies Press.*

TABLE 10.1 A Classification of Chronic Toxicities and their Human Health Effects

Toxicity	Adverse Health Effect
Germ cell mutagenicity	Mutation of adult germ cells resulting in heritable abnormal traits
Carcinogenicity	Chromosomal damage resulting in the development of malignant tumors (See Box 10.1)
Reproductive toxicity (teratogenicity)	Loss of adult fertility and abnormal development of fetus or children
Specific target organ toxicity	Loss of organ function

biochemical processes or genetic damage. Genetic damage can result in abnormal development, mutations, or cancer.

The following section reviews the paradigm employed by the U.S. federal government to assess human health hazards resulting from chemical pollution of the environment and to mitigate adverse health effects resulting from chemical pollution.

10.2. THE FEDERAL RISK ASSESSMENT PARADIGM

10.2.1. Risk Assessment

Scientific research supports human health risk assessment in three ways (see Figure 10.1): identifying hazardous chemical compounds, developing dose-response functions necessary to predict adverse effects based on low-level chronic doses delivered by the environment, and estimating uptake from various routes by representative human populations. Risk assessment draws from each of these scientific fields to characterize specific health risks, whether that is an incremental excess lifetime cancer risk or the probability of an adverse effect in a particular population.

10.2.2. Risk Management and Mitigation

Risk characterization is essentially a quantitative prediction of the adverse effect on human health that integrates our scientific understanding of toxicology and environmental exposure. Federal, state, and local government agencies assume ultimate responsibility for mitigating health hazards by establishing emission standards or action levels, both designed to reduce contaminant levels at various exposure points (air, water, and soil). Mitigation may involve removal and secure disposal of hazardous waste or excessively contaminated soil or, in many cases, management of contaminated sites by imposing institutional controls designed to limit human contact with contaminants that are impractical to remove. As indicated in Figure 10.1, risk management is an endeavor that attempts to balance the risk posed by pollution against the economic, social, and political costs of reducing risk. Not surprisingly, marginal costs increase substantially as risk declines.

10.3. DOSE-RESPONSE ASSESSMENT

From the outset, risk assessment follows different routes depending on whether a chemical compound is identified as a known (or potential) carcinogen. The assessment of noncarcinogens owes much to basic principles of pharmacology, while carcinogens owe their unique assessment to early studies of radiation-induced cancer. As our understanding of carcinogenesis improves, there is growing recognition within the scientific community that early distinctions between carcinogens and noncarcinogens were unwarranted. The regulatory community, however, has proven unwilling or unable to abandon practices developed 50 years ago.

10.3.1. Dose-Response Distributions

Toxicologists fit dose-response data to numerous mathematical response functions; some (but not all) are cumulative probability distributions (*cdf*) derived from processes with specific characteristics (e.g., a Poisson *cdf* represents the probability of random independent events such as radioactive decay). The primary distinguishing characteristic of these response functions is the presence or absence of a response threshold; the more common threshold response functions (Table 10.2) are plotted in Figures 10.2 to 10.4.

TABLE 10.2 The Most Common Threshold Response Functions

Model	Cumulative Distribution Function	Parameters
logistic	$P(D; \mu, \sigma) = \left(\dfrac{1}{1 + e^{-(D-\mu)/\sigma}} \right)$	*location* = $\mu > 0$, *scale* = $\sigma > 0$
Poisson	$P(D; \lambda) = \sum_{i=0}^{D} \left(\dfrac{\lambda^i \cdot e^{-\lambda}}{i!} \right)$	λ
Weibull	$P(D; \lambda, k) = 1 - e^{-(D/\lambda)^k}$	*scale* = $\lambda > 0$, *shape* = $k > 0$

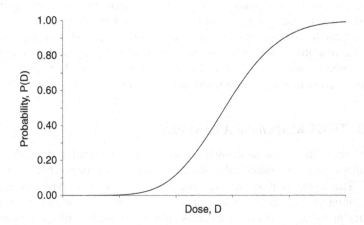

FIGURE 10.2 The Poisson cumulative probability distribution is a threshold response function.

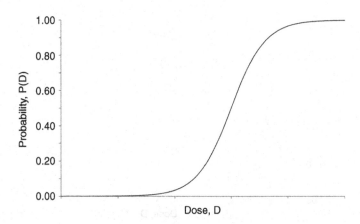

FIGURE 10.3   The logistic cumulative probability distribution is a threshold response function.

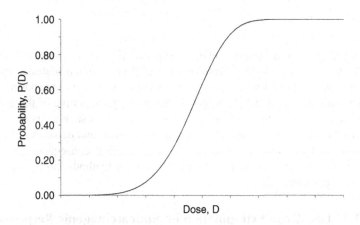

FIGURE 10.4   The Weibull cumulative probability distribution is a threshold response function.

10.3.2. The No-Threshold One-Hit Model

The *one-hit* model has a meaning that extends beyond a specific mathematical expression. The simplest one-hit *cdf* employs a single dose coefficient parameter λ.

$$P(D; \lambda) = 1 - e^{-\lambda \cdot D} \tag{10.1}$$

The most recognizable attribute of any one-hit model is the absence of a threshold dose D (Figure 10.5). "Carcinogen risk assessment models have generally been based on the premise that risk is proportional to cumulative lifetime dose. For lifetime human exposure scenarios, therefore, the exposure metric used

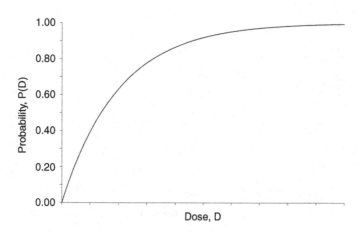

FIGURE 10.5 The cumulative probability distribution for the simplest one-hit model is a nonthreshold response function.

for carcinogenic risk assessment has been the lifetime average daily dose"
(EPA, 2005). The dose D is prorated over a 70-year human lifetime, scaling
the probability of carcinogenesis $P(D; \lambda)$ by continuous lifetime exposure.

The simplest no-threshold response function representative of the one-hit
model appears in Figure 10.5. While response functions such as those in Figures 10.2 to 10.5 are the subject of continuing research and debate by pharmacologists and toxicologists, they are of little practical consequence because
risk assessment focuses on low-dose extrapolation methods largely independent of a particular cdf.[1]

10.3.3. Low-Dose Extrapolation of Noncarcinogenic Response Functions

Several fundamental assumptions govern low-dose extrapolations when the
toxicant is a noncarcinogen. The first relates to the pharmacokinetics of
noncarcinogenic chemicals in humans and animals. The adverse effect of a
toxicant is a function of the steady-state concentration in the body—the *body
burden*. The steady-state body burden is a function of the intake and elimination rates from the body, quantified by the biological half-life $t_{1/2}$ of the
substance.

The dose plotted in a noncarcinogenic dose-response curve is the steady-state body burden $(mg \cdot kg^{-1})$ prorated on a daily basis: the *average daily dose*

---

1. Low-dose extrapolation using the benchmark–dose method does depend on the choice of cdf.

ADD $(mg \cdot kg^{-1} \cdot d^{-1})$. The cumulative dose ingested throughout the exposure duration is considered irrelevant.

10.3.4. Estimating the Steady-State Body Burden

The steady-state body burden of any chemical substance depends on intake and elimination rates, R_I and R_E. The elimination rate R_E depends on the elimination route, kinetics (typically first-order rate law), and storage (i.e., the bioconcentration factor BCF).

At steady state, the rate of intake equals the rate of elimination:

$$R_I = R_E$$

$$IR \cdot C_I = ER \cdot C_E \qquad (10.2)$$

The intake rate $R_I(mg \cdot d^{-1})$ is a function of the ingestion rate factor $IR(water: L \cdot d^{-1}, soil\ or\ food: kg \cdot d^{-1})$ (i.e., the volume of water ingested or the mass of a particular food ingested per day) times the concentration of the contaminant in the water or food being ingested C_I.

The elimination rate $R_E(mg \cdot d^{-1})$ is a function of the elimination rate factor—for example, the volume of urine eliminated per day $ER(L \cdot d^{-1})$—times the concentration of the contaminant in the urine C_E. The elimination rate constant k_E, a first-order constant, is a function of the biological half-life $t_{1/2}$ of the contaminant in the organism or, alternatively, the effective distribution volume of water in the body V_d:

$$k_E = \left(\frac{\ln(2)}{t_{1/2}}\right) = \left(\frac{ER}{V_d}\right) \qquad (10.3)$$

In the short term, the concentration being eliminated depends on the concentration in the food (or water), the relative rates of intake and elimination, and the biological half-life.

$$C_E = \left(\frac{IR \cdot C_I}{ER}\right) \cdot \left(1 - e^{-(ER/V_d) \cdot t}\right) = \left(\frac{IR \cdot C_I}{ER}\right) \cdot \left(1 - e^{-(\ln(2) \cdot t/t_{1/2})}\right) \qquad (10.4)$$

The steady-state concentration, however, is simply the concentration of the contaminant in the water or food being ingested C_I times the ratio of ingestion and elimination rates:

$$C_{E(SS)} = \frac{IR \cdot C_I}{ER} \qquad (10.5)$$

Figure 10.6 illustrates a low-dose extrapolation method for noncarcinogenic dose-response functions relative to three toxicological end-points: the lowest observable adverse effect level $LOAEL$, the no adverse effect level $NOAEL$, and the reference dose RfD. The error bars in Figure 10.6 illustrate the statistical distinction between $NOAEL$—the *highest* dose whose adverse effect is

FIGURE 10.6 Low-dose portion of an adverse effect dose-response curve for a noncarcinogen showing three toxicological end-points used in risk assessment: *LOAEL*, *NOAEL*, and *RfD*.

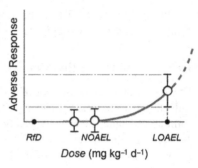

statistically insignificant—and *LOAEL*—the *lowest* dose whose adverse effect is statistically significant. This portion of the dose-response curve does not involve extrapolation. Relative to these two points is the *reference dose RfD*, an extrapolated toxicological end-point.

10.3.5. Reference Dose *RfD*

The reference dose *RfD* $(mg \cdot L^{-1} \cdot d^{-1})$ is a low-dose extrapolated end-point strictly limited to noncarcinogenic toxicants. The *RfD* end-point can be defined relative to either a *NOAEL* (10.6) or a *LOAEL* (10.7).

$$RfD = \frac{NOAEL}{UF \cdot MF} = \frac{NOAEL}{10^n \cdot 1} \tag{10.6}$$

$$RfD = \frac{LOAEL}{10 \cdot UF \cdot MF} = \frac{LOAEL}{10 \cdot 10^n \cdot 1} \tag{10.7}$$

The *RfD* end-point depends on the nature of the dose-response study, whether it involves humans or animals or whether it is a long- or short-term study.

> *Uncertainty Factors*
> $UF_H = 10^1 (long\text{-}term\ human\ studies)$
> $UF_A = 10^2 (long\text{-}term\ animal\ studies)$
> $UF_S = 10^3 (short\text{-}term\ animal\ studies)$
> *Modifying Factors* $(0 < MF \leq 1;\ default\ MF = 1)$

Interspecies extrapolation—extrapolating from animal-based dose-response studies to define human toxicological end-points—is a common practice involving much greater uncertainty than the type of extrapolation illustrated in Figure 10.6.

10.3.6. Low-Dose Extrapolation of Carcinogenic Response Functions

The earliest research on carcinogenesis came from scientists studying the health effects of ionizing radiation—high-energy alpha and beta particles emitted by radioactive elements and high-energy photons (gamma- and X-rays).

These studies led to the one-hit theory that postulates that the absorption of a single high-energy particle or photon by tissue is sufficient to initiate carcinogenesis. A key feature of the one-hit model is the absence of a threshold dose.

Scientists established certain polycyclic aromatic compounds (e.g., benzo [a]pyrene, Structure **1**, and naphthylamine, Structure **2**) as cancer-causing agents as early as 1949 (Goulden and Tipler, 1949; Cooper, 1954; Lefemine, 1954; Hoffmann, Masuda et al., 1969). The mechanisms of carcinogenesis were poorly understood at that time, leading toxicologists to adopt the one-hit theory of radiation-induced carcinogenesis.

Structure 1. Benzo[a]pyrene (CAS 50-32-8)

Structure 2. 1-Naphthylamine (CAS 134-32-7)

Carcinogenesis, regardless of the cause (Box 10.1), has a lengthy latency period where the duration of the latent period decreases as the dose (radiation or chemical) increases. Furthermore, carcinogens and radiation both produce mutations. Aside from these similarities, the actual interaction with tissue is profoundly different for radiation and chemicals.

The absorption and retention of any chemical compound, regardless of its toxicology (see Table 10.1), depend on its molecular properties. The anatomic path followed by a specific compound, the barriers (gastrointestinal, renal, placental, blood-brain, etc.) it crosses or cannot cross, also depends on their chemical properties. Unlike ionizing radiation, there are few universal carcinogens. Most carcinogens target specific organisms, a single gender, a particular age group, or individuals experiencing specific physiological stresses.

Figure 10.7 illustrates the low-dose extrapolation method for carcinogenic dose-response functions relative to three toxicological end-points: the effective dose resulting in a 10% incremental excess lifetime cancer risk ED_{10}, the lower-limit effective dose (i.e. the lower 95% confidence limit) associated with a 10% incremental excess lifetime cancer risk LED_{10}, and the *point-of-departure POD*. The error bars in Figure 10.7 illustrate the statistical uncertainty at the lower dose limit of the dose-response curve—the point where low-dose extrapolation begins.

The adverse response, given the extended latency of carcinogensis, is scaled as the *incremental excess lifetime cancer risk IELCR*. Figure 10.7 illustrates an assessment based on a 10% *IELCR* that intersects the dose-response curve at a dose designated ED_{10}, although empirical data may extend below

Box 10.1 The Knudson Two-Hit Model

Oncologists associate two regulatory gene classes with carcinogenesis (Gibbs, 2003): oncogenes and tumor suppressor genes. Proto-oncogenes promote cell proliferation in immature or stem cells, while tumor-suppressor genes repress cell division preparatory to further signals that initiate cell differentiation into mature tissue. Knudson (1971) suggested an alternative to the one-hit model of carcinogenesis that postulated the necessity of two disabling mutations, one for each allele of a tumor-suppressor gene, to initiate carcinogenesis—the so-called "two-hit" hypothesis. If the individual inherited a disabling mutation in a single tumor-suppressor allele, then exposure to a carcinogen resulting in a disabling mutation in the second allele—a "one-hit" scenario—would be sufficient to initiate carcinogenesis. Individuals susceptible to carcinogens presumably carry disabling mutations to a single tumor-suppressor allele.

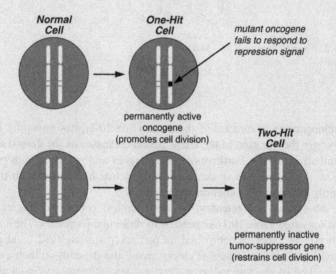

An enabling mutation of a proton-oncogene, perhaps a mutation to the promoter region disabling the binding of a repressor protein, is sufficient to initiate carcinogenesis. Mutant proto-oncogenes, rendering them unresponsive to repression, are known as oncogenes. A one-hit mutation targeting either proto-oncogene allele can result in carcinogenesis, although by a different mechanism than disabling mutations of tumor-suppressor genes.

ED_{10}. The statistical uncertainty associated with the ED_{10} end-point defines the LED_{10} dose associated with a 10% $IELCR$. The point-of-departure POD is best understood by reference to Figure 10.7.

The adverse effect of a carcinogen is assessed as a function of the cumulative lifetime body burden. The dose plotted in a carcinogenic dose-response curve is prorated over a human lifetime of 70 years (EPA,

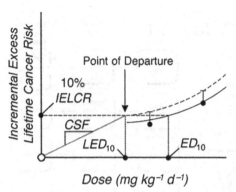

FIGURE 10.7 Low-dose portion of an adverse effect dose-response curve for a carcinogen showing three toxicological end-points used in risk assessment: ED_{10}, LED_{10}, and *POD*.

2005): the *lifetime average daily dose LADD* $mg \cdot kg^{-1} \cdot d^{-1}$. The assumption is that a high carcinogenic body burden accumulated over a short period of time is equivalent to a low body burden accumulated over a lifetime. For instance, daily consumption of 2 liters of water by an adult results in a average daily dose of 15 milligrams per kilogram per day if the exposure duration is 1 year and the concentration is 530 ppm but if this same exposure is prorated over 70 years the life time average daily dose is a mere 0.22 milligrams per kilogram per day.

The one-hit theory assumes there is no threshold and any dose has a finite probability of initiating carcinogenesis; performing a linear extrapolation from the *POD* to zero dose is the proper risk assessment protocol. Notice that the slope is steeper and the estimate of risk is more conservative if the extrapolation is made using a *POD* defined by LED_{10} rather than ED_{10}. The slope of this extrapolation is known as the cancer slope factor *CSF*.

Figures 10.6 and 10.7 demonstrate the profound significance of the one-hit theory on carcinogenic risk assessment. The one-hit theory determines the metrics used to quantify carcinogenic risk: reliance on cumulative rather than steady-state body burden and the low-dose extrapolation method (Figure 10.7). None of these explicitly depends on a specific dose-response function.

10.4. EXPOSURE PATHWAY ASSESSMENT

Figure 10.8 illustrates the elements of exposure pathway assessment. A receptor or exposed population receives a contaminant dose, provided that all elements in the exposure pathway are present.

10.4.1. Receptors

The exposure pathway terminates with a specific human population—the *receptor* in risk assessment terminology—that receives a contaminant dose from the environment (see Figure 10.8). The exposed or potentially exposed

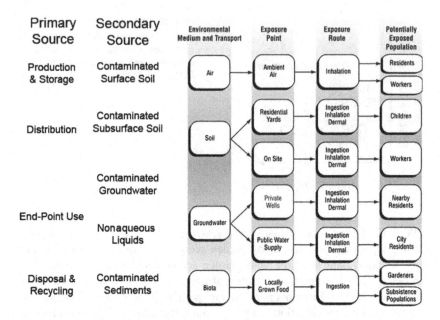

FIGURE 10.8 A complete exposure pathway begins with a contamination source and a mode of environmental transport and reaches an exposed population or receptors through some exposure point—contaminated environmental media—and an exposure (uptake) route. *Source: Reproduced with permission from U.S. Department of Health and Human Services, 1992. Public Health Assessment Guidance Manual. Agency for Toxic Substances and Disease Registry. Atlanta, GA.*

population is distinguished from the general population by some characteristic that brings this subpopulation into contact with contaminated environmental media. For instance, urban air pollution affects ambient air throughout the urban environment, placing the entire residential urban population in contact with contaminant gases and particulates, while certain manufacturing processes pollute a work site, exposing only workers at that site. One can further distinguish high-risk populations: a particular age group in a residential urban population or workers performing specific tasks at a work site.

The U.S. EPA *Exposure Factors Handbook* (Wood, Phillips et al., 1997) provides numerous examples of activity patterns that influence the type and duration of contact experienced by different subpopulations. Day and colleagues (2007), for instance, reported on the exposure pathway assessment at a factory producing copper-beryllium wire that distinguishes work area subpopulations.

10.4.2. Exposure Routes

Toxicologists employ pharmacological drug disposition models (Figure 10.9) to represent exposure routes. Contaminants entering the human body ultimately elicit an adverse effect by acting on a target organ. The magnitude

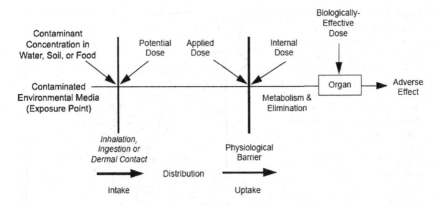

FIGURE 10.9   The generalized exposure route model is adopted from pharmacology drug disposition models. *Source: Reproduced with permission from Wood, P., Phillips, L., et al., 1997. Exposure Factors Handbook. U. S. E. P. Agency. Washington, DC, Office of Research and Development, National Center for Environmental Assessment, 1193.*

of the adverse effect is most closely linked to the contaminant concentration or dose at the target organ—the biologically effective dose. A complex internal pathway—the exposure route—connects the point where the contaminant enters the body—the exposure point—and the target organ.

The exposure route begins with either oral or dermal intake that delivers a potential dose to the human once the contaminant crosses an intake boundary—the lining of the sinus passages or lungs, at various points along the gastrointestinal tract, or the skin.[2] The potential dose is typically lower than the concentration at the exposure point because intake is usually less than 100% efficient. Following intake, the contaminant undergoes distribution throughout some portion of the human body. The distribution volume (Eq. (10.3)) and the possibility for clearance (i.e., elimination) reduce the applied dose that arrives at the biological barrier (vascular, renal, placental, blood-brain, etc.) that protects the organ from potentially toxic substances. Contaminant uptake is the passage of the compound across a biological barrier, resulting in an internal dose lowered by the transport properties of the barrier from the applied dose. Finally, metabolism and elimination processes (reverse transport across the biological barrier) further lower the internal dose to the biologically effective dose just mentioned.

Eukaryotes—organisms whose cells have a nucleus—rely on selective ion transporters to regulate uptake across the cell membrane into the cytoplasm. The selectivity of these ion transporters serves as a biological barrier to the uptake of inorganic toxic substances. Zinc homeostasis in mammalian cells

---

2. Skin and mucous membranes are considered nonspecific barriers.

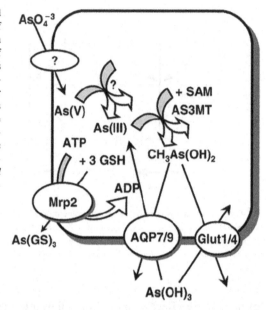

FIGURE 10.10 The generalized uptake and efflux pathways of arsenic in mammals include both the electrochemical reduction of *As(V)* to *As(III)* and transporters that export either the arsenic(III)-gluthathione complex *As(GS)₃* or arsenous acid. Arsenic uptake is believed to occur via phosphate ion transporters. *Source: Reproduced with permission from Rosen, B.P. Liu, Z., 2009. Transport pathways for arsenic and selenium: A minireview. Environment International 35, 512–515.*

is maintained by a combination of import and export transporters (Liuzzi and Cousins, 2004). There is no evidence that cadmium ions are exported from mammalian cells by any of the known *ZnT* zinc efflux transporters (Martelli, Rousselet et al., 2006), but cadmium export is possible by alternative routes: efflux by transporters of other divalent metal cations (e.g., Mn^{2+}) or export of cadmium complexes with ligands such as cysteine or glutathione. Figure 10.10 illustrates the cytoplasmic processes initiating the elimination of arsenate by mammalian cells; the intracellular reduction of As(IV) followed by the methylation and export of As(III).

10.4.3. Exposure Points

Potential exposed populations come into contact with media containing contaminants, known as the exposure point (see Figure 10.9). An exposure point can be a wide range of media (Wood, Phillips et al., 1997) ranging from ambient air, water, sediments, and soil to food and consumer products. Our primary concern here is with environmental media—air, water, sediments, and soil—that deliver contaminants to humans by inhalation, ingestion, and dermal contact (see Figure 10.8).

Table 10.3 lists possible exposure points. Risk analysts distinguish exposure points from environmental media at the source of the exposure pathway and mobile environmental media—air and water—that transport and disperse contaminants. Each exposure point implies potential exposure routes and exposed populations, the latter depending on specific activities.

TABLE 10.3 Possible Exposure Points for Contaminated Environmental Media

Environmental Media	Exposure Points
Soil	Surface soil at work sites, residential sites, and parks; subsurface soil exposed during excavation or other disturbance of the surface; excavated and transported soil used as fill
Sediment	Submerged or exposed stream and lake sediment, sandbars, overbank flood deposits, and drainage ditches; excavated and transported sediments used as fill or the result of ditch, drainage channel, canal, and watercourses maintenance
Groundwater	Wells and springs supplying water for municipal, domestic, industrial, agricultural, and recreational purposes
Surface Water	Lakes, ponds, streams, and storm drainage supplying water for municipal, domestic, industrial, agricultural, and recreational purposes
Air	Ambient air downwind of a contaminated site, indoor air containing migrating soil gases—e.g., flammable (methane) and asphyxiating (carbon dioxide) gas entering buildings on or adjacent to landfills
Biota	Plants (cultivated and gathered fruits and vegetables, plants used for medicinal purposes), animals (livestock, game, and other terrestrial or aquatic organisms), or other natural products that have contacted contaminated soil, sediment, waste materials, groundwater, surface water, or air
Other	Contaminated materials at commercial or industrial sites (e.g., raw materials, sludge from treatment processes, waste piles, radiation-laden metals) may provide a direct point of contact for on-site workers, visitors, or trespassers

Source: DHHS, 1992.

Establishing exposure points requires detailed knowledge of both the nature and extent of contamination through environmental testing and biological monitoring. Environmental testing analyzes all of the environmental media listed in Table 10.3 to identify and quantify potential contaminants. Biological monitoring collects and analyzes biological fluid and tissue samples from potential exposed populations to confirm and quantify exposure.

10.4.4. Fate and Transport

Continuing to work our way toward the contaminant source, the next component in the exposure pathway (see Figure 10.8) is contaminant fate and transport in environmental media. This is environmental chemistry's primary focus and the principal topic of this book. Most hazardous substances undergo profound biological and chemical transformation upon entering soils or aquifers, the atmosphere, or water; often the transformations are quite complex and depend on prevailing conditions.

Biological and chemical transformations acting in concert may degrade a hazardous substance, eliminating or reducing toxicity, or they can activate a substance, substantially enhancing toxicity. Examples of both are well known. Often biological and chemical transformations of inorganic substances that lower toxicity under prevailing conditions are reversed when chemical and biological conditions change over time and location.

Adsorption lowers toxicity by reducing biological availability (i.e., water solubility) and, in the process, leads to a substance's accumulation in soils, sediments, and aquifers. Contaminant storage and buildup in environmental solids contributes to legacy pollution, whose long-term risk is often difficult to assess.

Mobile environmental media—water (above- or belowground), solvents, and gases (above- and belowground)—disperse contaminants from their primary and secondary sources, expanding the area and volume of contaminated environmental media and increasing the opportunity for human exposure. Water percolating through soil, sediments, and aquifers transports contaminants slowly and over rather short distances belowground. Surface water flow and atmospheric currents disperse suspended particles rapidly and, in some cases, over considerable distances. The cost of environmental mitigation and the scope of the exposed population can increase dramatically with environmental transport.

Appendixes 10A–10D list chemical-specific and site-specific factors that influence the fate and transport of contaminants in the environment. Though lacking chemical and mechanistic detail, the tables in Appendixes 10A–10D give a broad perspective of the way risk analysts understand fate and transport processes.

10.4.5. Primary and Secondary Sources

Every hazardous substance that contaminates the environment has a history. The history begins with a production process that utilizes the compound of concern or that generates the compound as either a primary product or by-product. The history terminates with the compound of concern entering a waste stream that may or may not involve recovery or recycling. The agriculture, mining, manufacturing, and chemical industries in the United States are undergoing a transformation designed to minimize the use and production of hazardous substances, but this transformation will not entirely eliminate the release of hazardous substances into the environment at production and storage facilities.

Opportunities for environmental pollution occur during the distribution and end-point use. Often distributors or end-users fail to properly store or handle products that are hazardous. Waste management, especially when recovery and recycling are costly or inefficient, creates further opportunities for environmental contamination during waste storage, transport, and disposal.

Hazardous substances released into the environment from any of the primary sources just mentioned will contaminate soils, surface water bodies, groundwater, sediments, and aquifers. Contaminated environmental media become secondary sources upon the removal of the primary source—dismantling production, storage or distribution facilities, termination of end-point use, or closing of waste disposal sites—often leaving secondary sources as a continuing pollution legacy.

10.4.6. Exposure Assessment

The ultimate goal of exposure pathway assessment is identifying pathways of concern linking a contaminant source to an exposed population. A key factor in this process is exposure pathway elimination. If all components of an exposure pathway are present, then risk analysts identify this as a *completed exposure pathway*—a pathway of concern because a specific population is exposed via the pathway to a contamination source. If one or more elements of an exposure pathway are absent, then risk analysts identify this as an *eliminated exposure pathway*. A *potential exposure pathway* would be one where the presence or absence of one or more pathway elements is unverified. The "Exposure Pathway Elimination" (Box 10.2) illustrates this process with a real case.

10.5. INTAKE ESTIMATES

The general exposure route model (see Figure 10.9) begins with intake of contaminated media. This section discusses intake estimates for specific exposure routes. Intake equations (see Appendixes 10E–10H) estimate the dose delivered to an exposed population, specifically the body burden $(mg \cdot kg^{-1} \cdot d^{-1})$. Intake dose estimates for carcinogens must compute a cumulative body burden, while a steady-state body burden is computed for noncarcinogens. Intake dose estimates also depend on different criteria for setting the ingestion (or inhalation) rate *IR*, broadly known as exposure factors. Once we clearly understand the role of averaging time in computing body burden $(mg \cdot kg^{-1} \cdot d^{-1})$ and the statistical significance of exposure factors, we can fully comprehend and apply the various intake equations.

10.5.1. Averaging Time

OSWER Directive 9285.6-03 (EPA, 1991) provides the following guidance for setting the *averaging time* for intake equations:

Box 10.2 Exposure Pathway Elimination

In 1974, over opposition by surrounding landowners, the Dane County Zoning Board granted John DeBeck permission to open and operate a landfill in an abandoned limestone quarry near Black Earth Creek. The Board did not require DeBeck to install a secure liner or monitoring wells. DeBeck operated the Refuse Hideaway landfill from 1974 through 1987 in the face of continued opposition by landowners whose properties bordered the landfill. At one point homeowners filed a lawsuit that temporarily closed the landfill for 15 months.

In March 1988, the Wisconsin Department of Natural Resources (WDNR) detected cleaning solvents from the landfill in two domestic wells and contaminated runoff in a drainage ditch that emptied into Black Earth Creek. The WDNR asked the Dane County Zoning Board to rescind the landfill permit issued to DeBeck. After threatening to file a lawsuit, the Wisconsin attorney general, Don Hanaway, reached an agreement with DeBeck to stop accepting waste by May 16, 1988, and to install a clay cap and methane vents at the landfill. DeBeck filed for bankruptcy a year later, and in October 1992, the U.S. Environmental Protection Agency added Refuse Hideaway to the National Priorities List, designating it a Superfund site.

The following table lists the exposure pathway components and the exposure time frame for Refuse Hideaway. The exposure point (drinking water from private wells), the exposure route (ingestion of drinking water), and a potentially exposed population (landowners relying on private wells for drinking water) existed prior to the opening of Refuse Hideaway, but the exposure pathway is eliminated because a contaminant source and transport did not exist during that time frame.

Exposure Pathway Component	Time Frame of Exposure		
	Before 1974	*1974–1987*	*1988–Present*
Contaminant Source	No	Yes	Yes
Fate and Transport	No	Unknown	Yes
Exposure Point	Yes	Yes	Yes
Exposure Route	Yes	Yes	Yes
Potentially Exposed Populations	Yes	Yes	Yes
Conclusion	Eliminated Pathway	Potential Pathway	Completed Pathway

The presence of a potential exposure pathway linking Refuse Hideaway to domestic drinking water wells motivated the 53 landowners in the vicinity to oppose the opening and operation of the landfill. It provided a basis for the court judgment temporarily closing the landfill and for early efforts by the Wisconsin attorney general and the WDNR to terminate operations. Legal efforts to close the landfill were unsuccessful until the detection of solvents in domestic wells completed the exposure pathway.

Averaging time (AT) for exposure to noncarcinogenic compounds is always equal to
ED [exposure duration], whereas, for carcinogens, a 70-year AT is still used in order
to compare to [USEPA] Agency slope factors typically based on that value.

By specifying a human lifetime (70 years) as the averaging time, the carcinogen body burden $(mg \cdot kg^{-1} \cdot d^{-1})$ becomes effectively the cumulative
body burden, scaled to a daily intake $(mg \cdot kg^{-1} \cdot d^{-1})$. Noncarcinogen body
burden, which is also scaled to a daily intake $(mg \cdot kg^{-1} \cdot d^{-1})$, is averaged
over a much shorter time scale—the exposure duration *ED*. Equating averaging time *AT* to exposure duration *ED* for noncarcinogens implicitly assumes
any deviation from steady state can be neglected.

For noncarcinogens, the steady-state body burden $(mg \cdot kg^{-1} \cdot d^{-1})$ is
called the *Average Daily Dose (ADD)*. For carcinogens, the cumulative body
burden $(mg \cdot kg^{-1} \cdot d^{-1})$ is known as the *Lifetime Average Daily Dose*
(LADD), scaled using an averaging time *AT* equal to the mean lifetime:
$(70\ y) \cdot (365\ d \cdot y^{-1})$.

10.5.2. Exposure Factors

Key parameters influencing body burden estimates relate to the behavior and
physical characteristics of each potential exposed population: the ingestion rate
IR associated with a particular exposure route and the exposure duration *ED*.
Earlier we discussed the importance of probability distributions on statistical
estimates of central tendency and variation in any population. Toxicologists
studying exposure factors rely on statistical methods to estimate the most likely
exposure *MLE* and the reasonable maximum exposure *RME*, based on the mode
and the 95th percentile for the potential exposed population, respectively.

For example, the *Exposure Factors Handbook* (Wood, Phillips et al.,
1997) distinguishes several potential exposed populations, each with distinctive water ingestion rates: adults and children in the general population, lactat-
ing and pregnant females, and highly active individuals. Population studies
also distinguish ingestion rates *IR* and exposure durations *ED* for residential,
industrial, commercial, agricultural, and recreational activity.

Table 10.4 lists standard inhalation rates that distinguish ambient from
indoor air, where the key variable is not the breathing rate but exposure duration *ED*, a behavioral characteristic. The residential *RME* total inhalation rate
of 20 $m^3 \cdot d^{-1}$ is less than the total *RME* for the general population because
the typical residential population is "general population subgroups who would
be expected to spend the majority of their time at home: housewives; service and
household workers; retired people; and unemployed workers" (EPA, 1991).

The potable water ingestion rate in Table 10.4 corresponds to the 90th percentile for the general population and assumes that the general population is
absent from their residential water supply about 15 days per year. Potable
water ingestion in the workplace is assumed to be 50% of the total water
ingestion rate *IR*, a population behavior characteristic.

TABLE 10.4 Residential Exposure Factors for Selected Exposure Routes

Intake Route	Standard U.S. EPA Exposure Factor
Water Ingestion Rate	$2\ L \cdot d^{-1}$
Soil Ingestion Rate	Child (age 1–6 years): $200\ mg \cdot d^{-1}$ Adult (age > 6 years): $100\ mg \cdot d^{-1}$
Air Inhalation Rate	Total: $20\ m^3 \cdot d^{-1}$ Indoor: $15\ m^3 \cdot d^{-1}$
Homegrown Produce Ingestion Rate	Fruit: $42\ g \cdot d^{-1}$ Vegetables: $80\ g \cdot d^{-1}$
Subsistence Fish Ingestion Rate	$132\ g \cdot d^{-1}$

Source: EPA, 1991.

Soil ingestion includes actual "outdoor" soil and indoor dust. Soil ingestion covers both purposeful soil consumption (a behavior known as *pica*)
and incidental ingestion through hand-to-mouth contact or particulates entering the mouth or nostrils while breathing. Children younger than 6 years old
ingest significantly more soil than other age groups, yet another population
behavior characteristic.

Ingestion rates for certain foods identify subpopulations or food sources
associated with exposure pathways of concern: homegrown fruits and vegeta-
bles grown in contaminated soil, a subpopulation that subsists on fish caught
in contaminated lakes or streams, or nursing infants subsisting on breast milk.
Intake equations for water, soil, and food ingestion and air inhalation, along
with examples of each, appear in Appendixes 10E–10H.

10.6. RISK CHARACTERIZATION

Armed with equations for estimating contaminant intake based on concentrations
at the exposure point (Appendixes 10E–10H), we are now prepared to quantify
risk. Risk is a function of *intake* and *toxicity*. Quantifying risk from exposure to
a carcinogen is based on a definite statistical probability: the incremental excess
lifetime cancer risk *IELCR*. The risk factor for a noncarcinogen, on the other hand,
is not directly linked to the probability of an adverse effect on a population.

10.6.1. The Incremental Excess Lifetime Cancer Risk

The incremental excess lifetime cancer risk *IELCR* is the probability of developing cancer as the result of exposure to a specific carcinogen and appears
as an incremental increase in cancer cases in the exposed population over
what would occur in the absence of exposure. An acceptable *IELCR* can vary

but is often 1 to 10%, depending on variability in the population study.[3] The probability of developing cancer, based on the number of cancer cases in the population—regardless of age—is the lifetime cancer risk. Carcinogen toxicity is quantified by the cancer slope factor SF $\left(mg^{-1}_{toxicant} \cdot kg_{toxicant} \cdot d\right)$ of each substance (10.9), whose units are the inverse of the Lifetime Average Daily Dose $LADD$ because the ratio is a probability (i.e., unitless).

$$Risk = Dose \cdot Toxicity \tag{10.8}$$

$$IELCR = LADD \cdot SF \tag{10.9}$$

Example

An adult (70 kg) drinks water $\left(2\ L \cdot d^{-1}\right)$ containing 0.05 $\left(mg \cdot L^{-1}\right)$ benzene (see Appendix 10E, first example). The oral slope factor SF for the carcinogen benzene is 0.029 $\left(mg^{-1} \cdot kg \cdot d\right)$.

$$Water\ Intake = 5.9 \cdot 10^{-4}\ mg_{benzene} \cdot kg^{-1} \cdot d^{-1}$$

$$IELCR = LADD \cdot SF$$

$$IELCR = \left(5.9 \cdot 10^{-4}\ mg_{benzene} \cdot kg^{-1} \cdot d^{-1}\right) \cdot \left(2.9 \cdot 10^{-2}\ mg^{-1}_{benzene} \cdot kg \cdot d\right)$$

$$IELCR = 2 \cdot 10^{-5}$$

Example

A child (15 kg) ingests soil $\left(200\ mg \cdot d^{-1}\right)$ containing 1 $\left(mg \cdot kg^{-1}\right)$ benzene. The oral slope factor SF for the carcinogen benzene is 0.029 $\left(mg^{-1} \cdot kg \cdot d\right)$.

$$Soil\ Intake = 1.14 \cdot 10^{-6}\ mg_{benzene} \cdot kg^{-1} \cdot d^{-1}$$

$$IELCR = LADD \cdot SF$$

$$IELCR = \left(1.14 \cdot 10^{-6}\ mg_{benzene} \cdot kg^{-1} \cdot d^{-1}\right) \cdot \left(2.9 \cdot 10^{-2}\ mg^{-1}_{benzene} \cdot kg \cdot d\right)$$

$$IELCR = 3 \cdot 10^{-8}$$

Cumulative risk resulting from simultaneous exposure to several carcinogens is found by adding the $IELCR$ values for each carcinogen and applying a cumulative $IELCR$ to the total (see Appendix 10I). Generally, the

---

3. Variability in any population study of carcinogenesis ultimately determines the minimal $IELCR$ detectable with statistical confidence. A population study with low variability permits $IELCR$ detection as low as 1% with reasonable statistical confidence. Inherent variability typically limits $IELCR$ detection with reasonable statistical confidence to 5% or 10%. The corresponding $LADD$ is a toxicological end-point dose symbolized, respectively, as ED_{01}, ED_{05}, or ED_{10}.

acceptable cumulative *IELCR* should be no larger than the acceptable *IELCR* for exposure to a single carcinogen.

10.6.2. The Hazard Quotient

The hazard quotient *HQ* (the risk factor applied to noncarcinogens) relates the dose delivered at the exposure point (the Average Daily Dose *ADD*) to a toxicological end-point: the reference dose *RfD* (10.10). Earlier we described how the reference dose *RfD* is a low-dose end-point, extrapolated from either a *NOAEL* (Eq. (10.6)) or *LOAEL* (Eq. (10.7)). An acceptable hazard quotient *HQ* is less than 1.

$$Risk = \frac{Dose}{Toxicity}$$

$$HQ = \frac{ADD}{RfD} \qquad (10.10)$$

Example

An adult $(70 \; kg)$ drinks water $(2 \; L \cdot d^{-1})$ containing 0.5 $(mg \cdot L^{-1})$ toluene (see Appendix 10E, second example). The reference dose for toluene is $RfD = 0.2 \; mg_{toluene} \cdot kg^{-1} \cdot d^{-1}$.

$$Water \; ADD = 1.4 \cdot 10^{-2} \; mg_{toluene} \cdot kg^{-1} \cdot d^{-1}$$

$$HQ = \frac{ADD_{toluene}}{RfD_{oral, \, toluene}}$$

$$HQ = \frac{1.4 \cdot 10^{-2} \; mg_{toluene} \cdot kg^{-1} \cdot d^{-1}}{0.2 \; mg_{toluene} \cdot kg^{-1} \cdot d^{-1}}$$

$$HQ = 7 \cdot 10^{-2} < 1$$

Example

A child $(15 \; kg)$ ingests soil $(200 \; mg \; d^{-1})$ containing 10 $mg \; kg^{-1}$ toluene. The reference dose for toluene is $RfD = 0.2 \; mg_{toluene} \cdot kg^{-1} \cdot d^{-1}$.

$$Soil \; ADD = 1.33 \cdot 10^{-4} \; mg_{toluene} \cdot kg^{-1} \cdot d^{-1}$$

$$HQ = \frac{ADD_{toluene}}{RfD_{oral, \, toluene}}$$

$$HQ = \frac{1.33 \cdot 10^{-4} \; mg_{toluene} \cdot kg^{-1} \cdot d^{-1}}{0.2 \; mg_{toluene} \cdot kg^{-1} \cdot d^{-1}}$$

$$HQ = 6.7 \cdot 10^{-4} < 1$$

Cumulative risk resulting from simultaneous exposure to several noncarcinogens is found by adding the *HQ* values to obtain a Hazard Index *HI* (see Appendix 10J). An acceptable *HI* for cumulative exposure is the same as the acceptable *HQ* for exposure to a single noncarcinogen: *HI* < 1.

10.7. EXPOSURE MITIGATION

Federal and state agencies employ several practices to mitigate exposure. Mitigation typically involves specific actions taken to eliminate exposure pathways and to reduce risk to acceptable levels at exposure points. The following example illustrates actions taken by the Wisconsin DNR to mitigate exposure from contamination at Refuse Hideaway. The actions represent the selected remedy announced in the official Record of Decision (EPA, 1995) for Refuse Hideaway landfill.

The Source Control actions in Table 10.5 address contamination at the source of the exposure pathway and are designed to restrict access, eliminate primary and secondary contamination sources at the site, and restrict contaminant transport offsite. The Groundwater Treatment actions focus on off-site mitigation, primarily treating the groundwater plume to prevent further contaminant migration offsite. The Water Supply actions address the exposure point: domestic wells.

Mitigation also relies on setting and enforcing standards designed to limit risk in the exposed population. The selected remedies at Refuse Hideaway landfill refer to standards twice: extraction of groundwater when volatile organic compounds are "greater than 200 ppb" and installation of treatment systems for any private well "with concentrations exceeding NR 140 Enforcement Standards." The role of risk assessment in setting enforcement standards is the topic of this section.

You will encounter numerous acronyms used to identify *risk-based standards*; a few appear in Table 10.6. A RBCL, associated with a particular HQ or IELCR, or a MOE, are not enforceable standards but merely guidelines for setting enforceable standards. They would appear in the box labeled Regulatory Options in the Federal Risk Assessment Paradigm (see Figure 10.1). The setting of MCLs and MCLGs occurs at the level of Agency Action, representing a regulatory consensus designed to provide acceptable protection of public health while taking into account attendant economic, social, and political constraints.

An RBCL always relates to a particular exposure route because it estimates the contaminant concentration at the exposure point associated with a quantifiable risk. The following example estimates the RBCL for a carcinogen dissolved in water, the exposure point being ground or surface water and the exposure route water ingestion. Expression (10.11) is the LADD equation

TABLE 10.5 Selected Remedies for Risk Mitigation at Refuse Hideaway Landfill

Source Control	Deed restrictions and zoning modifications
	Warning signs posted around the perimeter of the property
	Maintenance of the existing single barrier (clay) cap, vegetation, and surface runoff controls
	Operation and maintenance of the existing landfill gas extraction and destruction system and leachate extraction and off-site treatment and disposal system
	Groundwater monitoring of selected monitoring wells and private home wells
Groundwater Treatment	Extraction of the most highly contaminated groundwater (greater than 200 ppb total [volatile organic compounds]) in the vicinity of the landfill and treatment of groundwater to meet applicable groundwater discharge standards
	Injection of the treated water into the aquifer upgradient of the landfill to stimulate *in situ* biodegradation of degradable components of the contamination
	Monitoring and evaluating the effectiveness of the groundwater extraction, treatment, and reinjection system in achieving progress toward cleanup standards
Water Supply	Supply a point-of-entry treatment system for any private well exhibiting contaminants originating at the Refuse Hideaway landfill with concentrations exceeding NR 140 Enforcement Standards (federal MCLs) or that are believed by the WDNR and EPA to be imminently at risk for exceeding those standards
	Construct a community water supply well if the number of homes requiring replacement water supplies makes [construction of a community well] cost effective

TABLE 10.6 Acronyms Used to Identify Air, Water, Soil, and Food Standards

Acronym	Definition
RBCL	A risk-based concentration level that meets an acceptable risk standard: $HQ < 1$ or IELCR in the range 10^{-5} to 10^{-3}
MCL	A maximum contaminant level subject to enforcement
MCLG	A maximum contaminant level goal that serves as a nonenforceable target for remediation
MOE	A margin of exposure (NOAEL $\div$ ADD) related to HQ (Eq. (10.10))

for water ingestion. Expression (10.12) quantifies LADD using the target IELCR.

$$\text{Intake Equation}: \quad LADD = \frac{C_W \cdot IR \cdot EF \cdot ED}{BM \cdot AT} \qquad (10.11)$$

$$\text{Risk Quantification}: \quad LADD = \frac{Target\ IELCR}{SF} \qquad (10.12)$$

The RBCL for a carcinogen in water is found by equating the two preceding expressions and solving for the concentration in water C_W:

$$C_{W,RBCL} \equiv \left(\frac{Target\ IELCR}{SF}\right) \cdot \left(\frac{BM \cdot AT}{IR \cdot EF \cdot ED}\right) \qquad (10.13)$$

Example

Compute the RBCL for the water ingestion of benzene by adults, assuming a very conservative target IELCR of 10^{-5}. The cancer slope factor SF for benzene is $0.029\ mg^{-1} \cdot kg \cdot d$. We will assume that the exposed population consists of adults who will drink contaminated water for 30 years.

$$RBCL = \left(\frac{10^{-5}}{0.029\ mg_{benzene}^{-1} \cdot kg \cdot d}\right) \cdot \left(\frac{BM \cdot AT}{IR \cdot EF \cdot ED}\right)$$

$$Adult\ RBCL = \left(\frac{10^{-5}\ mg_{benzene}}{0.029\ kg \cdot d}\right) \cdot \frac{(70\ kg) \cdot (70\ y \cdot 365\ d \cdot y^{-1})}{(2\ L_{water} \cdot d^{-1}) \cdot (350\ d \cdot y^{-1}) \cdot (30\ y)}$$

$$Adult\ RBCL = \left(\frac{17.9}{609}\right) \left[\frac{mg_{benzene}}{L}\right] = 0.029 \left[\frac{mg_{benzene}}{L}\right]$$

$$Adult\ RBCL = 29\ ppb$$

Example

Compute the RBCL for the water ingestion of toluene by adults, assuming the acceptable risk is $HQ \leq 1$. The reference dose RfD for toluene is $0.2\ mg \cdot kg^{-1} \cdot d^{-1}$. We will assume that the exposed population consists of adults who will drink contaminated water for 30 years.

$$ADD = \frac{C_W \cdot IR \cdot EF \cdot ED}{BM \cdot AT} = \frac{C_W \cdot IR}{BM}$$

$$ADD_{RB} = HQ \cdot RfD_{oral} = RfD_{oral}$$

$$C_{W,RBCL} = RfD_{oral} \cdot \left(\frac{BM}{IR}\right)$$

$$RBCL = \left(0.2 \; mg_{toluene} \cdot kg^{-1} \cdot d^{-1}\right) \cdot \left(\frac{BM}{IR}\right)$$

$$Adult \; RBCL = \left(\frac{0.2 \; mg_{toluene}}{kg \cdot d}\right) \cdot \left(\frac{70 \; kg}{2 \; L_{water} \cdot d^{-1}}\right)$$

$$Adult \; RBCL = \left(\frac{14}{2}\right) mg_{toluene} \cdot L^{-1}$$

$$Adult \; RBCL = 7 \; ppm$$

10.8. SUMMARY

Chemical contaminant risk assessment in the United States follows a paradigm (see Figure 10.1) that begins with hazard identification and dose-response assessment and reaches its completion with exposure assessment and risk characterization. Risk assessment results provide a scientific basis for risk management and mitigation. Dose-response assessment follows separate tracks for carcinogens and noncarcinogens, the principal difference being the absence of a threshold dose and response based on cumulative rather than steady-state dose when predicting carcinogenesis. Distinctions between carcinogens and noncarcinogens also determine the method for low-dose extrapolation and, ultimately, risk characterization.

A key element in all risk assessment studies of contaminants in the environment is exposure assessment, more specifically exposure pathway assessment. Exposure pathway assessment establishes links between the contamination source and potential exposed populations, evaluating all potential transport pathways, exposure points, and exposure routes that could connect a potential exposed population to a contamination source.

Risk characterization quantifies risk in exposed populations where an exposure pathway is completed. Risk characterization employs intake equations for each potential exposure route to estimate the potential dose. Risk characterization allows the setting of risk-based contaminant levels for each contaminant and exposure route. The enforcement standards—maximum contaminant levels MCLs—adopted by federal and state environmental agencies derive from risk-based contaminant levels RBCLs. The environmental chemist is a key participant in the risk assessment process, whether through research studies of fate and transport or collecting field data during environmental monitoring studies.

APPENDIX 10A. CHEMICAL- AND SITE-SPECIFIC FACTORS THAT MAY AFFECT CONTAMINANT TRANSPORT BY SURFACE WATER

Table 10A.1 lists the transport mechanisms and the chemical-specific and site-specific factors affecting contaminant fate and transport by surface water. It is excerpted from Appendix E of the *Public Health Assessment Guidance Manual* (EPA, 1992).

TABLE 10A.1 Chemical- and Site-Specific Factors Affecting Contaminant Transport: Surface Water

Transport Mechanism	Factors Affecting Transport	
	Chemical-Specific Considerations	*Site-Specific Considerations*
Surface Water		
Overland flow (via natural drainage or man-made channels)	• Water solubility • K_{OC}	• Precipitation (amount, frequency, duration) • Infiltration rate • Topography (especially gradients and sink holes) • Vegetative cover and land use • Soil/sediment type and chemistry • Use as water supply intake areas • Location, width, and depth of channel; velocity; dilution factors; direction of flow • Floodplains • Point and nonpoint source discharge areas
Volatilization	• Water solubility • Vapor pressure • Henry's Law constant	• Climatic conditions • Surface area • Contaminant concentration
Hydrologic connection between surface water and groundwater	• Henry's Law constant	• Groundwater/surface water recharge and discharge • Stream bed permeability • Soil type and chemistry • Geology (especially Karst conditions)
Adsorption to soil particles and particle sedimentation	• Water solubility • K_{OW}	• Particle size and density • Geochemistry of soils/sediments

Continued

TABLE 10A.1 Chemical- and Site-Specific Factors Affecting Contaminant Transport: Surface Water—Cont'd

Transport Mechanism	Factors Affecting Transport	
	Chemical-Specific Considerations	Site-Specific Considerations
	• K_{OC} • Density	• Organic carbon content of soils/sediment
Biologic uptake	• K_{OW} • BCF	• Chemical concentration • Presence of fish, plants, and other animals

Source: Reproduced with permission from United States Department of Health and Human Services, 1992. Public Health Assessment Guidance Manual. Agency for Toxic Substances and Disease Registry. Atlanta, GA.

APPENDIX 10B. CHEMICAL- AND SITE-SPECIFIC FACTORS THAT MAY AFFECT CONTAMINANT TRANSPORT BY GROUNDWATER

Table 10B.1 lists the transport mechanisms and the chemical-specific and site-specific factors affecting contaminant fate and transport by groundwater. It is excerpted from Appendix E of the *Public Health Assessment Guidance Manual* (EPA, 1992).

TABLE 10B.1 Chemical- and Site-Specific Factors Affecting Contaminant Transport: Groundwater

Transport Mechanism	Factors Affecting Transport	
	Chemical-Specific Considerations	Site-Specific Considerations
Groundwater		
Movement within and across aquifers or discharge to surface water	• Density (more or less dense than water) • Water solubility • K_{OC}	• Site hydrogeology • Precipitation • Infiltration rate • Porosity • Hydraulic conductivity • Groundwater flow direction • Depth to aquifer • Groundwater/surface water recharge and discharge zones

Continued

TABLE 10B.1 Chemical- and Site-Specific Factors Affecting Contaminant Transport: Groundwater—Cont'd

Transport Mechanism	Factors Affecting Transport	
	Chemical-Specific Considerations	Site-Specific Considerations
		• Presence of other compounds • Soil type • Geochemistry of site soils and aquifers • Presence and condition of wells (well location, depth, and use; casing material and construction; pumping rate) • Conduits, sewers
Volatilization (to soil gas, ambient air, and indoor air)	• Water solubility • Vapor pressure • Henry's Law constant • Diffusivity	• Depth to water table • Soil type and cover • Climatologic conditions • Contaminant concentrations • Properties of buildings • Porosity and permeability of soils and shallow geologic materials
Adsorption to soil particles or precipitation out of solution	• Water solubility • K_{OW} • K_{OC}	• Presence of natural carbon compounds • Soil type, temperature, and chemistry • Presence of other compounds
Biologic uptake	• K_{OW}	• Groundwater use for irrigation and livestock watering

Source: Reproduced with permission from United States Department of Health and Human Services, 1992. Public Health Assessment Guidance Manual. Agency for Toxic Substances and Disease Registry. Atlanta, GA.

APPENDIX 10C. CHEMICAL- AND SITE-SPECIFIC FACTORS THAT MAY AFFECT CONTAMINANT TRANSPORT INVOLVING SOILS OR SEDIMENTS

Table 10C.1 lists the transport mechanisms and the chemical-specific and site-specific factors affecting contaminant fate and transport from contaminated soils and sediments, including both particulate transport by erosion and mobilization of contaminants from soil by water. It is excerpted from Appendix E of the *Public Health Assessment Guidance Manual* (EPA, 1992).

TABLE 10C.1 Chemical- and Site-Specific Factors Affecting Contaminant Transport: Soil and Sediments

Transport Mechanism	Factors Affecting Transport	
	Chemical-Specific Considerations	Site-Specific Considerations
Soil (Surface and Subsurface) and Sediment		
Soil Erosion	• Particle density • Runoff velocity	• Plant cover • Soil erodibility • Rainfall intensity • Slope and slope length
Surface Water Runoff	• Water solubility • K_{OC}	• Plant cover • Infiltration rate and hydraulic conductivity of soil • Precipitation rate • Slope and slope length
Leaching	• Water solubility • K_{OC}	• Soil profile characteristics (e.g., impermeable horizons) • Soil porosity and permeability • Soil chemistry (especially acid/base) • Cation exchange capacity • Organic carbon content
Volatilization	• Vapor pressure • Henry's Law constant	• Physical properties of soil • Chemical properties of soil • Climatologic conditions
Biologic uptake	• BCF • Bioavailability	• Soil properties • Contaminant concentration

Source: Reproduced with permission from United States Department of Health and Human Services, 1992. Public Health Assessment Guidance Manual. Agency for Toxic Substances and Disease Registry. Atlanta, GA.

APPENDIX 10D. CHEMICAL- AND SITE-SPECIFIC FACTORS THAT MAY AFFECT CONTAMINANT TRANSPORT INVOLVING AIR AND BIOTA

Table 10D.1 lists the transport mechanisms and the chemical-specific and site-specific factors affecting contaminant fate and transport by the atmosphere and biota. It is excerpted from Appendix E of the *Public Health Assessment Guidance Manual* (EPA, 1992).

APPENDIX 10E. WATER INGESTION EQUATION

The water ingestion equation (10E.1) is a function of six variables listed in Table 10E.1. One parameter is the contaminant water concentration C_W

TABLE 10D.1 Chemical- and Site-Specific Factors Affecting Contaminant Transport: Air and Biota

Transport Mechanism	Factors Affecting Transport	
	Chemical-Specific Considerations	Site-Specific Considerations
Air		
Aerosolization	• Water solubility	• Chemicals stored under pressure
Atmospheric deposition	• Particle size	• Rainfall intensity and frequency • Wind speed and direction
Volatilization	• Henry's Law constant	• Presence of open containers, exposed surfaces, or leaking equipment
Wind	• NA	• Speed, direction, atmospheric stability
Biota		
Bioaccumulation	• K_{OW} • Biological half-life	• Presence of plants and animals • Consumption rate
Migration	NA	• Commercial activities (farming, aquaculture, livestock, dairies) • Sport activities (hunting, fishing) • Migratory species
Vapor sorption	NA	• Soil type • Plant species
Root uptake	NA	• Contaminant depth • Soil moisture • Plant species

Source: Reproduced with permission from United States Department of Health and Human Services, 1992. Public Health Assessment Guidance Manual. Agency for Toxic Substances and Disease Registry. Atlanta, GA.

TABLE 10E.1 Water Ingestion Equation Parameters

Parameter and Units	Symbol
Water Concentration $[mg \cdot L^{-1}]$	C_W
Ingestion Rate $[L \cdot d^{-1}]$	IR
Exposure Duration $[y]$	ED
Exposure Frequency $[d \cdot y^{-1}]$	EF
Body Mass $[kg]$	BM
Averaging Time $[d]$	AT

at the exposure point. The averaging time AT depends on the nature of the contaminant—carcinogen or noncarcinogen. The remaining parameters depend on the physical and behavioral characteristics of the potential exposed population.

$$Water\ Ingestion = \left(\frac{C_W \cdot IR \cdot (ED \cdot EF)}{BM}\right) \cdot \left(\frac{1}{AT}\right) \quad (10E.1)$$

The following two examples illustrate estimates of the body burden dose for a carcinogen—the lifetime average daily dose $LADD$—and a noncarcinogen—the average daily dose ADD. Notice that the product of exposure duration and exposure frequency $(EF \cdot ED)$ equals the averaging time AT for non-carcinogens.

Example

An adult (70 kg) drinks water (2 $L \cdot d^{-1}$) containing 0.05 $mg \cdot L^{-1}$ benzene. Estimate the benzene dose. Benzene is a known carcinogen.

$$Life\ time\ Average\ Daily\ Dose_{benzene} = \left(\frac{C_W \cdot IR \cdot (EF \cdot ED)}{BM}\right) \cdot \left(\frac{1}{AT_{carcinogen}}\right)$$

$$LADD_{benzene} = \left(\frac{(0.05\ mg \cdot L^{-1}) \cdot (2\ L \cdot d^{-1}) \cdot (350\ d \cdot y^{-1}) \cdot (30\ y)}{(70\ kg)}\right) \cdot \left(\frac{1}{(70\ y \cdot 365\ d \cdot y^{-1})_{mean\ life\ time}}\right)$$

$$LADD_{benzene} = \left(\frac{1.095 \cdot 10^3\ mg}{1.79 \cdot 10^6\ kg \cdot d}\right)$$

$$LADD_{benzene} = 5.9 \cdot 10^{-4}\ mg \cdot kg^{-1} \cdot d^{-1}$$

Example

An adult (70 kg) drinks water (2 $L \cdot d^{-1}$) containing 0.5 $mg \cdot L^{-1}$ toluene. Estimate the toluene dose. Toluene is classified as a noncarcinogen.

$$Average\ Daily\ Dose_{toluene} = \left(\frac{C_W \cdot IR \cdot (EF \cdot ED)}{BM}\right) \cdot \left(\frac{1}{AT_{noncarcinogen}}\right)$$

$$ADD_{toluene} = \left(\frac{C_W \cdot IR \cdot (EF \cdot ED)}{BM}\right) \cdot \left(\frac{1}{EF \cdot ED}\right) = \left(\frac{C_W \cdot IR}{BM}\right)$$

$$ADD_{toluene} = \left(\frac{\left(0.5\ \frac{mg_{toluene}}{L}\right) \cdot \left(2\ \frac{L}{d}\right)}{(70\ kg)}\right)$$

$$ADD_{toluene} = \left(\frac{1.0\ mg \cdot d^{-1}}{70\ kg}\right)$$

$$ADD_{toluene} = 1.4 \cdot 10^{-2}\ mg \cdot kg^{-1} \cdot d^{-1}$$

APPENDIX 10F. SOIL INGESTION EQUATION

The soil ingestion equation (10F.1) is a function of six variables and a conversion factor listed in Table 10F.1. One parameter is the contaminant soil concentration C_S at the exposure point. The averaging time AT, as always, depends on the nature of the contaminant—carcinogen or noncarcinogen. The remaining parameters depend on the physical and behavioral characteristics of the potential exposed population.

$$Soil\ Ingestion = \left(\frac{(C_S \cdot CF) \cdot IR \cdot (ED \cdot EF)}{BM}\right) \cdot \left(\frac{1}{AT}\right) \qquad (10F.1)$$

The following example estimates the body burden dose for a noncarcinogen.

Example

A child (15 kg) ingests soil (200 $mg \cdot d^{-1}$) containing 18.4 $mg \cdot kg^{-1}$ arsenic (Arain, Kazi et al., 2009). Estimate the arsenic dose using the averaging time for a noncarcinogen.

$$\underset{arsenic}{Average\ Daily\ Dose} = \left(\frac{(C_S \cdot CF) \cdot IR \cdot (EF \cdot ED)}{BM}\right) \cdot \left(\frac{1}{\underset{noncarcinogen}{AT}}\right)$$

$$\underset{arsenic}{ADD} = \left(\frac{(C_S \cdot CF) \cdot IR \cdot (EF \cdot ED)}{BM}\right) \cdot \left(\frac{1}{EF \cdot ED}\right)$$

$$\underset{arsenic}{ADD} = \frac{\left(\left(18.4\ \frac{mg_{arsenic}}{kg_{soil}}\right) \cdot \left(10^{-6}\ \frac{kg}{mg}\right)\right) \cdot \left(200\ \frac{mg_{soil}}{d}\right)}{(15kg)}$$

$$\underset{arsenic}{ADD} = \left(\frac{3.68 \cdot 10^{-3}\ mg_{arsenic}}{15\ kg \cdot d}\right)$$

$$\underset{arsenic}{ADD} = 2.45 \cdot 10^{-4}\ mg_{arsenic} \cdot kg^{-1} \cdot d^{-1}$$

TABLE 10F.1 Soil Ingestion Equation Parameters

Parameter and Units	Symbol
Soil Concentration $[mg \cdot kg^{-1}]$	C_S
Conversion Factor $[10^{-6}\ kg \cdot mg^{-1}]$	CF
Ingestion Rate $[mg \cdot d^{-1}]$	IR
Exposure Duration $[y]$	ED
Exposure Frequency $[d \cdot y^{-1}]$	EF
Body Mass $[kg]$	BM
Averaging Time $[d]$	AT

APPENDIX 10G. FOOD INGESTION EQUATION

The food ingestion equation is similar to the soil ingestion equation (10F.1), substituting food concentration C_F for soil concentration. The following example estimates the body burden dose for a noncarcinogen in food.

Example

An adult (60 kg) living in Bengal ingests spinach (100 $mg \cdot d^{-1}$) containing 0.9 $mg \cdot kg^{-1}$ arsenic (Arain, Kazi et al., 2009). Estimate the arsenic dose.

$$\underset{arsenic}{Average\ Daily\ Dose} = \left(\frac{(C_F \cdot CF) \cdot IR \cdot (EF \cdot ED)}{BM} \right) \cdot \left(\underset{noncarcinogen}{\frac{1}{AT}} \right)$$

$$\underset{arsenic}{ADD} = \left(\frac{(C_F \cdot CF) \cdot IR \cdot (EF \cdot ED)}{BM} \right) \cdot \left(\frac{1}{EF \cdot ED} \right)$$

$$\underset{arsenic}{ADD} = \frac{\left(\left(0.90\ \frac{mg_{arsenic}}{kg_{food}} \right) \cdot \left(10^{-6}\ \frac{kg}{mg} \right) \right) \cdot \left(100\ \frac{mg_{food}}{d} \right)}{(60 kg)}$$

$$\underset{arsenic}{ADD} = \left(\frac{9.0 \cdot 10^{-5}\ mg_{arsenic}}{60\ kg \cdot d} \right)$$

$$\underset{arsenic}{ADD} = 1.5 \cdot 10^{-6}\ mg_{arsenic} \cdot kg^{-1} \cdot d^{-1}$$

APPENDIX 10H. AIR INHALATION EQUATION

The air ingestion equation (10H.1) is a function of six variables listed in Table 10H.1. One parameter is the contaminant air concentration C_A at the exposure point. The averaging time AT, as always, depends on the nature of

TABLE 10H.1 Air Inhalation Equation Parameters

Parameter and Units	Symbol
Air Concentration $[mg \cdot m^{-3}]$	C_A
Inhalation Rate $[m^3 \cdot d^{-1}]$	IR
Exposure Duration $[y]$	ED
Exposure Frequency $[d \cdot y^{-1}]$	EF
Body Mass $[kg]$	BM
Averaging Time $[d]$	AT

the contaminant—carcinogen or noncarcinogen. The remaining parameters depend on the physical and behavioral characteristics of the potential exposed population.

$$Air\ Inhalation = \left(\frac{C_A \cdot IR \cdot (ED \cdot EF)}{BM}\right) \cdot \left(\frac{1}{AT}\right) \tag{10H.1}$$

The following example illustrates estimates of the body burden dose for radon-222, a carcinogen—the lifetime average daily dose.

Example

An adult (70 kg) living in Dane County, Wisconsin (Radon Zone 1), inhales indoor air (15 $m^3 \cdot d^{-1}$) containing radon-222. Dane County (EPA, 1993) is mapped as Zone 1 (predicted average indoor level > 0.148 $Bq \cdot L^{-1}$ indoor air). Estimate the radon-222 dose.

We can estimate the absorbed dose using the conventional air intake equation, assuming that the body absorbs all radioactive decays from inhaled radon-222.

$$\underset{radon-222}{Lifetime\,Average\,Daily\,Dose} = \left(\frac{C_A \cdot IR \cdot (EF \cdot ED)}{BM}\right) \cdot \left(\frac{1}{AT}\right)_{carcinogen}$$

$$\underset{radon-222}{LADD} = \left(\frac{(0.0148\,Bq \cdot L^{-1}) \cdot (10^3\,L \cdot m^{-3}) \cdot (15m^3 \cdot d^{-1}) \cdot (350 d \cdot y^{-1}) \cdot (30y)}{(70kg)}\right) \cdot \left(\frac{1}{AT}\right)_{carcinogen}$$

$$\underset{radon-222}{LADD} = \left(\frac{2.331 \cdot 10^6\,Bq}{(70\,kg) \cdot (2.45 \cdot 10^4\,d)}\right)$$

$$\underset{radon-222}{LADD} = 1.30\,Bq \cdot kg^{-1} \cdot d^{-1}$$

APPENDIX 10I. HAZARD INDEX—CUMULATIVE NONCARCINOGENIC RISK

Simultaneous exposure at subthreshold levels to several toxicants can result in adverse health effects. The Hazard Index HI (10I.1) estimates the cumulative risk from exposure to multiple noncarcinogens. Acceptable risk is defined as $HI < 1$.

$$HI = \sum_i HQ_i \tag{10I.1}$$

The following example lists three noncarcinogenic elements, each with a different average daily dose ADD based on typical groundwater concentrations in the United States (Newcomb and Rimstidt, 2002). The exposure route is drinking water consumption. The following table lists the hazard quotient HQ for each substance and the cumulative HI.

Toxicant	$ADD \left[\frac{mg_{toxicant}}{kg_{bodymass} \cdot d}\right]$	$RfD \left[\frac{mg_{toxicant}}{kg_{bodymass} \cdot d}\right]$	HQ
Cd	0.0003	0.001	0.30
Zn	0.0076	0.3	0.03
As	0.0004	0.0003	1.33
HI			1.66

APPENDIX 10J. CUMULATIVE TARGET RISK—CUMULATIVE CARCINOGENIC RISK

Simultaneous exposure at subthreshold levels to several toxicants can result in adverse health effects. The cumulative target risk CTR (10J.1) estimates the $IELCR$ resulting from exposure to two or more carcinogens. Acceptable risk is variable, depending on the acceptable number of excess lifetime cancer cases in the exposed population: $10^{-4} < CTR < 10^{-6}$.

$$CTR = \sum_i IELCR_i \qquad (10J.1)$$

The following example lists two known carcinogens, each with a different lifetime average daily dose $LADD$ based on exposure to untreated well water samples taken near the Refuse Hideaway landfill in 1987 (benzene = 24 $\mu g \cdot L^{-1}$; vinyl chloride = 20 $\mu g \cdot L^{-1}$). The exposure duration is calculated based on continuous exposure from 1987 to 2010. The cancer slope factor SF applies to all forms of cancer. The $IELCR$ for each substance and the cumulative target risk CTR both appear in the table. The excess cancer risk resulting from vinyl chloride exposure is predicted to be 180 cases per million, while 13 excess cases result from benzene exposure in drinking water contaminated by Refuse Hideaway landfill, resulting in a CTR of about 193 excess cancer cases in the exposed population.

Toxicant	$LADD \left[\frac{mg_{toxicant}}{kg_{bodymass} \cdot d}\right]$	$SF \left[\frac{kg_{bodymass} \cdot d}{mg_{toxicant}}\right]$	$IELCR$
Benzene	$2.3 \cdot 10^{-4}$	0.029	$1.3 \cdot 10^{-5}$
Vinyl chloride	$2.5 \cdot 10^{-4}$	0.72	$1.8 \cdot 10^{-4}$
CTR			$1.9 \cdot 10^{-4}$

Problems

1. The following "single-hit" model represents the probability of a toxic response from the ingestion of a contaminant. The probability of a toxic response $P(D;r)$ is plotted below the model using a logarithmic scale for dose D and probability r of damage to a single cell.

$$P(D;r) = 1 - e^{-r \cdot D}$$

Assume that the toxic response represented by the dose-response curve plotted below is the probability of a fatality caused by exposure to the contaminant. Draw on the graph (below) the lethal dose that represents a 50% fatality probability in the population: LD^{50}.

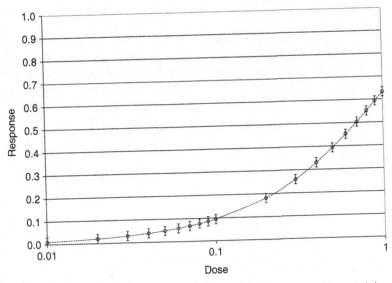

2. The "single-hit" dose-response curve plotted above represents the probability of a toxic response from the ingestion of a contaminant. NOTE: The error bars (plotted above) represent uncertainty in the observation of the response. Taking into account the uncertainty represented on the plot, determine a reasonable "no observable adverse effect level" (NOAEL) for this dose-response curve.

Solution
The NOAEL is the highest experimental measurement on the dose-response curve that is not statistically different from zero (the error bars for the response to this dose level overlap with "no response"). The LOAEL is the lowest experimental measurement on the dose-response curve that is statistically different from zero (the error bars for the response to this dose level do not span "no response").

3. A 15 kg child has been drinking water (0.6 L d^{-1} containing 0.05 mg L^{-1} benzene, a known carcinogen), for 6 years. Estimate the Lifetime Average Daily Dose $LADD$ for this case.

$$Lifetime\ Average\ Daily\ Dose = \frac{C_W \cdot IR \cdot EF \cdot ED}{BM \cdot AT}$$

Solution

Benzene is a carcinogen, so the averaging time AT is a lifetime, or 70 years, while the exposure duration ED is the length of time this adult consumed contaminated water, or 6 years.

$$Lifetime\ Average\ Daily\ Dose = \frac{C_W \cdot IR \cdot EF \cdot ED}{BM \cdot AT}$$

$$LADD_{benzene} = \frac{(0.05\ mg \cdot L^{-1}) \cdot (0.6\ L \cdot d^{-1}) \cdot (365\ d \cdot y^{-1}) \cdot (6\ y)}{(15\ kg) \cdot (70\ y \cdot 365\ d \cdot y^{-1})}$$

$$LADD_{benzene} = \left(\frac{65.7\ mg}{3.83 \cdot 10^5\ kg \cdot d} \right)$$

$$LADD_{benzene} = 1.71 \cdot 10^{-4}\ mg \cdot kg^{-1} \cdot d^{-1}$$

4. A 70 kg adult construction worker, age 50 years, ingests 330 $mg\ d^{-1}$ of soil containing 10 $mg\ kg^{-1}$ toluene (toluene has not been identified as a carcinogen). Work-related exposure frequency is 200 days per year. Assume that a typical construction worker has 30 years of work exposure. Average Daily Dose ADD of toluene in this case.

$$Average\ Daily\ Dose = \frac{(C_S \cdot CF) \cdot IR \cdot EF \cdot ED}{BM \cdot AT}$$

Solution

Toluene is not a carcinogen, so the Averaging Time AT is the duration of exposure, or 30 years.

$$Adult\ ADD_{toluene} = \frac{\left(\left(10\ \frac{mg_{toluene}}{kg_{soil}} \right) \cdot \left(10^{-6}\ \frac{kg}{mg} \right) \right) \cdot \left(330\ \frac{mg_{soil}}{d} \right) \cdot \left(200\ \frac{d}{y} \right) \cdot (30\ y)}{(70\ kg) \cdot \left(30\ y \cdot 200\ \frac{d}{y} \right)}$$

$$Adult\ ADD_{toluene} = 4.7 \cdot 10^{-5}\ mg \cdot kg^{-1} \cdot d^{-1}$$

5. What is the average daily dose ADD of methylmercury CH_3Hg^+ in a 70 kg adult who eats 500 g of tuna daily containing 0.2 ppm CH_3Hg^+? The half-life of CH_3Hg^+ in humans is 70 days.

$$Adult\ ADD_{methylHg} = \frac{\left(0.2\ \frac{mg_{methylHg}}{kg_{fish}} \right) \cdot \left(10^{-3}\ \frac{kg}{g} \right) \cdot \left(500\ \frac{g_{fish}}{d} \right)}{(70 \cdot kg)}$$

$$= 1.4 \cdot 10^{-3}\ mg \cdot kg^{-1} \cdot d^{-1}$$

6. Suppose a 70 kg adult consumes 100 micrograms a day of a substance x whose steady-state level in the body is later established to be 0.1 ppm. What is the half-life for elimination of this substance from the body?

Solution

$$t_{1/2} = \frac{BM \cdot C_{body} \cdot \ln(2)}{IR \cdot CF} = \frac{(70\ kg) \cdot (0.1\ mg \cdot kg^{-1}) \cdot \ln(2)}{(100\ \mu g \cdot d^{-1}) \cdot (10^{-3}\ mg \cdot \mu g^{-1})} \approx 50\ d$$

Once consumption of food containing this substance stops, how long does it take for the level of this substance to reach 0.025 $mg \; kg^{-1}$?

Solution
A steady-state body concentration of 0.025 $mg \; kg^{-1}$ is one-fourth of the original steady state, equivalent to 2 half-lives or 100 days.

7. In 1996, the USEPA developed a new reference dose *RfD* for methylmercury CH_3Hg^+ of 0.1 $\mu g \cdot kg^{-1} \cdot d^{-1}$. What mass of fish can a 60 kg woman safely eat each week if the average CH_3Hg^+ level in fish is 0.30 *ppm*?

Solution
First, estimate the acceptable mercury intake rate for this individual. Second, using the average CH_3Hg^+ level in fish, estimate the acceptable fish consumption rate.

$$Mercury \; Intake \; Rate = RfD \cdot BM$$

$$Mercury \; Intake \; Rate = \left(\frac{0.1 \; \mu g_{Hg}}{kg_{bodymass} \cdot d} \right) \cdot \left(60 \; kg_{bodymass} \right)$$

$$Mercury \; Intake \; Rate = 6 \; \mu g_{Hg} \cdot d^{-1}$$

$$Fish \; Consumption \; Rate = \frac{IR}{C_F}$$

$$Fish \; Consumption \; Rate = \frac{6 \; \mu g_{Hg} \cdot d^{-1}}{0.3 \mu g_{Hg} \cdot g_{fish}^{-1}} = 20 \; g_{fish} \cdot d^{-1}$$

8. According to an informal 1992 survey, the drinking water in about one-third of Chicago homes had lead levels of 10 *ppb*. Assuming that an adult drinks about 2 *L* of water a day, calculate the lead *ADD* an adult Chicago resident obtains daily from drinking water alone.

Solution
The drinking water *Pb* level of 10 *ppb* amounts to 10 micrograms per kilogram. Each liter of water weights 1 *kg*, yielding a daily *Pb* intake of 20 micrograms from drinking water alone. The *ADD* is this dose divided by the adult body mass: 0.29 $\mu g \cdot kg^{-1} \cdot d^{-1}$

References

Abbas, Z., Gunnarsson, M., et al., 2002. Corrected Debye-Hückel theory of salt solutions: Size asymmetry and effective diameters. J. Phys. Chem. B 106, 1406–1420.

Ahrens, L.H., 1954. The lognormal distribution of the elements: A fundamental law of geochemistry and its subsidiary. Geochim. Cosmochim. Acta 5, 49–73.

Ahrens, L.H., 1966. Element distributions in specific igneous rocks—VIII. Geochim. Cosmochim. Acta 30, 109–122.

Allison, L.E., Bernstein, L., et al., 1954. Diagnosis and Improvement of Saline and Alkali Soils. U. S. S. Laboratory. United States Government Printing Office, Washington, DC.

Altshuller, A.P., 1973. Atmospheric sulfur dioxide and sulfate: Distribution of concentration at urban and nonurban sites in United States. Environ. Sci. Technol. 7, 709–712.

Aluwihare, L.I., Repeta, D.J., et al., 2005. Two chemically distinct pools of organic nitrogen accumulate in the ocean. Science 308, 1007–1010.

Amend, J.P., Plyasunov, A.V., 2001. Carbohydrates in thermophile metabolism: Calculation of the standard molar thermodynamic properties of aqueous pentoses and hexoses at elevated temperatures and pressures. Geochim. Cosmochim. Acta 65, 3901–3917.

Amend, J.P., Shock, E.L., 2001. Energetics of overall metabolic reactions of thermophilic and hyperthermophilic Archaea and Bacteria. FEMS Microbiol. Rev. 25, 175–243.

Amirbahman, A., Reid, A.L., et al., 2002. Association of methylmercury with dissolved humic acids. Environ. Sci. Technol. 36, 690–695.

Ammari, T., Mengel, K., 2006. Total soluble Fe in soil solutions of chemically different soils. Geoderma 136, 876–885.

Anders, E., Grevesse, N., 1989. Abundance of the elements: meteoric and solar. Geochim. Cosmochim. Acta 53, 197–214.

Andren, A.W., Elzerman, A.W., et al., 1975. Chemical and physical aspects of surface organic microlayers in freshwater lakes. In: Proceedings of the First Specialty Symposium on Atmospheric Contributions to the Chemistry of Lake Waters. International Association for Great Lakes Research, Ann Arbor, MI.

Arain, M.B., Kazi, T.G., et al., 2009. Determination of arsenic levels in lake water, sediment, and foodstuff from selected area of Sindh, Pakistan: Estimation of daily dietary intake. Food Chem. Toxicol. 47, 242–248.

Argersinger, W.J., Davidson, A.W., et al., 1950. Thermodynamics and ion exchange phenomena. Trans. Kans. Acad. Sci. 53, 404–410.

Baas-Becking, L.G.M., Kaplan, I.R., et al., 1960. Limits of the natural environment in terms of pH and oxidation-reduction potentials. J. Geol. 68, 243–284.

Baisden, W.T., Amundson, R., et al., 2002. Turnover and storage of C and N in five density fractions from California annual grassland surface soils. Global Biogeochem. Cycles 16 (4), 1117.

Barak, P., Jobe, B.O., et al., 1997. Effects of long-term soil acidification due to nitrogen fertilizer inputs in Wisconsin. Plant Soil 197, 61–69.

Barger, J.R., Brown, G.E., et al., 1997. Surface complexation of Pb(II) at oxide-water interfaces: I. XAFS and bond-valence determination of mononuclear and polynuclear Pb(II) sorption products on aluminum oxides. Geochim. Cosmochim. Acta 61, 2617–2637.

449

Barger, J.R., Towle, S.N., et al., 1996. Outer-sphere Pb(II) adsorbed at specific surface sites on single crystal alpha-alumina. Geochim. Cosmochim. Acta 60, 3541–3547.

Benkovitz, C.M., Scholtz, M.T., et al., 1996. Global gridded inventories of anthropogenic emissions of sulfur and nitrogen. J. Geophys. Res. 101 (D22), 29239–29253.

Berkner, L.V., Marshall, L.C., 1965. History of major atmospheric components. Proc. Natl. Acad. Sci. U.S.A. 53, 1215–1226.

Boek, E.S., Coveney, P.V., et al., 1995. Monte Carlo molecular modeling studies of hydrated Li-, Na-, and K-smectites: Understanding the role of potassium as a clay swelling inhibitor. J. Am. Chem. Soc. 117, 12608–12617.

Bolt, G.H., 1957. Determination of the charge density of silica sols. J. Phys. Chem. 61, 1161–1169.

Borken, W., Matzner, E., 2004. Nitrate leaching in forest soils: an analysis of long-term monitoring sites in Germany. J. Plant Nutr. Soil Sci. 167, 277–283.

Bouwer, H., 1991. Simple derivation of the retardation equation and application to preferential flow and macrodispersion. Ground Water 29, 41–46.

Bower, C.A., Ogata, G., et al., 1968. Sodium hazard of irrigation water as influenced by leaching fraction and by precipitation or solution of calcium carbonate. Soil Sci. 106, 19–34.

Bower, C.A., Wilcox, L.V., et al., 1963. An index of the tendency of $CaCO_3$ to precipitate from irrigation waters. Soil Sci. Soc. Am. Proc. 29, 91–92.

Bragg, W.L., 1929. Atomic arrangement in the silicates. Trans. Faraday Soc. 25, 291–314.

Brown, I.D., Altermatt, K.K., 1985. Bond valence parameters obtained from a systematic analysis of the inorganic crystal-structure database. Acta Crystallogr. B 41, 244–247.

Budtz-Jørgensen, E., Keiding, N., et al., 2002. Estimation of health effects of prenatal methylmercury exposure using structural equation models. Environ. Health 1, 1–22.

Calace, N., Catrambone, T., et al., 2007. Fulvic acid in microlayer waters of the Gerlache Inlet Sea (Antarctica): Their distribution in dissolved and particulate phases. Water Res. 41, 152–158.

Cameotra, S.S., Bollag, J.M., 2003. Biosurfactant-enhanced bioremediation of polycyclic aromatic hydrocarbons. Critical Reviews in Environmental Science and Technology 33 (2), 111–126.

Chan, C.D.N., 2004. Bragg's law: according to the 2θ deviation, the phase shift causes constructive (left picture) or destructive (right picture) interferences. Retrieved August 27, 2009, from http://upload.wikimedia.org/wikipedia/commons/7/74/Loi_de_bragg.png.

Chaney, R.L., Brown, J.C., et al., 1972. Obligatory reduction of ferric chelates in iron uptake by soybean. Plant Physiol. 50, 208–213.

Chantigny, M.H., 2003. Dissolved and water-extractable organic matter in soils: A review on the influence of land use and management practices. Geoderma 113, 357–380.

Chavez-Paez, M., van Workum, K., et al., 2001. Monte Carlo simulations of Wyoming sodium montmorillonite hydrate. J. Phys. Chem. 114, 1405–1413.

Chen, M., Ma, L.Q., et al., 1999. Baseline concentrations of 15 trace elements in Florida surface soils. J. Environ. Qual. 28 (4), 1173–1181.

Chen, Y., Schnitzer, M., 1978. The surface tension of aqueous solutions of soil humic substances. Soil Sci. 125, 7–15.

Chiou, C.T., Peters, L.J., et al., 1979. A physical concept of soil-water equilibria for nonionic organic compounds. Science 206, 831–832.

Clapp, R.B., Hornberger, G.M., 1978. Empirical equations for some soil hydraulic properties. Water Resour. Res. 14 (4), 601–604.

Clark, C.J., McBride, M.B., 1984. Chemisorption of Cu(II) and Co(II) on allophane and imogolite. Clays Clay Miner. 32 (4), 300–310.

Contributors, 2008. Hematite in Scanning Electron Microscope, magnification 100x. Retrieved May 19, 2009, from http://upload.wikimedia.org/wikipedia/commons/1/17/Hematite_in_Scanning_Electron_Microscope%2C_magnification_100x.GIF.

Cooper, R.L., 1954. The determination of polycyclic hydrocarbons in town air. Analyst 79, 573–579.

Coulter, B.S., Talibudeen, O., 1968. Calcium-aluminum exchange equilibria in clay minerals and acid soils. J. Soil Sci. 19, 237–250.

Crandall, C.A., Katz, B.G., et al., 1999. Hydrochemical evidence for mixing of river water and groundwater during high-flow conditions, lower Suwannee River basin, Florida, USA. Hydrogeo. J. 7, 454–467.

Cunningham, D.P., Lundie, L.L., 1993. Precipitation of cadmium by Clostridium thernoaceticum. Appl. Environ. Microbiol. 59, 7–14.

Dai, Z., Liu, Y., et al., 1998. Changes in pH, CEC and exchangeable acidity of some forest soils in southern China during the last 32–35 years. Water Air Soil Pollut. 108 (3–4), 377–390.

Dapson, R.W., Kaplan, L., 1975. Biological half-life and distribution of radiocesium in a contaminated population of green treefrogs Hyla cinerea. Oikos 26, 39–42.

Davies, C.W., 1938. The extent of dissociation of salts in water. Part VIII. An equation for the mean ionic activity coefficient of an electrolyte in water, and a revision of the dissociation constants of some sulphates. J. Chem. Soc. (London), 2093–2098.

Day, G.A., Dufresne, A., et al., 2007. Exposure pathway assessment at a copper-beryllium alloy facility. Ann. Occup. Hyg. 51, 67–80.

Debye, P., Hückel, E., 1923. Zür Theorie der Elektrolyte. I. Gefrierpunktserniedrigung und verwandte Erscheinungen. Physikalische Zeitschrift 24, 185–206.

Deffeyes, K.S., 1965. Carbonate equilibria: A graphic and algebraic approach. Limnol. Oceanog. 10 (3), 412–426.

Delahay, P., Pourbaix, M., et al., 1950. Potential-pH diagrams. J. Chem. Educ. 27, 683–688.

Desai, J.D., Banat, I.M., 1997. Microbial production of surfactants and their commercial potential. Microbiol. Mol. Biol. Rev. 61 (1), 47–64.

de Vooys, A.C.A., Koper, M.T.M., et al., 2001. Mechanistic study of the nitric oxide reduction on a polycrystalline platinum electrode. Electrochimica Acta 46, 923–930.

de Vooys, A.C.A., Koper, M.T.M., et al., 2001. Mechanistic study on the electrocatalytic reduction of nitric oxide on transition-metal electrodes. J. Catalysis 202, 387–394.

DiChristina, T.J., Moore, C.M., et al., 2002. Dissimilatory Fe(III) and Mn(IV) reduction by Shewanella putrefaciens requires ferE, a Homolog of the pulE (gspE) Type II protein secretion gene. J. Bacteriol. 184, 142–151.

Dickson, A.G., Goyet, C., 1994. Handbook of methods for the analysis of the various parameters of the carbon dioxide system in sea water. C. D. S. S. T. Department of Energy. Department of Energy, Oak Ridge, TN.

Dingman, S.L., 1994. Physical Hydrology. Macmillan, New York.

Dodd, J.C., Burton, C.C., et al., 1987. Phosphatase activity associated with the roots and the rhizosphere of plants infected with vesicular-arbuscular mycorrhizal fungi. New Phytol. 107, 163–172.

Doucette, W., Andren, A., 1987. Correlation of octanol/water partition coefficients and total molecular surface area for highly hydrophobic aromatic compounds. Environ. Sci. Technol. 21, 821–824.

Doucette, W., Andren, A., 1988. Estimation of octanol/water partition coefficients: Evaluation of six methods for highly hydrophobic aromatic hydrocarbons. Chemosphere 17, 345–359.

Duce, R.A., Quinn, J.G., et al., 1972. Enrichment of heavy metals and organic compounds in the surface microlayer of Narragansett Bay, Rhode Island. Science 176, 161–163.

Dudka, S., Ponce-Hernandez, R., et al., 1995. Current level of total element concentrations in the surface layer of Sudbury's soils. Sci. Total Environ. 162, 161–171.

Environmental Science Center, 2006. PBT Profiler: Persistent, Bioaccumulative, and Toxic Profiles Estimated for Organic Chemicals On-Line. Office of Pollution Prevention and Toxics, U.S. Environmental Protection Agency.

Ferguson, A.D., Deisenhofer, J., 2002. TonB-dependent receptors—structural perspectives. Biochim. Biophys. Acta 1565, 318–332.

Foscolos, A.E., 1968. Cation-exchange equilibrium constants of aluminum-saturated montmorillonite and vermiculite clays. Soil Sci. Soc. Am. J. 32, 350–354.

Freeze, R.A., Cherry, J.A., 1979. Groundwater. Prentice-Hall, Englewood Cliffs, NJ.

Fuller, R.D., Simone, D.M., et al., 1988. Forest clearcutting effects on trace metal concentrations: Spatial patterns in soil solutions and streams. Water Air Soil Pollut. 40, 185–195.

Gaines, G.L., Thomas, H.C., 1953. Adsorption studies on clay minerals. II. A formulation of the thermodynamics of exchange adsorption. J. Chem. Phys. 21, 714–718.

Gapon, Y.N., 1933. On the theory of exchange adsorption in soils. Journal of General Chemistry of the U.S.S.R. 3, 144–160.

Gardner, W.K., Parbery, D.G., et al., 1982. The acquisition of phosphorus by Lupinus albus L. I. Some characteristics of the soil/root interface. Plant Soil 68, 19–32.

Garrels, R.M., Christ, C.L., 1965. Solutions, Minerals and Equilibria. Freeman, Cooper, San Francisco.

Garrels, R.M., Mackenzie, F.T., 1971. Evolution of Sedimentary Rocks. Norton, New York.

Gasparovic, B., Plavsic, M., et al., 2007. Organic matter characterization in the sea surface microlayers in the subarctic Norwegian fjords region. Mar. Chem. 105, 1–14.

Gast, R.G., 1969. Standard free energies of exchange for alkali metal cations on Wyoming bentonite. Soil Sci. Soc. Am. Proc. 33, 37–41.

Genomics: GTL, 2005. Simplified representation of the global carbon cycle. Retrieved May 19, 2010, from http://genomics.energy.gov/gallery/gtl/detail.np/RoadmapPics_CMYK/Pg034_fig7P.tif.

George, G.N., Gorbaty, M.L., 1989. Sulfur K-edge X-ray absorption spectroscopy of petroleum asphaltenes and model compounds. J. Am. Chem. Soc. 111 (9), 3182–3186.

Gibbs, W.W., 2003. Untangling the roots of cancer. Sci. Am. 289, 48–57.

Gillespie, R.J., Nyholm, R.S., 1957. Inorganic stereochemistry. Quar. Rev. Chem. Soc. 11, 339–380.

Goldschmidt, V.M., 1937. The principles of distribution of chemical elements in minerals and rocks. J. Chem. Soc. (London), 655–673.

Gorbaty, M.L., George, G.N., et al., 1990. Direct determination and quantification of sulfur forms in heavy petroleum and coals. 2. The sulfur-K edge x-ray absorption-spectroscopy approach. Fuel 69 (8), 945–949.

Goulden, F., Tipler, M.M., 1949. Experiments on the identification of 3:4-benzpyrene in domestic soot by means of the fluorescence spectrum. Br. J. Cancer 3, 157–160.

Gray, T., 2006. Dow-Popular Science periodic table. Retrieved December 3, 2008, from http://www.popsci.com/files/periodic_popup.html.

Greenwood, N.N., Earnshaw, A., 1997. Chemistry of the Elements. Butterworth-Heinemann, Boston.

Grieve, I.C., 1990. Variations in chemical composition of the soil solution over a four-year period at an upland site in southwest Scotland. Geoderma 46 (4), 351–362.

Griffin, R.A., Jurinak, J.J., 1973. Estimation of activity coefficients from the electrical conductivity of natural aquatic systems and soil extracts. Soil Sci. 116, 26–30.

Guetzloff, T.F., Rice, J.A., 1994. Does humic acid form a micelle? Sci. Total Environ. 152, 31–35.

Guntelberg, E., 1926. Untersuchungen über Ioneninteraktion. Z. Phys. Chem. 123, 199–247.

Hadley, A., Toumi, R., 2003. Assessing changes to the probability distribution of sulphur dioxide in the UK using a lognormal model. Atmos. Environ. 37, 1461–1474.

Hannam, K.D., Prescott, C.E., 2003. Soluble organic nitrogen in forests and adjacent clearcuts in British Columbia, Canada. Can. J. For. Res. 33, 1709–1718.

Hay, B.P., Dixon, D.A., et al., 2001. Structural criteria for the rational design of selective ligands. 3. Quantitative structure-stability relationship for iron(III) complexation by tris-catecholamide siderophores. Inorg. Chem. 40 (16), 3922–3935.

Hayase, K., Tsubota, H., 1983. Sedimentary humic acid and fulvic acid as surface active substances. Geochim. Cosmochim. Acta 47, 947–952.

Haynes, J.L., Thatcher, J.S., 1955. Crop rotation and soil nitrogen. Soil Sci. Soc. Am. Proc. 19, 324–327.

Haynes, R.J., 2000. Labile organic matter as an indicator of organic matter quality in arable and pastoral soils in New Zealand. Soil Biol. Biochem. 32, 211–219.

Helgeson, H.C., 1969. Evaluation of irreversible reactions in geochemical processes involving minerals and aqueous solutions: I. Applications. Geochim. Cosmochim. Acta 33, 455–481.

Helmke, P.A., 2000. The chemical composition of soils. In: Sumner, M.E. (Ed.), Handbook of Soil Science. CRC Press, Boca Raton, FL, pp. B3–B24.

Henin, S., Dupuis, M., 1945. Essai de bilan de la matiere organique du sol. Annales Agronomiques 15, 17–29.

Henrichs, S.M., Williams, P.M., 1985. Dissolved and particulate amino acids and carbohydrates in the sea surface microlayer. Mar. Chem. 17 (2), 141–163.

Herbert, B.E., Bertsch, P.M., 1995. Characterization of dissolved and colloidal organic matter in soil solution: a review. In: McFee, W.W., Kelly, J.M. (Eds.), Carbon Forms and Functions in Forest Soils. Soil Science Society of America, Madison, WI, pp. 63–88.

Higgins, G.H., 1959. Evaluation of the ground-water contamination hazard from underground nuclear explosions. J. Geophys. Res. 64, 1509–1959.

Hoffmann, D., Masuda, Y., et al., 1969. α-Naphthylamine and β-naphthylamine in cigarette smoke. Nature 221, 254–256.

Holtan, H.N., England, C.B., et al., 1968. Moisture-Tension Data for Selected Soils on Experimental Watersheds. A. R. Service. United States Department of Agriculture, Beltsville, MD.

Hopkins, C.G., Knox, W.H., et al., 1903. A quantitative method for determining the acidity of soils, Nineteenth Annual Convention of the Association of Official Agricultural Chemists. B. o. C. U. S. Department of Agriculture. Government Printing Office, Washington, DC, 114–119.

Hornberger, G.M., Raffensperger, J.P., et al., 1968. Elements of Physical Hydrology. Johns Hopkins University Press, Baltimore, MD.

Huang, C.P., Stumm, W., 1973. Specific adsorption of cations on hydrous gamma-Al2O3. J. Colloid Interface Sci. 43, 409–420.

Hubbert, M.K., 1940. The theory of ground-water motion. J. Geol. 48 (8), 785–944.

Hunter, K.A., Lee, K.C., 1986. Polarographic study of the interaction between humic acids and other surface-active organics in river waters. Water Res. 20, 1489–1491.

Ingham, E.R., Andrew, R.M., et al., 1999. The Soil Biology Primer. Natural Resources Conservation Service, Washington, DC.

Jackson, M.L., Tyler, S.A., et al., 1948. Weathering sequence of clay-size minerals in soils and sediments. 1. Fundamental generalizations. J. Phys. Colloid Chem. 52, 1237–1260.

Jenkinson, D.S., Andrew, S.P.S., et al., 1990. The turnover of organic carbon and nitrogen in soil. Philos. Trans. R. Soc. Lond. B Biol. Sci. 329 (1255), 361–368.

Jenkinson, D.S., Rayner, J.H., 1977. The turnover of soil organic matter in some of the Rothamsted Classical Experiments. Soil Sci. 123, 298–305.

Jenny, H., 1941. Factors of Soil Formation: A System of Quantitative Pedology. McGraw-Hill, New York.

Jensen, H.E., Babcock, K.L., 1973. Cation-exchange equilibria on a Yolo loam. Hilgardia 41, 475–487.

Johnson, J.F., Allan, D.L., et al., 1996. Root carbon dioxide fixation by phosphorus-deficient Lupinus albus: Contribution to organic acid exudation by proteoid roots. Plant Physiol. 112, 19–30.

Johnson, J.F., Vance, C.P., et al., 1996. Phosphorus deficiency in Lupinus albus: altered lateral root development and enhanced expression of phosphoenolpyruvate carboxylase. Plant Physiol. 112, 31–41.

Jones, D.L., Owen, A.G., et al., 2002. Simple method to enable the high resolution determination of total free amino acids in soil solutions and soil extracts. Soil Biol. Biochem. 34, 1893–1902.

Jones, D.L., Shannon, D., et al., 2004. Role of dissolved organic nitrogen (DON) in soil N cycling in grassland soils. Soil Biol. Biochem. 36, 749–756.

KAERI, 2009, 2000. Table of Nuclides. Retrieved September 4, 2009, from http://atom.kaeri.re.kr/.

Kaiser, K., Guggenberger, G., 2003. Mineral surfaces and soil organic matter. Euro. J. Soil Sci. 54 (2), 219–236.

Kapteyn, J.C., 1903. Skew frequency curves in biology and statistics. Astronomical Laboratory at Groningen and Noordhoff, Groningen.

Katchalsky, A., Spitnik, P., 1947. Potentiometric titrations of polymethacrylic acid. J. Polymer Sci. 2, 432–446.

Kerr, H.W., 1928. The nature of base exchange and soil acidity. J. Am. Soc. Agron. 20, 309–335.

Kile, D.E., Chiou, C.T., 1989. Water solubility enhancements of DDT and trichlorobenzene by some surfactants below and above the critical micelle concentration. Environ. Sci. Technol. 23 (7), 832–838.

Kim, K.Y., Jordan, D., et al., 1998. Effect of phosphate-solubilizing bacteria and vesicular-arbuscular mycorrhizae on tomato growth and soil microbial activity. Biol. Fertil. Soils 26 (2), 79–87.

Knudson, A., 1971. Mutation and cancer: Statistical study of retinoblastoma. Proc. Natl. Acad. Sci. U.S.A. 68, 820–823.

Knulst, J.C., Backlund, P., et al., 1997. Response of surface microlayer to artificial acid precipitation in a meso-humic lake in Norway. Water Res. 31, 2177–2186.

Knulst, J.C., Boerschke, R.C., et al., 1998. Differences in organic surface microlayers from an artificially acidified and control lake, elucidated by XAD-8/XAD-4 tandem separation and solid-state 13C NMR spectroscopy. Environ. Sci. Technol. 32, 8–12.

Kosman, D.J., 2003. Molecular mechanisms of iron uptake in fungi. Mol. Microbiol. 47 (5), 1185–1197.

Krishnamoorthy, C., Overstreet, R., 1950. An experimental evaluation of ion-exchange relationships. Soil Sci. 69, 41–55.

Lambert, J.B., 1964. Nitrogen-15 magnetic resonance spectroscopy. I. Chemical shifts. Proc. Natl. Acad. Sci. U.S.A. 51, 735–737.

Lambert, S.M., 1965. Movement and sorption of chemicals applied to soil. Weeds 13, 185–190.

Lambert, S.M., 1967. Functional relationship between sorption in soil and chemical structure. J. Agri. Food Chem. 15, 572–576.

Lambert, S.M., 1968. Omega: a useful index of soil sorption equilibria. J. Agri. Food Chem. 16, 340–343.

Lang, S., 2002. Biological amphiphiles (microbial biosurfactants). Curr. Opin. Colloid Interface Sci. 7 (1–2), 12–20.

Langelier, W.F., 1936. The analytical control of anticorrosion water treatment. J. Am. Water Works Assoc. 28, 1500–1521.

Langmuir, I., 1916. The constitution and fundamental properties of solids and liquids. Part I Solids. J. Am. Chem. Soc. 38, 2221–2295.

Langmuir, I., 1918. The adsorption of gases on plane surfaces of glass, mica and platinum. J. Am. Chem. Soc. 40, 1361–1403.

Langmuir, I., 1925. The distribution and orientation of molecules. In: Holmes, H.N. (Ed.), Third Colloid Symposium Monograph. The Chemical Catalog Company, New York, pp. 48–75.

Larson, R.B., Bromm, V., 2001. The first stars in the universe. Sci. Am. 285 (6), 4–11.

Lebron, I., Suarez, D.L., et al., 1994. Stability of a calcareous saline-sodic soil during reclamation. Soil Sci. Soc. Am. J. 58 (6), 1753–1762.

Lefemine, D.V., 1954. The identification of 3,4-benzpyrene in cigarette paper smoke and tars. Southeast Regional American Chemical Society Meeting, Birmingham, AL.

Lenhart, J.J., Saiers, J.E., 2004. Adsorption of natural organic matter to air-water interfaces during transport through unsaturated porous media. Environ. Sci. Technol. 38 (1), 120–126.

Lessard, G., 1981. Biogeochemical phenomena in quick clays and their effects on engineering properties. Department of Civil Engineering, University of California, Berkeley, CA.

Li, S., Matthews, J., et al., 2008. Atmospheric hydroxyl radical production from electronically excited NO2 and H2O. Science 319, 1657–1660.

Liang, C.C., Juliard, A.L., 1965. Reduction of oxygen at platinum electrode. Nature 207, 629–630.

Lide, D.R., 2005. Geophysics, Astronomy, and Acoustics: Abundance of Elements in the Earth's Crust and in the Sea. In: Lide, D.R. (Ed.), CRC Handbook of Chemistry and Physics. CRC Press, Boca Raton, FL, pp. 14–17.

Limpert, E., Stahel, W.A., et al., 2001. Log-normal distributions across the sciences: Keys and clues. Bioscience 51, 341–352.

Lindsay, W.L., 1979. Chemical Equilibra in Soils. John Wiley & Sons, New York.

Liss, P.S., 1975. Chemistry of the sea surface microlayer. In: Riley, J.P., Skirrow, G. (Eds.), Chemical Oceanography, vol. 2. Academic Press, New York, pp. 193–243.

Liu, C., Ya Suo, Y., et al., 1994. The energetic linkage of GTP hydrolysis and the synthesis of activated sulfate. Biochemistry 33, 7309–7314.

Liu, D., Wang, Z., et al., 2006. Spatial distribution of soil organic carbon and analysis of related factors in croplands of the black soil region, Northeast China. Agri. Ecosyst. Environ. 113, 73–81.

Liuzzi, J.P., Cousins, R.J., 2004. Mammalian zinc transporters. Ann. Rev. Nutrit. 24, 151–172.

Lovley, D.R., 1993. Dissimilatory metal reduction. Annual Reviews of Microbiology 47, 263–290.

Lovley, D.R., Holmes, D.E., et al., 2004. Dissimilatory Fe(III) and Mn(IV) reduction. Adv. Microb. Physiol. 49, 219–286.

L'vovich, M.I., 1979. World Water Resources. American Geophysical Union, Washington, DC.

Mackay, D., 1982. Correlation of bioconcentration factors. Environ. Sci. Technol. 16, 274–278.

Mackay, D., Fraser, A., 2000. Bioaccumulation of persistent organic chemicals: mechanisms and models. Environ. Pollut. 110, 375–391.

Mackin, J.E., Owen, R.M., et al., 1980. A factor analysis of elemental associations in the surface microlayer of Lake Michigan and Its fluvial inputs. J. Geophys. Res. 85, 1563–1569.

Maguire, R.J., Kuntz, K.W., et al., 1983. Chlorinated hydrocarbons in the surface microlayer of the Niagara River. J. Great Lakes Res. 9, 281–286.

Maidment, D.R., 1993. Hydrology. In: Maidment, D.R. (Ed.), Handbook of Hydrology. McGraw-Hill, New York, pp. 1.1–1.15.

Marcus, Y., 1987. The thermodynamics of solvation of ions. Part 2. The enthalpy of hydration at 298.15 K. J. Chem. Soc. Faraday Trans. I 83, 339–349.

Marcus, Y., 1988. Ionic radii in aqueous solutions. Chem. Rev. 88, 1475–1498.

Margitan, J.J., 1988. Mechanism of the atmospherlc oxidation of sulfur dioxide. Catalysis by hydroxyl radicals. J. Phys. Chem. 84, 3314–3318.

Martelli, A., Rousselet, E., et al., 2006. Cadmium toxicity in animal cells by interference with essential metals. Biochimie 88, 1807–1814.

Martin, A.J.P., Synge, R.L.M., 1941. A new form of chromatogram employing two liquid phases. Biochem. J. 35, 1358–1368.

Marty, J.C., Zutic, V., et al., 1988. Organic matter characterization in the northern Adriatic Sea with special reference to the sea-surface microlayer. Mar. Chem. 25, 243–263.

May, H.M., Kinniburgh, D.G., et al., 1986. Aqueous dissolution, solubilities and thermodynamic stabilities of common aluminosilicate clay minerals: Kaolinite and smectites. Geochim. Cosmochim. Acta 50, 1667–1677.

Meyers, P.A., Rice, C.P., et al., 1983. Input and removal of natural and pollutant materials in the surface microlayer on Lake Michigan. Ecological Bullitens (Stockholm) 35, 519–532.

Mikutta, R., Kleber, M., et al., 2006. Stabilization of soil organic matter: association with minerals or chemical recalcitrance? Biogeochemistry 77, 25–56.

Mooney, R.W., Keenan, A.G., et al., 1952. Adsorption of water vapor by montmorillonite. II. Effect of exchangeable ions and lattice swelling as measured by X-ray diffraction. J. Am. Chem. Soc. 74, 1371–1374.

Müller, B., 2009. ChemEQL V3.0: Summary and availability information. Retrieved August 27, 2009, from http://www.eawag.ch/research_e/surf/Researchgroups/sensors_and_analytic/chemeql.html.

Muller, P., 1994. Glossary of terms used in physical organic chemistry. Pure Appl. Chem. 66, 1077–1184.

Mulligan, C.N., 2005. Environmental applications for biosurfactants. Environ. Pollut. 133 (2), 183–198.

Myers, H.E., Hallsted, A.L., et al., 1943. Nitrogen and carbon changes in soils under low rainfall as influenced by cropping systems and soil treatment. Kansas Agricultural Experiment Station, Manhattan, KS.

Myers, J.M., Myers, C.R., 2002. Genetic complementation of an outer membrane cytochrome omcB mutant of Shewanella putrefaciens MR-1 requires omcB plus downstream DNA. Appl. Environ. Microbiol. 68, 2781–2793.

Naidenko, O., Leiba, N., et al., 2008. Bottled water quality investigation: 10 Major brands, 38 pollutants. Retrieved August 26, 2009, from http://www.ewg.org/reports/bottledwater.

Nam, K., Chung, N., et al., 1998. Relationship between organic matter content of soil and the sequestration of phenanthrene. Environ. Sci. Technol. 32 (23), 3785–3788.

National Academy of Science, 1983. Risk Assessment in the Federal Government: Managing the Process. National Academies Press, Washington, DC.

National Research Council, 1983. Risk Assessment in the Federal Government: Managing the Process. National Academies Press, Washington, D.C.

NDDC, 2009. Interactive Chart of Nuclides. Retrieved August 26, 2009, from http://www.nndc.bnl.gov/chart/.

Newcomb, W.D., Rimstidt, J.D., 2002. Trace element distribution in US groundwaters: a probabilistic assessment using public domain data. Appl. Geochem. 17, 49–57.

Nordstrom, D.K., Wilde, F.D., 1998. Reduction-oxidation potential (electrode method). In: Wilde, F.D., Radtke, D.B., Gibs, J., Iwatsubo, R.T. (Eds.), Geological Survey Techniques of Water-Resources Investigations, vol. 9. United States Geological Survey, Reston, VA, pp. 1–20.

Norrish, K., 1954. The swelling of montmorillonite. Discuss. Faraday Soc. 18, 120–134.

Ohtaki, H., Radnai, T., 1993. Structure and dynamics of hydrated ions. Chem. Rev. 93, 1157–1204.

Olness, A., Archer, D., 2005. Effect of organic carbon on available water in soil. Soil Sci. 170 (2), 90–101.

Oster, J.D., Shainberg, I., et al., 1980. Flocculation value and gel structure of sodium/calcium montmorillonite and illite suspensions. Soil Sci. Soc. Am. J. 44, 955–959.

Ott, W.R., 1990. A physical explanation of the lognormality of pollutant concentrations. J. Air Waste Manag. Assoc. 40, 1378–1383.

Parfitt, R.L., Russell, J.D., et al., 1976. Confirmation of the structure of goethite (a-FeOOH) and phosphated goethite by infrared spectroscopy. J. Chem. Soc. (London) Faraday Trans. I 72, 1082–1087.

Parks, G.A., de Bruyn, P.L., 1962. The zero point of charge of oxides. J. Phys. Chem. 66, 967–973.

Parton, W.J., Schimel, D.S., et al., 1987. Analysis of factors controlling soil organic matter levels in the Great Plains grasslands. Soil Sci. Soc. Am. J. 51, 1173–1179.

Pauling, L., 1929. The principles determining the structure of complex ionic crystals. J. Am. Chem. Soc. 51, 1010–1026.

Pauling, L., 1930a. The structure of the micas and related minerals. Proc. Natl. Acad. Sci. U.S.A. 16, 123–129.

Pauling, L., 1930b. The structure of chlorites. Proc. Natl. Acad. Sci. U.S.A. 16, 578–582.

Pauling, L., 1960. The Nature of the Chemical Bond and the Structure of Molecules and Crystals: An Introduction to Modern Structural Chemistry. Cornell University Press, Ithaca, NY.

Perminova, I.V., Frimmel, F.H., et al., 1998. Development of a predictive model for calculation of molecular weight of humic substances. Water Res. 32 (3), 872–881.

Petersen, F.F., Rhoades, J., et al., 1965. Selective adsorption of magnesium ions by vermiculite. Soil Sci. Soc. Am. Proc. 29, 327–328.

Philpott, C.C., 2006. Iron uptake in fungi: A system for every source. Biochim. Biophys. Acta 1763, 636–645.

Pierre, W.H., Banwart, W.L., 1973. Excess-base and excess-base/nitrogen ratio of various crop species and parts of plants. Agron. J. 65, 91–96.

Pietramellara, G., Ascher, J., et al., 2002. Soil as a biological system. Ann. Microbiol. 52, 119–131.

Pines, A., Gibby, M.G., et al., 1973. Proton-enhanced NMR of dilute spins in solids. J. Chem. Phys. 59, 569–590.

Pourbaix, M.J.N., 1938. Thermodynamique des solution aqueuses diluées. Le pH et le potentiel d'oxydo-réduction. Société chimique de Belgique, Bruxelles.

Qian, J., Skyllberg, U., et al., 2002. Bonding of methyl mercury to reduced sulfur groups in soil and stream organic matter as determined by X-ray absorption spectroscopy and binding affinity studies. Geochim. Cosmochim. Acta 66, 3873–3885.

Quideau, S.A., Bockheim, J.G., 1997. Biogeochemical cycling following planting to red pine on a sandy prairie soil. J. Environ. Qual. 26, 1167–1175.

Resendiz, D., 1977. Relevance of Atterberg limits in evaluating piping and breaching potential. In: Sherard, J.L., Decker, R.S. (Eds.), Dispersive Clay, Related Piping, and Erosion in Geotechnical Projects. Special Publication 623. American Society for Testing and Materials, Philadelphia, PA, pp. 341–353.

Rhoades, J., 1967. Cation exchange reactions of soil and specimen VERMICULITEs. Soil Sci. Soc. Am. Proc. 67, 361–365.

Rhoades, J.D., Krueger, D.B., et al., 1968. The effect of soil-mineral weathering on the sodium hazard of irrigation water. Soil Sci. Soc. Am. Proc. 32, 643–647.

Rice, J.A., MacCarthy, P., 1990. A model of humin. Environ. Sci. Technol. 24, 1875–1990.

Rice, J.A., MacCarthy, P., 1991. Statistical evaluation of the elemental composition of humic substances. Org. Geochem. 17 (5), 635–648.

Ritchie, J.D., Perdue, E.M., 2003. Proton-binding study of standard and reference fulvic acids, humic acids, and natural organic matter. Geochim. Cosmochim. Acta 67 (1), 85–96.

Ritchie, J.D., Perdue, E.M., 2008. Analytical constraints on acidic functional groups in humic substances. Org. Geochem. 39, 783–799.

Robarge, W.P., 2004. Precipitation and dissolution processes. In: Hillel, D., Hatfield, J.L. (Eds.), Encyclopedia of Soils in the Environment, vol. 3. Elsevier-Academic Press, Boston, MA, pp. 322–329.

Robson, A.D., Jarvis, S.C., 1983. The effects of nitrogen nutrition of plants on the development of acidity in Western Australian soils. 11 Effects of differences in cation/anion balance between plant species grown under non-leaching conditions. Aust. J. Agric. Res. 34, 355–365.

Roe, M., 2007. Well-crystallized kaolinite from the Keokuk geode, USA, a sample in the Macaulay Institute Collection. Retrieved May 19, 2009, from http://www.minersoc.org/pages/gallery/claypix/kaolinite/kao4_6big.jpg.

Römheld, V., Marschner, H., 1981. Rhythmic iron stress reactions in sunflower at suboptimal iron supply. Physiol. Plant. 53, 347–353.

Rosca, V., Koper, M.T.M., 2005. Mechanism of electrocatalytic reduction of nitric oxide on Pt (100). J. Phys. Chem. B 109, 16750–16759.

Rosen, B.P., Liu, Z., 2009. Transport pathways for arsenic and selenium: a minireview. Environ. Int. 35, 512–515.

Ruebush, S.S., Brantley, S.L., et al., 2006. Reduction of soluble and insoluble iron forms by membrane fractions of Shewanella oneidensis grown under aerobic and anaerobic conditions. Appl. Environ. Microbiol. 72, 2925–2935.

Russell, J.D., Parfitt, R.L., et al., 1974. Surface structures of gibbsite goethite and phosphated goethite. Nature 248, 220–221.

Ruvarac, A., Vesely, V., 1970. Simple graphical determination of thermodynamic constants of ion exchange reactions. Zeitschrift für physikalische Chemie (Frankfurt) 73, 1–6.

Saeki, K., Wada, S.I., et al., 2004. Ca2+-Fe2+ and Ca2+-Mn2+ exchange selectivity of kaolinite, montmorillonite, and illite. Soil Sci. 169, 125–132.

Salter, R.M., Green, T.C., 1933. Factors affecting the accumulation and loss of nitrogen and organic carbon in cropped soils. J. Am. Soc. Agron. 25, 622–630.

Sander, R., 2009. Henry's Law Constants (Solubilities). Retrieved January 6, 2010, from http://www.mpch-mainz.mpg.de/~sander/res/henry.html.

Schaefer, J., Stejskal, E.O., 1976. Carbon-13 nuclear magnetic resonance of polymers spinning at the magic angle. J. Am. Chem. Soc. 98 (4), 1031–1032.

Schindler, P.W., Furst, B., et al., 1976. Ligand properties of surface silanol groups. 1. Surface complex formation with Fe^{3+}, Cu^{2+}, Cd^{2+} and Pb^{2+}. J. Colloid Interface Sci. 55, 469–475.

Schindler, P.W., Kamber, H.R., 1968. Die Acidität von Silanolgruppen. Helv. Chim. Acta 51, 1781–1786.

Schwab, A.P., 2004. Chemical equilibria. In: Hillel, D., Hatfield, J.L. (Eds.), Encyclopedia of Soils in the Environment, vol. 1. Elsevier-Academic Press, Boston, MA, pp. 189–194.

Scott, W.D., Hobbs, P.V., 1967. The formation of sulfate in water droplets. J. Atmos. Sci. 24, 54–57.

Shacklette, H.T., Boerngen, J.G., 1984. Elemental concentrations in soils and other surficial materials in the conterminous United States. U. S. Geological Survey Professional Papers. United States Department of the Interior, Washington, DC, 1270.

Shanzer, A., Felder, C.E., et al., 2009. Natural and biomimetic hydroxamic acid based siderophores. In: Rappoport, Z., Liebman, J.F. (Eds.), The Chemistry of Hydroxylamines, Oximes and Hydroxamic Acids, Part 1. Wiley Interscience, Chichester, England, pp. 751–815.

Shinozuka, N., Lee, C., 1991. Aggregate formation of humic acids from marine sediments. Mar. Chem. 33, 229–241.

Shinozuka, N., Lee, C., et al., 1987. Solubilizing action of humic acid from marine sediment. Sci. Total Environ. 62, 311–314.

Shiu, W.Y., Doucette, W., et al., 1988. Bioaccumulation of persistent organic chemicals: Mechanisms and models. Environ. Sci. Technol. 22, 651–658.

Silva, I.A., Nyland, J.F., et al., 2004. Mercury exposure, malaria, and serum antinuclear/antinucleolar antibodies in Amazon populations in Brazil: a cross-sectional study. Environ. Health 3, 1–12.

Sinsabaugh, R.L., Lauber, C.L., et al., 2008. Stoichiometry of soil enzyme activity at global scale. Ecol. Lett. 11 (11), 1252–1264.

Six, J., Callewaert, P., et al., 2002. Measuring and understanding carbon storage in afforested soils by physical fractionation. Soil Sci. Soc. Am. J. 66 (6), 1981–1987.

Smith, R.M., Martell, A.E., 2001. NIST critically selected stability constants of metal complexes. United States Department of Commerce, National Institute of Standards and Technology, Standard Reference Data Program, Gaithersburg, MD.

Snyder, L.R., 1970. Petroleum nitrogen compounds and oxygen compounds. Acc. Chem. Res. 3 (9), 290–299.

Sposito, G., Holtzclaw, K.M., et al., 1983a. Sodium-calcium and sodium-magnesium exchange on Wyoming bentonite in perchlorate and chloride back-ground ionic media. Soil Sci. Soc. Am. J. 47, 51–56.

Sposito, G., Holtzclaw, K.M., et al., 1983b. Cation selectivity in sodium- calcium, sodium-magnesium, and calcium-magnesium exchange on Wyoming Bentonite at. 298 K. Soil Sci. Soc. Am. J. 47, 917–921.

Sprycha, R., 1989. Electrical double layer at alumina/electrolyte interface I. Surface charge and zero potential. J. Colloid Interface Sci. 127, 1–11.

Staff, 1987. Soil mechanics: Level I, Module 3, USDA Textural Soil Classification Study Guide. Retrieved July 2010, 2010, from http://www.wsi.nrcs.usda.gov/products/W2Q/H&H/docs/training_series_modules/soil-USDA-textural-class.pdf.

Stokes, G.G., 1851. On the effect of the internal friction on the motion of pendulums. Trans. Cambridge Phil. Soc. 9, 8–106.

Stumm, W., Hohl, H., et al., 1976. Interaction of metal ions with hydrous oxide surfaces. Croatcia Chemica Acta 48, 491–504.

Stumm, W., Huang, C.P., et al., 1970. Specific chemical interaction affecting stability if dispersed systems. Croatcia Chemica Acta 42, 223–245.

Stumm, W., Morgan, J.A., 1970. Aquatic Chemistry. Wiley-Interscience, New York.

Su, C., Suarez, D.L., 1997. In situ infrared speciation of adsorbed carbonate on aluminum and iron oxides. Clays Clay Miner. 45, 814–825.

Suarez, D.L., 1981. Relation between pHc and sodium adsorption ratio (SAR) and an alternative method of estimating SAR of soil or drainage waters. Soil Sci. Soc. Am. J. 45, 469–475.

Szulczewski, M.D., Helmke, P.A., et al., 2001. XANES spectroscopy studies of Cr(VI) reduction by thiols in organosulfur compounds and humic substances. Environ. Sci. Technol. 35 (6), 1134–1141.

Szytula, A., Burewicz, A., et al., 1968. Neutron diffraction studies of alpha-FeOOH. Physica Status Solidi B 42, 429–434.

Tambach, T.J., Bolhuis, P.G., et al., 2006. Hysteresis in clay swelling induced by hydrogen bonding: Accurate prediction of swelling states. Langmuir 22, 1223–1234.

Tambach, T.J., Hensen, E.J.M., et al., 2004. Molecular simulations of swelling clay minerals. J. Phys. Chem. B 108, 7586–7596.

Taylor, S.R., 1964. Abundance of chemical elements in the continental crust: A new table. Geochim. Cosmochim. Acta 28, 1273–1285.

Thauer, R., 1998. Biochemistry of methanogenesis: a tribute to Marjory Stephenson. Microbiology 144, 2377–2406.

Thomas, G.W., 1977. Historical developments in soil chemistry: Ion exchange. Soil Sci. Soc. Am. J. 41, 230–238.

Thompson, H.S., 1850. On the absorbent power of soils. J. Royal Agri. Soc. 11, 68–74.

Thorn, K.A., Cox, L.G., 2009. N-15 NMR spectra of naturally abundant nitrogen in soil and aquatic natural organic matter samples of the International Humic Substances Society. Org. Geochem. 40, 484–499.

Townsend, A.R., Vitousek, P.M., et al., 1997. Soil carbon pool structure and temperature sensitivity inferred using CO_2 and (13)CO_2 incubation fluxes from five Hawaiian soils. Biogeochemistry 38 (1), 1–17.

Trudinger, P.A., 1969. Assimilatory and dissimilatory metabolism of inorganic sulfur compounds by micro-organisms. In: Rose, A.H., Wilkinson, J.F. (Eds.), Advances in Microbial Physiology, vol. 3. Academic Press, New York, pp. 111–158.

Trumbore, S.E., Chadwick, O.A., et al., 1996. Rapid exchange between soil carbon and atmospheric carbon dioxide driven by temperature change. Science 272 (5260), 393–396.

Tschapek, M., Scoppa, C., et al., 1978. On the surface activity of humic acid. Zeitschrift für Pflanzenernährung und Bodenkunde 141 (2), 203–207.

Tschapek, M., Wasowski, C., et al., 1981. Humic acid as a colloidal surfactant. Plant Soil 63 (2), 261–271.

Udo, A.P., 1978. Thermodynamics of potassium-calcium and magnesium-calcium exchange reactions on a kaolinitic clay. Soil Sci. Soc. Am. J. 42, 556–560.

U.S. Congress. Senate, 1938. Federal Food, Drug, and Cosmetic Act. Statute, Federal Register 52, 1040.

U.S. Congress. Senate, 1954. Miller Pesticide Amendments. Statute 68, 511.

U.S. Congress. Senate, 1958. Food Additives Amendment of 1958. Statute 72, 1784.

U.S. Congress. Senate, 1960. Color Additives Amendment of 1960. Statute 74, 397.

U.S. Department of Health and Human Services, 1992. Public Health Assessment Guidance Manual. Agency for Toxic Substances and Disease Registry. United States Department of Health and Human Services, Public Health Service, Agency for Toxic Substances and Disease Registry, Atlanta, GA.

U.S. Environmental Protection Agency, 1991. Risk Assessment Guidance for Superfund. Volume 1. In: Human Health Evaluation Manual Supplemental Guidance. Office of Solid Waste and Emergency Response, United States Environmental Protection Agency, Washington, DC, p. 28.

U.S. Environmental Protection Agency, 1995. EPA Superfund Record of Decision: Refuse Hideaway Landfill. Wisconsin Department of Natural Resources and United States Environmental Protection Agency Region 5, Chicago, IL.

U.S. Environmental Protection Agency, 1998. U.S. EPA Announces Changes in Refuse Hideaway Superfund Site Cleanup Plan, United States Environmental Protection Agency, Office of Public Affairs Region 5, Chicago, IL, p. 6.

U.S. Environmental Protection Agency, 2005. Guidelines for Carcinogen Risk Assessment. R. A. Forum. United States Environmental Protection Agency, Washington, DC.

van Beusichem, M.L., Kirby, E.A., et al., 1988. Influence of nitrate and ammonium nutrition on the uptake, assimilation, and distribution of nutrients in Ricinus communis. Plant Physiol. 86, 914–921.

Vanselow, A.P., 1932. Equilibria of the base-exchange reactions of bentonites, permutites, soil colloids, and zeolites. Soil Sci. 33, 95–113.

Vermeulen, T., 1952. Ion-exchange chromatography of trace components: A design theory. Indust. Eng. Chem. 44, 636–651.

Vietch, F.P., 1902. The estimation of soil acidity and the lime requirements of soils. J. Am. Chem. Soc. 24, 1120–1128.

Vietch, F.P., 1904. Comparison of methods for the estimation of soil acidity. J. Am. Chem. Soc. 26, 637–662.

Villarreal, M.R., 2007. November 6 cross section view of the structures that can be formed by phospholipids in aqueous solutions. Retrieved May 12, 2009, from http://upload.wikimedia .org/wikipedia/commons/c/c6/Phospholipids_aqueous_solution_structures.svg.

Waksman, S.A., 1942. II. The microbiologist looks at soil organic matter. Soil Sci. Soc. Am. Proc. 7, 16–21.

Waldo, G.S., Carlson, R.M.K., et al., 1991. Sulfur Speciation in Heavy Petroleums—Information from X-Ray Absorption Near-Edge Structure. Geochim. Cosmochim. Acta 55 (3), 801–814.

Waldo, G.S., Mullins, O.C., et al., 1992. Determination of the chemical environment of sulfur in petroleum asphaltenes by x-ray absorption-spectroscopy. Fuel 71 (1), 53–57.

Way, J.T., 1850. On the power of soils to absorb manure. J. Royal Agri. Soc. 11, 313–379.

Way, J.T., 1852. On the power of soils to absorb manure. J. Royal Agri. Soc. 13, 123–143.

Wershaw, R.L., 1999. Molecular aggregation of humic substances. Soil Sci. 164 (11), 803–813.

Whittig, L.D., Janitzky, P., 1963. Mechanisms of formation of sodium carbonate in soils. I. Manifestations of biological conversions. J. Soil Sci. 14, 323–333.

Wild, A., Keay, J., 1964. Cation exchange equilibria with vermiculite. J. Soil Sci. 15, 135–144.

Wolock, D.M., Hornberger, G.M., et al., 1989. The relationship of catchment topography and soil hydraulic characteristics to lake alkalinity in the northeastern United States. Water Resour. Res. 25, 829–837.

Wood, P., Phillips, L., et al., 1997. Exposure Factors Handbook. U. S. E. P. Agency. Office of Research and Development, National Center for Environmental Assessment, Washington, DC.

Woodruff, C.M., 1949. Estimating the nitrogen delivery of soil from the organic matter determination as reflected by Sanborn Field. Soil Sci. Soc. Am. Proc. 14, 208–212.

Wosten, H.A.B., 2001. Hydrophobins: Multipurpose proteins. Annu. Rev. Microbiol. 55, 625–646.

Wosten, H.A.B., van Wetter, M.A., et al., 1999. How a fungus escapes the water to grow into the air. Curr. Biol. 9 (2), 85–88.

Wurl, O., Holmes, M., 2008. The gelatinous nature of the sea-surface microlayer. Mar. Chem. 110, 89–97.

Yang, H., Lu, R., et al., 2006. Goethite, [alpha]-FeO(OH), from single-crystal data. Acta Crystallogr. E62, i250–i252.

Yang, S.Y., Chang, W.L., 2005. Use of finite mixture distribution theory to determine the criteria of cadmium concentrations in Taiwan farmland soils. Soil Sci. 170, 55–62.

Yonebayashia, K., Hattori, T., 1987. Surface active properties of soil humic acids. Sci. Total Environ. 62, 55–64.

Yoon, S.J., Diener, L.M., et al., 2005. X-ray absorption studies of CH_3Hg-binding sites in humic substances. Geochim. Cosmochim. Acta 69 (5), 1111–1121.

Yoon, S.J., Helmke, P.A., et al., 2002. X-ray absorption and magnetic studies of trivalent lanthanide ions sorbed on pristine and phosphate-modified boehmite surfaces. Langmuir 18, 10128–10136.

Zhang, C., Manheim, F.T., et al., 2005. Statistical characterization of a large geochemical database and effect of sample size. Appl. Geochem. 20, 1857–1874.

Zhang, F.S., Ma, J., et al., 1997. Phosphorus deficiency enhances root exudation of low-molecular weight organic acids and utilization of sparingly soluble inorganic phosphates by radish (Rhaganus sativus L.) and rape (Brassica napus L.). Plant Soil 196, 261–264.

Zheng, D.Q., Guo, T.M., et al., 1997. Experimental and modeling studies on the solubility of CO2, CHClF2, CHF3, C2H2F4 and C2H4F2 in water and aqueous NaCl solutions under low pressures. Fluid Phase Equilibria 129, 197–209.

Zhou, Z., Gunter, W.D., 1992. The nature of the surface charge of kaolinite. Clays Clay Miner. 40 (3), 365–368.

Zimmermann, M., Leifeld, J., et al., 2007. Quantifying soil organic carbon fractions by infrared-spectroscopy. Soil Biol. Biochem. 39, 224–231.

Zumft, W.G., 1997. Cell biology and molecular basis of denitrification. Microbiol. Mol. Biol. Rev. 61 (4), 533–616.

Index

Note: Page numbers followed by '*b*' indicate boxes, '*f*' indicate figures and '*t*' indicate tables.

A

Acetealdehyde, 355
Acid, defined, 258, 259
Acid rain, 283
Acid strength, defining, 260–261
Acid-base chemistry
  alkalinity, 280
  Arrhenius model of acid-base reactions, 258
  atmospheric gas impacts, 269–275
  Brønsted-Lowery model, 259
  charge balancing, 278–283
  chemical weathering impacts, 263–269
  environmental perspective, 257–258
  nitrogen cycling in natural landscapes, 276–277
Acid-base chemistry principles
  aqueous carbon dioxide reference level, 262–263
  conjugate acids and bases, 259–260
  defining acid and base strength, 260–261
  dissociation, 258
  hydrogen ion transfer, 259
  water reference level, 261–262
Acidity and basicity
  from chemical weathering, 263–269
  criteria, 260–261
  environmental sources, 263–269
  reference point defining, 257–258, 261, 262, 280
  summary overview, 302–303
  See also exchangeable acidity
Acidometric titration, 234–235
Actinium, 1–2
Active acidity, 298
Activity and the equilibrium constant
  concentrations and, 153–154
  ion activity coefficient expressions, 154–156
  ionic strength I, 154
Activity coefficient validation, simulations, 191
Activity diagrams, 183–187

Activity index, 285–286
Activity product validation, simulations, 192
Adhesion, 371–372
Adsorbents, 371
  colloids as, 372–374, 387–388, 399
Adsorption, 371–372, 399
Adsorption edges, 389
Adsorption envelopes, typical
  anion, 390*f*
  cation, 389*f*
Adsorption experiment, pH-dependent, 388–392
Adsorption isotherm experiment, 374–378
  interpreting the
    ion exchange isotherms in, 382–383
    Langmuir adsorption model, 380–382
    linear adsorption or partitioning model, 383–386
    pH-dependent ion adsorption experiment vs., 388
Adsorption isotherm, parameters
  defining, 382
Adsorption isotherms
  C-type, 376–377, 377*f*, 379
    distribution coefficient, 383
  ion exchange, 382–383
  L-type, 375, 378*f*, 380–382, 381*f*
Adsorption processes, 371–373, 399
  site limiting, 380–382
Aerobic bacteria, 350–353, 351*t*, 352*t*
Affinity chromatography, 65–66
Agricultural soil
  acidification of, 280
  ammonia-based fertilizers, 276–277, 280
  fertility, 118
  soil organic carbon turnover, 219–220, 219*t*
Air, chemical- and site-specific factors
  affecting contaminant transport, 438
Air inhalation equation, 442–443
Alcohol fermentation, 354–355